Organic Spices

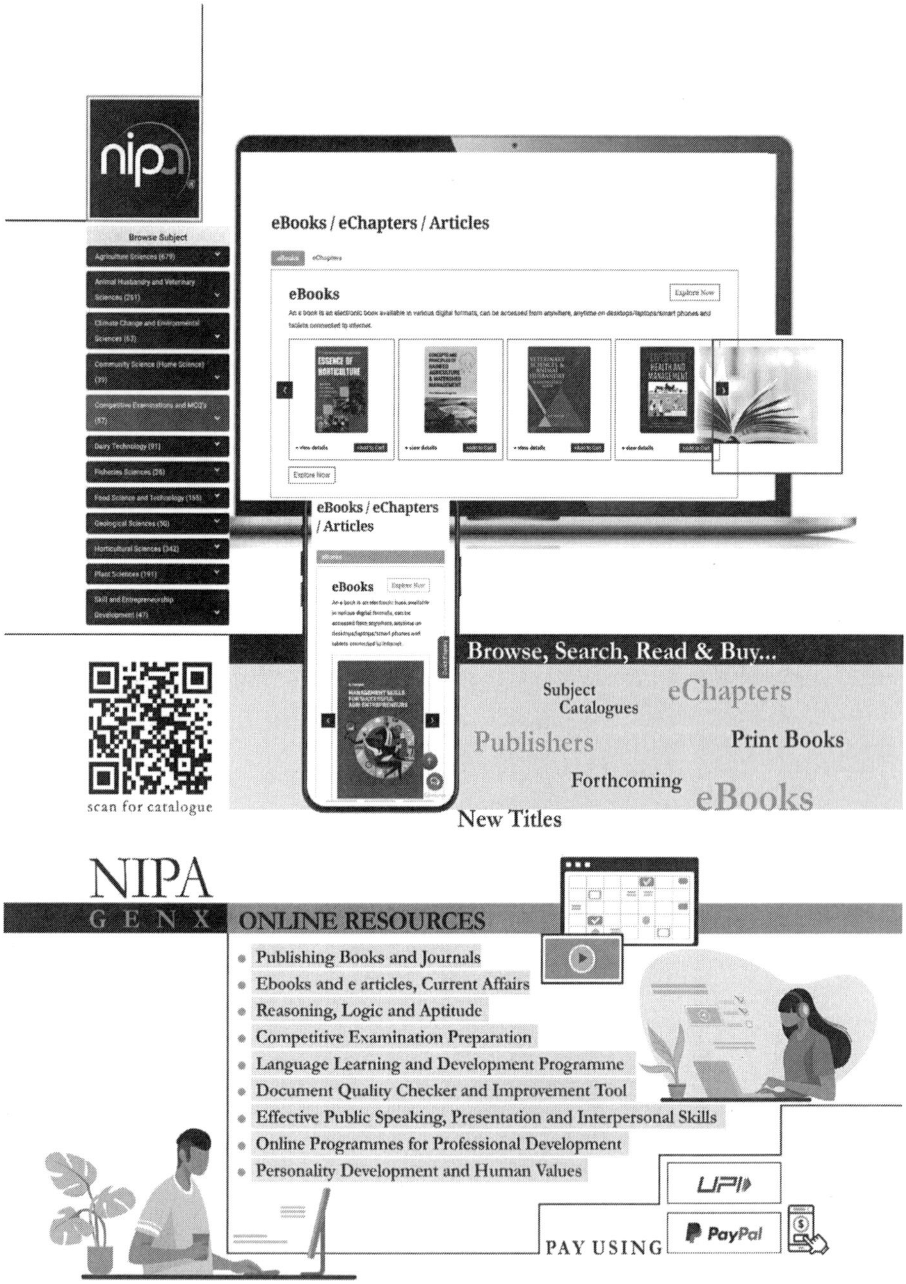
nipa
Browse Subject
eBooks / eChapters / Articles
eBooks
Explore Now
eBooks / eChapters / Articles
scan for catalogue
Browse, Search, Read & Buy...
Subject Catalogues
eChapters
Publishers
Print Books
Forthcoming
eBooks
New Titles
NIPA
GENX
ONLINE RESOURCES
Publishing Books and Journals
Ebooks and e articles, Current Affairs
Reasoning, Logic and Aptitude
Competitive Examination Preparation
Language Learning and Development Programme
Document Quality Checker and Improvement Tool
Effective Public Speaking, Presentation and Interpersonal Skills
Online Programmes for Professional Development
Personality Development and Human Values
PAY USING
UPI
PayPal

Organic Spices

Editors

V.A. PARTHASARATHY, Director

K. KANDIANNAN, Senior Scientist (Agronomy)

&

V. SRINIVASAN, Senior Scientist (Soil Science)

INDIAN INSTITUTE OF SPICES RESEARCH
Calicut - 673 012, Kerala (India)

New India Publishing Agency

Pitam Pura, New Delhi- 110 088

Published by
Sumit Pal Jain *for*

New India Publishing Agency

101, Vikas Surya Plaza, CU Block, L.S.C. Mkt.,
Pitam Pura, New Delhi- 110 088, (India)
Phone: 011-27341717, Fax: 011-27341616
E-mail: newindiapublishingagency@gmail.com
Web: www.bookfactoryindia.com

ISBN 978-81-89422-84-4

Composed and Designed by NIPA

Preface

Spices are important group of horticultural crops. The aroma and pungency of spices tasted by our ancestors are of the produce obtained from naturally fed plants. The art of growing spices became a business. The diversified use of spices demanded more production resulted in commercial ventures by skipping/supplementing traditional production system with modern production packages. But, this led to the use of certain harmful inputs which in turn affected flavour besides being toxic.

India is a "Land of Spices", each state or union territory in India cultivates one or other spice. Spices cultivated in an area of 2. 571 million hectare with a production of 3. 817 million tonnes. Spices exports have registered substantial growth during the last one decade. It has increased from 203398 tonnes valued MLN US $ 241 million in 1995-96 to 350363 tonnes valued MLN US $ 593 million in 2005-06, registering an annual average growth rate of 9.4% in value terms. During the year 2006-07, the spices export value has registered MLN US $ 793 million and the export has been 373750 tonnes. India commands a formidable position in the World Spice Trade with 47% share in volume and 40% in value.

Our traditional agricultural production systems were extensive in nature and relied on organic recycling of internal resources of a farm, as animals are part of the systems; their manures were used adequately to keep soil healthy. Green / green leaf manures, oil cakes, wood ash, crop residues are of great support in soil fertility maintenance. Crop rotation with legumes restored the soil vitality by promoting beneficial microbes. Nutrient imbalances and deficiency were seldom noticed in such a production system. Modern intensive systems depends on external inputs particularly fertilizer and pesticides. Enough evidences are documented to show indiscriminate use of these external inputs delivered negative effect on ecosystem at large ultimately human health and environment.

There is a growing demand for organic spices with an annual growth rate of 10 to 20 percent. Organic production systems prevailed in traditional farming of black pepper, cardamom, ginger, turmeric, chilli and paprika, coriander, cumin, fennel, fenugreek, nutmeg and vanilla and possible intervention without comprising quality and other issues like beneficial microbes, composting , biological control of pests and diseases and organic standards and procedure, agencies involved in certification are essential part of organic spices production.

Keeping the above in mind, we attempted to document all the work done on production of spices organically. This would serve as a reference book for researchers, teachers and students besides farmers, traders and consumers. We place on record our gratitude to all our colleagues at Indian Institute of Spices Research, Calicut for providing inputs.

Our Thanks are due to Mr. A. Sudhakaran, Photographer, Indian Institute of Spices Research for designing cover page and to Director, Indian Institute of Spices Research, Calicut for providing all possible help and photographs in this endeavor.

We also sincerely acknowledge the Food and Agricultural Organization of the United Nations for granting Authorization to use "Good Agricultural Practices for selected Agricultural Components" from the "Framework of Good Agricultural Practices" FAO (COAG/2003/6).

Editors

List of Contributors

V.A. PARTHASARATHY
Director
Indian Institute of Spices Research
Calicut - 673 012

M. ANANDARAJ
Project Coordinator (Spices)
Indian Institute of Spices Research
Calicut - 673 012

B. CHEMPAKAM
Head, Division of Crop Production
& Post Harvest Technology
Indian Institute of Spices Research
Calicut - 673 012

B.B. VASHISHTHA
Director
National Research Centre for
Seed Spices Tabiji, Ajmer - 305 206

C.K. THANKAMANI
Senior Scientist (Agronomy)
Indian Institute of Spices Research
Calicut - 673 012

D. PRASATH
Senior Scientist (Horticulture)
Indian Institute of Spices Research
Calicut - 673 012

M.N. VENUGOPAL
Head, Cardamom Research Centre
Indian Institute of Spices Research
Appangala, Heravanad
P.O. Madikeri - 571 201

K. KANDIANNAN
Senior Scientist (Agronomy)
Indian Institute of Spices Research
Calicut - 673 012

K.N. SHIVA
Senior Scientist (Horticulture)
Indian Institute of Spices Research
Calicut - 673 012

A. KUMAR
Senior Scientist (Plant Pathology)
Indian Institute of Spices Research
Calicut - 673 012

P.A. MATHEW
Head, Experimental Farm
Indian Institute of Spices Research
Peruvannamuzhi,
Calicut - 673 578

R. DINESH
Senior Scientist (Soil Science)
Indian Institute of Spices Research
Calicut - 673 012

R. SUSEELA BHAI
Senior Scientist (Plant Pathology)
Indian Institute of Spices Research
Calicut - 673 012

S. DEVASAHAYAM
Head, Division of Crop Protection
Indian Institute of Spices Research
Calicut - 673 012

S.K. MALHOTRA
Senior Scientist (Horticulture)
National Research Centre for Seed Spices
Tabiji, Ajmer - 305 206

SANTHOSH J. EAPEN
Senior Scientist (Nematology)
Indian Institute of Spices Research
Calicut - 673 012

T. JOHN ZACHARIAH
Senior Scientist (Bio Chemistry)
Indian Institute of Spices Research
Calicut - 673 012

V. SRINIVASAN
Senior Scientist
(Soil Science)
Indian Institute of Spices Research
Calicut - 673 012

S. HAMZA
Technical Officer
Indian Institute of Spices Research
Calicut - 673 012

Abbreviations

AGB	-	Alleppy Green Bold
AGL	-	Alleppy Green Light
AGS 1	-	Alleppy Green Shipment
AGS	-	Alleppy Green Superior
AICRP	-	All India Coordinated Research Project
AICRPS	-	All India Coordinated Research Project on Spices
AMF	-	Arbuscular Mycorrhizal Fungi
ANGRAU	-	Acharya N.G. Ranga Agricultural University
APEC	-	Asia Pacific Economic Cooperation
APEDA	-	Agricultural and Processed Food Products Export Development Authority
AQIS	-	Australian Quarantine and Inspection Service
ASM	-	Available Soil Moisture
ASTA	-	American Spice Trade Association
AVRDC	-	Asian Vegetable Research and Development Centre
BC	-	Before Christ
BHC	-	Benzene hexachloride (HCH)
BL	-	Bleached Light
BREB	-	Bowen ratio-energy balance
C/N	-	Carbon/ Nitrogen
CCS HAU	-	Chaudharay Charan Singh Haryana Agricultural University
CFA	-	Compost fungus activator
CFU	-	Colony forming units
CGAP	-	Code of Good Agricultural Practices
CGEB	-	Coorg Green Extra Bold

CMV	-	Cucumber Mosaic Virus
CND	-	Compositional Nutrient Diagnosis
CPE	-	Cumulative pan evaporation
DAP	-	Days After Planting
DARCOF	-	Danish Research Centre for Organic Food and Farming
DAS ELISA	-	Double Antibody Sandwich- Enzyme linked Immunosorbent Assay
DAS	-	Days after sowing
DASD	-	Directorate of Arecanut and Spices Development
DDT	-	Dichloro-Diphenyl-Trichloroethane
DRIS	-	Diagnosis and Recommendation Integrated System
DTPA	-	Diethylene Triamine Penta Acetic acid
ECOCERT	-	Ecological Certification, Germany
ELIZA	-	Enzyme linked immunosorbant assay
EM	-	Effective microorganisms
ESA	-	European Spice Association
ET_{C} -	-	Evapo - transpiration coefficient
ETL	-	Economic threshold level
EUREP	-	Retailer Produce Working Group
FAO	-	Food and Agricultural Organisation
FDA DALs	-	Food and Drug Administration Defect Action Levels
Fe	-	Iron
FiBL	-	Research Institute of Organic Agriculture
FYM	-	Farm yard manure
GA	-	Gibberellic acid
GAP	-	Good Agricultural Practices
GBPUAT	-	G.B. Pant University of Agriculture & Technology
GC	-	Gujarat Cumin
GCFV	-	Ginger Chlorotic Fleck Virus
GIS	-	Geographical Information System
GL	-	Garbled Light Pepper
GMO(GMOS)	-	Genetically Modified Organisms

GMP	-	Good Management Practice
IAA	-	ndole Acetic Acid IAA
IARI	-	Indian Agricultural Research Institute
ICRI	-	Indian Cardamom Research Institute
IFOAM	-	International Federation of Organic Agriculture Movements
IIHR	-	Indian Institute of Horticultural Research
IISR	-	Indian Institute of Spices Research
IMO	-	Institute of Marketecology, Switzerland
INDOCERT	-	Indian Organic Certification Agency
INDON-GAP	-	Indonesian Good Agricultural Practices
IPC	-	International Pepper Community
ISO	-	International Standard Organization
ITC	-	International Trade Center
ITF	-	International Task Force on Harmonization and Equivalence in Organic Agriculture
IU	-	International Units
IW	-	Irrigation water
JAS	-	Japanese Agricultural Standards
JNKV	-	Jawaharlal Nehru Krishi Vidyapeeth
KAU	-	Kerala Agricultural University
LDPE	-	Low Density Polyethylene
LLDPE	-	Li near low density polyethylene
LSD	-	Least Significant Difference
MDRIS	-	Modified Diagnosis and Recommendation Integrated System
MG	-	Malabar Garbled
Mn	-	Manganese
Mo	-	Molybdenum
MPKV	-	Mahatma Phule Krishi Vidyapeeth
MSL	-	Mean Sea Level
MT	-	Metric tonnes
MUG	-	Malabar Ungarbled
NAAS	-	National Academy of Agriculture Sciences

NAPP	-	National Accreditation Policy and Programme
NCM-ELISA	-	Nitrocellulose membrane Enzyme linked Immunosorbent Assay
NGO's	-	Non-Governmental Organizations
NOP	-	National Organic Programe
NPOP	-	National Programme for Organic Production
NPV	-	Nuclear polyhedrosis virus
NRCS	-	National Research Centre for Spices
NRCSS	-	National Research Centre for Seed Spices
NSKE	-	Neem Seed Kernel Extract
NUCG	-	Ungarbled Non-bleached Cochin Ginger
NUGK	-	Ungarbled Non-bleached Calicut Ginger
OA	-	Organic agriculture
OECD	-	Organisation for Economic Co-operation and Development
OUA & T	-	Orissa University of Agriculture and Technology
PAU	-	Punjab Agricultural University
PCR	-	Polymerase chain reaction.
PGPR	-	Plant growth promoting rhizobacteria
PH	-	Pin Heads
PM	-	Poultry manure
PMS	-	Postmenopausal syndrome
PROM	-	Phosphate rich organic manure
PSB	-	P-solubilising bacteria
PSM	-	Phosphate solubilising micro-organisms
PVBV	-	Pepper vein banding virus
PVMV	-	Pepper veinal mottle virus
PVX	-	Potato virus X
PVY	-	Potato virus Y
PYMV	-	Piper yellow mottle virus
RAJAU	-	Rajasthan Agricultural University
RAU	-	Rajendra Agricultural University
RCC	-	Reinforced concrete
SDAU	-	SardarKrushinagar Dhantiwada Agricultural University

SGS	-	Societe Generale de Surveillance, Switzerland
Skal	-	Certification Organic Production, Netherlands
SOEL	-	Stiftung Ökologie & Landbau (Foundation Ecology & Agriculture)
SOM	-	Soil organic matter
SQF	-	Safe Quality Food
SSP	-	Single Super Phosphate
TEV	-	Tobacco etch virus
TG	-	Tellicherry Garbled
TGEB	-	Tellicherry Garbled Extra Bold
TGSEB	-	Tellicherry Garbled Black Pepper Special Extra Bold
TMV	-	Tobacco mosaic virus
TNAU	-	Tamil Nadu Agricultural University
TQM	-	Total Quality Management
UAS	-	University of Agricultural Sciences
UAS-D	-	University of Agricultural Sciences-Dharwad
UGL	-	Ungarbled Light Pepper
UKROFS	-	UK Register of Organic Food Standards
UNDP	-	United Nations Development Programme
UNEP	-	United Nations Environment Programme
USDA	-	United States Department of Agriculture
USWB	-	United State Weather Bureau
VAM	-	Vesicular Arbuscular Mycorrhiza
WB	-	World Bank
WHO	-	World Health Organisation
ZECC	-	Zero Energy Cool Chamber
Zn	-	Zinc

Contents

1 Introduction

V.A. Parthasarathy, K. Kandiannan and V. Srinivasan

1.1 SPICE - DEFINITION AND USES

Spice is defined as "a strongly flavoured or aromatic substance of vegetable origin, obtained from tropical plants, commonly used as a condiment". A spice is a dried seed, fruit, root, rhizome, bark or vegetative substance or flower bud used in nutritionally insignificant quantities as a food additive for the purpose of flavoring. Spices are distinguished from herbs, which are leafy, green plant parts used for flavoring purposes. Herbs, such as basil or oregano, may be used fresh, and are commonly chopped into smaller pieces; spices, however, are dried and usually ground into a powder (http://www.reference.com/browse/wiki/Spice). American Spice Trade Association (ASTA) defines spices as "any dried plant product used primarily for seasoning purposes". Included are tropical aromatics (pepper, cinnamon, cloves, etc.), leafy herbs (basil, oregano, marjoram, etc.), spice seeds (sesame, poppy, mustard, etc.) and dehydrated vegetables (onions, garlic, etc.) (http://www.astaspice.org/spice/frame_spice.htm). Many spices have other uses like food preservation, medicine, religious rituals, cosmetics, and perfumery or as vegetables.

In ancient times, they were as precious as gold; and as significant as medicines, preservatives, and perfumes (http://www.indiatravelite.com/feature/worldofspice.htm). Spices are used primarily for flavoring, seasoning or for adding pungency and flavor to foods and beverages. "He, who controls the spice, controls the universe". Such were the words uttered by the main character of the movie Dune based on the Frank Herbert science fiction epic of the same name (http://asiapacificuniverse.com/pkm/spiceroutes.htm).

In the story, the spice was the lifeblood of a vast empire. For the leaders of this empire, it was essential that at all times 'the spice must flow'. The spice trade of the Dune movie was no doubt inspired by the historical trade in aromatics from ancient times to the present. According to a 15th century saying, "No man should die who can afford cinnamon". The fascinating history of spices is a story of adventure, exploration, conquest and fierce naval rivalry.

One of the first uses of spices was for religious purposes such as for incense, embalming, in sacrificial rites, and as charms, perfumes, or medicines. In the middle Ages, spice became a luxury item, a commodity so valued that it was accepted as currency. And is it too outlandish to claim that America was discovered due to spice-after all, it was the search for trade routes to India's valuable spices that inspired explorers like Columbus (Brown, 2003).

Spices have been used since time immemorial to enhance or vary the flavours of foods. Spices mask the taste of the otherwise tasteless food. Some spices are also used as food preservatives. Spices have medicinal properties. Spices are well known appetizers and digestives and considered essential in the culinary art all over the world. Some of them have anti-oxidant properties, while others possess strong anti-microbial and antibiotic activities. Spices increase the secretion of saliva rich in ptyalin, which facilitates starch digestion in the stomach, rendering the meals, which are rich in carbohydrates, more digestible. Spices possibly activate the adrenal-cortical function and fortify resistance and physical capacity. Stroke volume, blood pressure and stroke frequency can be markedly diminished or augmented by means of spices. Spices inhibit thrombus formation and accelerate thrombolysis (http:// indianfood.indianetzone.com/spices/).

The culinary history for spices is more interesting as it led to the development of "spice trade." Spices not only prized for their culinary aspects, but also for their monetary or barter value. Pepper, the most historically valuable spice was actually used as currency (http://www.suite101.com/ lesson.cfm/18003/1349). Parthasarathy *et al.* (2007) stated that references to pepper in ancient Sanskrit texts and Tamil literature indicate use of the spice over 3000 years ago (*Agananuru* and *Purananuru*). The word "spice" which is related to "species," originally meant merchandise, especially imported Asian products. A reason for 'spices' value as a trade commodity was the costly distance between the suppliers (Asia) and the market, (first the Mideast, then Europe). However, compared to other goods traded between the two regions, spices took up less cargo space, so their popularity was higher than bulkier products. The resultant spice trade is an integral part of the history of the rising and falling empires of India, China, Italian city states like Venice, Holland, England, and Portugal. The trade was anything but polite piracy and ruthlessness are key descriptors. Indeed, many current practices of modern international capitalism, both good and bad, were developed by the great (and unregulated) international spice trading companies like England's East India Company and Holland's Dutch East Indian Company. Gradually, as shipping became less expensive and herbs and spices could be grown in newly found territories such as the Americas, their culinary value came to the fore. Today, the U.S., Germany, Japan, and France are the biggest importers of spices, with India, Indonesia, Brazil, and the Malaysia the biggest exporters (http://www.suite101.com/lesson.cfm/18003/1349).

Spices have a profound influence on the course of human civilization. They permeate our lives from birth to death. In everyday life, spices succor us, cure us, relax us, and excite us. Ancient peoples such as the Egyptian, the Arab and the Roman made extensive uses of spices, not only to add flavor to foods and beverages, but as medicines, disinfectants, incenses, stimulants and even as aphrodisiac agents. No wonder they were sought after in the same manner gold and precious metals (Chomchalow, 1996).

The basic effects of spices when used in cooking and confectionary can be for flavouring, deodorizing/masking, pungency and colouring (Table 1).

Table 1. Uses of spices

Basic Function	Spices as major function	Spices as sub function
Flavouring	Parsley, cinnamon, allspice, dill, mint, tarragon, cumin, marjoram, star anise, basil, anise, mace, nutmeg, fennel, sesame, vanilla, fenugreek, cardamom, celery	Garlic, onion, bay leaves, clove, thyme, rosemary, caraway, sage, savory, coriander, pepper, oregano, horseradish, Japanese pepper, saffron, ginger, leek, mustard
Deodorizing/ masking	Garlic, savory, bay leaves, clove, leek, thyme, rosemary, caraway, sage, oregano, onion, coriander	
Pungency	Garlic, savory, bay leaves, clove, leek, thyme, rosemary, caraway, sage, oregano, onion, coriander, Japanese, pepper, mustard, ginger, horseradish, red pepper, pepper	Parsley, pepper, allspice, mint, tarragon, cumin, star anise, mace, fennel, sesame, cardamom, mustard, cinnamon, vanilla, horseradish, Japanese pepper, nutmeg, ginger
Colouring	Paprika, turmeric, saffron	

Ravindaran (2006)

1.2 SPICES - DISTRIBUTION

According to Ridley (1912), the history of the cultivation and use of spices is perhaps the most romantic story of any vegetable product. From the earliest known eras of civilization spices were eagerly sought in all parts of the world. It must be noted that the greater part of the spices that have been valued by man are derived from the Asiatic tropics, while the other quarters of the globe have produced comparatively few (Ridley, 1912). He also has mentioned centers of distribution or production of spices i.e., from Asia are derived black pepper, cardamoms, cinnamon (native of southern India and Ceylon, the present Sri Lanka), nutmegs, and mace; cloves, turmeric, ginger, greater galangal from the Malay Archipelago; cassia-bark and lesser galangal from China. Africa gave the grains of Paradise, Madagascar, while the American tropics gave vanilla, capsicums and pimento. The colder climate of northern Europe and Asia produced coriander, cumin, caraway, mustard seed and calamus roots.

1.3 SPICES - CLASSIFICATION

According to International Organization for Standardization (ISO) there are 109 spices which belonging to 31 families are grown around the world (Table 1). In India, about 52 spices are cultivated. They are classified in various ways. Based on botanical nomenclature, parts used (leaf/seed spices), quantity produced (major / minor spices), climate(temperate/tropical spices), duration (annual, biennial, perennial) and stature(herbs, shrubs, trees) etc., Based on the plant part used, they are classified as (I) rhizome and root spices, (II) bark spices, (III) leaf spices, (IV) flower spices, (V) fruit spices and (V) seed spices.

1.4 SPICES - HISTORY

From ancient times to the present, Asia has been well known as the 'Land of Spices'. The Maluku Islands of Indonesia, also known to English speakers as the Moluccas, are referred to as the 'Spice Islands'. It is the most popular of all terms used to describe the land of spices although this term could be applied to Sumatra as well as Sri Lanka, the latter also referred to as the 'Spice Isle'. Melaka (or Malacca), the most famous city known for its spice trade, is called the "Spice City", while the sea route that has linked Venice, Arabia, Sri Lanka, India, Malaysia (Malacca), Moluccas, and China through spice trading since the 16th century, is known as the 'Spice Route'(Chomchalow, 1996), ranking in importance next to the famous 'Silk Road'. For the Indians who produce and export the majority of spices for world markets, the term 'Spice Bowl of the World', is applied to the State of Kerala. Many of the world's finest spices like black pepper, cardamom, nutmeg, ginger and turmeric have been produced in Kerala (Chomchalow, 1996).

Table 2. Spices grown around the world

	Botanical name of the plant	**Family**	**Common name**	**Name of plant part used as spices**
1.	*Acorus calamus*	Araceae	Sweet flag, myrtle flag, calamus, flag root	Rhizome
2.	*Aframomum angustifolium*	Zingiberaceae	Madagascar cardamom	Fruit, seed
3.	*Aframomum hanburyi*	Zingiberaceae	Cameroon cardamom	Fruit, seed
4.	*Aframomum koranima*	Zingiberaceae	Korarima cardamom	Fruit, seed
5.	*Aframomum melegueta*	Zingiberaceae	Grain of paradise. Guinea grains	Fruit, seed
6.	*Allium ascalonicum*	Liliaceae	Shallot	Bulb
7.	*Allium cepa*	Liliaceae	Onion	Bulb
8.	*Allium cepa var. aggregatum*	Liliaceae	Potato onion	Bulb
9.	*Allium tuberosum*	Liliaceae	Indian leek, Chinese chive	Bulb, leaf
10.	*Allium fistulosum*	Liliaceae	Stony leek, Welsh onion, Japanese bunching onion	Leaf and bulb
11.	*Allium porrum*	Liliaceae	Leek, winter leek	Leaf and bulb
12.	*Allium sativum*	Liliaceaee	Garlic	Bulb
13.	*Allium schoenoprasum*	Liliaceae	Chive	Leaf
14.	*Alpinia galangal*	Zingiberaceae	Greater galangal, Longwas, Siamese ginger	Rhizome

Contd....

Table 2. Contd....

15.	*Alpinia officinarum*	Zingiberaceae	Lesser galangal	Rhizome
16.	*Amomum aromaticum*	Zingiberaceae	Bengal cardamom	Fruit, seed
17.	*Amomum kepulaga*	Zingiberaceae	Round cardamom, Chester cardamom, Siamese cardamom, Indonesian cardamom	Fruit, seed
18.	*Amomum krervanh*	Zingiberaceae	Cambodian cardamom	Fruit, seed
19.	*Amomum subulatum*	Zingiberaceae	Greater Indian cardamom, large cardamom, Nepalese cardamom	Fruit, seed
20.	*Amomum tsao-ko*	Zingiberaceae	Tsao-ko cardamom	Fruit, seed
21.	*Anethum graveolens*	Apiaceae (Umbelliferae)	Dill	Fruit, leaf, top
22.	*Anethum sowa*	Apiaceae (Umbelliferae)	Indian dill	Fruit
23.	*Angelica archangelica*	Apiaceae (Umbelliferae)	Garden angelica	Fruit, petiole
24.	*Anthriscus cereifolium*	Apiaceae (Umbelliferae)	Chervil	Leaf
25.	*Apium graveolens*	Apiaceae (Umbelliferae)	Celery, garden celery	Fruit, root, leaf

Contd....

Table 2. Contd....

26.	*Apium graveolens var. rapaceum*	Apiaceae (Umbelliferae) (Cniciferae)	Celeriac	Fruit, root, leaf
27.	*Armoracia rusticana*	Brassicaceae	Horse radish	Root
28.	*Artemisia dracunculus*	Asteraceae (Compositae)	Tarragon, estragon	Leaf
29.	*Averrhoa bilimbi*	Averrhoaceae	Belimbing, bilimbi cucumber tree	Fruit
30.	*Averrhoa carambola*	Averrhoaceae	Carambola, caramba	Fruit
31.	*Brassica junceae*	Brassicaceae	Indian mustard	Seed
32.	*Brassica nigra*	Brassicaceae	Black mustard	Seed
33.	*Bunium persicum*	Apiaceae (Umbelliferae)	Black caraway	Seed, tuber
34.	*Capparis spinosa*	Capparidaceae	Caper, common caper, caper bush	Floral bud
35.	*Capsicum annuum*	Solanaceae	Capsicum, chillies, paprika	Fruit
36.	*Capsicum frutescens*	Solanaceae	Chillies, bird's eye chilli	Fruit
37.	*Carum bulbocastanum*	Apiaceae (Umbelliferae)	Black caraway	Fruit, bulb
38.	*Carum carvi*	Apiaceae (Umbelliferae)	Caraway, blond caraway	Fruit
39.	*Cinnamomum aromaticum*	Lauracea	Cassia, Chinese cassia	Bark, leaves

Contd....

Table 2. Contd....

40.	*Cinnamomum burmanii*	Lauraceae	Indonesian cassia	Bark
41.	*Cinnamomum loureirii*	Lauraceae	Vietnamese cassia	Bark
42.	*Cinnamomum tamala*	Lauraceae	Tejpat, Indian cassia	Leaf, bark
43.	*Cinnamomum zeylanicum*	Lauraceae	Sri Lankan cinnamon, Indian cinnamon	Bark, leaf
44.	*Coriandrum sativum*	Apiaceae (Umbelliferae)	Coriander	Leaf, fruit
45.	*Crocus sativus*	Iridaceae	Saffron	Stigma
46.	*Cuminum cyminum*	Apiaceae (Umbelliferae)	Cumin	Fruit
47.	*Curcuma longa*	Zingiberaceae	Turmeric	Rhizome, leaf
48.	*Cymbopogon citratus*	Poaceae	West Indian lemongrass	Leaf
49.	*Cymbopogon nardus*	Poaceae	Sri Lankan citronella	Leaf
50.	*Elettaria cardamomum*	Zingiberaceae	Small cardamom	Fruit, seed
51.	*Elettaria cardamomum*	Zingiberaceae	Sri Lankan cardamom	Fruit, seed
52.	1 *Ferula assa-foetida* 2 *Ferula foetida* 3 *Ferula narthex*	Apiaceae (Umbelliferae)	Asafoetida	Rhizome
53.	*Foeniculum vulgare*	Apiaceae	Bitter fennel	Leaf, twig, fruit

Contd....

Table 2. Contd....

54.	*Foeniculum vulgare*	Apiaceae	Sweet fennel	Leaf, twig, fruit
55.	*Garcinia cambogia*	Clusiaceae	Garcinia, Camboge	Pericarp of the fruit
56.	*Garcinia indica*	Clusiaceae	Garcinia, Kokum	Pericarp of the fruit
57.	*Hyssopus officinalis*	Lamiaceae	Hyssop	Leaf
58.	*Illicum verum*	Dliciaceae	Star anise, Chinese anise	Fruit
59.	*Juniperus communis*	Cupressaceae	Common juniper	Fmit
60.	*Kaempferia galanga*	Zingiberaceae	Galangal	Rhizome
61.	*Laurus nobilis*	Lauraceae	Laurel, true laurel, bay leaf, sweet flag	Leaf
62.	*Levisticum officinale*	Apiaceae	Garden lovage, lovage	Fruit, leaf
63.	*1 Lippia graveolens* *2 Lippia berlandieri*	Verbenaceae	Mexican oregano	Leaf, terminal shoot
64.	*Mangifera indica*	Anacardiaceae	Mango	Immature fruit(rind)
65.	*Melissa officinalis*	Lamiaceae	Balm, lemon balm, melissa	Leaf, terminal shoot
66.	*Mentha arvensis*	Lamiaceae	Japanese mint, field mint, corn mint	Leaf, terminal shoot
67.	*Mentha citrate*	Lamiaceae	Bergamot	Leaf, terminal shoot
68.	*Mentha x piperita*	Lamiaceae	Peppermint	Leaf, terminal shoot
69.	*Mentha spicata*	Lamiaceae	Spearmint, garden mint	Leaf, terminal shoot
70.	*Murraya koenigii*	Rutaceae	Curry leaf	Leaf

Contd....

Table 2. Contd....

71.	*Myristica argentea*	Myristicaceae	Papuan nutmeg	Kernel
			Papuan mace	Aril
72.	*Myristica fragrans*	Myristicaceae	Indonesian type nutmeg, Indonesian type mace, Siauw type mace	Kernel Aril
73.	*Nigella damascene*	Ranunculaceae	Damas black cumin, love in a mist	Seed
74.	*Nigella sativa*	Ranunculaceae	Black cumin	Seed
75.	*Ocimum basilicum*	Lamiaceae	Sweet basil	Leaf, terminal shoot
76.	*Origanum majorana*	Lamiaceae	Sweet marjoram	Leaf, floral bud
77.	*Origanum vulgare*	Lamiaceae	Oregano, origan	Leaf, flower
78.	*Pandanus amaryllifolius*	Pandanaceae	Pandan wangi	Leaf
79.	*Papaver somniferum*	Papaveraceae	Poppy, blue maw, mawseed	Seed
80.	*Petroselinum crispum*	Apiceae	Parsley	Leaf, root
81.	*Pimenta dioica*	Myrtaceae	Pimento, allspice, Jamaica pepper	Immature fruit, leaf
82.	*Pimenta racemosa*	Myrtaceae	West Indian bay	Fruit, leaf
83.	*Pimpinella anisum*	Apiaceae	Aniseed	Fruit
84.	*Piper guineense*	Piperaceae	West African or Benin pepper	Fruit
85.	*Piper longum*	Piperaceae	Long pepper, Indian long pepper	Fruit
86.	*Piper nigrum*	Piperaceae	Black pepper, white pepper, green pepper	Fruit

Contd....

Table 2. Contd....

87.	*Punica granatum*	Punicaceae	Pomegranate	Seed (dried with flesh)
88.	*Rosmarinus officinalis*	Lamiaceae	Rosemary	Terminal shoot, leaf
89.	*Salvia officinalis*	Lamiaceae	Garden sage	Terminal shoot, leaf
90.	*Satureja hortensis*	Lamiaceae	Summer savory	Terminal shoot, leaf
91.	*Satureja Montana*	Lamiaceae	Winter savory	Leaf, twig
92.	*Schinus molle*	Anacardiaceae	American pepper, Califomian pepper tree	Fruit, wall (rind)
93.	*Schinus terebenthifolius*	Anacardiaceae	'Brazilian pepper'	Fruit
94.	*Sesamum indicum*	Pedaliaceae	Sesame, gingelly	Seed
96.	*Syzygium aromaticum*	Myrtaceae	Clove	Flower bud
97.	*Tamarindus indica*	Cesalpiniaceae	Tamarind	Fruit
98.	*Thymus serpyllum*	Lamiaceae	Mother of thyme, wild thyme, creeping thyme	Terminal shoot, leaf
99.	*Thymus vulgaris*	Lamiaceae	Thyme, common thyme	Terminal shoot, leaf
100.	*Trachyspermum ammi*	Apiaceae	Ajowan	Fruit
101.	*Trigonella foenumgracecum*	Fabaceae	Fenugreek	Seed, leaf

Contd....

Table 2. Contd....

102.	*Vanilla planifolia syn. Vanilla fragrans*	Orchidaceae	Vanilla	Fruit (pod)
103.	*Vanilla tahitensis*	Orchidaceae	Vanilla	Fruit (pod)
104.	*Vanilla pompona*	Orchidaceae	Pompona vanilla	Fruit (pod)
105.	*Xylopia aethiopica*	Annonaceae	Negro pepper, Guinean pepper	Fruit
106.	*Zanthoxylum bungei*	Rutaceae	Chinese prickly ash pepper, Sechuang pepper	Fruit
107.	*Zanthoxylum acanthopodium*	Rutaceae	Chinese pepper	Fruit
108.	*Zanthoxylum piperitum*	Rutaceae	Japanese pepper	Fruit
109.	*Zingiber officinale*	Zingiberaceae	Ginger	Rhizome

Source : Spices Board, Cochin, India (http://www.indianspices.com/html/iso_lis_spe.htm)

Spices have been prominent in human history virtually since its inception. Spices were some of the most valuable items of trade in the ancient and medieval world. In the story of Genesis, Joseph was sold into slavery by his brothers to spice merchants. In the biblical poem Song of Solomon, the male speaker compares his beloved to many forms of spices. Spices were the primary reason that Portuguese navigator Vasco Da Gama sailed to India. Around that same time, when Christopher Columbus happened upon the Americas, he was quick to describe to investors the many new spices available there (http://www.reference.com/browse/wiki/Spice). For more than three thousand years, elaborate trading routes that connected the Mediterranean World with the far off lands of Asia linked the ancient world. As caravans carried exotic goods on the Silk Road to burgeoning markets in the Roman Empire, a parallel maritime trade linked the markets of Eurasia. A complex network of sailing ships, dependent on seasonal monsoon winds, carried cargoes from India and Southeast Asia to ports throughout the ancient world. Arab, Indian, and Malay seafarers, who concealed the source of their goods eager to protect their lucrative trade, conducted this early trade. The demand for these fragrant spices and aromatic resins, worth their weight in gold, spurred a search for routes to the mythical *Golden Khersonese, Chryse* ("land of gold") and to *Agyre* ("land of silver") as geographers, astronomers, and merchant sailors soon charted new understandings of our world (http://ias.berkeley.edu/orias/Spice/textobjects/overview.htm).

European traders and merchants interested in the "fabulous and legendary Orient of silk, spices and pearls" wanted a sea route to the Orient. In 1498 this Portuguese navigator connected Europe to Asia via the African coastline. A tri-continental world of Europe, Africa and Asia entered a new phase of interaction, for better or for worse. They wanted access to the Indian Ocean, but the African land mass was in the way, a huge stumbling block to European greed (Mazrui, 1998).The famed spices which traveled from Africa to the Arabian traders and from thence to the markets of the classical Mediterranean world had their ultimate origin in 'Southeast Asia'. The aromatic trail known as the "Cinnamon Route" began somewhere in the Malay Archipelago, known as the "East Indies," and crossed the Indian Ocean to the southeastern coast of Africa. The spices may have landed initially at Madagascar and they eventually were transported to the East African trading ports in and around the city known in Greco-Roman literature as Rhapta. Merchants then moved the commodities northward along the coast. In Roman times, they traveled to Adulis in Ethiopia and then to Muza in Yemen and finally to Berenike in Egypt. From Egypt they made their way to all the markets of Europe and West Asia (Miller, 1969).

The cinnamon route started in the cinnamon and cassia-producing regions of northern Indochina and southern China and then likely proceeded from South China spice ports southward during the winter monsoon down the Philippine corridor. The route likely turned southeast at that point to Sumatra and/or Java to pick up different varieties of cinnamon and cassia along with aloeswood and benzoin. From southwestern Indonesia the voyage then took the Austronesian merchants across the great expanse of the Indian Ocean to Africa. According to Chau Ju-Kua, cloves and nutmeg were grown in two kingdoms found in Toupo, southeast of southern China. Ships coming from Toupo to China sailed for twenty-five days on a northwest course before arriving at Sanfotsi, and from there proceeded either due to north to reach Tsu'an-chou (Fuzhou or Xiamen) or a bit northwest to reach Canton. So it is clear and logical that the 'clove route' went through the Philippines, where Sanfotsi was located and this also was the most direct course for the trade between the clove and nutmeg-bearing regions with the Chinese coast (http://asiapacificuniverse.com/pkm/spiceroutes.htm).

The fame of Indian spices is older than the recorded history. The story of Indian Spices is more than 7000 years old. Centuries before Greece and Rome had been discovered, sailing ships were carrying Indian spices, perfumes and textiles to Mesopotamia, Arabia and Egypt. It was the lure of these that brought many seafarers to the shores of India. Theophrasus described black pepper and long pepper (*P. longum* L.) in the 4th century BCE. The powerful Chola Kings before 100 BCE supposedly took pepper to Indonesia (Parthasarathy *et al.* 2007). Long before Christian era, the Greek merchants thronged the markets of South India, buying many expensive items amongst which spices were one. Epicurean Rome was spending a fortune on Indian spices, silks, brocades, Dhaka Muslin and cloth of gold, etc. It is believed that the Parthian wars were being fought by Rome largely to keep open the trade route to India. It is also said that Indian spices and her famed products were the main lure for crusades and expeditions to the East. Today when spices cost so little, it seems unbelievable that they were once a royal luxury and that man were willing to risk their lives in quest of them. Though it was the Dark Ages, but there were rich people who had gold to exchange for pepper and cinnamon. It was in the year 1492 A.D., that Christopher Columbus discovered the New World. Five years later, four tiny ships sailed southward from the port of Lisbon, Portugal, under the guidance of Captain Vasco De Gama. Like Columbus, Vasco De Gama too was searching for a new route to the spice lands of Asia. While Columbus failed to achieve the goal, De Gama succeeded. In a two year, 24,000 miles round trip, he took his ships around the continent of Africa to India and back to Lisbon. Only two of the four ships survived to reach their homeport. These two ships brought back a cargo of spices and other products worth 60 times the cost of the said voyage (http://indianfood.indianetzone.com/spices/1/history_spices.htm). The spices of the East were valuable in those times, During these Middle Ages, a pound of ginger was worth a sheep, a pound of mace worth three sheeps or half a cow. Pepper, the most valuable spice of all, was counted out in individual peppercorns, and a sack of pepper was said to be worth a man's life. De Gama's successful voyage intensified an international power struggle for control over the spice trade. For three centuries afterwards the nations of Western Europe - Portugal, Spain, France, Holland, and Great Britain - fought bloody sea-wars over the spice-producing colonies.

By the year 1000AD Arabians had conquered the Indus valley, what is now India. They brought the cumin and coriander that mixed with Indian pepper, ginger and turmeric make up the base of so many South Asian dishes. It was this combination of spices that centuries later British sailors spread throughout the world as curry powder. In India, Arabian traders got the rare and exotic spices of the Far East from local spice merchants. India had spent the previous two millennia spreading its culture to the Spice Islands of the east (http://indianfood.indianetzone.com/spices/1/history_spices.htm). Arabian traders were able to make good money supplying these spices, even with the high prices paid to the Indian middle men, not only to their countrymen back home, but to Europe as well. These traders of spices paid for the Art and Education for which Arabia became famous in the present day. In many ways the culture of Arabia loved studying and learning different things. Many great Greek and Roman plays translated in Arabic, so too were the geographic writings of Pliny and Ptolemy telling of the general location of the tabled Spice Islands.

As Arabian astronomers charted the stars in order to study them and understand mans relation with them, they realized these same charts could be used for navigation. And then Arabian traders invented the technology and knew the odds. Soon they were sailing to what is now Indonesia and Malaysia to purchase spices directly, bypassing the Indian middlemen. By the middle of the 13th century Arabian merchants were regularly visiting Sumatra for cassia from the slopes of Mount Korintje. While traveling they would stop at little villages and towns that had fresh water resource to refill their water supplies. At these stops the merchants would barter their cumin, coriander and saffron and speak and preach of their religion as well. The tropical climate did not suit the saffron but coriander now plays an integral part in so many dishes across Indonesia. (http://indianfood.indianetzone.com/spices/).

Hippalus showed that it was possible to make the round trip between Berenice, on the Egyptian coast of the Red Sea, and the pepper-producing Malabar Coast of India in less than a year, a voyage that had previously taken at least 2 years. Since the spices cardamom and turmeric, indigenous to India, were cultivated as early as the 8th century BC in the gardens of Babylon, the propagating material or seeds of these spices must have been introduced to Babylonia from India far back in time(Rosengarten, 1969). Although small physic (medicinal-plant) gardens had been known in Europe for several centuries before his time, Charlemagne (742-814), king of the Franks (France) and emperor of the West, was an important figure in the development of herbs. A patron of literature, art, and science, he organized for the first time in his realm larger systematic plantings of temperate condimental plants. In 812 Charlemagne ordered a number of useful plants, among them anise, fennel, fenugreek, and flax, to be grown on the imperial farms in Germany. With the advent of Arabian travelers and geographers more information became available concerning the areas where spices were produced. Ali al-Masudi, a 10th-century historian, described in his *Meadows of Gold* the vast cultivated plantings of the *maharaj*, or king of the East Indian Archipelago, whose produce included cloves, nutmegs, mace, cardamom, sweet aloes, and camphor(http://www.hort.purdue.edu/newcrop/history/lecture26/r_26-1l.html) original source (Rosengarten, 1969).

In 1519 the Spanish conquistador Hernán Cortés was given command of an expedition that culminated in the conquest of Mexico. His soldiers also made a great contribution to the history of spices while on military reconnaissance in the moist, shady coastal rain forests of southeastern Mexico. They found vanilla. This youngest and newest of the tropical spices wag being used by the Aztecs as a delicate, flavoring to season their chocolate beverages (http://www.hort.purdue.edu/newcrop/history/lecture26/r_26-1l.html) original source (Rosengarten, 1969). During the 16th and 17th centuries ginger plantings successfully established in the West Indies with propagating material brought from the East Indies. Since ginger was probably the easiest of the Oriental tropical spices to grow, its simple cultivation requirements soon led to its being produced on a large scale in Santo Domingo, Jamaica, and Barbados. Early Jamaican records report that in 1547 one thousand tons of ginger was exported from the West Indies to Spain. The Portuguese under Albuquerque had won several decisive naval battles over the Muslims by 1511, thereby gaining control of many ancient Oriental spice producing areas-the Malabar Coast of India, Ceylon, Java, Sumatra, and the great Malayan spice trading center of Malacca. By 1514 they dominated Ternate and the other Spice Islands. Indonesia's abundant spices first brought Portuguese merchants to the key trading port of Malacca in 1511. Prized for their flavor, spices such as cloves, nutmeg and mace were also believed to cure everything from the plague to venereal disease, and were literally worth their weight in gold. The Dutch eventually wrested control of the spice trade from Portuguese, and the tenacious Dutch East India Company (known by initials VOC) established a spice monopoly which lasted well into the 18th century (http://www.indo.com/indonesia/history.html).

Cloves were introduced in 1818, and flourished in the tropical climate and fertile soil of the western areas of both Zanzibar and Pemba. By the middle of the century, the Zanzibar archipelago was the world's largest producer of cloves, and the largest slave trading centre on the East African coast. Slaves were used for the cultivation and harvesting of cloves, and the Sultan occupied so many plots that by his death in 1856, he had 45 plantations. Plots were also acquired by his children, and many concubines and eunuchs from the royal harem. Over time, several other spices such as cinnamon, cumin, ginger, pepper and cardamom were introduced. Their rich fragrance became synonymous with Zanzibar, which became known as the 'Spice Islands'. Slaves, spices and ivory provided the basis of considerable prosperity, and Zanzibar became the most important entrepôt in the Western Indian Ocean. All other East African coastal centres were subject to it and almost all trade passed through Zanzibar (http://zanzibar.net/zanzibar/history/spice_trade).

1.5 SPICES PRODUCTION IN THE WORLD

The International Standard Organisation (ISO) recognized 109 plants as spice, but still many plants may be there in nature with of spicy characters. Many highly valued spices like black pepper (King of Spices), Cardamom (Queen of Spices), Vanilla (Prince of Spices), ginger and turmeric are of tropical habitat in adaptation. Many herbal spices are in temperate origin. According to their ecological niche, they distributed in different countries of the world. However, their area of production varies with country. Compared to many other field and horticultural crops, spices area and production is limited. The FAO database gives area and production only limited number of spices (Table 3). Spices cultivated in an area of 7587.02 ('000ha) with a production of 31859.69 ('000t) during 2005. The world export of spices during 2005 was 3592. 48 ('000t) and import was 3454.40 ('000t) (FAO database 2007).

Table 3. World area and production of important spices during 2005

Name of spice(s)	Area (' 000 ha)	Production ('000t)
Anise, badian, fennel, coriander	661.16	467.86
Chillies and peppers (dry)	2004.81	2662.73
Chillies and peppers(green)	1725.54	24803.01
Cinnamon (canella)	176.98	134.8
Cloves	466.08	145.18
Ginger	338.9	1119.74
Pepper (*Piper* sp)	473.55	407.41
Nutmeg, mace and cardamom	222.89	74.02
Spices, nec*	1440.67	2034.58
Vanilla	76.44	10.36
Total	7587.02	31859.69

* Other spices; *Source*: FAO database 2007.

1.6 CONVENTIONAL AGRICULTURE

The conventional agriculture in present day usages means - agriculture involving external inputs particularly fertilizer and pesticides. Besides these, growth regulators, mechanization, animal medicines are part of the system of modern agriculture.

1.6.1 Fertilizer

Before 19th century, most food in the world was organically produced using organic manure and human and animal power (White, 1970). The agricultural revolution began in England in the early 19th century, when Jethro Tull invented a horse–drawn hoe and a seed drill (Prasad, 2005). It was the German chemist, Justus von Liebig, (during 1830-1840) applied scientific principles to the study of plant and animal nutrition. It was discovered that mineral salts present in the humus and manure only were absorbed by plants and not the organic matter. Sir Humphery Davy and Justus von Leibig were the key founders of this theory and published their ideas (Davy, 1813; von Leibig, 1840). Later Lawes and Gilbert initiated both laboratory and field plot studies that led to the establishment of the Rothamsted Experimental Station at Harpenden, England. First fertilizer (Single Super Phosphate (SSP)) industry started in 1842 at Rothamsted, England. Fertilizer industry developed gradually as result consumption also increased (Fig. 1). After 1990's fertilizer consumption declined in developed countries might be due to change of consumer's preference to organic food. The developing countries continue to use fertilizer in large quantities (Fig. 1).

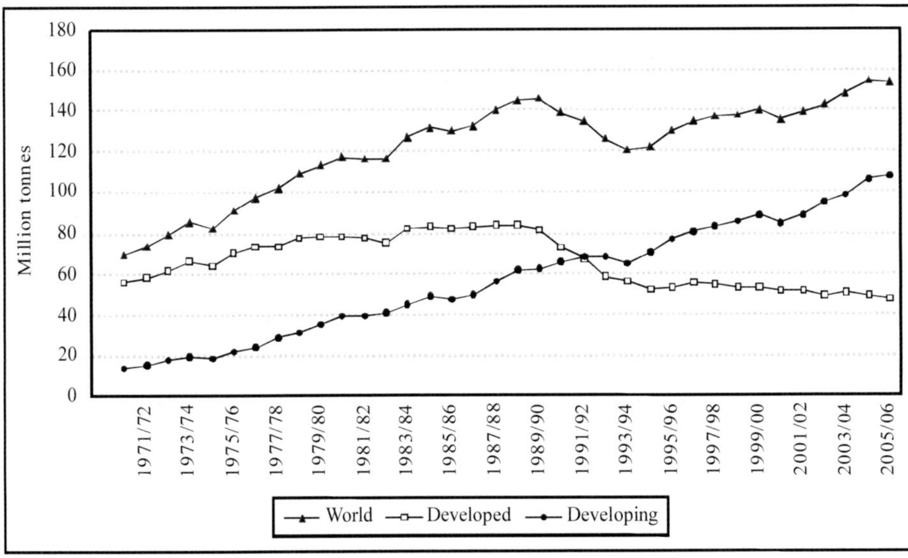

Fig. 1 Total fertilizer nutrient consumption - Million tones nutrients, N + P_2O_5 + K_2O
Source: IFA 2007. http://www.fertilizer.org/ifa/statistics/indicators/tablenpk.asp (online) Graph prepared by authors from the data provided from this site.

In countries like India, fertilizer is an important component along with short duration high yielding varieties and pesticides during 'green revolution'. There is a steady increase in fertilizer consumption in India after 1960's (Fig.2.) could contribute for self sufficient in food production.

The world consumption of nutrients N, P_2O_5, K_2O put together during 2005-06 was 154.07 million tones. Out of which the maximum (90.86 million tones) consumed was nitrogen nutrient. Indiscriminate use of nitrogen nutrient leads to excess NO_3 and NO_2 in water and food. Methemoglobinemia cancer is due to pollution of excess of N. Eutrophication in water bodies also due to surplus N present in water. The fertilizer use efficiency is not more than 50% and the remaining is huge loss Yadav (2003). Duram (2005) quoted those high levels of nitrate in drinking water than standard levels recorded, can cause "blue baby syndrome" in which oxygen levels in an infant's blood drop dangerously low.

1.6.2 Pesticides

Unintentional poisonings kill an estimated 3,55,000 people globally each year (WHO, 2003). In developing countries - where two thirds of these deaths occur - such poisonings are associated strongly with excessive exposure to, and inappropriate use of, toxic chemicals. In many such settings, toxic chemicals may be emitted directly into soil, air, and water - from industrial processes, pulp and paper plants, tanning operations, mining, and unsustainable forms of agriculture - at levels or rates well in excess of those tolerable to human health (Yáñez *et al.* 2002; UNDP, 1998; WB, 2002). Acute exposure to pesticides can lead to death or serious illness (OECD, 2001). Chronic pesticide exposure is most often a problem in the occupational setting, particularly among poor rural populations where men, women, and children all work and live in close proximity to fields and orchards where chemicals are applied and stored (UNDP, 1998; WHO, 1990).

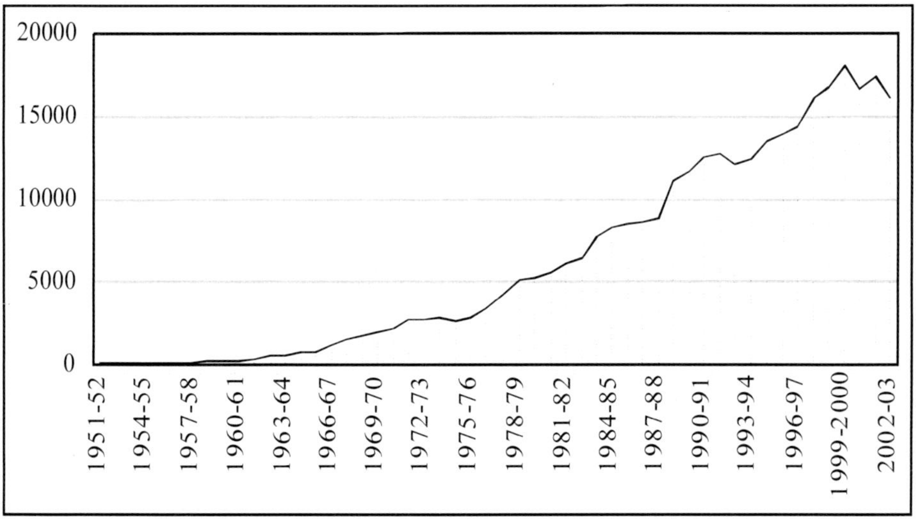

Fig 2. Fertilizer consumption in India ('000 tones). *Source:* http://dacnet.nic.in/eands/PORTRAT-G.pdf (graph drawn by authors from data provided in the website)

Long-term exposure to pesticides can increase the risk of developmental and reproductive disorders, immune-system disruption, endocrine disruption, impaired nervous-system function, and development of certain cancers. Children are at higher risk from exposure than are adults (OECD, 2001; WHO, 1990, Yáñez *et al.* 2002; UNDP 1998).

Pesticides, when not judiciously used, may disrupt natural biological pest control mechanisms. More vigorous pest attacks may result, along with heavier chemical use, and increased health exposures. Pesticides, as well as fertilizers, can infiltrate water sources - contaminating drinking water and animal species, e.g. fish, upon which humans rely for nutrition. Such contamination can lead to a range of secondary public health impacts (UNEP, 2004; UNEP, 1999; WB, 2003).

Around 30% of pesticides marketed in developing countries with an estimated market value of US$900 million annually do not meet internationally accepted quality standards. They are posing a serious threat to human health and the environment. The global market value for pesticides is estimated at US$32 billion in 2000, with the share of developing countries around US$3 billion. In developing countries, pesticides are mainly used for agriculture, but also for public health, such as insecticides for controlling insects spreading malaria (WHO, 2001).

1.6.2.1 Pesticides use in India – An example

Use of pesticides in India began in 1948 when DDT imported for malaria control and BHC for locust control. India started pesticide production with manufacturing plant for DDT and benzene hexachloride (BHC) (HCH) in the year 1952. According to Gupta (2004), in 1958, India was producing over 5000 metric tonnes of pesticides. Currently, there are approximately 145 pesticides registered for use, and production has increased (Fig.3) to

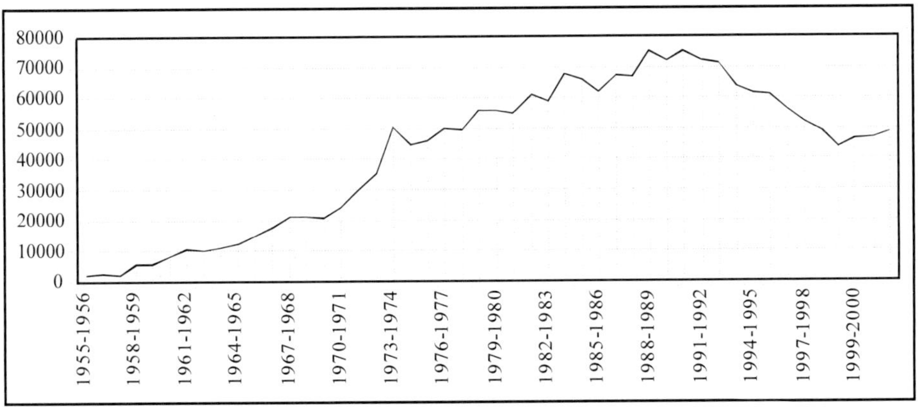

Fig. 3. Pesticides consumption (Metric tones) technical grade in India
Source: Graph prepared by authors from the data provided from this site. (http : / /www.ncipm.org.in/ asps/ Display/ Pesticides.asp)

approximately 85,000 metric tonnes. Rampant use of these chemicals has given rise to several short-term and long-term adverse effects of these chemicals. The first report of poisoning due to pesticides in India came from Kerala in 1958 where, over 100 people died after consuming wheat flour contaminated with parathion. Subsequently several cases of pesticide-poisoning including the Bhopal disaster have been reported. Despite the fact that the consumption of pesticides in India is still very low, about 0.5 kg/ha of pesticides against 6.60 and 12.0 kg/ha in Korea and Japan, respectively, there has been a widespread contamination of food commodities with pesticide residues, basically due to non-judicious use of pesticides. In India, 51% of food commodities are contaminated with pesticide residues and out of these, 20% have pesticides residues above the maximum residue level values on a worldwide basis. It has been observed that their long-term, low-dose exposure are increasingly linked to human health effects such as immune-suppression, hormone disruption, diminished intelligence, reproductive abnormalities, and cancer. In this light, problems of pesticide safety, regulation of pesticide use, use of biotechnology, and biopesticides, and use of pesticides obtained from natural plant sources such as neem extracts are some of the future strategies for minimizing human exposure to pesticides (Gupta, 2004). Among the pesticides, use of insecticides is maximum followed by fungicides and weedicides (Fig. 4).

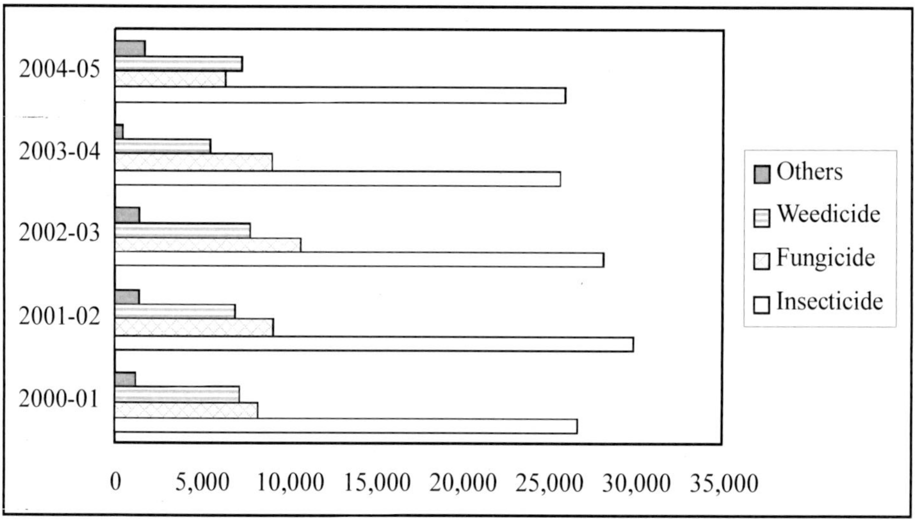

Fig. 4. Pesticides consumption in India (Metric tones)

The levels of HCH and DDT residues detected in rice, maize, turmeric, coriander, black pepper, and all the vegetables and fruits also lower than those found in wheat flour, oil, and fat samples analyzed and suggested that a restricted and controlled use of such persistent pesticides may be useful for decreasing their contamination levels in different food items (Kaphalia *et al.* 1990). Acute and chronic diseases like cancer, lymphoma, birth defects, infant death associated with pesticide exposure. According to Duram (2005), about 350 pesticides commonly used in US on food production system, in homes and on pets. Impacts of pesticides noticed in every location on the earth even in southern and northern arctic areas. The inert materials used in pesticides are not studied well and not enough data on it. Water pollution by pesticides and fertilizers are equally disturbing. Industrial agricultural chemical soared from 1965 to 1990, the estimated crop losses from insects, diseases, and weeds actually increased from about 35 percent to 42 percent world wide. This indicates that while we are poisoning our environment and ourselves we are also failing to control crop pests. Industrial agriculture reduces diversity on relying on just a few crops. This, in turn, simplifies and intensifies the pest complex present.

FAO (2001) noted that during the second half of the 1990s, a strong and steady growth in the sales of organic foods has provided a viable and sometimes value added market niche. Changes in dietary habits among many segments of the population of developed countries - resulting from increased health awareness and the increasing demand for a wider variety of products, including convenience food - have contributed to this growth. Due to major food scares, which hit many countries in western Europe in the late 1990s and early years of this century, consumers in general have become more critical when purchasing food. Moreover, they have become more demanding regarding information on production and processing aspects (including tradability of the product). The sale of organic horticultural products has been expanding rapidly in many of the major organic markets (e.g. the United States, countries in the European Community and Japan). However, the market share of organic products in total food sales is still small, with shares ranging between one and three percent.

1.7 ORGANIC AGRICULTURE MOVEMENT

The human civilization from 'Hunting and Gathering' to present day 'Global Village' under went several changes. The fast changes witnessed depletion of production bases in many fertile grounds coupled with contamination of natural resources and human health hazards. That warranted the change in directions of agriculture for sustainable development. Modern agriculture and food systems, including organic agriculture, are undergoing a technological and structural modernization and faced with a growing globalization. Organic agriculture (OA) seen as pioneering efforts to create sustainable based on other principles than mainstream agriculture (Kristensen, 2005). The mainstream agriculture is 'conventional agriculture' where, externally industry produced inputs such as fertilizers and pesticides are used. The degree of use of these inputs varies among farmers or countries. Still there are farmers in developing countries produce crops without the support of external inputs.

Organic farming is a way of agriculture that relies on ecosystem management rather than external agricultural inputs. This approach excludes the use of synthetic inputs, such as synthetic fertilizers and pesticides, veterinary drugs, and genetically modified organisms (GMOS), and usually subscribes to the principles of sustainable agriculture. Its theoretical basis emphasizes soil health as the foundation for successful production. Its proponents believe that healthy soil, maintained without the use of man-made fertilizers and pesticides, and livestock raised without drugs, yields higher quality food than conventional, chemical-based agriculture. In many countries, including the US and in the EU, organic farming is also defined by law, making the commercial use of the term *organic* to describe farming and food products regulated by the government.

Codex Alimentarius Commission a joint body of FAO/WHO defines "organic agriculture as holistic food production management systems, which promotes and enhances agro-ecosystem health, including biodiversity, biological cycles, and soil biological activity. It emphasizes the use of management practices in preferences to the use of off-farm inputs, taking into account that regional conditions require locally adapted systems. This is accomplished by using, where possible, agronomic, biological and mechanical methods, as opposed to using synthetic materials, to fulfill any specific function within the system"(FAO, 1999).

Health is a wealth; modern life is under immense stress particularly for urban dwellers. All our natural resources - air, water, and soil are polluted, and some places it is beyond reclaim. Duram, (2005) states that organic farming promoted as

- environmentally friendly and people saw it as a solution to the social and environment problems of industrial agriculture.
- consumers turned to organic foods because of food scares such as mad cow disease and concerns over genetically modified organisms (GMOS).
- consumers were becoming less accepting of mass-produced food and inhumane treatment of animals in industrial production.
- a lifestyle choice, perhaps related to socioeconomic status, influenced consumer purchases of organic food,
- organic farming was becoming more accepted within rural areas and among farmers themselves.

Organic crop nutritionally superior evidenced from findings of USA and Europe. Organic food supplies more antioxidants, primary and secondary nutrients, and secondary metabolites. Evidence show that livestock prefers organic food. Organic food strengthens immune system. Synthetic pesticides depress defense mechanism of plants in non-organic production system. Experiment evidence shows that Organically grown straw berries, black berries and corn had 19 per cent, 50 per cent and 58 per cent higher anti-oxidant, respectively than conventional crop. Nutritional superiority, food safety, fresher taste, and environmental concern, consumer demand is clearly strong and growing as people are willing to pay more organically grown food. Commonly organic food priced 10 to 30 % higher over non-organic food. However, organic research receives very little support. Out of 8,85,863 acres of land grant universities, in US only 151 acres (0.02%) are certified organic system. Incredibly, of the thirty thousands USDA research studies, only thirty four have a 'strong organic' focus and these represents 0.1% (Duram, 2005). Organic farmers have their own experiment, developed pest management and soil fertility techniques, and share their findings with other organic farmers. The human health benefit of organic farming is due to overall improvement of ecosystem particularly soil was recorded by Reganold *et al.* (1993). However, it varies with socio-economic and eco-system setup of a country or a region, consumer acceptance and market access etc., According to Kristiansen (2006), the European form of organic agriculture, especially its current market-driven style, is not necessarily the most appropriate system for other countries. The principles of OA are guides to tailor organic practices to each individual farming location.

In ancient times, particularly in orient, maintenance of soil productivity was by addition of every possible material that the farmer considered would contribute nutrients to his crops. They are various kinds of organic matter, including human and animal manures, crop residues, canal muck, and vegetation from hill lands and canals. All ashes saved and returned to the land directly or indirectly. Fish fertilizers, bean, and peanut cakes also used and agriculture were self-sustaining basis in all of the oriental countries for centuries (King, 1911; Allison, 1973). Composting was the channel by which many of the plant nutrients prepared for application to the soil. Fussell (1966) states that the Greeks and Romans realized the value of marl, pigeon, poultry dung, green manures, legumes, and mulches. Printing invented during the fifteenth century and for the first time in world history it was possible for agricultural knowledge easily and widely disseminated and rapidly during following one or two centuries. Prior to this only handwritten books were available and these found only in the libraries of monasteries and the great landowners.

1.7.1 History of organic movement

The first 40 years of the 20th century saw simultaneous advances in biochemistry and engineering that rapidly and profoundly changed farming. The introduction of the gasoline internal combustion engine ushered in the era of the tractor, and made possible hundreds of mechanized farm implements. Research in plant breedling led to the commercialization of hybrid seed. And a new manufacturing process made nitrogen fertilizer - first synthesized in the mid-1800s - affordably abundant (Prasad, 2005).

In the early 1900s, American agronomist F.H. King toured China, Korea, and Japan, studying traditional fertilization, tillage, and general farming practices. He published his findings in *Farmers of Forty Centuries* (1911). King (1911) probably did not view himself as part of a movement, organic or otherwise, but in later years his book became an important organic reference.

In Germany, Rudolf Steiner, the founder of the philosophy of 'Anthroposophy' and development biodynamic agriculture was probably the first comprehensive organic farming system (the apparent beginning of which was a lecture Steiner presented in 1924). The Steiner's philosophy different from organic agriculture principally as it has spiritual, mystical and astrological aspects (Kristiansen, 2006). The first organic certification and labelling system, 'Demeter', created in 1924.

Sir Albert Howard, often referred to as the father of modern organic agriculture. From 1905 to 1924, he worked as an agricultural adviser in Pusa, Bengal, India, where he documented traditional Indian farming practices, and came to regard them as superior to his conventional agriculture science. His research and further development of these methods recorded in his writings, notably, his 1940 book, *An Agricultural Testament*, which influenced many scientists and farmers of the day.

In 1939, influenced by Sir Howard's work, Lady Eve Balfour launched the Haughley Experiment on farmland in England and compared organic and conventional farming. She published *The Living Soil*, based on the initial findings. Widely read, it led to the formation of a key international organic advocacy group, the Soil Association.

The coinage of the term *organic farming* was by Lord Northbourne, in his book, *Look to the Land* (1940), wherein he described a holistic, ecologically-balanced approach to farming (Northbourne, 1940).

In the late 1930's in rural Pennsylvania, USA, J.I. Rodale realised the importance and protecting the natural health of the soil to preserve and improve human health. In 1947, he founded the Soil and Health Foundation that later become The Rodale Institute. The central message and philosophy is 'healthy soil, equals healthy food, equals healthy people'.

In Japan, Masanobu Fukuoka, a microbiologist working in soil science and plant pathology, began to doubt the modern agricultural movement. In the early 1940s, he quit his job as a research scientist, returned to his family's farm, and devoted the next 30 years to developing a radical no-till organic method for growing grain, now known as Fukuoka farming.

Technological advances during World War II accelerated post-war innovation in all aspects of agriculture, resulting in big advances in mechanization (including large-scale irrigation), fertilization, and pesticides. In particular, two chemicals that had been produced in quantity for warfare, were repurposed to peace-time agricultural uses. Ammonium nitrate, used in munitions, became an abundantly cheap source of nitrogen. And a range of new pesticides appeared: DDT, which had been used to control disease-carrying insects around troops, became a general insecticide, launching the era of widespread pesticide use.

At the same time, increasingly powerful and sophisticated farm machinery allowed a single farmer to work ever larger areas of land. In 1944, an international campaign called the Green Revolution launched in Mexico with private funding from the US. It encouraged the development of hybrid plants, chemical controls, large-scale irrigation, and heavy mechanization in agriculture around the world.

During the 1950s, sustainable agriculture was a topic of scientific interest, but research tended to concentrate on developing the new chemical approaches. In 1962, Rachel Carson, a prominent scientist and naturalist, published *Silent Spring*, chronicling the effects of DDT and other pesticides on the environment. *Silent Spring* is widely considered as being a key factor in the US government's 1972 banning of DDT. The book and its author often credited with launching the worldwide environmental movement. Similar book by Duram, L. A. (2005) 'Good Growing - Why Organic Farming Works', University of Nebraska Press, Lincoln and London also documented the ill effects of chemical farming.

In the 1970s, global movements concerned with pollution and the environment increased their focus on organic farming, which was promoted through slogans like *"Know Your Farmer, Know Your Food"*. In 1972, the International Federation of Organic Agriculture Movements, widely known as IFOAM, was founded and dedicated to the diffusion and exchange of information on the principles and practises of organic agriculture of all schools and across national and linguistic boundaries. In 1975, Fukuoka released his first book, *One Straw Revolution*, with a strong impact in certain areas of the agricultural world. His approach to small-scale grain production emphasized a meticulous balance of the local farming ecosystem, and a minimum of human interference and labor.

In the 1980s, around the world, various farming and consumer groups began seriously pressuring for government regulation of organic production. This led to legislation and certification standards enacted through the 1990s and to date. Since the early 1990s, the retail market for organic farming in developed economies has been growing by about 20% annually due to increasing consumer demand. Concern for the quality and safety of food, and the potential for environmental damage from conventional agriculture, are apparently responsible for this trend.A large number of terms used as an alternative to organic farming. These are: biological, ecological, bio-dynamic, organic-biological and natural agriculture.

1.8 PRINCIPLES OF ORGANIC AGRICULTURE

IFOAM (2002) set four principles - The principle of health; The principle of ecology; The principle of fairness and The principle of care. The principles are to be used as a whole. They are composed as ethical principles to inspire action.

Principle of health : Organic Agriculture should sustain and enhance the health of soil, plant, animal and human as one and indivisible.

Principle of ecology : Organic Agriculture should be based on living ecological systems and cycles, work with them, emulate them and help sustain them.

Principle of fairness : Organic Agriculture should build on relationships that ensure fairness with regard to the common environment and life opportunities.

Principle of care : Organic Agriculture should be managed in a precautionary and responsible manner to protect the health and well-being of current and future generations and the environment.

1.9 ORGANIC FARMING - A GLOBAL PERSPECTIVE

According Helga and Yussefi (2007), organic agriculture is developing rapidly and is now practiced in more than 120 countries of the world. Its share of agricultural land and farms continues to grow in many countries. Furthermore, it can reasonably be assumed that uncertified organic farming is practiced in even more countries.

According to the SOEL-FiBL Survey 2007, almost 31 million hectares are currently managed organically by more than 600'000 farms worldwide (Table 4). This constitutes 0.7 percent of the agricultural land of the countries covered by the survey. The continent with most organic land is Australia/ Oceania with almost 11.9 million hectares, followed by Europe with almost 7 million hectares, Latin America (5.8 million hectares), Asia (almost 2.9 million hectares), North America (2.2 million hectares) and Africa (almost 0.9 million hectares). On a continent level, the share or organic land in proportion to all agricultural land is highest in Australia/Oceania (2.6 percent), followed by Europe. It should be noted, though, that some countries in Europe exhibit a much higher percentage; some countries have reached shares of more than ten percent of agricultural land (Austria, Switzerland). In the European Union, the share of organic land is almost four percent. However, countries like India growth is slow (Table 5).

Table 4 Organic area in the world

Continent	Organic land area(hectares)	Share of total agricultural area	Organic farms
Africa	890'504	0.11%	124'805
Asia	2'893'572	0.21%	129'927
Europe	6'920'462	1.38%	187'697
Latin America	5'809'320	0.93%	176'710
North America	2'199'225	0.56%	12'063
Oceania	11'845'100	2.59%	2'689
Total	30'558'183	0.74%	633'891

Source: SOEL-FiBL Survey 2007

Table 5. Organic farming status in India

Year	Organic Area (Ha)		% share to Agri area in India	No. of Organic Farms	
	World	India		World	India
2003	22811267	41000	0.03	398804	5661
2004	24070010	37050	0.03	462475	5147
2005	26458270	76326	0.05	558449	5147
2006	31502786	114037*	0.06*	622782	5147

Source: SOEL Survey 2003 to 2006. (Yussefi, M. Foundation Ecology & Agriculture SOEL, Germany, www.soel.de)

Global sales of organic food and drink are soaring, however supply-related problems are expected to increase as production becomes international. Exceptionally high market growth rates are causing demand to outpace supply. The North American region is the most adversely affected with food manufacturers and retailers grappling to find adequate supply of organic products. American companies are currently scouring the globe for organic ingredients. The European organic food industry is also experiencing supply shortages. Even sectors that have been suffering from overproduction have fallen victim. Significant volumes of organic meat products were sold as conventional meats between 2002 and 2004 because demand fell short of supply. The European organic dairy sector also suffered from oversupply with over a third of all organic milk produced in countries like Denmark, the UK, and France going into the conventional market. Both sectors are now experiencing undersupply with retailers struggling to find new sources. One British retailer is marketing transitional organic milk, whilst other retailers have resorted to importing organic milk. A growing number of European supermarkets are offering fixed contracts to livestock farmers to encourage them to convert to organic practices.

Product shortages in North America and Europe are resulting in organic food imports from across the globe. Organic ingredients like beans, seeds and nuts are increasingly coming from countries like China, Turkey and Brazil. Organic herbs and spices are being imported from India, Paraguay and Ethiopia. Increasing volumes of organic fresh fruit & vegetables are coming in from African and Asian countries. Latin America and Australasia are already established sources of organic meat products.

Organic food production has indeed become global, however the disparity between producer and consumer countries is growing. Demand for organic products is concentrated in affluent countries where production of organic foods is increasing at a relatively slow rate. The largest increases in organic food production are in developing countries which have very small internal markets for organic products. For instance, the amount of organic farmland in Africa, Asia and Latin America has reported triple-digit growth since 2000 compared to double-digit growth in other regions. However, production in these regions is highly export-geared with relatively few certified organic products sold in producer countries.

Most sales of organic foods are from countries where consumers have high disposable incomes. The G7 countries account for over 80% of total sales, whereas their share of international organic farmland is a mere 12%. This over-concentration of demand puts the global organic food industry in a potentially fragile condition. Most organic food is produced for two regions. Protectionism measures or sluggish demand from one or both regions would close up markets for producers in developing countries.

Supply of organic foods has become global, however demand has not. Producers in developing countries are advised to develop internal markets for organic products to spread business risk. The challenge is creating demand in countries where large sections of the population cannot afford to pay the price premium for organic foods. Producers can learn from Latin American producers who have successfully developed regional markets for their organic products. The Eco Vida Network in Brazil is a fine example of how producers have developed local markets by building strong relationships with consumers. By thinking local and undertaking such initiatives, producers can ensure the global organic food industry continues to go from strength to strength (http://www.organicmonitor.com/700240.htm).

According to the Fair-trade Foundation, global sales of fair-trade certified products topped the €1.1 billion mark in 2005 - an increase of 37% from 2004. The USA and the UK are leading the fair-trade crusade, with markets worth € 344 million and € 277 million respectively, but other countries are experiencing exponential year on year growth, notably Australia/New Zealand (178%), Canada (99%), Finland (73%), Sweden (69%) and Austria (62%). The absence of a pan-European organic certification body makes it harder to get a clear consensus on the size of the organic food and drink market. UK organic certification body, the Soil Association, estimated the global market for organic food and drink at £16.7 billion (€24.2 billion) in 2005, while the just-food report 2006 Global market review of the whole, natural, organic and ethical food markets valued it at US$36.7 billion. Within Europe, Germany is the largest market for organic products, followed closely by the fast-growing UK market, which is worth £ 1.6 billion (€ 2.2 billion) and expanding by 30% annually. Given that overall food industry growth in most European markets is somewhere between 2 and 6%, the ripe ethical food and drink market offers rich pickings for food manufacturers. If only it were that straightforward. One barrier which is thwarting the new product development efforts of some manufacturers is securing fair-trade or organic certified ingredients in large enough quantities. Ben and Jerry's, for example, has just replaced its regular vanilla ice cream with a fair-trade version in the UK, The Netherlands and Belgium, and admits that tracking down enough fair-trade certified vanilla was a problem (Searby, 2007).

The demand for organic product is high and growth rate of organic industry is fast. However, countries like India where population is ever increasing; the sustainability is based site-specific balanced and adequate fertilization with conjunctive use of organic and inorganic sources of plant nutrients (NAAS, 2005) to meet the food requirement. However, it was suggested to develop a organic farming to met the export demand of a specialized crops only.

1.10 MARKET OPPORTUNITY FOR ORGANIC SPICES

According to Export-Import Bank of India (Anon, 2003), total global organic spice exports in 2001-2002 were 933.5 metric tons valued at approx. US$ 3.2 million. Organic spice exports have been growing rapidly in recent years, recorded 25 percent increase in quantity, and 11 percent increase in value terms during 2001-02. The share of organic spice is negligible compared to conventional spices trade. The main organic spices traded by volume are pepper (39 percent), ginger (19 percent), turmeric (11 percent), cardamom (8 percent), chilli (7 percent), cloves (7 percent), nutmeg (7 percent), vanilla (1 percent) and others (7 percent). The largest producers and exporters of organic spices are Sri Lanka (28 percent), India (12 percent), Indonesia (10 percent), Madagascar (8 percent), Tanzania (10 percent) and Guatemala (8 percent). Sri Lanka is the largest producer of organic pepper followed by India. Guatemala is the largest producer of organic cardamom and Madagascar is the main player in the organic vanilla trade. Organic spice production country wise and crop wise given in Fig 5a & b.

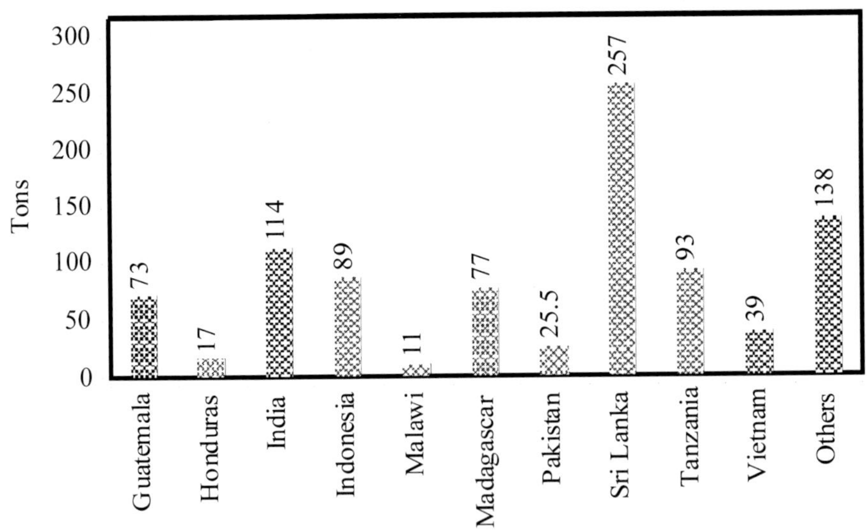

Fig. 5 (a) Country wise production of organic spices during 2001-02

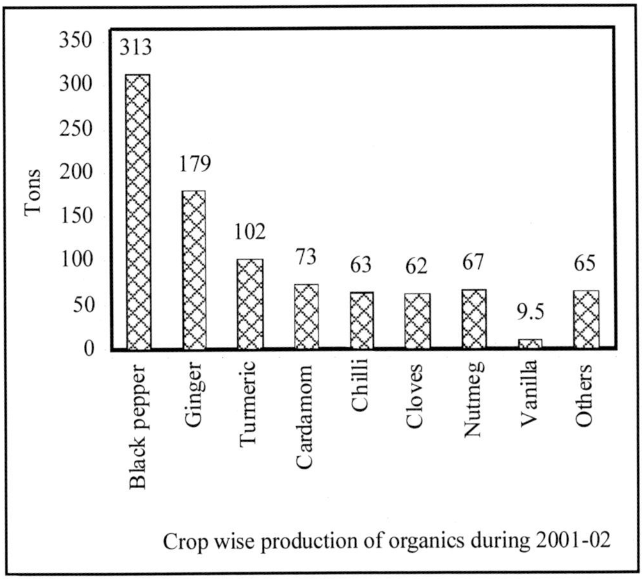

Fig 5 b. Supply of organic spices during 2001-02

The main consumers of organic spices are Germany, UK, France, Japan and USA. The organic spice imported and consumed was approximately 875 metric tons valued 3.0 million. Pepper, ginger and turmeric were the main spices (over 60 percent) consumed, followed by cardamom, chilli, cloves and nutmeg (Fig. 6 a & b).

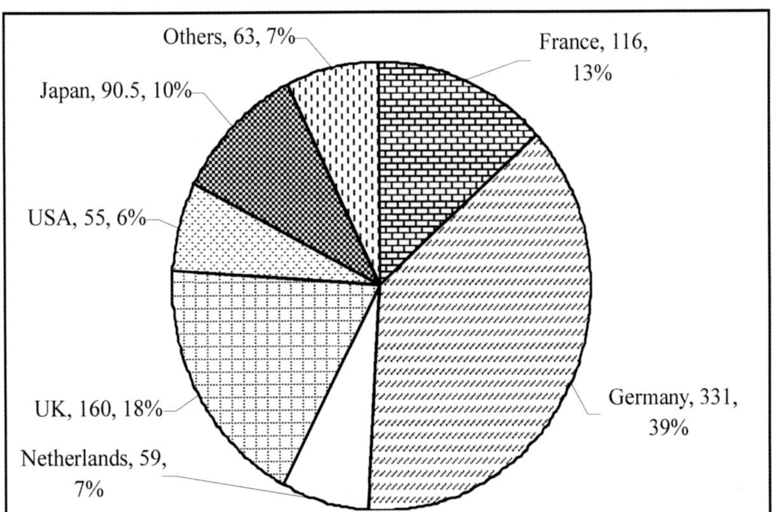

Fig 6 a. Country wise consumption of organic spices (tons, %) during 2001-02

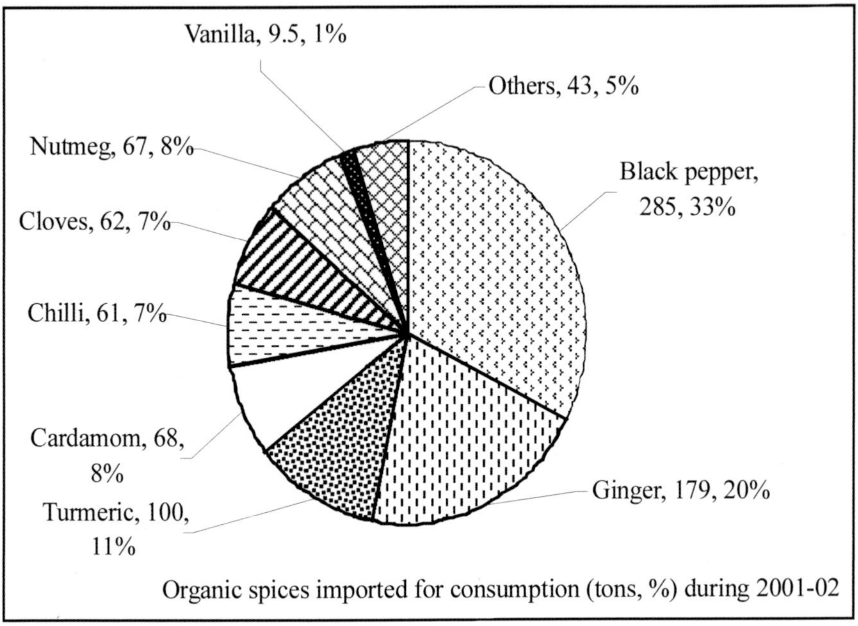

Fig. 6 b. Import of organic spices and consumption during 2001-02

Ethnic cuisine is becoming very popular in the west where people are increasingly experimenting with spices that were once considered exotic. This trend leads to a growing demand for both conventional and organic spices in the market. The important factors that driving the demand for organic spices are

i) *Health*: Health consciousness is fast becoming an important determinant in western lifestyles.

ii) *Medicinal*: Essential oils and spices that are used for phyto-medicinal purposes are generally required to be residue free. Certified organic spices are only 100 percent chemical free.

iii) *Food processing*: Most food processing industries have come under stringent standards, testing inputs for various aflatoxins and residues are switching to organic spice inputs.

iv) *Life style and diet*: With the changing culinary habits more consumers either purchase spices for at-home meal preparation, or frequent restaurants that serve spicy cuisines.

v) *Competition*: Competition from artificial flavourings is also an important factor affecting the demand for natural spices.

REFERENCES

Allison, F.E. (1973) *Soil Organic Matter and Its Role in Crop Production*. Elsevier Scientific Publishing Company, Amsterdam. P.637.

Anonymous, (2003) *Export of organic products from India : Prospects and Challenges. Occasional Paper No. 97*. Export-Import Bank of India, Mumbai.

Brown, Michele R. (2003) Spice It Up (http://www.answerpoint.org/columns2.asp?column_id=681&column_type=feature) (online accessed on 09-3-3-2007).

Chomchalow, N. (1996) Spice Production in Asia - An Overview. Unpublished paper presented at the IBC's Asia Spice Markets '96 Conference, Singapore, 27-28, May 1996 (www.journal.au.edu/au_techno/2001/oct2001/article6.pdf) online accessed on 08-3-2007.

Davy, H. (1813) Elements of Agricultural Chemistry. In a course of Lectures for the Boards of Agriculture. W. Bulwer, London.

Duram, L. A. (2005) Good Growing – Why Organic Farming Works. University of Nebraska Press, Lincoln and London. p. 250.

FAO. (1999) Organic Agriculture. Food and Agriculture Organization of the United Nations, Rome. http://www.fao.org/unfao/bodies/COAG/COAG15/X0075E.htm (online accessed on 09-3-2007).

FAO. (2001) World Markets for Organic Fruit and Vegetables-Opportunities for Developing Countries in the Production and Export of Organic Horticultural Products 2001 International Trade Centre, Technical Centre For Agricultural And Rural Cooperation, Food And Agriculture Organization Of The United Nations Rome, http://www.fao.org/docrep/004/Y1669E/y1669e02.htm#bm02 (online) (online accessed on 09-3-2007).

FAO. (2007) *FAO Database*. Food and Agriculture Organization of the United Nations, Rome.

Fussell, G.E. (1966) *Farming Technique from Prehistoric to Modern Times*. Pergamon, New York, P. 269.

Gupta P. K. (2004) Pesticide exposure—Indian scene. *Toxicology*. 198(1-3): 83-90.

Helga, W. and Yussefi, M. (2007) *The World of Organic Agriculture. Statistics and Emerging Trends 2007*. 9th edition, International Federation of Organic Agriculture Movements IFOAM, Bonn, Germany & Research Institute of Organic Agriculture FiBL, Frick, Switzerland.

http://asiapacificuniverse.com/pkm/spiceroutes.htm (online accessed on 09-3-2007).

http://www.astaspice.org/spice/frame_spice.htm (online accessed on 09-3-2007).

http://dacnet.nic.in/eands/PORTRAT-G.pdf (online accessed on 09-3-2007).

http://ias.berkeley.edu/orias/Spice/textobjects/overview.htm (online accessed on 09-3-2007).

http://indianfood.indianetzone.com/spices/ (online accessed on 09-3-2007).

http://www.ncipm.org.inlasps/Display/Pesticides - asp (online accessed on 9-3-2007).

http://studentsfriend.com/sf/part1see/part1-3.html (online accessed on 09-3-2007).

http://www.fertilizer.org/ifa/statistics/indicators/tablenpk.asp (online accessed on 09-3-2007).

http://www.indo.com/indonesia/history.html. (online accessed on 09-3-2007).

http://www.organicmonitor.com/700240.htm (online accessed on 09-3-2007).

http://www.reference.com/browse/wiki/Spice (online accessed on 09-3-2007).

http://www.reference.com/browse/wiki/Spice. (online accessed on 09-3-2007).

http://www.suite101.com/lesson.cfm/18003/1349 (online accessed on 09-3-2007).

http://zanzibar.net/zanzibar/history/spice_trade (online accessed on 09-3-2007).

IFOAM, (2002) Norms for organic production and processing, section B, General Principles, Recommendations and Standards. International Federation of Organic Agriculture Movement, http://www.ifoam.org.

Kaphalia, B. S., Takroo, R., Mehrotra, S., Nigam, U., Seth, T. D. (1990) Organochlorine pesticide residues in different Indian cereals, pulses, spices, vegetables, fruits, milk, butter, Deshi ghee, and edible oils. *J Assoc off Anal Chem.* 73(4):509-12.

King, F. H. (1911) *Farmers of Forty Centuries*. Democrat. Madison, Wisc., P. 438.

Kristensen, E. S. (2005) *Organic Farming in a Global Perspective – Globalisation, Sustainable Development and Ecological Justice*. Danish Research Centre for Organic Food and Farming (DARCOF), Denmark.

Kristiansen, P. (2006) Overview of organic agriculture. In: (Eds. P. Kristiansen, A.Taji and J. Reganold) *Organic Agriculture : A Global Perspective*. CSIRO Publishing, Collingwood Victoria, Australia.

Mazrui, A. A. (1998) *A tale of two continents*: Africa, Asia & the dialectic of Globalization (http://tcdc.undp.org/CoopSouth/1998_2/cop98213.pdf) (online accessed on 09-3-2007).

Miller, J. I. (1969) *The spice trade of the Roman Empire*, 29 B.C. to A.D. 641. Clarendon Press, Oxford.

NAAS, (2005) *Organic farming: Approaches and possibilities in the context of Indian agriculture.* Policy Paper 30. National Academy of Agriculture Sciences, New Delhi.

Northbourne, Lord. (1940) *Look to the Land*. Basis Books, London.

OECD, (2001) OECD *Environmental outlook for the chemicals industry*. Organisation for Economic Co-operation and Development, Environment Directorate, Paris.

Parthasarathy, V.A., Sasikumar, B., Nair, R. R. and Johnson George, K. (2007) Black pepper: Botany and Horticulture. *Horticulture review*. 33: 173 – 266.

Prasad, R. (2005) Organic farming *vis-à-vis* modern agriculture. *Current Sci.* 89 (2): 252-254.

Ravindaran, P. N. (2006) Spices: Definition, Classification, History, Properties, Uses and Role in Indian Life. In: (P.N.Ravindran, K.Nirmal Babu, K.N.Shiva, Johny, A.K. (Eds.)) *Advances in Spices Research – History and Achievements of Spices Research in India since Independence*, Agrobios (India), Jodhpur.

Reganold, J. P., Palmer, A. S., Lockhart, J. C. and Macgregor, A. N. (1993) Science 260: 344-349.

Ridley, R. N. (1912) *Spices*. Macmillan and Co. Limited, London (Reprinted at Ram Printograph (India) New Delhi).

Rosengarten, Jr. F. (1969) *The Book of Spices*, Jove Publ., Inc., New York, p. 23-96.

Searby, L. (2007) Sourcing Ethical & Organic Ingredients. http://www.thefoodworld.com/show_article/36/ (Online) (accessed on 05-5-2007).

UNDP, (1998) *Human development report – consumption for human development*. United Nations Development Programme, New York/Oxford.

UNEP. (1999) *The state of the environment: freshwater*. GEO-2000: global environment outlook. United Nations Environment Programme, Nairobi.

UNEP. (2004) Food and Agriculture Organization of the United Nations/United Nations Environment Programme/World Health Organization. *Childhood pesticide poisoning: information for advocacy and action*. United Nations Environment Programme, Geneva.

von Leibig, J. (1840) *Organic Chemistry in its Application to Agriculture and Physiology*. Taylor and Walton, London.

WB, (2002) Toxics and poverty: the impact of toxic substances on the poor in developing countries. World Bank, Washington, DC.

WB, (2003) *Persistent organic pollutants: a legacy of environmental harm and threats to health*. May 2003 (Environment Strategy Series, No. 6). World Bank, Washington, DC.

White, K. D. (1970) *Agricultural History* 44: 281-290.

WHO. (1990) *Public health impact of pesticides used in agriculture*. World Health Organization, Geneva.

WHO. (2001) FAO/WHO: amount of poor-quality pesticides sold in developing countries alarmingly high Press Release WHO/04, 1 February 2001. http://www.who.int/inf-pr-2001/en/pr2001-04.html (accessed on 09-03-07).

WHO. (2003) *The world health report 2003 – shaping the future.* World Health Organization, Geneva.

Yadav, J. S. P. (2003) Managing soil health for sustained high productivity. *J. Indian Soc. Soil Sci.* 51(4): 448-465.

Yáñez, L., Ortiz, D., Calderón, J., Batres, L., Carrizales, L., Mejía, J., Martínez, L., García-Nieto, E. and Díaz-Barriga, F. (2002) Overview of human health and chemical mixtures: problems facing developing countries. *Environmental Health Perspectives,* 110(6): 901–909.

2

Organic Manures

R. Dinesh, V. Srinivasan, K. Kandiannan, S. Hamza

1.0 INTRODUCTION

Soil fertility can be considered to be a measure of the soil's ability to sustain satisfactory crop growth, both in the short and longer-term. Organic farming recognizes the soil as being central to a sustainable farming system. It could be argued that organically managed soils require a higher level of baseline fertility because shortfalls cannot be made up by applications of soluble fertilizers. Soil fertility is determined by a set of interactions between the physical and chemical environments of the system and by biological activity. Organic matter is linked intrinsically to soil fertility, because it is important in maintaining good soil physical conditions (e.g. soil structure, aeration and water holding capacity) and is an important nutrient reserve. Organic matter contains most of the soil reserve of N and large proportions of other nutrients such as phosphorus and sulphur. Soil fertility is thus markedly affected by not only *quantity*, but also *quality* of the organic matter. To maintain soil fertility, organic farming relies on sound crop rotations to include fertility building and fertility depleting stages, returns of crop residues, N-fixation by legumes/ rhizobium, nutrient retention by green manures and effective use of manures/ composted wastes.

It is well known that organic farming system implies a farming system that primarily aims at cultivating land and raising crops under ecologically favourable condition. It emphasizes restricting the use of chemical inputs whether it is inorganic fertilizers or pesticides and instead, relies more on an integrated approach of crop management practices making use of cultural, biological and natural inputs. Addition of organic manures such as FYM, recycling of organic wastes through composting, green manures and biological inputs like vermicompost and biofertilizers etc. constitute important components for plant nutrient management in organic farming. Soil organic matter is the key to soil fertility and productivity in organic farms. In the absence of organic matter, the soil is a mixture of sand, silt and clay. Organic matter induces life into this inert mixture and promotes biological production. Soil organic matter consists of a variety of components. These include, in varying proportions and many intermediate stages:

- Raw plant residues and microorganisms (1 to 10 per cent)
- "Active" organic traction (10 to 40 per cent)
- Resistant or stable organic matter (40 to 60 per cent) also referred to as humus.

Raw plant residues, on the surface, help reduce surface wind speed and water runoff. Removal, incorporation or burning of residues predisposes the soil to serious erosion. The "active" and some of the resistant soil organic components, together with microorganisms (especially fungi) are involved in binding small soil particles into larger aggregates. Aggregation is important for good soil structure, aeration, water infiltration and resistance to erosion and crusting. The resistant or stable fraction of soil organic matter contributes mainly to nutrient holding capacity (cation exchange capacity) and soil color. This fraction of organic matter decomposes very slowly and therefore has less influence on soil fertility than the "active" organic fraction.

Organic matter in soil serves several functions. From a practical agricultural standpoint, it is important for two main reasons. First as a "revolving nutrient bank account"; and second, as an agent to improve soil structure, maintain tilth, and minimize erosion.

2.0 FUNCTIONS OF SOIL ORGANIC MATTER (SOM)

Soil organic matter is recognized as a fundamental indicator of soil quality and sustainability (Tiessen *et al.* 1994; Reeves 1997), through its improvement of soil aggregation, permeability and soil water holding capacity (Schjønning *et al.* 2002). Increasing organic matter and improving soil structure also reduces runoff and wind and water erosion in most soils (Lal 2002). As a consequence, many systems that have been managed to enhance organic matter show less long-term yield variability and are less sensitive to drought than conventionally managed systems (Lotter *et al.* 1993). Organic matter inputs through organic amendment, in addition to supplying nutrients, improve soil aggregation (Caravaca *et al.* 2002; Ferreras *et al.* 2006) and pore size distribution (Tejada and Gonzalez 2003), which markedly eases root penetration and flow of water and gases, directly related to plant growth (Marinari *et al.* 2000) and seedling emergence (Ferreras *et al.* 2000). Organic inputs also stimulate microbial diversity and activity (Carpenter-Boggs *et al.* 2000; Dinesh *et al.* 2000), replenishes essential micronutrients in soil, increases the utilisation efficiency of applied fertilizers and favors nutrient availability to the plant (Lee *et al.* 2004). The role of composted organic manures in disease suppression (Noble and Coventry 2005; Termorshuizen *et al.* 2006), adsorption and transport of heavy metals (Kaschl *et al.* 2002; De Volder *et al.* 2003) and the potential to adsorb or transform hazardous organic pollutants and persistent biological molecules (Moeller and Reeh 2003; Löser *et al.* 2004) is well recognized.

Long-term experiments have shown the benefit of manure, adequate fertilization, and crop rotation on maintaining agronomic productivity by increasing C input into the soil. Reduced and zero tillage, set-aside, perennial crops are other recommended management practices commonly recognized as preserving or increasing SOC (Paustian *et al.* 1997a; Paustian *et al.* 1997b; Smith 2004; Lal 2004; Freibauer *et al.* 2004).

As a revolving nutrient bank account, organic matter serves two main functions:

- Since soil organic matter is derived mainly from plant residues, it contains all of the essential plant nutrients. Accumulated organic matter, therefore, is a storehouse of plant nutrients. Upon decomposition, the nutrients are released in a plant-available form.
- The stable organic fraction (humus) adsorbs and holds nutrients in a plant available form.

Organic matter does not add any "new' plant nutrients but releases nutrients in a plant available form through the process of decomposition. In order to maintain this nutrient cycling system, the rate of addition from crop residues and manure must equal the rate of decomposition. If the rate of addition is less than the rate of decomposition, soil organic matter will decline and, conversely if the rate of addition is greater than the rate of decomposition, soil organic matter will increase. The term steady state has been used to describe a condition where the rate of addition is equal to the rate of decomposition.

Even in well-fertilized soils, soil fertility is reduced by loss of soil organic matter (Aref and Wander 1997; Martin-Rueda *et al.* 2007). Accompanying these losses in productive potential are losses in agroecosystem efficiency. Crop response to inputs is increased in soils where organic matter status and biological and physical properties influenced by SOM are enhanced (Avnimelech 1986; Martin-Rueda *et al.* 2007). Enhancements in SOM status and crop performances are reported for a variety of management practices, including organic (Wander *et al.* 1994; Olesen *et al.* 2007), compost amended (Stone *et al.* 2001), mixed-crop and cover-cropped (Stevenson *et al.* 1998; Hutchinson *et al.* 2007) and no-till systems (*Ouédraogo et al.* 2006; Lal *et al.* 2007).

3.0 ORGANIC MANURES

The regular incorporation and recycling of organic wastes to the soil is the most efficient method of maintaining optimum levels of soil organic matter. In the traditional agriculture, followed over generations in India, the use of plant and animal wastes as source of plant nutrients was the accepted practice.

Organic manure in a broad sense includes composts from rural and urban wastes, crop residues, agro industrial bio wastes and green manures, apart from the commonly used FYM. The availability of on farm wastes/ by products in the plantation sector of India is given in Table 1. Farmyard manure made from cattle dung, excreta of other animals, animal tissues and excretory products, and compost from rural and urban wastes, crop residues and green manures are collectively termed as bulky organic manures because of their low contents of major plant nutrients (< 2% N), while materials like oil cakes, fish meal, animal meal, poultry manures, slaughter house wastes containing comparatively higher contents of plant nutrients (> 2% N) are termed as concentrated organic manures.

Table 1. Availability of on farm wastes/ by-products in plantation sector in India

Crop/ Crop waste	Quantity
Coconut (excluding coir pith)*	11.2 million tons
Areca leaves	0.13 million tons
Areca rachis	0.08 million tons
Areca husk	0.22 million tons
Cocoa shed leaves	360.03 t
Cocoa pruning	12056.33 t
Cocoa pod husk	32900 t
Coir pith	7.5 million tons
Coffee husk	0.18 million tons
Tea waste	0.22 million tons

* includes spadices, bunch wastes, sheath, inflorescences and husks

Source: Bidappa *et al.* 1996

It is well established that FYM plays an additional role than its capacity to contribute NPK. Unlike chemical fertilizers that supply only the major nutrients, FYM is a storehouse of several plant nutrients and acts as a good soil conditioner. Green manuring has been under practice by our farmers for decades. Estimates suggest that a 40-50 days old green manure crop can supply up to 80-100 kg N ha^{-1}. Even if half of this N is crop utilizable, a green manure crop can be a substitute to 50-60 kg fertilizer N ha^{-1}. Some of the potential legumes for green manuring are dhanicha, sunhemp, cowpea, mung, bean, guar and berseem etc. Dhanicha, sunhemp, mung bean and guar grown during kharif season as green manure crops have been reported to contribute 8-21 tons of green matter and 42-95 kg of N ha^{-1}. Similarly, Khesari, cowpea and berseem grown during rabi season can contribute 12-29 tons of green matter and 67-68 kg of N ha^{-1}.

To be sustainable, organic farming needs to be self-sufficient in nitrogen (N) (Berry *et al.* 2002; Shepherd, 2002) through the fixation of atmospheric di-nitrogen (N_2) by legumes, recycling of crop residues (green manures) and the application of animal manure, or compost. Only fixed N represents a true import of N on to the farm in this situation (see review by Cuttle *et al.* 2004) unless manure is imported. As N is most often the limiting nutrient in organic systems, the aim has to be to maximize N_2 fixation within the system. Despite this reliance on legumes for N, much remains to be understood about how to maximize N_2 fixation and to find ways to make the most efficient use of it.

3.1 Plant and animal by-products

Assorted by-products of the food and fiber industries are occasionally used as soil amendments, returning to the land nutrients that might otherwise be wasted. Many of these products are far too expensive to justify their use in other than very specialized horticultural applications. However, certain agro wastes obtained from shade trees provide ample scope for use as organic amendment. These wastes have good nutritive value (Table 2) and can be used as an input during composting.

3.1.1 Plant by-products

3.1.1.1 Alfalfa meal (or pellets)

It contains around 3% nitrogen and is commonly used as an animal feed. It is an excellent fertilizer material in horticulture, and is said to contain unknown growth factors which make its mineral content more effective as plant nutrients.

Table 2. Nutrient concentration of shade trees and other byproducts in plantations*

Common Name	Botanical Name	Nutrient concentration									
		N	P	K	Ca	Mg	Fe	Mn	Zn	Cu	Mo
		g/kg					mg/kg				
Ebony leaf	*Diospyrose ebenum* J. König	15.1	1.8	21.6	5.2	3.0	268	303	47	2	0.43
Ben teak (Nandi) leaf	*Lagerstroemia microcarpa* Wight	11.1	1.9	15.9	11	4.4	189	123	10	10	0.66
Jack leaf	*Artocarpus integrifolia* L. f.	8.3	1.3	11.8	8.3	2.3	202	148	22	30	0.76
Pink or red cedar leaf	*Aerocarpus fraxinifolius* L.	8.5	1.5	12.1	8.4	1.6	439	73	38	20	0.70
Indian Gooseberry (Aonla)	*Phyllanthus emblica L.*	9.0	1.2	12.9	9.7	4.0	148	62	10	20	0.82
Dragon's Eye leaf	*Dimocarpus longan Lour.*	5.0	1.4	17.4	8.5	1.8	418	47	25	20	0.70
Beleric myrobolon leaf	*Terminalia bellarica Roxb.*	6.0	1.2	8.4	8.8	2.5	337	791	17	30	0.77
Sambul leaf	*Syzygium cumini* (L.) Skeels	11	1.2	15.8	.9.4	3.5	375	265	16	2	0.86
Chocolate tree leaf	*Theobroma cacao* L	8.7	0.7	11	10	2.6	573	257	24	10	1.1
Chempaka Merah leaf	*Michelia champaca* L	16	1.8	23	13	3.2	202	193	32	10	1.9
Malabar tallow or Piney warmish leaf	*Vateria indica* L	6.4	2.0	9.2	5.9	2.1	244	152	21	10	0.32
Silver oak leaf	*Grevillea robusta* Cunn.	15.4	0.4	8.1	18	1.7	725	412	16	11	-
Indian coral tree leaf	*Erythrina indica* L.	24	2.3	20	9.0	3.0	370	160	22	28	-
Kikar or Kareyam leaf	*Garuga pinnata* Roxb	15	2.3	18	7.9	2.2	346	44	19	34	-
Gliricidia leaf	*Gliricidia sepium (Jacq.) Steud.*	28	2.2	22	9	2.9	976	121	30	19	-
Cardamom leaf	*Elettaria cardamomum* (L.) Maton.)	10.1	0.7	15	19	4.1	411	586	18	11	-

Black Pepper leaf	*Piper nigrum* L.	18	1.3	16	31	4.1	102	237	20	15	-
Coffee leaf	*Coffee arabica* L.	25	1.2	17	29	3.7	110	45	20	12	-
Coffee pulp	*Coffee arabica* L.	28	2.3	27	5.6	1.6	272	35	13	36	-
Arecanut (Betel palm) leaf	*Areca catechu* L.	23	1.5	8	9.1	2.3	364	112	19	92	-
Arecanut leaf sheath	*Areca catechu* L.	20	0.8	11	7.0	1.9	325	175	15	77	-
Ficus (Indian fig) fruit	*Ficus indica* (L.) Miller	25	2.2	25	9.6	38	123	22	23	16	-
Coconut leaf	*Cocos nucifera* L.	8.2	1.0	5.0	6.1	2.5	112	26	16	10	-
Coconut husk	*Cocos nucifera* L.	3.1	0.3	4.6	1.5	0.6	58	36	4	3	-

* Approximately 8-14 tons/ha in mixed cropping system

3.1.1.2 Cottonseed meal

It is a rich source of nitrogen (7%). Unfortunately, a substantial percentage of the insecticides used in the U.S. are applied to cotton, and some of these tend to leave residues in the seeds. Most organic certification programs restrict or prohibit the use of cottonseed meal.

3.1.1.3 Fruit pomaces

These are what remain after the juice is extracted. They are heavy, wet products normally available only locally, and best composted before use.

3.1.1.4 Leaf compost

It is increasingly available as more and more municipalities' compost urban and suburban leaves. In principle, the product is a good one, but it is often contaminated with "impurities" ranging from transmission fluid to trash bags.

3.1.1.5 Soybean meal

It is, like alfalfa, most commonly used as a protein supplement for animal feeds. With about 7% nitrogen it can be a useful, but expensive, fertilizer material.

3.1.1.6 Wood ash

It contains about 2% phosphate and 6% potash, but may be contaminated with heavy metals or plastic and typically has a high salt content. Wood ash is rather alkaline, and excessive use can be quite damaging to many soils. Some organic programs restrict its use.

3.1.2 Animal by-products

3.1.2.1 Blood meal

It is dried slaughterhouse waste containing about 12% nitrogen. Unless used carefully, it can burn plants with ammonia, lose much of its nitrogen through volatilization, or encourage fungal growth. In view of the extremely high cost of blood meal, farmers should be sure that it really is the best source of nitrogen in a given situation.

3.1.2.2 Feather meal

It is a common by-product of the poultry slaughter industry. Although total nitrogen levels are fairly high (7 to 10%), the nature of feathers is such that they break down and release their nitrogen much more slowly than many products of similar price.

3.1.2.3 *Fish meal* and *fish emulsion*

These are, like most animal by-products, rich in nitrogen. Fishmeal contains about 10% nitrogen, along with about 6% phosphate. It is most frequently used as a feed additive, but can be used as a fertilizer. The fertilizer analysis of fish *emulsion* varies with preparation method. Whole fish and fish parts must be digested to form slurry, a process accomplished with the aid of either phosphoric acid or special enzymes. Acid-digested fish emulsion usually has an analysis around 4-4-1, while enzyme-digested fish emulsion is usually measured as 4-1-1. Fish emulsion may be fortified with chemical fertilizer, so organic farmers should be suspicious of any product with nitrogen content in excess of 5%.

3.1.2.4 *Leather meal*

It is ground tannery waste with 10% nitrogen. Unfortunately, most leather meal also contains about 3% added chromium (a toxic heavy metal), and is thus prohibited in organic agriculture.

3.2 MANURE- AND COMPOST-BASED PRODUCTS

One of the most common types of prepackaged alternative soil amendments is the manure- or compost-based blended fertilizer. Several of these products have national distribution, and many more enjoy a loyal regional following. Such products are typically analyzed at 2 to 5% for each nutrient. Dried compost is used as a bulking agent, source of nutrients, and organic matter. It is blended with several of the materials including rock minerals and plant and animal by-products. Nearly all products of this class sell for prices about three times greater than their conventional fertilizer value, but may be quite effective in farm situations. However, farmers with access to other sources of manure or compost can realize substantial savings by relying on local manure resources. The nutritive value of different organic compost/ manures suited for spice based cropping systems and the organic manure recommended for spice crops are given in Tables 3 and 4 respectively.

3.2.1 Farmyard manure (FYM)

FYM is composed of dung, urine, bedding and straw. Dung comes mostly from undigested material and the urine from the digested material. More than 50 per cent of the organic matter that is present in dung is in the form of complex products composed of lignin and protein which are resistant to further decomposition and therefore the nutrients present in dung are released slowly. Contrarily, the nutrients from urine become readily available. Straw, saw dust or other bedding materials are used in cattle sheds to reduce the loss

Table 3. Nutritive value of different organic compost/ manures

Organic	N	P	K	Ca	Mg	S	Fe	Mn	Zn	Cu	Mo
	(g/ kg)						(mg/ kg)				
Neem cake	30	1.0	15	10	3.2	2.3	2085	65	17	10	2.5
Cotton cake	35	2.9	13	4.0	3.9	1.5	169	14	37	9	1.6
Mustard cake	22	2.1	12	6.0	2.1	1.5	1062	27	21	18	2.0
Groundnut cake	68	3.6	12	2.0	3.3	2.1	667	37	48	16	5.4
Gingelly cake	60	3.9	11	10	5.0	1.9	1262	50	64	25	11.0
Castor cake	58	11	9.2	3.6	2.4	-	197	22	36	13	-
Farmyard manure	5	2.5	3.0	11	5.5	1.8	2745	466	59	29	11.0
Goat manure	31	1.8	13	18	7.0	1.5	1874	711	120	67	12.0
Poultry manure	30	2.3	17	23	6.0	3.1	2411	234	194	40	10.0
Leaf compost	10	1.0	16	21	3.1	2.1	656	177	12	11	6.0
Vermi compost	20	8.0	12	33	11	8.6	6130	221	44	36	1.6
Coir compost	12	0.6	12	5.0	4.2	1.0	950	50	16	6.0	1.0

Source: Hamza and Srinivasan (2003)

Table 4. Recommended organic manure for spice crops

S. No.	Spice crops	Organic manure	Quantity
1	Black pepper	Farm yard manure (FYM)	5-10 kg / vine
2	Small cardamom	Neem cake/ FYM/ Vermi compost/Poultry manure	4 -10 kg/plant 4 -5 kg/plant
3	Large cardamom	Cattle manures/ Organic cakes	2 kg/plant
4	Vanilla	FYM/ Vermicompost	4 -5 kg/plant
5	Chilli	FYM / Sheep manure/ Neem cake	4 -5 t/ ha 4-5 t ha 3 -5 q/ ha 3 – 4 q/ ha
6	Ginger	FYM/ Neem cake	5 -6 t/ ha 2 t/ ha

Contd...

Table 4. Contd...

7	Turmeric	FYM/ Neem cake	5 -6 t/ ha 2 t/ ha
8	Fennel	FYM	10 -12 t/ ha
9	Coriander	FYM	4 t/ ha
10	Cumin	FYM	4 -5 t/ ha
11	Fenugreek	FYM	4 -5 t/ ha
12	Celery	FYM	10 – 12 t/ ha
13	Clove	FYM	15 – 40 kg/ plant
14	Nutmeg	FYM	15 – 40 kg/ plant

Source: Guidelines for production of Organic Spice in India, Spices Board. 2001

of urine and to increase the bulk of manure. On an average, about 3-5 kg bedding material per animal is used by farmers. When bedding material, as for example, straw with a high C to N ratio is mixed with dung and urine, immobilization of N derived from both the manure itself and from soil may be even more pronounced (Christensen 1986; Thomsen, 2005).

Approximately 70-80% of the nitrogen, 60-85% of the phosphorus and 80-90% of the potassium in feeds are excreted in the manure. The amount of nutrients available for recycling to plants varies widely being dependent upon the composition of the feed ration, the amount of bedding and water added or lost, the method of manure collection and storage, the method of land application, and characteristics of the soil, crop and climate. Manure contains all the plant nutrients needed for crop growth including trace elements. The availability or efficiency of manure utilization by a crop is determined by the method of application, time to incorporation and the rate of manure decomposition by microorganisms in soil. Breland (1994) reported higher N mineralization of crop residues when these were added as discrete clumps compared with residues mixed evenly into the soil. The intensity of soil tillage applied when incorporating manure may correspondingly affect the following turnover.

The varying pattern in immobilization and mineralization calls for an examination of possibilities in optimizing the synchronization between crop N uptake and N mineralization by adjusting the application strategy. Manure contains stable and unstable forms of nitrogen. Unstable N occurs in urine as urea and may account for more than 50% of the total N in manure. Urea changes rapidly to ammonium ions then quickly to ammonia as pH increases and manure begins to dry. Ammonia is extremely volatile resulting in N loss. Nearly all the ammonium N can be lost from surface applied manure if it is not incorporated within a few days. The more intimate contact between soil and manure obtained where the FYM had been mixed into the soil was assumed to give a more extensive incorporation of nutrients into the microbial biomass resulting in lower net N mineralization (Breland 1994; Sørensen and Jensen, 1995).

When manure is applied at a rate to supply the N need of a crop, the P and K will likely be in excess of the crop requirement. Essentially all of the K is available for plant growth the year manure is applied. However, some of the P may be in the form of insoluble inorganic compounds or as organic P and, like stable organic N must be mineralized before it is available. Timing and method of manure application determine the efficiency of nutrient recycling. Incorporating manure immediately minimizes odors and ammonia loss. If manure supplies more N than needed then some ammonia loss is unimportant. However, it is better to apply manure to more acres than to apply in excess to fewer acres. Manure must be spread uniformly to get consistent results. Applying and incorporating manure too early for the crop, and in the fall or early winter in high rainfall areas, could result in N loss and groundwater contamination. Likewise surface runoff and soil erosion must be controlled to protect surface waters. A cover crop planted early (late August to mid September) can be effective in reducing nitrogen leaching through plant uptake

FYM is prepared by the pit method and the heap method. The pit method is followed in areas where the annual precipitation is less than 1000 mm. The pit where the FYM is prepared is generally 1 m deep and 1 m wide and of convenient length. In the pit an absorbent layer is created at the bottom by spreading straw at the rate of 3 to 5 kg per animal kept. The substrate containing well mixed dung, cattle urine and mixed with greens and refuse straw is spread over the absorbent layer daily to form a layer of 30 cm thick and the process is continued until the pit is filled. Each days layer should be pressed, moistened if dry and covered with a 305 cm layer of well ground fertile soil to hasten the decomposition and to absorb ammonia. The heap is turned every three months and in about a years time digested FYM is available with 0.5% N, 0.2% P and 0.5% K. In the heap method, the daily collections from the cattle shed are spread in uniform layers until the heap attains a maximum height of one metre above ground. The top of the heap is rounded of and plastered with dung and mud mixture. In both the methods, aeration is allowed in the beginning and later on anaerobic conditions set in and continue for a long period. The manure is ready for use in 5-6 months.

3.2.2 Poultry manure (PM)

Poultry waste comprises waste feed, solid and liquid droppings, litter, eggshell, diseased and dead birds, feathers and other wastes from poultry sheds. PM is a good source of nutrients, particularly for vegetables and other crops requiring instant N and the benefits of PM and litter in crop production is well documented (Edwards and Daniel 1993; Adams *et al.* 1994; Adegbidi and Briggs 2003; Motavalli *et al.* 2003). Poultry litter (manure plus bedding materials) can improve soil tilth, reduce soil compaction, and add organic matter and nutrients to increase soil fertility and productivity (Edmisten *et al.* 1992). Poultry litter can be applied either as fresh or composted manure (Eghball and Power 1999; Kelling *et al.* 1995).

The nutrient content of PM from different sources is given in Table 5. It has been reported that approximately 74% of total P and 40% of total N in PM are in available form (Shepherd and Withers 1999). N mineralization from poultry manure applied to soils is lower in composted PM compared to non-composted PM. Tyson and Cabrera (1993) observed organic N mineralization rates of 0.4–5.8% and 25.4–39.8% for composted and non-composted PM, respectively, after 8 weeks of incubation at 25°C while Hadas and Portnoy (1994) reported N mineralization rates of 11–29% for poultry and dairy manure derived composts incubated during 32 weeks at 30°C. When applied to soils, PM undergoes rapid decomposition due to fine size particles (Cabrera *et al.* 1993). They observed that fine size particles contributed to increased decomposition of poultry litter. They also observed that 50–60% of total N applied in fresh poultry litter was lost through volatilization during the first five weeks of incubation. Adegbidi and Briggs (2003) estimated that composted poultry manure provided approximately 360 kg ha^{-1} of plant available N with 32% gross mineralization rate. The other animal manures (Table 6) that could possibly be used are manures from horse, swine, sheep/goat, rabbit etc.

Table 5. Nutrient levels in poultry waste

Source	Nitrogen	Ammonia	Phosphorus	Potassium
	----------(%)----------			
Fresh chicken manure	3.7-8.8	0.4-1.1	1.2-2.9	1.2-2.7
Poultry litter manure	1.4-6.8	0.5-1.1	0.5-3.5	1.1-2.7
Broiler litter	2.3-6.0	-	0.6-3.9	0.7-5.2

Source: Sims and Wolf, 1994

Table 6. Approximate NPK values of various animal manures

Animal	Nitrogen	Phosphoric acid	Potash
	--------------------%--------------------------		
Dairy cow	0.57	0.23	0.62
Beef steer	0.73	0.48	0.55
Horse	0.70	0.25	0.77
Swine	0.49	0.34	0.47
Sheep/Goat	1.44	0.50	1.21
Rabbit	2.40	1.40	0.60
Chicken	1.00	0.80	0.39

Source: Anonymous, 1998

3.2.3 Cover crops and green manures

Cover crops are considered as integral components of organic farming systems and are termed as living mulches. A living mulch system is a mixed cropping system in which one partner acts chiefly as a live soil cover for a considerable part of the life cycle of the main crop. Green manuring involves the soil incorporation of any field or forage crop while green or soon after flowering, for the purpose of soil improvement. A cover crop is any crop grown to provide soil cover, regardless of whether it is later incorporated. Cover crops are grown primarily to prevent soil erosion by wind and water. Cover crops and green manures can be annual, biennial, or perennial herbaceous plants grown in a pure or mixed stand during all or part of the year. In addition to providing ground cover and, in the case of a legume, fixing nitrogen, they also help suppress weeds and reduce insect pests and diseases. When cover crops are planted to reduce nutrient leaching following a main crop, they are often termed catch crops.

A winter cover crop is planted in late summer or fall to provide soil cover during the winter. Often a legume is chosen for the added benefit of nitrogen fixation. In northern states, the plant selected needs to possess enough cold tolerance to survive hard winters. Hairy vetch and rye are among the few selections that meet this need. These cool-season legumes include clovers, vetches, medics, and field peas. They are sometimes planted in a mix with winter cereal grains such as oats, rye, or wheat. Winter cover crops can be established by aerial seeding into maturing cash crops in the fall, as well as by drilling or broadcasting seed immediately following harvest.

A summer green manure occupies the land for a portion of the summer growing season. These warm-season cover crops can be used to fill a niche in crop rotations, to improve the conditions of poor soils, or to prepare land for a perennial crop. Legumes such as cowpeas, soybeans, annual sweet clover, sesbania, guar, crotalaria, or velvet beans may be grown as summer green manure crops to add nitrogen along with organic matter. Non-legumes such as sorghum-sudan grass, millet, forage sorghum, or buckwheat are grown to provide biomass, smother weeds, and improve soil tilth.

3.2.3.1 Principal uses of cover crops and green manures

The cornerstone of the organic philosophy is the use of legume-based leys to build fertility. Indeed, the organic standards require the inclusion of legumes in the rotation. Fertility building crops and green manures are therefore key components of organic rotations. Cover crops reduce soil erosion and therefore soil nutrient and organic matter losses (Lal *et al.* 1991). Besides, they fix atmospheric N, recycle nutrients and affect soil nutrient availability (Lehmann *et al.* 2000). Advantages of cover crops include efficient control of erosion, reduction of nitrate leaching, better soil structure, suppression of weeds, improved self-regulation of pests and diseases (Abdin *et al.* 2000; Feil and Liedgens 2001; Mäder *et al.* 2002; Liedgens *et al.* 2004; den Hollander *et al.* 2007; Hiltbrunner *et al.* 2007), increased soil C sequestration (Hutchinson *et al.* 2007), enhanced soil biochemical and microbial activity (Dinesh *et al.* 2004; Dinesh *et al.* 2006). While legumes may supply N and increase crop yields, nonlegumes may reduce N leaching compared with no cover crops (Kuo *et al.* 1997; Sainju *et al.* 2006). Biomass production, important characteristics and nutrient additions to soils by a few cover crops is given in Table 7.

Thus far, the living mulch concept has been tested with a number of typical row crops, e.g. maize (*Zea mays* L.), soybeans (*Glycine max* (L.) Merr.), cabbage (*Brassica oleracea* L. var. capitata L.), broccoli (*Brassica oleracea* L. var. italica Plenck), asparagus (*Asparagus officinal*is L.), strawberry (*Fragaria x ananassa* Duch.), coconut (*Cocos nucifera* L.), arecanut (*Areca catechu* L.) and rubber (*Hevea brasiliensis* Muell.- Arg.) plantations. Numerous species and mixtures of species have been investigated for their suitability as ground cover. Like cover cropping, green manuring is also considered an important management practice with potential to reduce the dependence on mineral fertilizers and to maintain soil organic matter content. The importance of leguminous green manure crops in improving soil fertility and soil physical properties has received increasing attention (Whitbread *et al.* 2000; Mandal *et al.* 2003) in organic farming. The improvement in soil physical conditions as a result of build up of organic matter by incorporation of green manure or crop residue is associated with a decrease in bulk density, increase in total pore space, water stable aggregates and hydraulic conductivity of the soil (Tirlok Singh *et al.* 1980; Boparai *et al.* 1992). A major objective in the application of fertility-building crops (Cover crops and green manures) is to provide the N required for optimal performance of the subsequent exploitative crops in the rotation (Watson *et al.* 2002). They also have effects on other nutrients and are important sources of organic matter to the soil.

Table 7. Biomass, organic and inorganic characteristics and nutrient additions by cover crops in a coconut plantation

Cover crop	Biomass[a] (Mg ha^{-1})	Composition (g kg^{-1})			C/N	Additions (kg ha^{-1})		
		Total C	Total N	Carbohydrate		Total C	Total N	Carbohydrate
Calopogonium mucunoides	37.6	416.4	26.8	103.2	15.5	15.7	1.01	3.88
Pueraria phaseoloides	102.4	433.3	32.0	113.2	13.5	44.4	3.28	11.60
Centrosema pubescens	43.6	418.6	25.4	122.4	16.5	18.2	1.11	5.34
Atylosia scarabaeoides	89.6	424.0	32.4	118.6	13.1	38.0	2.90	10.63

[a]Sum of 10 years, included both above- and below-ground biomass
Source: Dinesh *et al.* 2004

As other organic amendments, green manure provides nutrients for plant growth and the organic carbon serves as an energy supply for heterotrophic soil microorganisms. Green manure commonly replaces farmyard manure amendments on stockless farms and a question of great concern is whether green manure can be considered to be equal to farmyard manure as a carbon source for improving and sustaining soil biological properties and fertility. Inputs of green manure or crop residues can increase the size and activity of soil microbial communities (Bolton *et al.* 1985, Martens *et al.* 1992; Fauci and Dick, 1994, Dinesh *et al.* 2000; Kautz *et al.* 2004; Elfstrand *et al.* 2007).

In tropical Asia, cover crops are frequently planted in oil palm (Broughton, 1976) and coconut plantations (Aldaba, 1995). The total benefit of leguminous cover intercropped with oil palm was found to be 239 kg N ha^{-1} $year^{-1}$ (Agamuthu and Broughton, 1985). Similarly, a mixture of cover crops (calopo, centrosema, pueraria, atylosia) grown in the interspaces of a 19-year-old coconut plantation was found to contribute 3.8–10.2 Mg ha^{-1} $year^{-1}$ of biomass. Annual incorporation of this biomass over a period of 10 years (Table 6) increased soil organic C by 1.9–4.6 g kg^{-1}, total N by 0.52–0.78 g kg^{-1}, Bray P by 1.3–2.3 mg 100 g^{-1} and extractable K by 3.2–4.2 mg 100 g^{-1} (Dinesh 2004; Dinesh *et al.* 2004). Similar to cover crops, the contribution of organic matter to the soil from a green manure crop is comparable to the addition of 9 to 13 tons per acre of farmyard manure or 1.8 to 2.2 tons dry matter per acre (Schmid and Klay 1984). The potential for leguminous cover crops which are managed as N sources (green manures) to supplement, or possibly replace, inorganic N fertilizer has been broadly discussed in the agricultural literature (Singh *et al.* 1992; Peoples *et al.* 1995, Thorup-Kristensen *et al.* 2003, Tonitto *et al.* 2006). N-fixation measurements demonstrate that clover and vetch varieties commonly used as green manures in annual cropping systems can supply 50–370 kg N ha^{-1} (Peoples *et al.* 1995). The nitrogen fertilizer equivalence, an index used to compare green manure efficiency with chemical fertilizer, has been found to range from 40-200 kg ha^{-1} of nitrogen in different crops but more typically is between 75 and 100 kg ha^{-1} (Smith *et al.* 1987). The cover crops/ green manure crops suited for different cropping systems in India are given in Table 8.

Apparently, much larger contributions of N are to be expected from crops planted solely as green manures and a wide range of legumes have been used for a wide range of subsequent crops (Lathwell 1990). One problem is that legume residues can decompose quickly as soon as they are incorporated into the soil and may release organic acids or other compounds detrimental to germination and seedling growth of the following crop.

Furthermore, a large proportion on the N in green manure residues may be released before the roots of the next crop are established to absorb it (Singh *et al.* 1992) and large losses of N via leaching and even volatilization from such residues have been reported (Costa *et al.* 1990; Janzen and McGinn 1991). There are several solutions to this problem; one is to use legumes with a high fibre, lignin or polyphenol content which degrade slowly, releasing N at the time when the subsequent crop requires it. The rate of release of N from crop residues is closely related to their C-to-N ratio and content of lignin and polyphenol such that a close linear relationship of N release (mineralisation) with the (lignin + polyphenol)-to- N ratio has been reported by several authors

(Thomas and Asakawa 1993; Handayanto *et al.* 1994; Dinesh *et al.* 2001). In all cases, however, it is necessary that farmers are willing to invest time or money on planting a crop with no immediate economic return and this is rarely the case if they can use the land for a grain or fodder crop at this time. Better acceptance is likely in regions where it is too dry for a grain or forage crop to be planted.

Table 8. Cover crops/ green manure crops suited for different cropping systems in India

	Cropping system	Cover crop/ Green manure
1.	Rice (Monocrop)	*Tephrosia purpurea* *Crotalaria juncea, Indigofera tinctoria, Azolla* sp.
2.	Rice (Double crop)	*Sesbania aculeata, Phaseolus trilobus, S. rostrata, Azolla* sp.
3.	Wheat	*Vigna radiata*
4.	Rice-wheat	*Sesbania aculeata*
5.	Sorghum	*Crotolaria juncea, Leucaena leucocephala*
6.	Cotton	*Trifolium alexandrium, Vigna radiata, Vigna mungo, Cyamopsis tetragonoloba, Crotalaria juncea, Sesbania aculeata*
7.	Potato	*Crotalaria juncea, Sesbania aculeata, Vigna ungiculata, Dolichos biflorus, Cyamopsis tetragonoloba*
8.	Castor rainfed	*Vigna ungiculata* (intercropped)
9.	Plantations crops (Black pepper, coconut, banana)	*Calopogonium mucunoides, Pueraria phaseoloides*

Source: Verma, 1995

3.2.4 Organic wastes from plantations

Unlike other crops, spices are perennial in nature and explore a limited volume of soil continuously throughout its lifespan. Hence, it is imperative that a substrate rich environment is provided in the root zone to obtain sustainable yields. As the pressure on land is increasing, area expansion for spice cultivation is a remote possibility. An ideal option would be to cultivate spices as a component crop in the existing coffee, tea, coconut and arecanut plantations. Mixed cropping of black pepper with rubber, cocoa, orange etc are common in countries like Brazil.

There is a great scope of bringing tea and coffee plantations under black pepper cultivation by using all the shade tree for trailing pepper. Area under coffee and tea in South India can accommodate about 3.8 and 2.0 million tree supports of pepper, respectively. Plantation crops including black pepper and cardamom and various shade trees contribute large quantities of waste by-products, which by composting and recycling will meet the nitrogen requirement of crop and partly other nutrients (Tables 1 & 2). Leaves and bunch wastes of coconut, arecanut, oil palm, coffee husk and tea spent have been found to give very high bioconversion efficiency. The multiple cropping system results in the continuous uptake, addition of biomass and higher level of nutrient recycling that would lead to a positive influence on the physico-chemical and biological properties of soil. Leaves and bunch wastes of coconut, arecanut, oil palm, coffee husk and tea spent have been found to give very high bioconversion efficiency. In south India, where coconut is a major plantation crop, abundance of waste material, especially 7.5 million tons of coir dust annually from coir-based industries provides a good scope for recycling these organic materials into the system. Compost made of leaf and sheath wastes and husks of these plantation crops are valuable sources of plant nutrients (Parthasarathy *et al.* 2006).

4.0 QUALITY OF RAW AND COMPOSTED ORGANIC MANURES

Organic manure is traditionally a key fertilizer in sustainable soil management. It is most effectively used in combination with other sustainable practices. These include crop rotation, cover cropping, green manuring, liming, and the addition of other natural or biologically friendly fertilizers and amendments. In organic production, manure is commonly applied to the field in either a raw (fresh or dried) or composted state.

4.1 Raw manure

Raw manure is an excellent resource for organic crop production. It supplies nutrients and organic matter, stimulating the biological processes in the soil that help to build fertility. Still, a number of cautions and restrictions are in order, based on concerns about produce quality, food contamination, soil fertility imbalances, weed problems, and pollution hazards.

Some manure may contain contaminants such as residual hormones, antibiotics, pesticides, disease organisms, and other undesirable substances. Since many of these can be eliminated through high-temperature aerobic composting, this practice is recommended where low levels of organic contaminants may be present. Organic resources can also potentially stimulate harmful pests and diseases (Vanlauwe and Giller 2006). Research has demonstrated that *Salmonella and E. coli* bacteria appear to survive the composting process much better than previously thought. (Droffner and Brinton 1995; Cekmecelioglu *et al.* 2005) The possibility of transmitting human diseases discourages the use of fresh manures (and even some composts) as preplant or side-dress fertilizers on vegetable crops especially crops that are commonly eaten raw. (Pederson 1998).

Unlike conventional farmers, who have only safety guidelines regarding manure use, certified organic farmers must follow stringent protocols. Raw manure may NOT be applied to food crops within 120 days of harvest where edible portions have soil contact (i.e., most vegetables, strawberries, etc.); it may NOT be applied to food crops within 90 days of harvest where edible portions do not have soil contact (i.e., grain crops, most tree fruits). Such restrictions do not apply to feed and fiber crops (Neal 1994)

Organic substances are not the only contaminants found in livestock manures. Heavy metals can be a problem (Nicholson *et al.* 1995; Krapac *et al.* 2002; Guerra-Rodríguez *et al.* 2006), especially where industrial-scale production systems are used. Concerns over heavy-metal and other chemical contamination have dogged the use of poultry litter as an organically acceptable fertilizer. Heavy-metal contamination is also a concern with composted sewage sludge (Barros *et al.* 2007; Speir *et al.* 2007) —a major reason for its being prohibited from certified organic production.

4.1.1 Quality of produce

It has long been acknowledged that improper use of raw manure can adversely affect the quality of such vegetable crops as potatoes, cucumbers, squash, turnips, cauliflower, cabbage, broccoli, and kale. As it breaks down in the soil, manure releases chemical compounds such as skatole, indole, and other phenols. When absorbed by the growing plants, these compounds can impart off-flavors and odors to the vegetables (Eghball and Lesoing 2000). For this reason, raw manure should not be directly applied to vegetable crops; it should instead be spread on cover crops planted the previous season.

4.1.2 Effect on soil fertility

Raw manure use has often been associated with imbalances in soil fertility. Continual manure use tends to acidify soil. As manure breaks down it releases various organic acids that assist in making soil minerals available—a benefit of manure that is often unrecognized. Over time, however, this process depletes the soil of calcium and causes pH levels to fall below the optimum for most crops. Manures do supply some calcium, but not enough to counterbalance the tendency toward increased acidity (Eghball and Lesoing 2000). Possible exceptions include caged-layer manure (when oyster-shell or similar calcium supplements are fed) and manure from dairy operations where lime is used.

When fresh manure containing large amounts of nitrogen and salts is applied to a crop, it can have the same effects as excessive applications of soluble commercial fertilizers—it can burn seedling roots, reduce immunity to pests, and shorten produce shelf life. To avoid manure-induced imbalances, continuous soil quality assessment through appropriate soil tests and application of lime to ensure soil pH balance is essential.

Understanding the soil's needs is only part of the equation. The nutrient content of the applied manure also needs to be known. Standard fertilizer values (such as those shown in Table 6) should be used only for crude approximations. The precise nutrient content of any manure is dependent not only on livestock species, but also on the ration fed, the kind of bedding used, amount of liquid added, and the kind of capture and handling system employed. Also, some traditional assumptions about manure composition may need to be updated. Because of the abundance of sulfur in rations, manure has long been recognized as a good source of sulfur. However, less sulfur is applied to crops in contemporary high-analysis fertilizers, and atmospheric deposition has been decreased by pollution controls. Sulfur deficiencies are appearing in many soils, and levels in manure may also be diminished (Eghball and Lesoing 2000). It is always advisable to test manure in order to assess its fertilizer value.

4.1.3 Weed problems

Use of raw manures has often been associated with increased weed problems. Some manure contains weed seed, often from bedding materials like small-grain straw and old hay. High-temperature aerobic composting can greatly reduce the number of viable weed seeds. (Larney and Blackshaw 2003; Larney and Hao 2007). In many cases, however, the lush growth of weeds that follows manuring does not result from weed seeds in the manure, but from the stimulating effect manure has on seeds already present in the soil (as demonstrated through studies at Auburn University using broiler litter) (Haapala 1997). The flush of weeds may result from enhanced biological activity, the presence of organic acids, an excess of nitrates, or some other change in the fertility status of the soil. Depending on the weed species that emerge, the problem may be related to the sort of fertility imbalances described above. Excesses of potash and nitrogen in particular can encourage weeds (Eghball and Lesoing 2000).

4.1.4 Pollution

When the nutrients in raw or composted manure are eroded or leached from farm fields or holding areas, they present a potential pollution problem, in addition to being a resource lost to the farmer. Leached into groundwater, nitrates and P load (Dunne *et al.* 2005; Ekholm *et al.* 2005) from manure and fertilizers have been linked to a number of human health problems. Flushed into surface waters, nutrients can cause eutrophication of ponds, lakes, and streams. The manner in which manure is collected and stored prior to field use affects the stabilization and conservation of valuable nutrients and organic matter. Composting is one means of good manure handling and is discussed in more detail in section 5.0.

Reducing manure run-off and leaching losses from fields is a matter of both volume and timing. Manure application far in excess of crop needs greatly increases the chances of nutrient loss, especially in high-rainfall areas. Manure spread on bare, frozen or erodable ground is subject to run-off, especially where heavy winter rains are common. Under some conditions, however, manure can actually slow run-off and erosion losses from fields (Ramos and Martínez-Casasnovas 2006) via improvements in the physical properties of a soil (Oades 1984; Hamblin 1991; Ouédraogo *et al.* 2001). For this reason the application of organic waste or compost could be beneficial for soil conservation (Pinamonti and Zorzi 1996), especially in degraded soils and soils susceptible to erosion.

Sheet-composting manure (tilling it into the soil shortly after spreading) or applying it to growing cover crops are two advisable means of conserving manure nutrients. Grass cover crops, such as rye and ryegrass, are especially good as "catch crops"-cover crops grown to absorb soluble nutrients from the soil profile to prevent them from leaching (Jackson *et al.* 2004). It is a sound strategy, therefore, to apply manure to growing catch crops or just prior to planting them. However, both sheet composting and applying to cover crops have trade-offs. Sheet composting improves the capture of ammonia nitrogen from manure, but requires tillage, which leaves the soil bare and vulnerable to erosion and leaching losses. Surface-applying to cover crops (with no soil-incorporation) eliminates most leaching and erosion losses but increases ammonia losses to the atmosphere.

5.0 COMPOSTING

Large quantity of fresh crop residues, on application directly to soil, causes extremely severe nitrogen immobilization (Sørensen and Amato 2002; Chaves *et al.* 2005) and development of excessive reduced condition of the soil (Bernal *et al.* 1998a; Lesturgez *et al.* 2006). The fresh residues are, therefore, piled up, moistened, churned periodically to aerate and allowed to decompose partially. This process is called composting.

As composting proceeds, the C/N ratio gradually decreases from 30:1 to 10-15:1 for the finished product (Dickson *et al.* 1991). This occurs because each time the organic compounds are consumed by microorganisms, two-thirds of the carbon is given off as carbon dioxide. The remaining third is incorporated along with nitrogen into microbial cells, and later released for further use once those cells die (http://www.css.cornell.edu/compost/chemistry.html; accessed on 02 February 2007).

Composting is the process of producing compost through aerobic decomposition of biodegradable organic matter. The decomposition is performed primarily by aerobes, although larger creatures such as ants, nematodes, and oligochaete worms also contribute. This decomposition occurs naturally in all but the most hostile environments, such as within landfills or in extremely arid deserts, which prevent the microbes and other decomposers from thriving. Although attaining a C/N ratio of roughly 10-15:1 is a useful goal in planning composting operations, this ratio may need to be adjusted according to the bioavailability of the materials in question. Most of the nitrogen in compostable materials is readily available. Some of the carbon, however, may be bound up in compounds that are highly resistant to biological degradation. Newspaper, for example, is slower than other types of paper to break down because it is made up of cellulose fibers sheathed in lignin, a highly resistant compound found in wood. Corn stalks and straw are similarly slow to break down because they are made up of a resistant form of cellulose. Although all of these materials can still be composted, their relatively slow rates of decomposition mean that not all of their carbon will be readily available to microorganisms, so a higher initial C/N ratio can be planned. Particle size also is a relevant consideration; although the same amount of carbon is contained in comparable masses of wood chips and sawdust, the larger surface area in the sawdust makes its carbon more readily available for microbial use. However, the use of compost in soils requires that it achieves an adequate degree of maturity, which implies stable organic matter content and the absence of phytotoxic compounds and plant or animal pathogens (Bernal *et al.* 1998b; Gomez 1998).

Composting can be divided into the two areas of home composting and industrial composting. Both scales of composting use the same biological processes, however techniques and different factors must be taken into account. Composting is the controlled decomposition of organic matter. Rather than allowing nature to take its slow course, a composter provides an optimal environment in which decomposers can thrive. To encourage the most active microbes, a compost pile needs the correct mix of carbon, nitrogen, oxygen (from the air) and water. Decomposition happens even in the absence of some of these ingredients, but not as quickly or as pleasantly. (For example, vegetables in a plastic bag will decompose, but the absence of air encourages the growth of anaerobic microbes, which produce disagreeable odors).

The goal in a composting system is to provide a healthy environment and nutrition for the rapid decomposers, the bacteria. The most rapid composting occurs with the ideal carbon to nitrogen ratio of between 25 and 30 to 1 by dry chemical weight. In other words, the ingredients placed in the pile should contain 25 to 30 times as much carbon as nitrogen. For example, grass clippings average about 19-to-1 and dry autumn leaves average about 55-to-1. Mixing equal parts by volume approximates the ideal range. The organic refuse used for composting may be of rural and urban origin and may include straw, leaves, paddy husk, groundnut shells, sugarcane trash, bagasse, cattle dung, urine, crop residues, city garbage, night soil, sewage, kitchen and vegetable wastes, hedge clippings, water hyacinth and all other residues containing organic matter.

Given enough time to die, all biodegradable material will compost. However, not all compost feed stocks are appropriate for backyard composting. Most backyard systems will not reach high enough temperatures to kill pathogens and deter vermin, so pet droppings, non-vegetarian animal manure, meat scraps, and dairy products are best left to operators of high-rate, thermophylic composting systems. Certain substances should not be composted by the average homeowner, as they require more sophisticated systems, competent management, and more efficient, cost-competitive, environmentally sound technology. These substances include non-vegetarian animal manures and bedding, by-products of food production and processing, restaurant grease and cooking oils, and residuals from the treatment of wastewater and drinking water. Composting will also break down petroleum hydrocarbons (Namkoong *et al.* 2002; Antizar-Ladislao *et al.* 2006) and some toxic compounds for recycling and beneficial reuse. The use of composting for such purposes is most commonly referred to as a form of bioremediation (Jørgensen *et al.* 2000; Rojas-Avelizapa *et al.* 2007).

High-carbon sources provide the cellulose needed by the composting bacteria for conversion to sugars and heat, while high-nitrogen sources provide the most concentrated protein, which allow the compost bacteria to thrive. Some ingredients with higher carbon content are dry, straw-type material, such as cereal straws, autumn leaves, sawdust and wood chips, some paper and cardboard (such as corrugated cardboard or newsprint with soy-based inks). Some ingredients with higher nitrogen content are green plant material (fresh or wilted) such as crop residues, hay, grass clippings, weeds, animal manures, fruit and vegetable trimmings, seaweeds, used coffee grounds, poultry manure provides lots of nitrogen (Walker and Bernal 2007) but little carbon. Horse manure provides both (Moral *et al.* 2005). Sheep and cattle manure don't drive the compost heap to as high a temperature as poultry or horse manure (Martins and Dewes 1992; Wolter *et al.* 2004), so the heap takes longer to produce the finished product (http://en.wikipedia.org/wiki/Composting; accessed on 12 February 2007).

Mixing the materials as they are added increases the rate of decomposition (Illmer and Schinner 1997), but it can be easier to place the materials in alternating layers, approximately 15 cm (6 in) thick, to help estimate the quantities. Keeping carbon and nitrogen sources separated in the pile can slow down the process, but decomposition will occur in any event. Greasy food waste and wastes from meat, dairy products, and eggs tend to attract unwanted vermin and take longer to decompose. However, eggshells are a good source of nutrients for the compost pile and the soil although they typically take more than one year to decompose. Manure from non-vegetarian animals should never be used, and neither should human or pet waste.

5.1 Science and Engineering of Composting

5.1.1 Invertebrates of the compost pile

In small-scale outdoor composting systems, such as backyard compost piles, soil invertebrates are likely to contribute to the decomposition process. Together with bacteria, fungi, and other microbes, these organisms make up a complex food web or energy pyramid with primary, secondary, and tertiary level consumers. The base of the pyramid, or energy source, is made up of organic matter including plant and animal residues.

Tertiary consumers: Includes organisms that eat secondary consumers like centipedes, predatory mites, rove beetles, fomicid ants, carabid beetles

Secondary consumers: Includes organisms that eat primary consumers like springtails, some types of mites, feather-winged beetles, nematodes, protozoa, rotifera, soil flatworms

Primary consumers: Includes organisms that eat organic residues like bacteria, fungi, actinomycetes, nematodes, some types of mites, snails, slugs, earthworms, millipedes etc.

5.1.2 Microorganisms of the compost pile

Microbial activity plays a leading role on transformations occurring during humification processes (Beffa *et al.* 1996). Thus, inoculation with proper microorganisms may activate the biodegradation of organic matter (Elorrieta *et al.* 2002) or improve the final characteristics of the compost (Requena *et al.* 1996; Gaind *et al.* 2005).

5.1.2.1 Bacteria

Bacteria are the smallest living organisms and the most numerous in compost; they make up 80 to 90% of the billions of microorganisms typically found in a gram of compost. Bacteria are responsible for most of the decomposition and heat generation in compost. They are the most nutritionally diverse group of compost organisms, using a broad range of enzymes to chemically break down a variety of organic materials. At the beginning of the composting process (0-40°C), mesophilic bacteria predominate. Most of these are forms that can also be found in topsoil. As the compost heats up above 40°C, thermophilic bacteria take over. The microbial population during this phase is dominated by members of the genus *Bacillus*. The diversity of bacilli species is fairly high at temperatures from 50-55°C but decreases dramatically at 60°C or above. When conditions become unfavorable, bacilli survive by forming endospores, thick-walled spores that are highly resistant to heat, cold, dryness, or lack of food. They are ubiquitous in nature and become active whenever environmental conditions are favorable.

At the highest compost temperatures, bacteria of the genus *Thermus* have been isolated. Composters sometimes wonder how microorganisms evolved in nature that can withstand the high temperatures found in active compost. *Thermus* bacteria were first found in hot springs in Yellowstone National Park and may have evolved there. Other places where thermophilic conditions exist in nature include deep sea thermal vents, manure droppings, and accumulations of decomposing vegetation that have the right conditions to heat up just as they would in a compost pile. Once the compost cools down, mesophilic bacteria again predominate. The numbers and types of mesophilic microbes that recolonize compost as it matures depend on what spores and organisms are present in the compost as well as in the immediate environment. In general, the longer the curing or maturation phase, the more diverse the microbial community it supports.

5.1.2.2 Actinomycetes

The characteristic earthy smell of soil is caused by actinomycetes, organisms that resemble fungi but actually are filamentous bacteria. Like other bacteria, they lack nuclei, but they grow multicellular filaments like fungi. In composting they play an important role in degrading complex organics such as cellulose, lignin, chitin, and proteins. Their enzymes enable them to chemically break down tough debris such as woody stems, bark, or newspaper. Some species appear during the thermophilic phase, and others become important during the cooler curing phase, when only the most resistant compounds remain in the last stages of the formation of humus. Actinomycetes form long, thread-like branched filaments that look like gray spider webs stretching through compost. These filaments are most commonly seen toward the end of the composting process, in the outer 10 to 15 centimeters of the pile. Sometimes they appear as circular colonies that gradually expand in diameter.

5.1.2.3 Fungi

Fungi include molds and yeasts, and collectively they are responsible for the decomposition of many complex plant polymers in soil and compost. In compost, fungi are important because they break down tough debris, enabling bacteria to continue the decomposition process once most of the cellulose has been exhausted. They spread and grow vigorously by producing many cells and filaments, and they can attack organic residues that are too dry, acidic, or low in nitrogen for bacterial decomposition.

Most fungi are classified as saprophytes because they live on dead or dying material and obtain energy by breaking down organic matter in dead plants and animals. Fungal species are numerous during both mesophilic and thermophilic phases of composting. Most fungi live in the outer layer of compost when temperatures are high. Compost molds are strict aerobes that grow both as unseen filaments and as gray or white fuzzy colonies on the compost surface.

5.1.2.4 Protozoa

Protozoa are one-celled microscopic animals. They are found in water droplets in compost but play a relatively minor role in decomposition. Protozoa obtain their food from organic matter in the same way as bacteria do but also act as secondary consumers ingesting bacteria and fungi.

5.1.2.5 Rotifers

Rotifers are microscopic multicellular organisms also found in films of water in the compost. They feed on organic matter and also ingest bacteria and fungi.

5.1.3 Microbes and heating the pile

An effective compost pile is kept about as damp as a well wrung-out sponge. This provides the moisture that all life needs to survive. Bacteria and other microorganisms fall into a variety of groups in terms of what their ideal temperature is and how much heat they generate as they do their work. *Mesophilic* bacteria enjoy midrange temperatures, from about 20 to 40 °C (70 to 110 °F). As they decompose the organic matter, they generate heat, and the inner part of a compost pile heats up the most.

The heap should be about 1 m (3 ft) wide, 1 m (3 ft) tall, and as long as is practicable. This provides a suitable insulating mass to allow a good heat build-up as the material decays. The ideal temperature is around 60 °C (140 °F), which kills most pathogens and weed seeds while providing a suitable environment for *thermophilic* (heat-loving) bacteria, which are the fastest acting decomposers. It is important to mention that a large variety of mesophilic, thermotolerant and thermophilic aerobic microorganisms, including bacteria, actinomycetes, yeasts and various other fungi have been extensively reported in composts and other self-heating organic materials (Amner *1* 1988; Faure and Deschamps 1991; Beffa *et al.* 1996; Hassen *et al.* 2001). However, many factors determine the microbial community during composting. Under aerobic conditions, temperature is the major factor that determines the types of microorganisms, species diversity, and the rate of metabolic activities.

The centre of the heap can get too warm, possibly hot enough to burn a bare hand. If this fails to happen, common reasons include the following:

- The heap is too wet, thus excluding the oxygen required by the compost bacteria
- The heap is too dry, so that the bacteria do not have the moisture needed to survive and reproduce
- There is insufficient protein (nitrogen-rich material)
- The solution is to add material, if necessary, and/or to turn the pile to aerate it.

On the other hand, the composting process can, if not properly managed, induce the proliferation and dispersion of potentially pathogenic and/or allergenic thermotolerant/thermophilic fungi and bacteria (Millner *et al.* 1994). Among the fungi, the mold *Aspergillus fumigatus* is predominant because of its cellulolytic and thermotolerant properties (Beffa *et al.* 1996). Among the bacteria, *Salmonella*, *Shigella*, *Escherichia coli*, *Enterobacter*, *Yersinia*, Streptococci and *Klebsiella* can emerge and cause infections for compost handlers and agricultural users (Strauch 1996).

Depending on how quickly the compost is required, the heap can be turned one or more times to bring the outer layers to the inside of the heap and vice versa, as well as to aerate the mixture. Adding water at this time helps keep the pile as damp. One guideline is to turn the pile when the high temperature has begun to drop, indicating that the food source for the fastest-acting bacteria (in the center of the pile) has been largely consumed. When the temperature stops rising after the pile has been turned, there is no further advantage in turning the pile. When all the material has turned into dark brown or nearly black crumbly matter, it is ready to use.

5.1.4 Phases of composting

In the process of composting, microorganisms break down organic matter and produce carbon dioxide, water, heat, and humus, the relatively stable organic end product. The temperature within a composting mass determines the rate at which many of the biological processes take place and plays a selective role on the evolution and the succession of the microbiological communities (Mustin 1987). In biological terms, the operating temperature ranges are as follows: >55°C to maximize sanitation, 45–55°C to maximize the biodegradation rate, and 35–40°C to maximize microbial diversity (Stentiford 1996). The process of aerobic composting can be divided into three major steps, a mesophilic-heating phase, or moderate-temperature phase, which lasts for a couple of days, 2) the thermophilic, or high-temperature phase, which can last from a few days to several months, and finally, 3) a several-month cooling and maturation phase (Mustin 1987; Leton and Stentiford 1990).

The findings of a study conducted in a semi-industrial pilot plant using a moderate aeration during the composting of municipal solid waste (Hassen *et al.* 2001) indicated that:

(i) Auto-sterilization induced by relatively high temperatures (60–55°C) caused a significant change in bacterial communities. For instance, *Escherichia coli* and faecal Streptococci populations decreased, respectively, from 2×10^7 to 3.1×10^3 and 10^7 to 1.5×10^3 cells/g waste dry weight (WDW); yeasts and filamentous fungi decreased from 4.5×10^6 to 2.6×10^3 cells/g WDW and mesophilic bacteria were reduced from 5.8×10^9 to 1.8×10^7 bacteria/g WDW. On the other hand, the number of bacterial spores increased at the beginning of the composting process, but after the third week their number decreased notably.

(ii) *Salmonella* disappeared completely from compost by the 25th day as soon as the temperature reached 60°C; and

(iii) The bacterial population increased gradually during the cooling phase. While Staphylococci seemed to be the dominant bacteria during the mesophilic phase and at the beginning of the thermophilic phase, bacilli predominated during the remainder of the composting cycle.

5.2 COMPOST CHEMISTRY

5.2.1 C/N Ratio

Of the many elements required for microbial decomposition, carbon and nitrogen are the most important. Carbon provides both an energy source and and the basic building block making up about 50 percent of the mass of microbial cells. Nitrogen is a crucial component of the proteins, nucleic acids, amino acids, enzymes and co-enzymes necessary for cell growth and function.

To provide optimal amounts of these two crucial elements, you can use the carbon-to-nitrogen (C/N) ratio for each of your compost ingredients. C/N ratio is one of the important factors affecting composting process and compost quality. It is considered that a C/N ratio of 25–30: 1 is the initial optimum ratio for composting (Fong *et al.* 1999; Zhu, 2007). At lower ratios, nitrogen will be supplied in excess and will be lost as ammonia gas, causing undesirable odors. Higher ratios mean that there is not sufficient nitrogen for optimal growth of the microbial populations, so the compost will remain relatively cool and degradation will proceed at a slow rate.

Typical C/N ratios for common compost materials can be looked up in published tables such as Appendix A, On-Farm Composting Handbook (http://compost.css.cornell.edu/OnFarmHandbook/ apa.taba1.html; accessed on 06.02.2007). In general, materials that are green and moist tend to be high in nitrogen, and those that are brown and dry are high in carbon (Table 9). As composting proceeds, the C/N ratio gradually decreases from 30:1 to 10-15:1 for the finished product.

Table 9. C/N ratio of organic materials used for composting

Materials High in Carbon	C/N*
Autumn leaves	30-80:1
Straw	40-100:1
Wood chips or sawdust	100-500: 1
Bark	100-130:1
Mixed paper	150-200:1
Newspaper or corrugated cardboard	560:1

Contd...

Table 9. Contd...

Materials High in Nitrogen	
Vegetable scraps	15-20:1
Coffee grounds	20:1
Grass clippings	15-25:1
Manure	5-25:1

Source: Dickson *et al.* 1991

5.2.2 Oxygen

The oxygen concentration plays a dominant role in composting (Van Ginkel *et al.* 2001). As microorganisms oxidize carbon for energy, oxygen is used up and carbon dioxide is produced. Without sufficient oxygen, the process will become an1aerobic and produce undesirable odors, including the rotten-egg smell of hydrogen sulfide gas.

Although the atmosphere is 21% oxygen, aerobic microbes can survive at concentrations as low as 5%. Oxygen concentrations greater than 10% are considered optimal for maintaining aerobic composting. Some compost systems are able to maintain adequate oxygen passively, through natural diffusion and convection. Other systems require active aeration, provided by blowers or through turning or mixing the compost ingredients.

5.2.3 Nutrient balance

Adequate phosphorus, potassium, and trace minerals (calcium, iron, boron, copper, etc.) are essential to microbial metabolism. Normally these nutrients are not limiting because they are present in ample concentration in the compost source materials.

5.2.4 pH

A pH between 5.5 and 8.5 is optimal for compost microorganisms. As bacteria and fungi digest organic matter, they release organic acids. In the early stages of composting, these acids often accumulate and cause the pH to drop, which encourages the growth of fungi and the breakdown of lignin and cellulose. Usually the organic acids become further broken down during the composting process. If the system becomes anaerobic, however, acid accumulation can lower the pH to 4.5, severely limiting microbial activity. In such cases, aeration usually is sufficient to return the compost pH to acceptable ranges.

5.3 ON- FARM COMPOSTING TECHNIQUES

There are a number of different techniques for composting all employing the two primary methods of aerobic composting:

5.3.1 Active (or hot) composting

This allows aerobic bacteria to thrive, kills most pathogens and seeds, and rapidly produces usable compost. Aerobic bacteria produce less odour and fewer destructive green house gases than their anaerobic counterparts. In addition, they are usually faster at breaking down material and the faster material is broken down, the faster compost is created for your garden.

5.3.2 Passive (or cold) composting

This lets nature take its course in a more leisurely manner and leaves many pathogens and seeds dormant in the pile. Cold composting is the type of composting done in most domestic garden compost bins where temperatures never reach above 30 °C (86 °F). Cold composting is typically people placing their kitchen scraps in the garden compost bin and forgetting about them. In this circumstance, a garden compost bin becomes a storage container for scraps and has very high moisture content, which means it turns quite anaerobic and smelly. This type of composting generates significant adverse green house gas emissions. When composting this way, a gardener can improve the process by adding some wood chip or small pieces of bark, leaves, twigs or a combination of these materials through the mixture. This material helps to improve drainage and airflow.

Composting is gaining increased attention for treating food wastes with various agricultural by-products in different systems (Droffner and Brinton, 1995; Elwell *et al.* 1996; Donahue *et al.* 1998; Laos *et al.* 1998; Seymour *et al.* 2001; Filippi *et al.* 2002; Das *et al.* 2003). Composting systems are either enclosed (home container composting, industrial in-vessel composting) or in piles (industrial windrow composting). The selection of the system is dependent on the nature of the waste to be composted, available manpower, and economic conditions (Cekmecelioglu *et al.* 2005).

5.3.3 Traditional techniques

5.3.3.1 The Indian Bangalore method (FAO, 1980)

This method was developed at Bangalore, India by Acharya. This method is recommended when night soil and refuse are used for preparing the compost. This mehtod overcomes many of the disadvantages of the Indore method such as problem of heap protection from adverse weather, nutrient losses due to high winds/ strong sunrays, frequent turning requirements, fly nuisance etc. However, the time involved in production of finished compost is much longer. The method suits areas with less rainfall.

Pit preparation and filling: Trenches or pits about one meter deep are dug. The dimensions of the pit, however, vary depending on the availability of land and the type of material to be composted. The trenches should preferable have sloping walls and a floor of 90 cm slope to prevent water logging For filling the pits, organic residues and night soil are put in alternate layers and after filling, the pit is covered with a 15-20 cm thick layer of refuse. The materials are allowed to remain in the pit without turning and watering for three months. During this period, the material settles down due to reduction in volume and additional night soil and refuse are placed on top in alternate layers and plastered or covered with mud or earth to prevent moisture loss. After the initial aerobic composting which is for about eight to ten days, the material undergoes anaerobic decomposition at a very slow rate and it takes about six to eight months to obtain the finished product.

5.3.3.2 Passive composting of manure piles (Robert, 1992)

This involves stacking the materials in piles to decompose over a long time period with little agitation and management. The process has been used for composting of animal wastes. Without considerable bedding material, the moisture content of manure exceeds the levels, which enables an open porous structure to exist in the pile. Under these circumstances, anaerobic microorganisms dominate the pile. All of the undesirable effects with anaerobic degradation occur, including low temperature, unpleasant odor, release of H_2S etc. This passive method of composting is essentially windrow composting but with a much less frequent turning schedule. It is a common method used for composting leaves. It demands minimal labour and equipment. It is a slow process because of its low aeration rate.

5.3.3.3 The Indian Indore method (FAO, 1980)

Sir Albert Howard at Indore, Madhya Pradesh (India) developed this method during 1924-26. The raw materials used are mixed plant residues, animal dung and urine, earth, wood ash and water. All organic wastes available on a farm such as weeds, stalks, stems, fallen leaves, pruning, chaff, fodder leftovers and so on, are collected and stacked in a pile. Woody materials like cotton or pigeon pea stalks and stubbles are first spread on the farm road and crushed under vehicles such as tractors or bullock carts before being piled. Such hard materials should not exceed ten per cent of the total plant residues. Green materials, which are soft and succulent are allowed to wilt for two to three days to remove excess moisture before stacking. While stacking, each type of material is spread in layers about 15 cm thick until the heap is about 1.5 meters high. The heap is then cut into vertical slices and about 20-25 kg are put under the feet of cattle in the shed as bedding for the night. The next morning the bedding, along with the dung and urine-earth, is taken to the pits where the composting is to be done. The composting is done through pit method or heap method.

In the former, pits 1 m deep and 1.5-2 m wide and of suitable length are made. The material brought from the cattle shed is spread evenly in the pit in layers of 10-15 cm. On each layer, slurry made with 4.5 kg dung, 3.5 kg urine-earth and 4.5 kg of inoculum taken from a 15 day old composting pit is spread. Sufficient quantity of water is sprinkled over the material in the pit. The pit is filled layer by layer and should not take longer than one week to fill. Compaction should be avoided. The material is turned three times during the whole period of composting; the first time after filling the pit, the second after another 15 days and the third after one month. The material is mixed thoroughly, moistened with water and replaced in the pit.

In the heap method, the organic wastes are piled in heaps about 2m wide at the base, 1.5 m high and 2 m long. The sides are tapered so that the top is about 0.5 m narrower in width than the base. The heap is usually built with a 20 cm layer of carbonaceous material such as leaves, hay, straw, saw dust, wood chips and chopped corn stalks. This is then covered with 10 cm nitrogenous material such as fresh grass, weeds or garden plant residues, fresh or dry manure or digested sewage sludge. The pattern of 20 cm carbonaceous material and 10 cm of nitrogenous material is followed until the pile is 1.5 m high and the material is wetted so that it becomes damp. The pile is covered with soil or hay to retain heat and is turned at six- and twelve-week intervals.

5.3.3.4 Container composting

There are sound reasons for composting in containers, although there is debate as to whether slatted or closed sided bins are preferable, for this will affect air circulation within the compost pile, as well as the potential for heat loss. The *Indore* method developed by Sir Albert Howard and the *Shewell Cooper* method favour slats, while the *New Zealand Box* method advocates the use of closed sides.

There are also differences between these techniques in terms of activators (that is, high nitrogen content organic substances to stimulate high bacterial activity within the heap, e.g., urine, grass mowings etc.) and materials used.

However, most agree that a good mixture of carbon and nitrogenous materials, usually created in layers and on a base consisting of rougher, stemmy material (to encourage air circulation) that is in contact with the soil are essential to all successful composting processes. For those who do not have a lot of space, composting can be carried out with good results by using cylindrical bins provided that attention is paid to the all-important issues of aeration and C:N ratios.

5.3.3.5 *In-vessel composting*

It is an industrial form of composting biodegradable waste that occurs in enclosed reactors. These generally consist of metal tanks through which air flow and temperature can be controlled. Offensive odours are caused by excess nitrogen or moisture. This may be controlled with a higher C:N ratio or increased aeration by ventilation, mixing, or by using a coarser grade of carbon material. Insects may be controlled by keeping the bin enclosed with the minimum size vents necessary for adequate air exchange. Insects outside the bin enter through the vents and tend to stay inside where the food is. A relatively dry layer of carbon material on top of the compost filters odors and, given adequate ventilation, maintains a moisture gradient that helps keep insects inside where the moisture is. Despite the high initial costs, the in-vessel systems require less space and provide better control than windrows for uniformly handling mixtures, and for manipulating gas emissions and polluting leachates. Other advantages of in-vessel systems over windrow composting involve shortened mesophilic (25–45°C) and thermophilic (45–75°C) stages, higher process efficiency resulting in a decreased number of pathogenic microorganisms, and thus more valuable final product (Nair *et al.* 2006). Donahue *et al.* (1998) demonstrated that food waste was successfully composted with sawdust and mulch chips in an in-vessel system within 14 days, after which the product was placed in windrows for the curing stage.

5.3.3.6 *Windrow composting*

It is the production of compost by piling organic matter or biodegradable waste, like animal manure and crop residues, in long rows called *windrows* (Cekmecelioglu *et al.* 2005). This method is suited to producing large volumes of compost. These piles are generally turned to improve porosity and oxygen content, mix in or remove moisture, and redistribute cooler and hotter portions of the pile. Typically, the windrows are initially from 3 ft high for dense materials like manures to 12 ft high for fluffy materials like leaves. The width varies from 10 to 20 ft. By the end of the first week, the windrow height diminishes appreciably and by the end of the second week it may be as low as

2 ft. In this method, the active composting stage lasts for three to nine weeks depending on the nature of the materials used and the frequency of turning. Windrow composting is the most commonly used of farm scale composting methods. Process control parameters include the initial ratios of carbon and nitrogen rich materials, the amount of bulking agents added to assure air porosity, the pile size, moisture content, and turning frequency. However, the lower capital investments, high environmental-friendliness and the less need for intensive knowledge and skills during operation make windrow systems more attractive to farmers for routine use (Tai and He, 2007).

5.3.4 Rapid composting methods

While traditional composting procedures take as long as 4-8 months to produce finished compost, rapid methods offer possibilities for reducing the composting period up to three weeks. A variety of approaches and their combinations have been used to hasten the process, which include:

5.3.4.1 The Berkley rapid composting method (Raabe, 2001)

This method rectifies some of the problems associated with the traditional methods. Compost can be made in two to three weeks with this method. The important factors associated with rapid composting are:

The material will compost best if it is between ½ to 1 ½ inches in size. Woody materials should be put through a grinder. The materials to be composted should have a C: N ratio of 30: 1. Materials like soil, ashes from a fireplace, manure from carnivorous animals should not be added to the pile. Once a pile is started, do not add anything. Composting works best if the moisture content of the materials in the pile is about 50%. Too much moisture will make a soggy mass and decomposition will be slow. Also, if the material is too dry, decomposition will not occur at all or will be very slow. Heat is very important for quick composting. To prevent heat loss and build up heat, a pile of atleast 36″ x 36″ x 36″ is recommended. The compost pile needs to be turned to prevent it from getting too hot. By turning the pile it will not overheat, and it will provide aeration also.

5.3.4.2 North Dakota State University hot composting (Smith, 1995)

Compost piles with a height of 6 feet 91.8 m) are raised. The maximum size of the organic matter pieces should be 6-9 inches (15-23 cm) long. If bins are constrcuted, the dimensions of 5 x 5 x 6 feet will yield 150 cubic feet (4.3 cu m) of compost. This is a good volume of compost to produce in a four-to-six week time period. To keep the aerobic bacterial population high and active, 0.12 kg of actual nitrogenous fertilizer should be added per cubic foot of dry matter and holes punched (four to five) into the centre of the pile. This is best done in phases or stages as the compost pile is building up. For example for a 150 cubic feet volume of dry matter, if the pile is built up over a period of three stages- at 2 feet, 4 feet and 6 feet, 5.7 kg of nitrogenous fertilizer should be added at each step. The total should be about 17-18 kg of fertilizer for the entire pile.

In this high temperature, bacterially active system, it is best to turn the composting material every three to four days. Once activated, expect the temperature to range between 120-160^{0}F (49-71^{0}C). The decomposition will be faster in summer and take more time in the winter. Once the compost is no longer hot and is odour free, crumbling material, it is ready for use.

5.3.4.3 Effective microrganism (EM) based quick composting production process (Hiraoka, 2002)

Since 1999, seven small-scale organic fertilizer plants, using this EM based technique have been in operation in Myanmar. They are owned and operated by Women's Income Generation Groups. A unit plant consists of nine pits of 6 feet (length) x 4 feet (width) x 3 feet (depth), enclosed by low walls and covered with roof.

The raw materials used in this technique are as follows:

Cow dung:	2 portions
Rice husk:	1 portion
Rice husk/ charcoal:	1 portion
Rice bran, milled:	1 portion
Accelerator:	33 litres of EM solution or *Trichoderma* solution per pit

The EM solution or accelerator is prepared by mixing 10 ml EM, 40 ml molasses and 950 ml water and leaving it for five to seven days depending on the temperature. The solution is then added to one litre of molasses and 98 litres of water to obtain 100 litres of ready-to-use EN solution. The amount is sufficient for three pits and reduced the composting period from three months to one month.

For composting, mix all the ingredients, except the accelarator. Then make 0.5 feet layer of mixture in the pit and sprinkle accelerator over the layer. Repeat the procedure until the pit is full. Cover with plastic sheet. Two or three weeks later, miz the whole pit to boost aerobic decomposition. The fertilizer is ready to use a couple of weeks later. A pit turns out 900 kg of final product per batch. Assuming that it takes 30 days on an average to produce a batch and only eight pits are being used, the annual production works out to 86.4 tons.

5.3.4.4 IBS Rapid composting technology (Virginia, 1997)

The IBS (Institute of Biological Sciences, University of the Philippines, Los Baños) technology involves inoculating the plant substrates used for composting with cultures of *Trichoderma harzianum*, a cellulose decomposer fungus. The fungus grown in a medium of sawdust mized with leaves of ipil ipil (*Leucaena leucocephala*), a legumnous tree, is termed as compost fungus activator (CFA). The technology is a development of windrow type of composting. The composting time, using this procedure, ranges from 21 to 45 days, depending on the plant substrates used.

The procedure consists of two parts: the production of the compost fungus activator and the composting process. To start with, agrowastes like rice straw, weeds and grasses should be chopped, moistened with water. For the compost mixture, carbonaceous substrates should be mixed with nitrogenous ones at a ratio of 4:1 or less, but never lower than 1:1 (on a dry weight basis). Some posiible combinations are:

3 parts rice- 1part ipil-ipil

4 parts rice straw- 1 part poultry manure

4 parts grasses-1 part legume materials + 1 part manure

4 partsgrasses- 1part *Chromolaena odorata* or *Mikania cordata* + 1 part animal manure.

For composting, the substrates should be piled loosely in a compost pen to provide aeration and the CFA is broadcastt onto the substartes during piling. The amount of CFA used is 1% of the total weight of the substrates (i.e. aboit 1 kg compost activator per 100 kg substrate). For faster decomposition, larger amount of CFA may be used. The heap should be covered over completely. This maintains the heat for compsoting and minimizes moisture loss. The compost heap usually heats up in 24-48 hours. Heat should be maintained at 50^{0}C or higher, and the heap should be turned once every 5-7 days for the first two weeks, and thereafter once every two weeks. After the first week, the volume of the pile should be reduced by one-third. After two-weeks, the volume of the pile should be reduced to one half of the original. The mature compost should be removed from the pen, and dried in the sun for two days. It should then be put in sacks and stored in a dry area.

Decomposition will continue until the substrate is finely fragmented and the texture becomes powdery, after which it is sun-dried again until the moisture content is 10-20%. If mature compost is needed at once, it should be sun-dried for one day, as soon as its temperature drops to 30^{0}C. Although the compost retains some fibres, it can be applied immediately as fertilizer.

5.3.5 Vermicomposting

Much attention has been paid in recent years to manage different organic waste resources at low-input as well as eco-friendly basis. Vermicomposting, utilizing earthworms, is an eco-biotechnological process that transforms energy-rich and complex organic substances into a stabilized humus-like product (Benitez *et al.* 2000). Vermitechnology has been proposed globally as potential tool to stabilize the natural and anthropogenic wastes, such as sewage sludge, industrial sludge, plant-derived wastes, agro-industrial solid waste, household waste, animal dung, etc. Although microbes are responsible for the biochemical degradation of the organic matter, earthworms are the important drivers of the process, conditioning the substrate and altering biological activity (Aira *et al.* 2002). There has been extensive research into using earthworm to manage the various industrial wastes (Elvira *et al.* 1998; Maboeta and van Rensburg 2003; Garg and Kaushik 2005). The nutrient contents of different types of vermicompost prepared using *Eudrilus eugeniae* at maturity period is given in Table (Jeyabal and Kuppuswamy 2001)

Vermicomposts have been reported to enhance plant germination, growth and yields in greenhouse crops (Atiyeh *et al.* 2000; Atiyeh *et al.* 2001; Edwards and Arancon 2004a; Edwards and Arancon 2004b; Edwards *et al.* 2004) and field crops (Buckerfield and Webster 1998; Venkatesh *et al.* 1998; Vadiraj *et al.* 1998; Arancon *et al.* 2003a; Arancon *et al.* 2004) including bush pepper (Srinivasan *et al.* 2002). Such increased productivity of crops, in response to vermicomposts amendments, have been attributed to greater availability of mineral nutrients (Edwards and Burrows 1988; Werner and Cuevas 1996), as well as their rich microbial populations (Arancon *et al.* 2006; Pramanik *et al.* 2007). The nutrient content of different types of vermicompost (*Eudrilus eugeniae)* at maturity period is given in Table 10. The presence of plant growth-influencing substances, such as plant growth hormones and humic acids in vermicomposts has also been suggested as a possible factor contributing to increased plant growth and yields (Arancon *et al.* 2003b).

Table 10. Nutrient contents of different types of vermicompost (*Eudrilus eugeniae*) at maturity period

Organic materials	C:N ratio	Organic N (%)	NH_4-N (%)	P (%)	K (%)	Cu (mg kg^{-1})	Mn (mg kg^{-1})	Zn (mg kg^{-1})	Fe (mg kg^{-1})
Biodigested slurry + coir pith	17.6	0.68	0.84	0.74	0.52	8	120	51	182
Biodigested slurry + weeds	13.7	0.78	1.25	0.75	0.54	11	115	55	195
Cowdung + coir pith	17.3	0.61	0.63	0.76	0.51	8	132	57	225
Cowdung + weeds	12.4	0.87	0.98	0.79	0.55	12	125	50	222
Sugarcane press mud + coir pith	16.2	0.60	0.66	1.20	0.62	13	119	53	198
Sugar cane press mud + weeds	12.5	0.90	0.96	1.39	0.60	10	123	56	229
Lsd (0.05)	3.5	0.16	0.19	0.21	0.04	NS	NS	NS	NS

Source: Jeyabal and Kuppuswamy, 2001

Several epigeic earthworms have been identified as detritus feeder, which can be reared on large numbers in organic waste resources. Some of these epigeics, e.g. *Eisenia foetida* (Savigny), *Perionyx excavatus* (Perrier), and *Eudrilus eugeniae* Kinberg, have been appeared as key candidates for organic waste recycling practices (Gajalakshmi *et al.* 2002; Garg and Kaushik 2005; Suthar 2006). Also, *Lumbricus rubellus* and *Dendrovena veneta,* which are very common in temperate soils, have been suggested in the literature to combat the variety of organic wastes (Elvira *et al.* 1997; Edwards 1998). Recent studies revealed that some epigeics show excellent results in beddings amended with sheep manure (Nogales *et al.* 1999; Garg *et al.* 2005). In Indian soils *Lampito mauritii, P. excavatus, Dichogaster bolaui, Amynthas morrisi,* and *Drawida willsi* are very common and their composting potential is well established (Kale *et al.* 1982; Singh 1997; Gajalakshmi *et al.* 2002; Tripathi and Bhardwaj 2004a).

Studies conducted at the Indian Institute of Spices Research, Calicut, Kerala (India) showed that growing earthworms in bush black pepper pots improved the soil physico chemical properties and yield of pepper. It has been estimated that earthworms can produce three tons of compost from ten tons of feeding material. Different types of soils have different species of earthworms and hence choosing local or native species is ideal. Local species like *Lampito mauritii, Octochaetona serrata, Perionyx excavatus* are found to be good. Indigenous species like *Eudrilus eugeniae* and *Eisenia foetida* are also good. The Central Plantation Crops Research Institute, Kasargod, Kerala, (India) has identified efficient local worm species to compost coconut leaves and fronds.

5.3.5.1 *Small scale vermiculture*

Small-scale systems may use a wide variety of bins. Often, small-scale composters build their own bins. Companies also sell such bins. Commonly, bins are made of old plastic containers, wood, Styrofoam containers, or metal containers. Bins should have holes in the sides to allow air to flow, and a spout that can be opened or closed or holes in the bottom to drain into a collection tray. Plastic bins require more drainage than wooden ones because they are non-absorbent. The design of a small bin usually depends on where an individual wishes to store the bin and how they wish to feed the worms. Most small bins can be grouped into three categories:

Non-continuous: an undivided container. A layer of bedding materials is placed in the bin, lining the bottom. Worms are added and organic matter for composting is added in a layer above the bedding. Another layer is added on top of the organic matter and the worms will start to compost the organic matter and bedding. This type of bin is often used because it is small and easy to build. But it is relatively difficult to harvest because all the materials and worms must be emptied out when harvesting.

Continuous vertical flow: a series of trays stacked vertically. The bottom-most tray is filled first, in a similar fashion to any other bin, but is not harvested when it is full. Instead, a thick layer of bedding is added on top and the tray above is used for adding organic material. Worms finish composting the bottom tray and then migrate to the one above. When a sufficient number of worms have migrated, the bottom tray can be collected and should be relatively free of worms. These bins provide an easier method of harvesting.

Continuous horizontal flow: a series of trays lined horizontally. This method too relies on the earthworms migrating towards a food source in order to ease the process of harvesting. The bin is usually constructed to be similar to a non-continuous bin but longer horizontally. It is divided in half, usually by a large gauge screen of chicken wire. One half is used until it becomes full, then the other half is filled with bedding and organic matter. In time, the worms migrate to the side with the food and the compost can be collected. These bins are larger than a non-continuous system but still small enough to be used indoors, with the added bonus of being easier to harvest.

5.3.5.2 Large scale vermiculture

Most large-scale vermiculture systems do not incorporate a physical bin, because it is impractical. A large system usually uses a windrow, which consists of bedding materials for the earthworms to live in (see bedding below) and acts as a large bin; organic material is added to it. Although the windrow has no physical barriers to prevent worms from escaping, in theory they should not due to an abundance of organic matter for them to feed on. Often windrows are used on a concrete surface to prevent predators from killing the worm population. Another property of large scale windrows is that when fed on one side consistently a wave motion is generated over time.

5.3.5.3 Vermicomposting in bins

When beginning a vermicomposting bin, put moist bedding into the bin, and add as many composting worms as available. In hot climates place the bin in the shade or away from midday direct sun. Quantities of kitchen waste appropriate for the worm population can be added to the bin daily or weekly. At first, feed the worms approximately one-half their body weight in kitchen scraps a day, maximum. That is, if you have 1 kg of worms, feed them about 1/2 kg of kitchen scraps a day. After they have established themselves, you can feed them up to their entire body weight. It is best not to add new food until the old food has been processed by the worms.

Hartenstein and Hartenstein (1981) in their lab-scale (10 cm-diameter petri-dishes) experiments on vermicomposting of activated sludge observed that approximately 1 g worm could convert 4 g of activated sludge in 5 days. This converts to a feeding rate of 0.8 kg-feed/kg-worm/day. Riggle and Holmes (1994) noted that worms can consume their own weight in 24 h, while earlier studies conducted by Wright (1972) on *Lumbricus terrestris* on feeding rates of worms showed that the feeding rate was dependent on the feed as well as the feed preparation or feed pretreatment. Reviewing the work of numerous researchers, Edwards and Bohlen (1996) also demonstrated that the feed rates varied greatly not only with earthworm species but also with the feed type. Further, Edwards and Bohlen (1996) observed that the amount of litter that the earthworms ingest seems to depend more on the total amount of suitable organic matter than on other factors, an observation which is corroborated by the earlier work of Neuhauser *et al.* (1980), in their studies on vermicomposting of activated sludge.

A stocking density of 1.60 kg-worms/m^2 (0.33 lb-worms/ft^2) and a feeding rate of 1.25 kg-feed/kg-worm/day resulted in the highest bioconversion of the substrate into earthworm biomass. The best vermicompost was obtained at the same stocking density and a feeding rate of 0.75 kg-feed/kg-worm/day (Ndegwa *et al.* 2000). Earlier, Neuhauser *et al.* (1980) indirectly studied the effect of population density on growth and production of *E. foetida* and using linear regressions showed that growth declined with increase in population density, while production increased with population density. Using this same approach, the potential stocking density of *E. foetida* was estimated to be approximately 0.8 kg-worms/m^2 on horse manure and 2.9 kg-worms/m^2 on activated sludge. Similar studies by Dominguez and Edwards (1997) on the effect of stocking rate on growth and maturation of *Eisenia andrei* concluded that, whereas individual worms grew more and faster at the lowest population density, the total biomass production was maximum at the highest population density in their test. At higher rates, they noted that the worms sexually matured faster than in the lower stocking rates.

5.3.5.4 Bedding

Bedding is the living medium for the worms but also a food source. It is material high in carbon and made to mimic dried leaves on the forest floor, the worms' natural habitat. The bedding should be moist (often similar to the consistency of a wrung-out sponge) and loose to enable the earthworms to breathe and to facilitate aerobic decomposition. A wide variety of bedding materials can be used including newspaper, sawdust, hay, cardboard, burlap coffee sacks, peat moss, pre-composted (aged) manure, and dried leaves. Also agriculture waste (crop residues and cattle waste), farmyard manure obtained from cattle yard, consisting of cattle excreta, discarded cattle feed stock and urban solid wastes like vegetable market waste and household waste are used as bedding material (Tripathi and Bhardwaj 2004b; Suthar 2007).

5.3.5.5 Temperature

Worms used in composting systems prefer temperatures of 55 to 70 degrees Fahrenheit (12-21 degrees Celsius). The temperature of the bedding should not drop below freezing or above 85 °F (29 °C). Maintaining moderate bed temperatures for vermicomposting systems during low ambient temperatures can significantly increase earthworm density. For instance, studies on the epigeic species *Eudrilus eugeniae, Perionyx excavatus* and *Eisenia foetida* by Reinecke *et al.* (1992) showed that *Eisenia foetida* has a wider tolerance for temperatures than *E. eudrilus eugeniae* and *P. excavatus* which allows this species to be cultivated in areas with higher temperatures (often as high as 43°C) as well as areas with lower soil temperatures (often below 5°C). Therefore, the other two species will have limited outdoor application in vermiculture systems. Fayolle *et al.* (1997) found that *Dendrobaena veneta* kept at 10^{0}C produce around eight times fewer cocoons worm^{-1} than those maintained at 15^{0}C, while Neuhauser (1988) reported that *D. veneta* failed to produce cocoons at 15^{0}C. Viljoen *et al.* (1992) found that cocoon production for *D. veneta* was 0.28 cocoon worm^{-1} day^{-1} at 25^{0}C but only 0.17 cocoon worm^{-1} day^{-1} at 15^{0}C. Frederickson and Howell (2003) demonstrated that vermicomposting systems operating at low ambient temperatures (6.3 ± 2.3 °C for 34 coldest weeks) have very low rates of worm reproduction. However, by maintaining moderate processing temperatures (13.7 ± 0.8 °C for the same period), significantly more cocoons and hatchlings were produced during 60 weeks. Total earthworm biomass was almost five-fold greater at the end of the study period for the heated beds compared with unheated. Apparently reproduction rates of the worms appear to be highly temperature dependent.

5.3.5.6 Worm feeding

There are two methods of feeding:

Top feeding: organic matter is placed directly on top of the existing layer of bedding in a bin and then covered with another layer of bedding. This is repeated every time the bin is fed.

Pocket feeding: a top layer of bedding is maintained and food is buried beneath. The location of the food is changed each time and often the bin is fed in more than one location. As bedding runs low more is added.

Vermicomposters often use a combination of both methods. Sometimes unburied food can attract fruit flies. For this reason, food should be buried at least one inch under the surface of the bedding material.

5.3.5.7 Preparing vermicompost (Hemangee, 2002)

The bed should be 2 ½ feet (height) x 4 feet (width) x any desired length. Apply worms for every part of the waste. Decomposition can be accelerated by shredding the raw materials into small pieces.Carbonaceous materials like sawdust, paper and straw can be mixed with nitrogen rich materials such as sewage sludge, biogas slurry and fish scraps to obtain a near optimum C/ N ratio of 30:1/ 40:1. A varied mixture of substances produces good qulaity compost, rich in major and micronutrients. The raw materials should be kept in piles and the temperature allowed reaching 50-55^{0}C.

The piles should remain at this temperature for 7 to 10 days. The optimum moisture level for manintaining aerobic conditions is 40-45%. Proper moisture and aeration can be maintained by mixing fibrous with nitrogen rich materials. The temperature of the piles should be within 28-30^{0}C. The pH of the raw material should not exceed 6.5-7.0. After a month, the compost is ready. It will be black, granular, lightweight and humus rich. To facilitate separating the compost, stop watering two or three days before emptying the beds. This will force about 80% of the worms to the bottom of the bed. The rest of the worms can be removed by hand. The vermicompost is then ready for application.

5.3.6 Coir pith compost

Coir pith, a highly lignocellulosic material, has wider C: N ratios ranging from 58:1 to 112: 1 (Nagarajan *et al.* 1985). Such wide C: N ratios coupled with low N content, presence of soluble tannin related phenolic compounds (8-12%), low biodegradability etc are some of the problems associated with raw coir pith. In such cases, the best way to utilize coir pith is after suitable composting. The plantation sector generates upto 7.5 million tons of coir pith (Table 1), which on composting serves as a rich source of nutrients.

5.3.6.1 Converting coir pith into compost

The Tamil Nadu Agricultural University, Coimbatore (Tamil Nadu State, India) and the Coir Board (India), have developed a technology for converting the coir pith into compost by using a fungus call *Pleurotus sajor caju* and urea. For converting one ton of coir pith into compost, 5 bottles of spawn culture and 5 kg of urea are required. The raw coir pith (100 kg) is spread uniformly over a hard floor of cement, stone, slab or brick in a layer of size 5 x 3 m with a thickness of 10 cm. Then 1 bottle (350 g.) of the fungus culture (spawn) is spread over the coir pith. Another layer of 100 kg coir pith is spread over it and 1 kg urea is added over the second layer. This process is repeated by adding the fungus spawn and urea alternatively with 100 kg coir pith till heap reaches up to a height of one meter. The heap is kept as such with constant watering and covering with a thin layer of coir waste to conserve moisture. After 30 days of decomposition, the coir pith turns into a black mass of compost with reduced lignin, cellulose, organic carbon and C: N ratio. The volume of material is also reduced by 40 percent. In order to ensure moisture retention and protection from heavy rainfall and wind adequate shelter should be provided to the heap. The major nutrients present in the coir pith compost are given in Table 11 (http://www.cpcri.ernet.in/faqnutrition1.htm; accessed on 08/02/2007).

Table 11. Nutrient content of coir pith compost

Nutrient	Level (%)
a. Nitrogen	4.02
b. Phosphorus	0.06
c. Potassium	1.20
d. Calcium	0.50
e. Magnesium	0.48
f. Organic carbon	24.50
g. C:N ratio	24:1

Source: http://www.cpcri.ernet.in/faqnutrition1.htm; accessed on 08/02/2007.

6.0 CONCLUSION

In a crop production system relying solely on organic inputs, the key to sustainable crop yield is the organic matter level in the soil. Apparently, higher soil organic matter enhances soil quality through marked improvements in physical and chemical characteristics. Besides, numerous studies indicate that addition of organic manures to soils also enhance the levels of substrates (dissolved organic C, dissolved organic N, labile organic N, light fraction organic matter-C & -N, carbohydrates etc), microbial biomass-C, -N and –P and microbial activity. Apparently, this suggests that addition of organic manures makes the soil biologically more active; a parameter, which is vital to the success of organically- managed cultivation systems.

In spice based cropping systems involving a mixture of annual and perennial crops, generation of agro wastes is a continuous process. While addition of these wastes in their raw state is beset with problems, suitable composting techniques do exist to convert them into nutrient rich manures. However, it is for the farmer to optimize the broad framework of steps he must take in order to generate sufficient organic waste from his own farm coupled with the ideal composting technique. A perfect balance of green manures, cover crops, and shade trees coupled with a sizeable population of poultry birds and livestock ably supported by scientific management would provide the necessary impetus. However, our experience suggests that it is easier said than done because such efforts are often constrained by the size of the farm holdings, the economic state of the farmer, his level of skill and more importantly the prevailing market price for organically produced spices. Be that as it may, in every way it seems that the mantra 'back to nature' has broadened dramatically from organic farming to nature inspired farming. For the land that gave birth to ayurveda this may come as no surprise.

REFERENCES

Abdin, O. A., Zhou, X. M., Cloutier, D., Coulman, D. C., Faris, M. A. and Smith, D. L. (2000) Cover crops and interrow tillage for weed control in short season maize *(Zea mays)*. *Eur.J. Agron.*, 12: 93-102.

Adams, P. L., Daniel, T. C., Edwards, D. R., Nichols, D. J., Pote, D. H. and Scott, H. D. (1994) Poultry litter and manure contribution to nitrate leaching through the vadose zone. *Soil Sci. Soc. Am. J.*, 58: 1206–1211.

Adegbidi, H. G. and Briggs, R. D. (2003) Nitrogen mineralization of sewage sludge and composted poultry manure applied to willow in a greenhouse experiment. *Biomass and Bioenergy*, 25: 665-673

Agamuthu, P. and Broughton, W. J. (1985) Nutrient cycling within the developing oil palm–legume ecosystem. *Agric. Ecosyst. Environ.*, 13: 111–123.

Aira M, Monroy F, Dominguez J. and Mato, S. (2002) How earthworm density affects microbial biomass and activity in pig manure. *Eur. J. Soil Biol.*, 38:7–10.

Aldaba, F. R. (1995) Coconut production in the Philippines: problems and prospects, *Plantations, Rech. Dev.*, 15–18.

Amner, W., McCarthy, A. J. and Edwards, C. (1988) Quantitative assessment of factors affecting the recovery of indigenous and release thermophilic bacteria from compost. *Appl. Environ. Microbiol.*, 54: 3107–3112.

Anonymous (1998) Fertilizer values of some manures. *Countryside & Small Stock Journal*. September-October. p. 75.

Antizar-Ladislao, B., Lopez-Real, J. and James Beck, A. (2006) Degradation of polycyclic aromatic hydrocarbons (PAHs) in an aged coal tar contaminated soil under in-vessel composting conditions. *Environmental Pollution*, 141: 459-468.

Arancon, N. Q., Edwards, C. A. and Bierman, P. (2006) Influences of vermicomposts on field strawberries: Part 2. Effects on soil microbiological and chemical properties. *Bioresour. Technol*, 97:831-840.

Arancon, N. Q., Edwards, C. A., Bierman, P., Welch, C. and Metzger, J. D. (2003a) Effects of vermicomposts applied to tomatoes and peppers grown in the field, and strawberries grown under high plastic tunnels. *Pedobiol.*, 47: 731–735.

Arancon, N. Q., Edwards, C. A., Bierman, P., Welch, C. and Metzger, J. D. (2004) The influence of vermicompost applications to strawberries: Part 1. Effects on growth and yield. *Bioresour. Technol*, 93: 145–153.

Arancon, N. Q., Lee, S., Edwards, C. A., Atiyeh, R. M. (2003b) Effects of humic acids and aqueous extracts derived from cattle, food and paper-waste vermicomposts on growth of greenhouse plants, *Pedobiol.*, 47: 741–744.

Aref, S. and Wander, M. M. (1997) Long-term trends of corn yield and soil organic matter in different crop sequences and soil fertility treatments. *Adv. Agron.*, 62:153-197.

Atiyeh, R. M., Arancon, N. Q., Edwards, C. A. and Metzger, J. D. (2001) The influence of earthworm-processed pig manure on the growth and productivity of marigolds. *Bioresour. Technol*, 81:103–108.

Atiyeh, R. M., Edwards, C. A., Subler, S. and Metzger, J. D. (2000) Earthworm processed organic wastes as components of horticultural potting media for growing marigolds and vegetable seedlings. *Compos. Sci. Util.*, 8: 215–223.

Avnimelech, Y. (1986) "Organic residues in modern agriculture". In: Chen Y. and Avnimelech, Y. (eds.) *The role of organic matter in modern agriculture*. Martinus Nijhoff, Dordrecht, pp. 1-10.

Barros, A. J. M., Prasad, S., Leite, V. D. and Souza, A. G. (2007) Biosorption of heavy metals in upflow sludge columns. *Bioresour. Technol* 98: 1418-1425.

Beffa, T., Blanc, M., Marilley, L., Fischer, J. L., Lyon, P. F. and Aragno, M. (1996) "Taxonomic and metabolic microbial diversity during composting". In: De Bertoldi, M., Sequi, P., Lemmes, B. and Papi, T. (eds.) *The science of composting*. Part I, Chapman and Hall, London, pp. 149–161.

Benitez, E., Nogales, R., Masciandro, G. and Ceccanti, B. (2000) Isolation by isoelectric focusing of humic-urease complexes from earthworm (*Eisenia foetida*) processed sewage sludges. *Biol. Fert. Soils*, 31: 489–493.

Bernal, M. P., Paredes, C., Sánchez-Monedero, M. A. and Cegarra, J. (1998b) Maturity and stability parameters of composts prepared with a wide range of organic wastes. *Bioresour. Technol*, 63: 91–99.

Bernal, M. P., Sánchez-Monedero, M. A, Paredes, C. and Roig, A. (1998a) Carbon mineralization from organic wastes at different composting stages during their incubation with soil. *Agric. Ecosyst. Environ.*, 69: 175–189.

Berry, P M., Sylvester-Bradley, R., Phillips, L., Hatch, D. J., Cuttle, S., Raynes, F. and Gosling, P. (2002) Is the productivity of organic farms restricted by the supply of available nitrogen?, *Soil Use Manage*, 15:137–143.

Biddappa, C. C., Upadhyay, A. K., Hegde, M. R. and Palaniswami, C. (1996) Organic matter recycling in plantation crops. *J. Plantn. Crops*, 24: 71-85.

Bolton, H., Elliott, L. F., Papendick, R. I. and Bezdicek, D. F. (1985) Soil microbial biomass and selected soil enzyme activities: effect of fertilization and cropping practices. *Soil Biol. Biochem.*, 17: 297–302.

Boparai, B. S., Yadvinder Singh and Sharma, B. D. (1992) Effect of green manure (*Sesbania aculeata*) on physical properties of soil and growth of rice-wheat and maize-wheat cropping system. *Int. Agrophys.*, 6: 95–101.

Breland, T. A. (1994) Enhanced mineralization and denitrification as a result of heterogeneous distribution of clover residues in soil. *Plant Soil*, 166: 1–12.

Broughton, W. J. (1976) Effects of various covers on the performance of Elaeis guineensis (Jacq.) on different soils. In: Earp D.A. and Newall, N. (eds.), International Oil Palm Developments, *Proc. International Oil Palm Conference*, Kuala Lumpur, 501–525.

Buckerfield, J. C. and Webster, K. A. (1998) Worm-worked waste boosts grape yields: Prospects for vermicompost use in vineyards. The Australian and New Zealand *Wine Industry Journal*, 13: 73–76.

Cabrera, M.L., Chiang, S.C., Merka, W.C., Thompson, S.A. and Pancorbo, O.C. (1993) Nitrogen transformations in surface-applied poultry litter: effect of litter physical characteristics. *Soil Sci. Soc. Am. J.* 57: 1519–1525.

Caravaca, F., Masciandaro, G. and Ceccanti, B. (2002) Land use in relation to soil chemical and biochemical properties in a semiarid Mediterranean environment. *Soil Till. Res.* 68: 23–30.

Carpenter-Boggs, L., Kennedy, A.C. and Reganold, J.P. (2000) Organic and biodynamic management: Effects on soil biology. *Soil Sci. Soc. Am. J.*, 54:1651–1659.

Cekmecelioglu, D., Demirci, A., Graves, R.E., Davitt, N.H. (2005) Applicability of Optimised In-vessel Food Waste Composting for Windrow Systems. *Biosystems Engineering*, 91: 479-486.

Chaves, B., Neve, S. D., Boeckx, P., Cleemput, O. V. and Hofman, G. (2005) Screening organic biological wastes for their potential to manipulate the N release from N-rich vegetable crop residues in soil. *Agric. Ecosyst. Environ.*, 111: 81-92

Christensen, B.T. (1986) Barley straw decomposition under field conditions: effect of placement and initial nitrogen content on weight loss and nitrogen dynamics. *Soil Biol. Biochem.*, 18: 523–529.

Costa, F. J. S. A., Bouldin, D. R. and Suhet, A. R. (1990) Evaluation of N recovery from mucuna placed on the surface or incorporated in a Brazilian oxisol. *Plant Soil*, 124: 91-96.

Cuttle, S., Shepherd, M. and Goodlass, G. (2004) A review of leguminous fertility-building crops. *Defra project OF0316*. Available from: http://www.organicsoilfertility.co.uk. [Accessed 02 February 2007).

Das, K.C., Tollner, E.W. and Eiteman, M.A. (2003) Comparison of synthetic and natural bulking agents in food waste composting. *Comp. Sci. & Util.*, 11: 27–35.

Den Hollander, N.G., Bastiaans, L. and Kropff, M.J. (2007) Clover as a cover crop for weed suppression in an intercropping design: II. Competitive ability of several clover species. *Eur. J. Agron.*, 26: 104-112.

De Volder, P.S., Brown, S.L., Hesterberg, D. and Pandya, K. (2003) Metal bioavailability and speciation in a wetland tailings repository amended with biosolids compost, wood ash, and sulfate. *J. Environ. Qual.*, 32: 854–864.

Dickson, N., Richard, T., Kozlowski, R. (eds.) (1991) *Composting to Reduce the Waste Stream: A Guide to Small Scale Food and Yard Waste Composting*. Northeast Regional Agricultural Engineering Service, Cornell University, 152 Riley-Robb Hall, Ithaca, NY 14853; 607-255-7654.

Dinesh, R. (2004) Long-term effects of leguminous cover crops on microbial indices and their relationships in soils of a coconut plantation of a humid tropical region. *J. Plant Nutr. Soil Sci.,* 167:189–195.

Dinesh, R., Dubey, R. P., Ganeshamurthy, A. N. and Shyam Prasad, G. (2000) Organic manuring in rice based cropping system: Effects on soil microbial biomass and selected enzyme activities. *Curr. Sci.,* 79, 1716-1720.

Dinesh, R., Suryanarayana, M. A., Ghoshal Chaudhuri, S., Sheeja, T. E. and Shiva, K. N. (2006) Long-term effects of leguminous cover crops on biochemical and biological properties in the organic and mineral layers of soils of a coconut plantation. *European J. Soil Biol.,* 42: 147-157.

Dinesh, R., Suryanarayana, M. A., Ghoshal Chaudhuri, S. and Sheeja, T. E. (2004) Influence of leguminous cover crops on the general and specific biochemical characteristics in soils of a plantation. *Soil Till. Res.,* 77: 69-77.

Dinesh, R., Suryanarayana, M.A., Nair, A.K. and Ghoshal Chaudhuri, S. (2001) Leguminous Cover Crop Effects on nitrogen mineralization rates and kinetics in soils *J. Agron. Crop Sci.,* 187: 161–166.

Dominguez, J. and Edwards, C.A. (1997) Effects of stocking rate and moisture content on the growth and maturation of *Eisenia andrei* (Oligochaeta) in pig manure. *Soil Biol. Biochem.,* 29:743–746.

Donahue, D.W., Chalmers, J.A. and Storey, J.A. (1998) Evaluation of in-vessel composting of university postconsumer food wastes. *Compost Sci. Util.,* 6: 75–81.

Droffner, M.L. and Brinton, W.F. (1995) Survival of E. coli and Salmonella population in aerobic thermophilic composts as measured with DNA gene probes. *Zentralbl Hygiene,* 197:387–397.

Dunne, E.J., Culleton, D., O'Donovan, G., Harrington, R. and Olsen, A.E. (2005) An integrated constructed wetland to treat contaminants and nutrients from dairy farmyard dirty water. *Ecological Engineering,* 24: 219-232.

Edmisten, K.L., Mitchell, C.C., Wood, C.W. and Delaney, D.D. (1992) Questions and answers about fertilizing with poultry litter. Alabama Cooperative Extension System. *Circular ANR-763*. pp. 1–6.

Edwards, C.A. and Bohlen, P.J. (1996). *Biology and ecology of earthworms,* 3rd ed., NY, Chapman and Hall.

Edwards, C.A. and Arancon, N.Q. (2004a) "The use of earthworms in the breakdown and management of organic wastes to produce vermicomposts and fees protein".

In: Edwards, C.A. (ed.), *Earthworm Ecology* (second ed.), CRC Press, Boca Raton, FL, pp. 345-379.

Edwards, C.A. and Arancon, N.Q. (2004b) "Interactions among organic matter, earthworms and microorganisms in promoting plant growth". *In*: Edwards, C.A. (ed.), *Functions and Management of Soil Organic Matter in Agroecosystems*, vol. 11. CRC Press, Boca Raton, FL, pp. 327-376.

Edwards, C.A. and Burrows, I. (1988) "The potential of earthworm composts as plant growth media". *In*: Edwards, C. A. and Neuhauser, E. F. (eds.), *Earthworms in Environmental and Waste Management*, SPB Academic Publ. b.v., The Netherlands, pp. 211-220.

Edwards, C.A., Dominguez, J. and Arancon, N.Q. (2004) "The influence of vermicomposts on plant growth and pest incidence". In: Mikhail, W.Z.A. and Shakir, S.H. (eds.), *Soil Animals and Sustainable Development*, pp. 397-420.

Edwards, CA., Dominguez, J. and Neuhauser, E.F. (1998) Growth and reproduction of *Perionyx excavatus* (Perr.) (Megascolecidae) as factors in organic waste management. *Biol. Fert. Soils*, 27: 155-161.

Edwards, D.R. and Daniel, T.C. (1993) Effects of poultry litter application rate and rainfall intensity from fescue grass plots. *J.Environ.Qual.*, 22: 361-365.

Eghball, B. and Power, J.F. (1999) Composted and noncomposted manure application to conventional and no-tillage systems: corn yield and nitrogen uptake. *Agronomy J.* 91: 819-825.

Eghball, E. and Lesoing, G.W. (2000) Viability of weed seeds following manure windrow composting. Compost Science & Utilization. Winter. p. 46-53.

Ekholm, P., Turtola, E., Grönroos, J., Seuri, P. and Ylivainio, K. (2005) Phosphorus loss from different farming systems estimated from soil surface phosphorus balance. *Agric. Ecosyst. Environ.*, 110: 266-278.

Elfstrand, S., Hedlund, K. and Mårtensson, A. (2007) Soil enzyme activities, microbial community composition and function after 47 years of continuous green manuring. *Appl. Soil Ecol.*, 35: 610-621.

Elorrieta, M.A., López, M.J., Suárez-Estrella, F., Vargas-García, M.C. and Moreno, J. (2002) "Composting of different horticultural wastes: effect of fungal inoculation". In: Insam, H., Riddech, N. and Klammer, S. (eds.), *Microbiology of composting*, Springer, Berlin, pp. 119-132.

Elvira C., Sampedro, L., Benitez, E. and Nogales, R. (1998) Vermicomposting of sludges from paper mill and dairy industries with Eisenia andrei: A pilot scale study. *Bioresour. Technol*, 63: 205-211.

Elvira, C., Sampedro, L., Domingnez, J. and Mato, S. (1997) Vermicomposting of water stable sludge from paper pulp industry with nitrogen rich material. *Soil Biol. Biochem.*, 29: 759-762.

Elwell, D.L., Keener, H.M. and Hansen, R.C. (1996) Controlled high rate composting of mixtures of food residuals, yard trimmings and chicken manure. *Comp. Sci. & Util.*, 4: 6–15.

FAO (1980) A manual of rural composting. *FAO/UNDP Regional project RAS/75/004 Field Document* 15.

Fauci, M.F. and Dick, R.P. (1994) Soil microbial dynamics—short-term and long-term effects of inorganic and organic nitrogen. *Soil Sci. Soc. Am. J.* 58: 801–806.

Faure, D. and Deschamps, A.M. (1991) The effect of bacterial inoculation on the initiation of composting grape pulps. *Bioresour. Technol.* 37: 235–238.

Fayolle, L., Michaud, H., Cluzeau, D. and Stawieck, J. (1997) Influence of temperature and food source on the life cycle of the earthworm *Dendrobaena veneta* (*oligochaeta*). *Soil Biol. Biochem.* 29: 747–750.

Feil, B. and Liedgens, M. (2001) Crop production in living mulches - A review. *Pflanzenbauwissenschaften*, 5:15-23.

Ferreras, L., Gomez, E., Toresani, S., Firpo, I., Rotondo, R. (2006) Effect of organic amendments on some physical, chemical and biological properties in a horticultural soil. *Bioresour. Technol.* 97: 635-640.

Ferreras, L.A., Costa, J.L., Garcý´a, F.O. and Pecorari, C. (2000) Effect of no-tillage on some soil physical properties of a structural degraded Petrocalcic Paleudoll of the southern Pampa of Argentina. *Soil Till. Res.* 54: 31–39.

Filippi, C., Bedini, S., Levi-Minzi, R., Cardelli, R. and Saviozzi, A. (2002) Cocomposting of olive oil mill by-products: chemical and microbiological evaluations. *Comp. Sci. & Util.*, 10: 63–71.

Fong, M., Wong, J.W.C. and Wong, M.H. (1999) Review on evaluation of compost maturity and stability of solid waste, *Shanghai Environ. Sci.* 18: 91–93.

Frederickson, J., and Howell, G. (2003) Large-scale vermicomposting: emission of nitrous oxide and effects of temperature on earthworm populations: The 7th international symposium on earthworm ecology, Cardiff, Wales, 2002. *Pedobiol.*, 47: 724-730.

Freibauer, A., Rounsevell, M.D.A., Smith, P. and Verhagen, J. (2004) Carbon sequestration in the agricultural soils of Europe. *Geoderma*, 122: 1–23.

Gaind, S., Pandey, A.K. and Lata (2005) Biodegradation study of crop residues as affected by exogenous inorganic nitrogen and fungal inoculants, *J Basic Microbiol.*, 4: 301–311.

Gajalakshmi, S., Ramasamy, E. V. and Abbasi, S. A. (2002) Vermicomposting of different forms of water hyacinth by earthworm Eudrilus eugeniae, Kinberg. *Bioresour. Technol.* 82: (2002), pp. 165–169.

Garg, V.K., Chand, S., Chhillar and Yadav, A. (2005) Growth and reproduction of Eisenia foetida in various animal waste during vermicomposting. *Appl. Ecol. Environ. Res.,* 3: 51–59.

Garg, V.K. and Kaushik, P. (2005) Vermistabilization of textile mill sludge spiked with poultry droppings by an epigeic earthworm *Eisenia foetida. Bioresour. Technol.,* 96: 1063–1071.

Gomez A (1998) The evaluation of compost quality. Trends Anal. Chem. 10: 310–314.

Guerra-Rodríguez, E., Alonso, J., Melgar, M.J. and Vázquez, M. (2006) Evaluation of heavy metal contents in co-composts of poultry manure with barley wastes or chestnut burr/leaf litter. *Chemosphere,* 65: 1801-1805.

Haapala, J.J. (1997) Risks of chicken manure as fertilizer. *In: Good Tilth.* June. p. 10, 12.

Hadas, A., Portnoy, R. (1994) Nitrogen and carbon mineralization rates of composted manures incubated in soil. *J. Environ. Qual.* 23:1184–1189.

Hamblin, A. (1991) Sustainable agricultural systems: what are the appropriate measures for soil structure?, Aust. J. Soil Res., 29:709–715.

Hamza, S. and Srinivasan, V. (2003) Organic manures for spices. *Planter's chronicle,* 99 (5): 4-5.

Handayanto E., Cadisch, G. and Giller K. E. (1994) Nitrogen release from prunings of legume hedgerow trees in relation to quality of the prunings and incubation method. *Plant and Soil* 160, 237-248.

Hartenstein, R. and Hartenstein, F. (1981) Physicochemical changes affected in activated sludge by the earthworm Eisenia foetida. J. Environ. Qual., 10: 372–376.

Hassen, A., Belguith, K., Jedidi, N., Cherif, A., Cherif, M. and Boudabous, A. (2001) Microbial characterization during composting of municipal solid waste. *Bioresour. Technol* 80: 217-225.

Hemangee, J. (2002) Vermiculture in India-online training material, Maharashtra *Agricultural Bioteks,* Pune, India.

Hiltbrunner, J., Liedgens, M., Bloch, L., Stamp, P. and Streit, B. (2007) Legume cover crops as living mulches for winter wheat: Components of biomass and the control of weeds. *Eur. J. Agron.* 26: 21-29.

Hiraoka, H. (2002) Biofertilizer production plant, Myanmar (FAO/ UNDP Project), *Back to office report,* February 2002, FAO-UNDP, Bangkok.

Hutchinson, J.J., Campbell, C.A. and Desjardins, R..L. (2007) Some perspectives on carbon sequestration in agriculture. *Agricultural and Forest Meteorology,* 142: 288-302

Illmer, P. and Schinner, F. (1997) Compost turning — A central factor for a rapid and high-quality degradation in household composting. *Bioresour. Technol,* 59: 157-162.

Jackson, L.E., Ramirez, I., Yokota, R., Fennimore, S.A., Koike, S.T., Henderson, D.M., Chaney, W.E., Calderón, F.J. and Klonsky, K. (2004) On-farm assessment of organic matter and tillage management on vegetable yield, soil, weeds, pests, and economics in California. *Agric. Ecosyst. Environ.*, 103: 443-463.

Janzen, H. H. and McGinn, S. M. (1991) Volatile loss of nitrogen during decomposition of legume green manure. *Soil Biol. Biochem.*, 23: 291-297.

Jeyabal, A. and Kuppuswamy, G. (2001) Recycling of organic wastes for the production of vermicompost and its response in rice–legume cropping system and soil fertility. *Eur. J. Agron.*, 15: 153-170.

Jørgensen, K.S., Puustinen, J. and Suortti, A–M. (2000) Bioremediation of petroleum hydrocarbon- contaminated soil by composting in biopiles. *Environmental Pollution*, 107: 245-254.

Kale, R.D., Bano, K. and Krishnamoorthy, R.V. (1982) Potential of Perionyx excavatus for utilizing organic wastes. *Pedobiol.*, 23: 419–425.

Kaschl, A., Romheld, V. and Chen, Y. (2002) The influence of soluble organic matter from municipal solid waste compost on trace metal leaching in calcareous soils, *Science of the Total Environment*, 291: 45–57.

Kautz, T., Wirth, S. and Ellmer, F. (2004) Microbial activity in a sandy arable soil is governed by the fertilization regime. *Eur. J. Soil Biol.*, 40: 87–94.

Kelling, K.A., Hero, D. and Rylant, K.E. (1995) "Effectiveness of composted manure for supplying nutrients". *In*: *Fertilizer, Aglime, and Pest Management Conference*, 17–18 January. pp. 77–81

Krapac, I.G., Dey, W.S., Roy, W.R., Smyth, C.A., Storment, E., Sargent, S.L. and Steele, J.D. (2002) Impacts of swine manure pits on groundwater quality. *Environmental Pollution*, 120: 475-492.

Kuo, S., Sainju, U.M. and Jellum, E.J. (1997) Winter cover cropping influence on nitrogen in soil. *Soil Sci. Soc. Am. J.*, 61:1392–1399.

Lal, R. (2002) "Why carbon sequestration in agricultural soils". *In*: Kimble, J.M., Lal, R. and Follet, R. F. (eds.), *Agricultural Practices and Policies for Carbon Sequestration in Soil*, Lewis Publishers, Boca Raton, 21–30.

Lal, R. (2004) Soil carbon sequestration to mitigate climate change. *Geoderma*, 123: 1–22.

Lal, R., Regnier, E., Eckert, D.J., Edwards, W.M. and Hammond, R. (1991) Expectations of cover crops for sustainable agriculture. In: Hargrove, W.L. (ed.), *Cover Crops for Clean Water*, Proc. of the Conference of the Soil and Water Conservation Society, 1–11.

Lal, R., Reicosky, D.C. and Hanson, J.D. (2007) Evolution of the plow over 10,000 years and the rationale for no-till farming. *Soil Till. Res.*, 93:1-12.

Laos, F., Mazzarino, M.J., Walter, I. and Roselli, L. (1998) Composting of fish waste with wood by-products and testing compost quality as a soil amendment: experiences in the Patagonia Region of Argentina. *Comp. Sci. Util.*, 6:59–66.

Larney, F.J. and Blackshaw, R.E. (2003) Weed seed viability in composted beef cattle feedlot manure. *J Environ. Qual.*, 32: 1105–1113.

Larney, F.J. and Hao, X. (2007) A review of composting as a management alternative for beef cattle feedlot manure in southern Alberta, Canada. *Bioresour. Technol*, 98: 3221-3227.

Lathwell, D.J. (1990) Legume green manures: principles for management based on recent research, Trop. *Soils Bulletin, No. 90-01*. North Carolina State University, NC, USA.

Lee, J.J., Park, R.D., Kim, Y.W., Shim, J.H., Chae, D.H., Rim, Y.S., Sohn, B.K., Kim, T.H. and Kim, K.Y. (2004) Effect of food waste compost on microbial population, soil enzyme activity and lettuce growth, *Bioresour. Technol*, 93: 21–28.

Lehmann, J., da Silva, J.P., Trujillo, L. and Uguen, K. (2000) Legume cover crops and nutrient cycling in tropical fruit tree production. *Acta Hort.*, 531: 65–72.

Lesturgez, G., Poss, R., Noble, A., Grünberger, O., Chintachao, W. and Tessier, D. (2006) Soil acidification without pH drop under intensive cropping systems in Northeast Thailand. *Agric. Ecosyst. Environ.* 114: 239-248.

Leton, T.G. and Stentiford, E.I. (1990) Control of aeration in static pile composting. *Waste Manag. Res.*, 8: 299–306.

Liedgens, M., Soldati, A., Stamp, P. (2004) Interactions of maize and Italian ryegrass in a living mulch system: (1) Shoot growth and rooting patterns. *Plant Soil*, 262: 191-203.

Löser, C., Ulbricht, H. and Seidel, H. (2004) Degradation of polycyclic aromatic hydrocarbons (PAHs) in waste wood. *Compos. Sci. Util.*, 12: 335–341.

Lotter, D.W., Seidel, R. and Liebhardt, W. (1993) The performance of organic and conventional cropping systems in an extreme climate year. *Am. J. Altern. Agr.*, 18: 146–154.

Maboeta, M. S. and van Rensburg, L. (2003) Vermicomposting of industrially produced woodchips and sewage sludge utilizing *Eisenia fetida, Ecotoxicol. Environ. Safety*, 56: 256–270.

Mäder, P., Fließbach, A., Dubois, D., Gunst, L., Fried, P. and Niggli, U. (2002) Soil fertility and biodiversity in organic farming. *Science*, 296:1694-1697.

Mandal, U.K., Singh, G., Victor, U.S. and Sharma, K.L. (2003) Green manuring: its effect on soil properties and crop growth under rice-wheat cropping system. *Eur. J. Agron.*, 19: 225-237.

Marinari, S., Masciandaro, G., Ceccanti, B. and Grego, S. (2000) Influence of organic and mineral fertilisers on soil biological and physical properties. *Bioresour. Technol*, 72: 9–17.

Martens, D.A., Johanson, J.B., Frankeberger, Jr. W.T. (1992) Production and persistence of soil enzymes with repeated additions of organic residues. *Soil Sci.*, 153: 53–61.

Martin-Rueda, I., Muñoz-Guerra, L.M., Yunta, F., Esteban, E., Tenorio, J.L., Lucena, J.J. (2007) Tillage and crop rotation effects on barley yield and soil nutrients on a Calciortidic Haploxeralf. *Soil Till. Res.*, 92: Pages 1-9.

Martins, O., Dewes, T. (1992) Loss of nitrogenous compounds during composting of animal wastes. *Bioresour. Technol*, 42: 103-111.

Millner, P.D., Olenchock, SA., Epstein, E., Rylander, R., Walker, J., Ooi, B.L., Horne, E. and Maritato, M.(1994) Bioaerosols associated with compost facilities. *Comp. Sci. Util.*, 2: 6–57.

Moeller, J., Reeh, U. (2003) Degradation of DEHP, PAHs and LAS in source separated MSW and sewage sludge during composting. *Comp. Sci. Util.*, 11: 370–378.

Moral, R., Moreno-Caselles, J., Perez-Murcia, M.D., Perez-Espinosa, A., Rufete, B., Paredes, C. (2005) Characterisation of the organic matter pool in manures. *Bioresour. Technol*, 96: 153-158.

Motavalli, P.P., Stevens, W.E. and Hartwig, G. (2003) Remediation of subsoil compaction and compaction effects on corn N availability by deep tillage and application of poultry manure in a sandy-textured soil. *Soil Till. Res.*, 71: 121-131.

Mustin, M. (1987) Le compost, *gestion de la matière organique*. F. Dubusc Eds., Paris, pp. 954.

Nagarajan, R., Manickam, T. S., Kothandaraman, G. V., Ramaswamy, K. and Palaniswamy, G. V. (1985) Manurial value of coir pith. *Madras Agric. J.*, 72: 533-535.

Nair, J., Sekiozoic, V. and Anda, M. (2006) Effect of pre-composting on vermicomposting of kitchen waste. *Bioresour. Technol*, 97: 2091-2095.

Namkoong, W., Hwang, E-Y., Park, J-S., Choi, J-Y. (2002) Bioremediation of diesel-contaminated soil with composting. *Environmental Pollution*, 119: 23-31.

Ndegwa, P.M., Thompson, S.A. and Das, K.C. (2000) Effects of stocking density and feeding rate on vermicomposting of biosolids. *Bioresour. Technol.* 71: 5-12.

Neal, K. (1994) Manure: The good, the bad, the ugly & how it works with your soil. *Acres USA*. October. p. 8, 10, 11, 13.

Neuhauser, E. F., Hartenstein, R. and Kaplan, D. L. (1980) Growth of the earthworm *Eisenia foetida* in relation to population density and food rationing. *OIKOS* 35: 93–98.

Neuhauser, E.F., Loehr, R.C. and Malecki, M.R. (1988) "The potential of earthworms for managing sewage sludge". *In*: Edwards, C. A. and Neuhauser, E. F. (eds.), *Earthworms in waste and environmental management*, SPB Academic Publishing, The Hague, pp. 9–20.

Nicholson, F.A., Chambers, B.J., Williams, J.R. and Unwin, R.J. (1995) Heavy metal contents of livestock feeds and animal manures in England and Wales. *Biores. Technol*, 70: 23-31.

Noble R. and Coventry, E. (2005) Suppression of soil-borne plant diseases with composts: a review. Biocontrol Sci. and Technol., 15: 3–20.

Nogales, R., Elvira, C., Benitez, E., Thompson, R. and Gomez, M. (1999) Feasibility of vermicomposting dairy biosolids using a modified system to avoid earthworm mortality. *J. Environ. Sci. Health* B., 34: 151–169.

Oades, J.M .(1984) Soil organic matter and structural stability: mechanisms and implications for management. *Plant* Soil, 76: 319–337.

Olesen, J.E., Hansen, E.M., Askegaard, M. and Rasmussen, I.A. (2007) The value of catch crops and organic manures for spring barley in organic arable farming. *Field Crops Research*, 100: 168-178.

Ouédraogo, E., Mando, A. and Stroosnijder, L. (2006) Effects of tillage, organic resources and nitrogen fertilizer on soil carbon dynamics and crop nitrogen uptake in semi-arid West Africa. *Soil Till Res.*, 91: 57-67.

Ouédraogo, E., Mando, A.M. and Zombré, N.P. (2001) Use of compost to improve soil properties and crop productivity under low input agricultural system in West *Africa. Agric. Ecosyst. Environ.*, 84: 259–266.

Parthasarathy, V.A., Srinivasan, V. and Dinesh, R. (2006) Organic spices- present status and scope. *Souvenir of National seminar on appropriate technologies for sustainable horticulture*, Department of Horticulture, Faculty of Agriculture, Annamalai University, Tamil Nadu (India), 34-42.

Paustian, K., Andren, O., Janzen, H., Lal, R., Smith, P., Tian, G., Tiessen, H., Van Noordwijk M. and Woomer P. (1997a) Agricultural soil as a C sink to offset CO_2 emissions. *Soil Use Manage*. 13: 230–244.

Paustian K., Collins H.P. and Paul E.A. (1997b) Management controls on soil carbon. *In*: Paul, E. A., Paustian, K., Elliott, E. T. and Cole, C. V. (eds.), *Soil Organic Matter in Temperate Agroecosystems: Long-Term Experiments in North America*, CRC Press, Boca Raton, FL, USA, 15–49.

Pederson, L. (1998). Prevent pathogens. *American Agriculturist*, May. p. 26.

Peoples, M.B., Herridge, D.F. and Ladha, J.K. (1995) Biological nitrogen fixation: an efficient source of nitrogen for sustainable agricultural production? *Plant Soil*, 174:3–28.

Pinamonti, F. and Zorzi, G. (1996) Experiences of compost use in agriculture and in land reclamation projects. *In*: De Bertoldi, M. (ed.), *The Science of Composting*: Part I, Blackie, Glasgow, UK, pp. 517–527.

Pramanik, P., Ghosh, G.K., Ghosal, P.K. and Banik, P. (2007) Changes in organic - C, N, P and K and enzyme activities in vermicompost of biodegradable organic wastes under liming and microbial inoculants. Bioresour. Technol (in Press).

Raabe, R.D. (2001) *The rapid composting method.* Co-operative extension, Division of Agriculture and Natural resources, University of California.

Ramos, M.C. and Martínez-Casasnovas, J.A. (2006) Erosion rates and nutrient losses affected by composted cattle manure application in vineyard soils of NE Spain. *CATENA*, 68: 177-185.

Reeves, D.W. (1997) The role of organic matter in maintaining soil quality in continuous cropping system. *Soil Till. Res.* 43: 131–167.

Reinecke, A.J., Viljoen, S.A., Saayman, R.J. (1992) The suitability of Eudrilus eugeniae, perionyx excavatus and Eisenia fetida (Oligochaeta) for vermicomposting in southern Africa in terms of their temperature requirements. *Soil Biol. Biochem.*, 24: 1295-1307.

Requena, N., Azcón, R. and Baca, M.T. (1996) Chemical changes in humic substances from compost due to incubation with ligno-celulolytic microorganisms and effects on lettuce growth. *Appl. Microbiol. Biotechnol.*, 45: 857–863.

Riggle, D. and Holmes, H. (1994). New horizons of commercial vermiculture. Biocycle, 35: 58-62.

Robert, R. (ed.) (1992) *On-farm composting*, Natural Resource, Agriculture, and Engineering Service, Cooperative Extension, Ithaka, New York.

Rojas-Avelizapa, N.G., Roldán-Carrillo, T., Zegarra-Martínez, H., Muñoz-Colunga, A.M., Fernández-Linares, L.C.A. (2007) Field trial for an ex-situ bioremediation of a drilling mud-polluted site. *Chemosphere*, 66: 1595-1600.

Sainju, U.M., Whitehead, W.F., Singh, B.P. and Wang, S. (2006) Tillage, cover crops, and nitrogen fertilization effects on soil nitrogen and cotton and sorghum yields. *Eur. J. Agron.*, 25: 372-382.

Schjønning, P., Elmholt, S., Munkholm, L.J. and Debosz, K. (2002) Soil quality aspects of humid sandy loams as influenced by organic and conventional long-term management. *Agric. Ecosyst. Environ.*, 88:195–214.

Schmid, O. and Klayz R. (1984) "Green Manuring: Principles and Practice". *In*: Brinton, Jr. W. F. (Translator) Publication of the Research Institute for Biological Husbandry, Switzerland and Woods End Agricultural Institute, Mt. Vernon, Maine 50 p.

Seymour, R.M., Donahue, D., Bourdon, M., Evans, J.R. and Wentworth, D. (2001) Intermittent aeration for in-vessel composting of crab processing waste. *Comp. Sci. Util.*, 9: 98–106.

Shepherd, M. (2002) Is the productivity of organic farms restricted by the supply of available nitrogen? *Soil Use Manage.*, (Special Edition on Organic Farming), 18: 248–255.

Shepherd, M.A. and Withers, P.J. (1999) Applications of poultry litter and triple superphosphate fertilizer to a sandy soil: effect on soil phosphorus status and profile distribution. *Nutr. Cycl. Agroecosyst.*, 54: pp. 233–242.

Sims, J.T. and Wolf, D.C. (1994) Poultry waste management: Agricultural and environmental issues. *Adv. Agron.*, 52:1-83.

Singh, J. (1997) Habitat preference of selected Indian earthworm species and their efficiency in reduction of organic materials. *Soil Biol. Biochem.*, 29: 585–588.

Singh, Y., Singh, B. and Khind, C.S. (1992) Nutrient transformations in soils amended with green manures. *Adv. Soil Sci.* 20: 237–309.

Smith, R.C.1995 *Composting Practices*. NDSU Extension Service, North Dakota State University of Agriculture and Applied Science, and USDA

Smith, M.S., Frye, W.W. and Varco, J.J. (1987) Legume winter cover crops. *Adv. Soil Sci.*, 7: 95-139.

Smith, P. (2004) Carbon sequestration in croplands: the potential in Europe and the global context. *Eur. J. Agron.*, 20: 229–236.

Sørensen P, Amato M (2002) Remineralisation and residual effects of N after application of pig slurry to soil. *Eur. J. Agron.* 16: 81–95.

Sørensen, P. and Jensen, E.S. (1995) Mineralization–immobilization and plant uptake of nitrogen as influenced by the spatial distribution of cattle slurry in soils of different texture. *Plant Soil*, 173: 283–291.

Speir, T.W., van Schaik, A.P., Hunter, L.C., Ryburn, J.L. and Percival, H.J. (2007) Attempts to derive EC_{50} values for heavy metals from land-applied Cu-, Ni-, and Zn-spiked sewage sludge. *Soil Biol. Biochem.*, 39: 539-549.

Srinivasan, V., Rubina, M.R., Hamza, S. and Kavitha Ramachandran (2002) Forms of soil potassium as influenced by organic fertilizers in bush black pepper (*Piper nigrum L.*). *J. Potassium Res.*, 18: 99-104.

Stentiford, E.I. (1996) "Composting control: principles and practice". *In*: De Bertoldi, M., Sequi, P., Lemmes, B., Papi, T. (eds.), *The Sciences of Composting*, Blackie Academic and Professional, Glasgow, UK, pp. 49–59.

Stevenson, F., Walley, F.L. and van Kessel, C. (1998). Direct vs. indirect nitrogen-15 approaches to estimate nitrogen contributions from crop residues. *Soil Sci. Soc. Am. J.*, 62, 1327–1334.

Stone, A.G., Traina, S.J. and Hoitink, H.A.J. (2001). Particulate organic matter composition and Pythium damping-off of cucumber. *Soil Sci. Soc. Am. J.*, 65:761-770

Strauch, D. (1996) Occurrence of microorganisms pathogenic for man and animals in source separated biowaste and compost importance, control, limits and epidemiology. In: De Bertoldi, M., Sequi, P., Lemmes, B. and Papi, T. (eds.), *The Sciences of Composting*, Blackie Academic and Professional, Glasgow, UK, pp. 224–232.

Suthar, S. (2006) Potential utilization of guar gum industrial waste in vermicompost production. *Bioresour. Technol.*, 97: 2474-2477.

Suthar, S. (2007) Nutrient changes and biodynamics of epigeic earthworm Perionyx excavatus (Perrier) during recycling of some agriculture wastes. *Bioresour. Technol.*, 98: 1608-1614.

Tai, H-S., He, W-H. (2007) A novel composting process for plant wastes in Taiwan military barracks. Resources, Conservation and Recycling, 51: 408-417.

Tejada, M. and Gonzalez, J.L. (2003) Effects of the application of a compost originating from crushed cotton gin residues on wheat yield under dryland conditions. *Eur. J. Agron.*, 19: 357–368.

Termorshuizen, A.J., van Rijn, E., van der Gaag, D.J., Alabouvette, C., Chen, Y., Lagerlöf, J., Malandrakis, A.A., Paplomatas, E.J., Rämert, B., Ryckeboer, J., Steinberg, C., Zmora-Nahum, S. (2006) Suppressiveness of 18 composts against 7 soil borne plant pathogens. *Soil Biol. Biochem.*, 38: 2461–2477.

Thomas, R. J. and Asakawa, N. M. (1993) Decomposition of leaf litter from tropical forage grasses and legumes. *Soil Biol. Biochem.*, 25: 1351-1361.

Thomas, R.J., Fisher, M.J., Ayarza, M.A., Sanz, J.I. (1995) "The role of forage grasses and legumes in maintaining the productivity of acid soils in Latin America". In: Lal R. and Stewart, B.A. (eds.), Soil Management: Experimental Basis for Sustainability and Environmental Quality. *Adv. Soil Sci. Series*, Lewis, Boca Raton, FL, pp. 61-81.

Thomsen, I.K. (2005) Crop N utilization and leaching losses as affected by time and method of application of farmyard manure. *Eur. J. Agron.*, 22: 1-9.

Thorup-Kristensen, K., Magid, J. and Jensen, L.S. (2003) Catch crops and green manures as biological tools in nitrogen management in temperate zones. *Adv. Agron.*, 79: 227–302.

Tiessen, H., Cuevas, E. and Chacon, P, (1994) The role of soil organic matter in sustaining soil fertility. *Nature*, 371: 783 - 785

Tirlok Singh, Nagarajarao, Y. and Sadaphal, M.N. (1980) Effect of legumes on physical properties of soil in mixed cropping with maize. *Indian J. Agron.*, 25: 592–599.

Tonitto, C., David, M.B. and Drinkwater, L.E. (2006) Replacing bare fallows with cover crops in fertilizer-intensive cropping systems: A meta-analysis of crop yield and N dynamics. *Agric. Ecosyst. Environ.*, 112: 58-72.

Tripathi, G. and Bhardwaj, P. (2004a) Decomposition of kitchen waste amended with cow manure using epigeic species (Eisenia fetida) and an anecic species (Lampito mauritii), *Bioresour. Technol.*, 92 : 215–218.

Tripathi, G. and Bhardwaj, P. (2004b) Comparative studies on biomass production, life cycles and composting efficiency of Eisenia fetida (Savigny) and Lampito mauritii (Kinberg). *Bioresour. Technol.*, 92: 275-283.

Tyson, S.C. and Cabrera, M.L. (1993) Nitrogen mineralization in soils amended with composted and uncomposted poultry litter. *Commun. Soil Sci. Pl. Anal.*, 24: 2361–2374.

Vadiraj, B.A., Siddagangaiah and Potty, S.N. (1998) Response of coriander (Coriandrum sativum L.) cultivars to graded levels of vermicomposts. *J. Spices Aromatic Crops*, 7: 141–143.

Van Ginkel, J.T., Van Haneghem, I.A. and Raats, P.A.C. (2001) SE–structures and environment: Physical properties of composting material: gas permeability, oxygen diffusion coefficient and thermal conductivity. *Biosystems Engineering*, 81: 113-125.

Vanlauwe B. and Giller K.E. (2006) Popular myths around soil fertility management in sub-Saharan Africa. *Agric. Ecosyst. Environ.*, 116: 34-46.

Venkatesh, Patil, P.B., Patil, C.V. and Giraddi, R.S. (1998) Effect of in situ vermiculture and vermicomposts on availability and plant concentration of major nutrients in grapes. *Karnataka J. Agric. Sci.*, 11: 117–121.

Verma, L.N. (1995) "Conservation and efficient use of organic sources of plant nutrients." In: Thampan, P.K. (ed.), *Organic Agriculture*, Peekay Tree Crops Development Foundation, Cochin, Kerala, India, p. 127.

Viljoen, S.A., Reinecke, A.J. and Hartman, L. (1992) The influence of temperature on the life cycle of *Dendrobaena veneta* (Oligochaeta). *Soil Biol. Biochem.*, 24: 141–1344.

Virginia, C.C. (1997) Rapid composting technology in Philippines: its role in producing good qulaity fertilizers. *Extension Bulletin*, FFTC, Taiwan.

Walker, D.J. and Bernal, M.P. (2007) The effects of olive mill waste compost and poultry manure on the availability and plant uptake of nutrients in a highly saline soil. *Bioresour. Technol.* (in press). (doi: 10. 1016/J.biortech. 2006.12.2006).

Wander, M.M., Traina, S.J., Stinner, B.R. and Peters, S.E. (1994) Organic and conventional management effects on biologically active soil organic matter pools. *Soil Sci. Soc. Am. J.*, 58:1130-1139.

Watson, C.A., Atkinson, D., Gosling, P., Jackson, L.R. and Rayns, F. (2002). Managing soil fertility in organic farming systems. *Soil Use and Manage.*, 18: 239-247.

Werner, M. and Cuevas, R. (1996) *Vermiculture in Cuba*. Biocycle, 37, JG Press, Emmaus, PA, 61–62.

Whitbread, A.M., Blair, G.J. and Lefroy, R.D.B. (2000) Managing legume leys, residues and fertilisers to enhance the sustainability of wheat cropping systems in Australia 1. The effects on wheat yields and nutrient balances. *Soil Till. Res.* 54: 63–75.

Wolter, M., Prayitno, S. and Schuchardt, F. (2004) Greenhouse gas emission during storage of pig manure on a pilot scale. *Bioresour. Technol.*, 95: 235-244.

Wright, M.A. (1972). Factors governing ingestion by the earthworm *Lumbricus terrestris* (L.) with special reference to apple leaves. *Ann. Appl. Biol.*, 70: 175-188.

Young, R.A. and Holt, R.F. (1977) Winter-applied manure: Effects on annual runoff, erosion, and nutrient movement. *J. Soil Water Conserv.*, September-October. p. 219–222.

Zhu, N. (2007) Effect of low initial C/N ratio on aerobic composting of swine manure with rice straw. *Bioresour. Technol.*, 98: 9-13.

3

Use of Microbes for Spices Production

M. Anandaraj and R. Dinesh

1.0. INTRODUCTION

The plant nutrient cycling depends on micro organisms that constitute major components of the biosphere. Although less than one per cent is cultured, a tremendous diversity exists in their populations. Transformation of nutrients from soil to plants is brought about by soil microorganisms. Latest molecular techniques enable scientists to understand and appreciate the diversity and help to utilize the beneficial effects of microbial diversity for increased productivity. Among soil organisms intense activity is found in and around rhizosphere of plants because of the presence of the secreted nutrients such as carbohydrates, amino acids and other cellular components present at the growing tip of roots. These root colonizing micro organism form symbiotic, associative, or parasitic relationship with the plants. These organisms have evolved mechanisms to occupy and reside in the specialized niche. These mechanisms are often utilized both for growth enhancement and disease suppression. The host also plays an important role in selecting its own microflora for the rhizosphere similar to the probiotic microflora of human gastrointestinal tract (Haas and Keel 2003).

2.0. USEFUL MICROBES

The activities of plant-associated microbial organisms are intense at the rhizosphere due to the presence of root exudates. Major genera associated in the rhizosphere are; *Bacillus, Burkholderia, Pseudomonas, Streptomyces* and *Trichoderma* besides symbiotic mycorriza (McSpadden and Fravel 2002). It is reported that often the soil type and host plant determine the structure of microbial communities (Garbeva *et al* 2004). The beneficial organisms screened for spices belong to the major groups namely, mycorrhiza, Plant growth promoting rhizobacteria (PGPR) and *Trichoderma*. These organisms were selected for the twin purpose of growth promotion and disease suppression. The principal mechanisms involved in these organisms are discussed followed by the results of application of these organisms for spice crops.

2.1. Mycorrhiza

Among symbiotic associations of plants with fungi, the most important is the mycorrhizal association which is highly developed form for mutual benefit. It has been reported that such symbiosis existed from Devonian period (Taylor 1990). The plant supplies the fungus its carbohydrate and the fungus provides nutrients especially Phosphorus, Zinc, Copper Sulphur etc. (Powell and Bagyaraj 1984). Mycorrhizae are also reported to alleviate transplantation shock and moisture stress (Davies *et al* 1993). There are two major types of associations namely; ectomycorrhiza, and endomycorrhiza. In ectomycorrhiza the fungus grows and forms an external mantle and grows intercellurly in the cortex and never grows intracellularly. Whereas in endomycorrhiza the fungus grows both inter and tra cellular. In this type the fungal mycelium penetrate in to cortex and inside the cortical cells forms a highly branched structure called arbuscles. Some of them also produce enlarged mycelium in the form of vesicles that store food reserves and later transformed in to resting spores. Because of the presence thse structures these organisms were referred to as vesicular arbuscular mycorrhiza (VAM) or Arbuscular Mycorrhizal Fungi (AMF). In AMF association, part of the mycelium is present outside the root and form extensive mycelial network. The external mycelium functions as extended root hair and takes part in absorption of water and nutrients and transports the same to the plant root. Ectendomycorrhizae is structurally different from ectomycorrhiza. There are also ericoid and orchid mycorrhiza colonizing members of ericaceae and orchids (Sieverding 1991). In most spice crops association of endomycorrhiza is reported (Manjunath and Bagyaraj 1982, Anandaraj *et al* 1991, 1996).

2.1.1. Beneficial effects of Mycorrhiza

As the external mycelium function as root hair, the plant is able to obtain fixed nutrients like Phosphorus beyond the zone of depletion. Increased uptake of K, S, Cu and Zn are also reported (Bolan 1991, Jacobson *et al* 2001). The main benefit of mycorrhizal association is that it ameliorates moisture stress and prevents transplantation shock in several transplanted crops (Davis *et al*

1993). The rhizosphere communities present in a mycorrhizal plant is different and hence the term Mycorhizosphere and it is often regarded as first line of defense against soil borne pathogens (Linderman 1988). Mycorrhizal inoculations are useful in the transplanted crops as the initial inoculation persists and multiplies on the newly formed roots (Bellguard 1992, Kaya *et al* 2003).

2.2.0. Plant growth promoting rhizobacteria (PGPR)

Among the root inhabiting organisms, fluorescent pseudomonads constitute a major successful colonizers followed by bacilli. The commonly occurring species include *Pseudomonas fluorescens, P. putida* and *P. aeruginosa.* Kloepper and Schroth (1978) first reported that certain root-colonizing bacteria could promote radish growth and named the bacteria "plant growth-promoting rhizobacteria" (PGPR). Since then a large number of scientists have been involved in the research on PGPR all over the world.

Soil bacteria in the genera *Rhizobium, Bradyrhizobium, Sinorhizobium, Azorhizobium, Mesorhizobium* and *Allorhizobium* (family Rhizobiaceae) collectively referred to as rhizobia invade plant root systems and form root nodules (Martinez-Romero and Wang, 2000) and are called iPGPR. iPGPRs are mostly Gram-negative and rod-shaped, with a lower proportion being Gram-positive rods, cocci and pleomorphic forms. The primary mechanism by which rhizobia increase plant growth is N_2 fixation. ePGPR do not form nodules, but increase plant growth through a variety of mechanisms; these include genera such as *Bacillus, Pseudomonas, Erwinia, Caulobacter, Serratia, Arthrobacter, Micrococcus, Flavobacterium, Chromobacterum, Agrobacterium, Hyphomycrobium* and free-living nitrogen-fixing bacteria (Prithiviraj *et al.,* 2003}. Rhizobacteria that establish inside plant roots, forming more intimate associations, are endophytes. These include a wide range of soil bacteria forming less formal associations than the rhizobia–legume symbiosis; endophytes may stimulate plant growth, directly or indirectly (Kobayashi *et al.,* 1995) and include the rhizobia. In general, a greater proportion of endophytes are PGPR than is the case for bacteria inhabiting the rhizoplane or rhizosphere (Nowak, 1998). Nodulating (rhizobia) and other N_2-fixing rhizobacteria are also endophytes, living in specially developed root organs. They promote plant growth through N fixation and these bacteria are rhizobia and the woody plant associated *Frankia* (Gardner *et al.,* 1984, Reddell and Spain, 1991, Rosbrook and Reddell, 1995, Nickel et al., 2001 and Rojas et al., 2001) and also the cyanobacterial N_2-fixing symbionts of the cycads are also called as PGPR.

PGPR can affect plant growth either indirectly or directly; indirect promotion of plant growth occurs when PGPR lessen or prevent the deleterious effects of one or more phytopathogenic organisms; while direct promotion of plant growth by PGPR involves either providing plants with a compound synthesized by the bacterium or facilitating the uptake of certain nutrients from the environment. Thus, in the broadest sense, PGPR include the N_2-fixing rhizobacteria that colonize the rhizosphere, providing N to plants, including the rhizobia of the well-characterized legume–rhizobia symbiosis. Regardless of the mechanisms of plant growth promotion, PGPR must colonize the rhizosphere around the roots, the rhizoplane or the root itself (Glick, 1995).

2.2.1. Classification of PGPR

Free-living soil and rhizosphere bacteria that are beneficial to plants are often referred to as plant growth-promoting rhizobacteria or PGPR (Kloepper et al., 1989; Glick, 1995a). A number of different bacteria may be considered to be PGPR including *Azotobacter, Azospirillum,* Pseudomonads, *Acetobacter, Burkholderia,Enterobacter* and bacilli (Brown, 1974; Kloepper et al., 1989; Bashan and Levanony, 1991; Tang, 1994; Glick, 1995b. PGPR can affect plant growth either directly or indirectly. The direct promotion of plant growth by PGPR for the most part entails either providing the plant with a compound that is synthesized by the bacterium or facilitating the uptake of certain nutrients from the environment. The direct effects of PGPR include providing the host plant with i) fixed nitrogen, ii) phosphorus and iron solubilized from the soil and iii) phytohormones such as auxins, cytokinins and gibberelins that are synthesized by the bacterium. The indirect promotion of plant growth occurs when PGPR lessen or prevent the deleterious effects of one or more phytopathogenic organisms. It was recently suggested that soil bacteria that indirectly stimulate plant growth be referred to as biocontrol-PGPR or biocontrol plant growth-promoting bacteria while bacteria that directly stimulate plant growth be referred to as PGPR. Gray and Smith (2005) have introduced two simple terms to aid in the conceptualization of PGPR; intracellular PGPR—(iPGPR) bacteria residing inside plant cells, producing nodules and being localized inside those specialized structures, and extracellular PGPR (ePGPR)—those bacteria living outside plant cells and not producing nodules, but enhancing plant growth through production of signal compounds that directly stimulate plant growth, improve plant disease resistance, or improve mobilization of soil nutrients. ePGPR can be subdivided into three types, based on degree of association with plant roots: those living near, but not in contact with, the roots; those colonizing the root surface; and those living in the spaces between cells of the root cortex during infection and

(ii) in the plant cytoplasm); and ePGPR ((i) living near, but not in contact with roots; (ii) colonizing the root surface; (iii) living in spaces between cells of the root cortex (Gray and Smith (2005).

2.2.2 Rhizosphere colonization

The activity of microbes is intense where nutrients are available. The rhizosphere of plants is one such active zone. There is stiff competition for this niche both by saprophytes and pathogens. Rhizosphere colonization is considered as the basic step for both root invading soil borne pathogens and surface colonizers. The understanding of root colonization is important for various applications such as biofertilization, biocontrol, biostimulation and bioremediation. Along root colonizing bacteria *Pseudomonas* has been rated at the top. Root colonization is studied using several techniques. The important being confocal laser scanning microscopy, green fluorescent pigment (gfp) in gnotobiotic systems (Troxler *et al* 1997, Zahir *et al* 2004). Fluorescent pseudomonads being efficient colonizers have been studied intensively. Competitions among fluorescent pseudomonads also occur as revealed by studies involving mutants.(Carrol *et al* 1995, Lugtenberg *et al* 2001).

2.3 *Trichoderma*

The genus *Trichoderma* has been exploited for two reasons; one as a source of cellulolytic enzymes and secondly for their ability to antagonize plant pathogenic fungi (Hjelford and Tronsmo 1998). Species of *Trichoderma* are conspicuous for their ubiquitous distribution, dominance over other soil fungal communities. Their nutritional requirement is minimal and they produce abundant propagules which include mycelium, conidia and quiescent resting spores which grow upon availability of nutritional source (Howell 2003). They are adapted to survive in varied habitats such as decaying wood, cotton fabrics and aerial and subterranean plant parts. In order to survive in such varied habitats they are endowed with superior competitive nature and possess an arsenal of cellulolytic enzymes. Their ability to produce various volatile and non-volatile metabolites makes them efficient antagonists for several pathogens of agricultural crop plants. The high degree of ecological adaptability, the ability to grow on inexpensive substrates makes *Trichoderma* amenable for mass multiplication and manipulation for various formulations and delivery systems (Nordlund 1996). Recent reports suggest that besides being an efficient biocontrol agent, they also promote growth by nutrient mobilization and also by triggering defense genes in plants through induced systemic resistance (Altomare *et al* 1999, Howell 2003).

3.0. MECHANISM OF GROWTH PROMOTION

The plant surface provides a specialized niche for various microbial communities that range from symbiotic or associative relationship among themselves and with the plants (Andrews and Harris 2000). When the association is for mutual benefit it often results in enhanced growth of plants and have developed various mechanisms.

3.1 Growth promotion by Mycorrhiza

The mycorrhizal fungi extend beyond the root hairs and mine nutrients especially P and in association with other beneficial organisms like P solubilizers, nitrogen fixers the uptake of nutrients is enhanced as a result the root mass and consequently the entire biomass is enhanced (Powell and Bagyaraj 1984, Sieverding 1991). The altered physiology and accumulation of sulfur containing amino acids prevent the molting of root knot nematode and preventing nematode infestation (Safir 1987). Protection from pathogens further promotes growth. Increased mobilization and uptake of other plant nutrients such as S, N, Cu and Zn are also reported (Cooper, 1984, Cox and Tinker 1976, Mosse et al 1981, Majunath and Habte 1988, Jacobson *et al* 2001). The altered physiology and synthesis of growth promoting substances contribute for the further growth enhancement in mycorrhizal plants (Allen *et al* 1980, Barea and Azcon-Aguilar 1982, Sieverding 1991). Mycorrhizal inoculation in the rooting medium has enhanced rooting of black pepper and the effect was comparable to rooting hormone used. This criteria was followed for selecting effective strains for black pepper (Anandaraj and Sarma 1994). Further inoculation of different species of mycorrhiza resulted in enhanced rooting, growth and nutrition uptake and biomass production (Thanuja *et al* 2002).

3.2 Growth promotion by PGPR

In the present day agriculture, PGPR, including rhizobia and other root endophytic bacteria, are being increasingly exploited because of their role in biocontrol, biofertilization and phytostimulation. These may be manifested through accelerated growth and development due to: production of growth stimulating phytohormones, enhanced availability of minerals, for example, Fe, through production of bacterial siderophores, or phosphorus through enhanced solubilization (Kloepper *et al.,* 1980); or increased resistance to pathogens and/or abiotic stresses such as frost damage (Sturz *et al.,* 2000). Glick *et al.* (1994) have shown that some PGPR can promote growth by modulating the levels of plant signals (ethylene). In some cases there can be favorable interactions between iPGPR and ePGPR; many ePGPR strains, as well as some activators produced by them, promote legume nodulation and nitrogen fixation in association with rhizobia (Zhang and Smith, 1997). They have been added to rhizobial inoculants, promoting legume nodulation and nitrogen fixation.

3.3 Growth promotion by *Trichoderma*

The observed growth promotion by *Trichoderma* is brought about by various ways; by suppressing the soil borne pathogens, by mobilizing nutrients and by inducing systemic resistance. Hence these organisms are termed as opportunistic avirulent plant symbionts (Harman *et al.*, 2004). Growth promotion by *Trichoderma* involves nutrient mobilization and production of growth substances (Benitez *et al* 2004). Production of siderophores for chelation of iron by *Pseudomonas fluorescens* is well documented (Meyer and Abdallah, 1978, Bakker *et al* 1993). A large number of fungi are also reported to produce iron chelating siderophores for sequestration and Hydroxamate type siderophores have been described that store iron (Winkleman 1992). Reports by Altamore *et al* (1999) have indicated that *Trichoderma harzianum* could solubilize rock phosphate, MnO2, Fe2O3 and Zn at very slightly acidic or alkaline pH. The mechanism of chelation appears to be different as chelation does not depend up on Fe deficiency in the surroundings. Micro organisms solubilize P by releasing organic acids (Cunningham and Kuiack 1992). In fungi other mechanisms are also reported. As dissolved P is continuously absorbed by the fungus, the disturbance in the equilibrium between insoluble P and soluble P account for P solubilization by fungi (Halvorson *et al* 1990, Illmer and Schinner 1992). The P could be stored in the mycelium and released at the lysis of mycelium. As *Trichoderma* mycelium grows continuously along the root, it ensures adequate supply of P to roots (Sivan and Harman 1992). In addition to organic acids metabolites such as peptides, proteins, phenolics and chitin are involved in metal complexing (Manulis *et al* 1987, Dori *et al* 1990). The genes involved in specific peptide synthesis in *Trichoderma* are reported (Wilhite *et al* 2001). Thus the growth promotion by *Trichoderma* is attributed to solubilization and sequestration of many plant nutrients such as P, Mn, Fe and Zinc and supply to the plant (Altamare *et al* 1999).

4.0 PGPR AS BIOFERTILIZERS

Although the molecular basis for the interactions is not always well known, several basic principles of molecular interplay between the PGPR and plants have been successfully unraveled (Table 1). The most important biofertilizers are the nitrogen-fixing rhizobia that enhance the supply of nitrogen to the plant. Unlike other bacteria, these form a host-specific symbiosis with leguminous plants. The molecular interaction between the plants (providing the carbon source) and the microorganisms (providing the nitrogen supply) is highly complex and involves many factors. For example, the enhanced rate of nitrogen-uptake stimulated by N-fixing *Azospirillum* is largely indirect and is linked to the impact of phytohormones on root development (Steenhoudt *et al.* 2000). Other PGPR biofertilizers influence the availability of phosphate by secreting phosphatases for mineralization of organic phosphorus, or by releasing organic acids for the solubilization of inorganic phosphates (Rodríguez and Fraga, 1999). The release of siderophores that chelate iron and make it available to the plant root is another example of biofertilization (Bloemberg and Lugtenberg, 2001).

4.1 Production of hormones by PGPR

The production of phytohormones, namely auxin, cytokinin and gibberellin are the most commonly invoked mechanisms of plant growth promotion by ePGPR; with auxin being of primary interest (Garcia de Salamone *et al.*, 2001). The auxins are well characterized class of plant hormones, and this is particularly so of indole-3-acetic acid (IAA), known to stimulate both short and long term responses in plants (Hagen, 1990). The various secretions by PGPR are reported in the literature that are involved in the growth promotion are reviewed periodically (Ping and Boland 2004, Zahir *et al* 2004). In addition to growth regulators other substances that regulate growth are also reported to be secreted by PGPR. One such example being 1-aminocyclopropane-1-carboxylic acid (ACC) deaminase. This enzyme regulates the endogenous level of ethylene and promotes root growth (Glick *et al* 1994). Some of the common growth factors involved in growth enhancement in PGPR are listed in Table 1.

Tien *et al.* (1979) showed that *Azospirillum brasilense* produce IAA as well as indole lactic acid when exposed to tryptophan. Further, IAA production was shown to increase with the age of the culture. *Bacillus* spp., isolated from the rhizosphere of *P. vulgaris* produce significant amounts IAA (Srinivasan *et al.*, 1996). IAA producing *Azospirillum brasilense* can stimulate formation of wheat seedling lateral roots, while a mutant strain, synthesizing low levels of IAA, does not (Barbieri *et al.*, 1993). Further, 72 h after tryptophan addition, *Azospirillum brasilense* strain Sp 245 produces 15 mg/l of IAA (Zakharova *et al.*, 1999). *Acetobacter diazotrophicus* produced 32 and 21 ng IAA ml^{-1} when cultured with 10 and 15% sucrose (Bastian *et al.*, 1998). In this same study, *Herbaspirillum seropediacae* produced less IAA, 7 ng IAA ml^{-1}, when cultured in NFb medium containing no sucrose. *P. fluorescens* BSP53a also produces IAA and a mutant overproducing IAA stimulates root development of black currant softwood cuttings and inhibits cherry (Dubeikovesky et al., 1993). These results indicate that the response of plants treated with an IAA-secreting PGPR is affected by both the amount of IAA produced by the bacteria and the characteristics of the plants.

Table 1. Production of growth regulators by PGPR

Growth substance	Organism	Reference
Indole -3-acetic acid	*Azotobacter chroococcum* *Azotobacter* sp. *Azospirillum* sp.	Muller *et al.* (1989) Zahir *et al.* (1998), (2000); Khalid *et al.* (2001) Lambrecht *et al.* (2000) Dobbelere *et al.* (2001)
Gibberillic acid	*Azospirillum* sp. *Rhizobium*	Lucangeli and Bottini (1997) Yanni *et al.* (2001)
Indole -3-acetic acid and Gibberin like substances	*Azotobacter* sp.	Mahmaud *et al.* (1984)
Cytokinins	*Rhizobium*	Persello-Cartieaux *et al.* (2003), Bloemberg and Lutenberg (2001)
Cytokinin like substances	*Azotobacter beijernickii* *Azotobacterlipoferum* *Azospirillum* sp.	Nieto and Frankenberger (1989) Bottini *et al.* (1989) Hubbel *et al.* (1979)
Zeatin	*Azotobacter chroococcum* *Pseudomonas fluorescens* G20-18	Nieto and Frankenberger (1989) Garcia de Salamone *et al.* (2001)
Ethylene	*Aeromonas* sp. *Azospirillum* sp. *Bacillus licheniformis* *Enterobacter aerogens* *Pseudomonas fluorescens* *Pseudomonas tabaci*	Billington *et al.* (1979) Strzelezyk *et al.* (1994) Fukodo *et al.* (1989) Mansouri and Bunch (1989) Pazout *et al.* (1981) Swanson *et al.* (1979)
ACC deaminase	*Enterobacter* *Pseudomonas putida*	Persello-Cartieaux *et al.* (2003) Glick *et al.* 1994

Cytokinins influence plant growth and development such as cell division, seed germination, root elongation (Stenlid, 1982), chlorophyll accumulation (Weidhase *et al.*, 1987), leaf expansion and delay senescence (Hackett, 1985). Cytokinins are N6-substituted aminopurines and it has been postulated that cytokinins produced by rhizobacteria influence plant growth and development. Some *P. fluorescens* strains producing cytokinins promote seedling emergence and increase root length of several crop species under controlled conditions (Gacia de Salamone *et al.*, 2001). Recent evidence suggests that some bacteria produce volatile compounds that influence plant growth (Ryu *et al.*, 2004). They proposed that a blend of airborne chemicals is responsible for increased growth in *Arabidopsis*. Further, the potent growth promoting strains *B. subtilis* GB03 and *Bacillus amyloliquefaciens* IN937a produced volatiles of 3-hydroxy-2-butanone (acetoin) and 2,3 butanediol (Ramos *et al.*, 2000).

Bacillus spp. produces high levels of gibberellins: A_1, A_3, A_4 and A_{20}, inducing positive effects on stem and shoot elongation in *A. glutinosa* (Gutierrez Munero *et al.*, 2001). *Acetobacter diazotrophicus* and *Herbaspirillum seropediacae* produce indole-3-acetic acid and gibberellins, A_1 and A_3, enhancing growth of members of the Graminaceae (Bastian *et al.*, 1998).

PGPR that are able to affect plant physiology directly often synthesize and exudate plant hormones or compounds that mimic their action because of structural similarities. Some PGPRs produce auxin (indole-3-acetic acid), cytokinin(s), gibberellin(s) (Steenhoudt and Vanderleyden, 2000; Yanni *et al.* 2001), octadecanoids and compounds that mimic the action of jasmonate. Others produce salicylate (Parker, 2003; Persello-Cartieaux *et al.* 2003; Bloemberg and Lugtenberg, 2001) or control the biosynthesis of ethylene via an ACC deaminase (1-amino-cyclopropyl-carboxylic acid) that reduces the available pool of ACC for ethylene biosynthesis by the plant (Persello-Cartieaux , 2003). Because ethylene often acts either synergistically or antagonistically with other plant hormones, manipulation of this pathway has an impact on many response complexes, for example, ISR, SAR, or salicylic acid- and jasmonic acid-linked gene expressions (Walling, 2000). The phenomenon of phytostimulation profits particularly from manipulating the complex and balanced network of plant hormones or hormone-like compounds that directly influence growth or stimulate root formation. For example, *Azospirillum* spp. produce auxins, cytokinins and gibberellins that stimulate root development, which can lead to improved crop yield (Steenhoudt and Vanderleyden, 2000). Although the view that plants can produce cytokinins is controversial (Holland, 1997), it is interesting that many PGPR from the phyllo- and rhizosphere can produce cytokinins that exert a pronounced stimulatory effect.

5.0 MECHANISM OF BIOCONTROL

Biological control of plant diseases involve reduction in inoculum density or disease causing activities of pathogens by one or more organisms accomplished naturally through manipulation of environment, host or antagonist (Baker and Cook 1974). Microbial antagonists against soil borne pathogens bring about suppression of diseases either by indirect means by improved plant nutrition or by increased resistance. The mechanisms vary among the biocontrol agents and some times a combination of mechanisms are involved.

5.1 Disease suppression by Mycorrhiza

Mycorrhizae have been reported to protect plants against soil borne diseases of several crop plants. The protection offered is both due to enhanced rooting and better uptake of nutrients as well as modified physiology of the mycorrhizal roots (Krishna and Bagyaraj 1983, Tello *et al* 1987, Jalali 1991). The mechanism appears to be both competitions for infection court and due to altered physiology of roots (Garcia- Garrido *et al* 1993.) In case of soil borne pathogens like *Phytophthora* and plant pathogenic nematodes, the preferred site of infection is the zone of elongation. As the site is already occupied by mycorrhizal fungi, site elimination appears to be one of the basic mechanisms for the suppression root rot and the enhanced nutrition (Hussey and Rancodri 1982, Cordier *et al* 1996). In bell pepper, higher endogenous levels of phenolics in mycorrhizal plants were attributed to the reduced disease against *Verticillium* wilt (Karagiannidis *et al* 2002; Garmendia *et al* 2004). In case of black pepper, mycorrhizal plants could suppress the damages caused by three root pathogens namely, *Phytophthora capsici, Meloidogyne incognita* and *Radopholus similis* due to enhanced rooting, uptake of nutrients and also by site exclusion (Anandaraj *et al* 1996).

5.2 Disease suppression by PGPR

According to their mode of action, PGPR have been divided into two large classes, namely the Biocontrol-PGPR, which suppress plant pathogens, thereby benefiting the plant indirectly, and the PGPR that directly affect plant metabolism resulting in increased plant growth, seed emergence or improved crop yields (Bashan and Holguin, 1997). The mode of action involves production of metabolites that reduce the populations of pathogens (Kloepper 1996, Glick 1995a). The bacterial metabolites and other secretions involved in biological control are listed in Table 2.

5.3 Antagonistic mechanisms of *Trichoderma*

The genus *Trichoderma* comprises of several antagonistic species with multiple mechanisms. The antagonistic mechanisms are both indirect and direct. Indirect methods includes competition for nutrition and space, modifying environmental conditions, promoting plant growth and enhancing defense mechanism of plants. Direct mechanisms are mycoparasitism (Benetez *et al* 2004).

Table 2. Production of bacterial metabolites for disease suppression

Bacterial metabolite	Organism	Effect on plant	Reference
Salicylic acid	*Serratia*	Systemic acquired resistance (SAR)	Persello-Cartieaux *et al.* (2003)
Flagellins	*Pseudomonas*	Induced Systemic resistance (ISR) or	Parker (2003)
Lipopolysaccharides	*Pseudomonas*	Host plant protection	Persello-Cartieaux *et al.* (2003)
Antibiotics- Pyoluteorin, Phenazine, Phloroglucinols	*Pseudomonas*	Disease suppression ISR	Persello-Cartieaux *et al.* (2003); Bloemberg and Lugtenberg
Lytic enzymes	*Pseudomonas fluorescens*	Growth promotion, Disease suppression	Velazhahan *et al.* (1999)
Volatiles- butane-diol	*Bacillus* sp.	ISR	Ryu *et al.* 2004
Volatile and non-volatile metabolites-HCN	*Pseudomonas fluorescens*	Disease suppression	Diby *et al.* (2005a; 2005b)
Metabolites	*Pseudomonas fluorescens*	ISR	Viswanathan and Samiyappan (1999)

5.3.1 Mycoparasitism

Mycoparasitism is one of the methods in which biocontrol mechanism operates. Due to this the biocontrol agents destroy the target fungi. The sequence of events at molecular level is reviewed by Chet and Inbar (1994). An early event in parasitism by *Trichoderma* spp is coiling of the *Trichoderma* spp around its host (pathogen). Later, the antagonist penetrates into the cells of the host, causing cell wall alterations, plasma membrane disorganization and aggregation of the cytoplasm. Benhamou & Chet (1993) reported that damage to *R. solani* hyphae began soon after coiling of *T. harzianum* around its host, followed by the rapid collapse and loss of turgor of host hyphae, suggesting that lytic enzymes were produced by *Trichoderma* spp in response to contact between both the partners. A *Fusarium chlamydosporum* strain, a mycoparasite of groundnut rust, *Puccinia arachidis* produces an endo-chitinase that inhibits germination of uredospores of the rust fungus indicating a significant antifungal activity of chitinase (Mathivanan *et al.*, 1998, Alexandra *et al* 2002). Higher levels of antifungal activity were recorded when more chitinolytic/glucanolytic enzymes of *T. harzianum* are tested together. Combining a 42-kDa endochitinase and a 40-kDa chitobiosidase from *T. harzianum* strain P1 resulted in a synergistic increase in antifungal activity (Lorito *et al.*, 1993; 1994). A variety of synergistic interactions have been found when different enzymes were combined or associated with biotic or abiotic antifungal agents (Chernin & Chet 2002).

Inglis & Kawchuk (2001) tested 14 fungi primarily representing mycoparasitic and biocontrol fungi for their ability to grow on and degrade cell walls (CWs) of an oomycete (*Pythium ultimum*), ascomycete (*Fusarium equisetii*), and basidiomycete (*Rhizoctonia solani*), and their hydrolytic enzymes were characterized. Protein was detected in the cultural medium of eleven of the test isolates, and these fungi significantly degraded CWs over the 14-day duration of the experiment. In general, a greater level of CW degradation occurred for *F. equisetii* and *P. ultimum* than for *R. solani*. Fungi that degraded *F. equisetii* CWs were *Coniothyrium minitans, Gliocladium roseum, Myrothecium verrucaria, Talaromyces flavus,* and *Trichoderma harzianum*. Taxa degrading *P. ultimum* CWs included *Chaetomium globosum, Coniothyrium minitans, M. verrucaria, Seimatosporium* sp., *Talaromyces flavus, Trichoderma hamatum, T. harzianum,* and *T. viride*. Production of extracellular protein was highly correlated with CW degradation.

5.3.2 Chemotropic growth and recognition

Trichoderma is reported to have chemotactic stimulus from the host fungus and grow towards it. Then the recognition process is by lectin carbohydrate binding. Lectins from *Sclerotium rolfsii* has been purified and demonstrated to agglutinate *Trichoderma* spores. (Inbar and Chet 1994). After recognition *Trichoderma* hyphae forms coils round the host mycelium and produces appresorium like structures. This is followed by dissolution of cell wall by lytic activity. The dissolution of cell wall is brought about by chitinases, glucanases and proteases (Lorito 1998). The wild type strains are reported to secrete endochitinase; chitobiosidase, b-N- acetyl hexosaminidase, N-acetyl - b - glalactosaminidase, b, 1,3, glucanase, b 1-6, glucanase, protease, DNase, &-amylase, cellulase, lipase, mannanase xylanase, urease, RNase, Pectinase, pectin lyase, laccase and peroxidase. The different enzyme system operating in various species, their synergistic activities with enzyme system has been reviewed (Lorito 1998, Benitez *et al* 1998).

5.3.3 Competition

Microorganisms compete for nutrition available in the vicinity for their growth and survival. Ecologists have coined the terms exploitation, competition and interference competition. Interference competition involves chemical mechanisms by which one organism limits another's access to substrate which results in mycelial interaction (Wicklow 1992). In exploitation competition, the nutritional resources are depleted by a given organism without directly limiting other organisms access to that resource. Competitions also occur for necrotic tissues, plant exudates and to wound sites. Preemptive colonization of senescing tissues by antagonists provides better biocontrol against such fungi (Hjelford and Tronsmo 1998).

5.3.4 Interference with pathogenesis

Microbial pathogens of crop plants use an array of enzymes to breach the mechanical barriers of cell walls and destabilize the integrity and gain entry. The enzymes such as cutinase, which hydrolyse ester linkages of the cutin polymer, pectolytic enzymes exo-and endo polygalacturonases, pectate lyase, pectin methyl esterase and cellulolytic enzymes. The proteins and secondary metabolites secreted by antagonistic *Trichoderma* are reported to interfere in these processes thereby affecting the pathogenic process. Many fungal diseases caused by *Fusarium, Colletotrichum* and bacterial diseases caused by *Pseudomonas* have been reported (Kubicek *et al.*, 2001, Brunner *et al.* 2003, 2005).

6.0 USE OF MICROBIAL CONSORTIA

In a study on effect of combined inoculation of AMF, *Glomus intraradices* Schenck & Smith, a PGPR, *Bacillus subtilis*, and a filamentous soil fungus, *Aspergillus niger*, on growth of lettuce plants and on indicators of biological soil quality, it was revealed that water-soluble carbohydrates and microbial biomass increased in the rhizosphere soil of *G. intraradices*-plants and the soil from all microbial inoculation treatments had significantly higher dehydrogenase activity, particularly in the soil inoculated with *B. subtilis* (about 21% higher than control soil). Inoculation with *A. niger* or *B. subtilis* increased significantly the urease, protease and phosphatase activities of the rhizosphere soil of the lettuce plants. The foliar P and K contents increased significantly with the *B. subtilis* or *G. intraradices* inoculation, alone or in combination. The most effective co-inoculation was observed in the combined treatment of inoculation with *G. intraradices* and *B. subtilis*, which synergistically increased plant growth compared with singly inoculated (Kohler *et al.*, 2007)

Two promising plant growth promoting rhizobacteria identified by 16S rDNA sequencing *viz., Pseudomonas putida* CC-FR2-4 and *Bacillus subtilis* CC-pg104, were inoculated to *Lectuca sativa* L. seedlings as free cells and entrapped in beads. Significant increase in shoot height after 21 days of growth was observed with encapsulated CC-pg104 inoculated plants. Highest increase in root length was observed with CC-pg104 free-cell inoculated plants, followed by plants inoculated with encapsulated CC-pg104. Results clearly demonstrated that inoculation of the encapsulated bacterial isolates promoted plant growth similar to their respective free cells and could be a novel and feasible technique for application in agricultural industry (Rekha et al., 2007).

The presence of *Pseudomonas* BA-8, *Bacillus* OSU-142 and BA-8 + OSU-142 stimulated plant growth and resulted in significant yield increase in sweet cherry (*Prunus avium* L.). Floral and foliar applications of BA-8, OSU-142 and BA-8 + OSU-142 in sweet cherry significantly increased yield per trunk cross-sectional area (16.3, 10.9 and 21.7%), fruit weight (4.15, 5.37 and 1.24%) and shoot length (11.3, 11.8 and 29.6%), respectively, compared with the control. In addition, N, P and K contents of sweet cherry leaves with BA-8, OSU-142 and BA-8 + OSU-142 treatments, Fe and Zn contents of leaves with BA-8 + OSU-142 treatment and Mn content of the leaves of sweet cherry with BA-8 and OSU-142 applications significantly increased (Esitken *et al.*, 2006).

Nine different isolates of PGPR from a pool of 233 rhizobacterial isolates obtained from the peanut rhizosphere on the basis of ACC-deaminase activity were selected out of which four of these isolates, viz. PGPR1, PGPR2, PGPR4 and PGPR7 (all fluorescent pseudomonads), were the best in producing siderophore and indole acetic acid (IAA). Besides, *Pseudomonas fluorescens* PGPR1 also possessed the characters like tri-calcium phosphate solubilization, ammonification and inhibited *Aspergillus niger* and *A. flavus* in vitro. In addition to the traits exhibited by PGPR1, PGPR4 showed strong in vitro inhibition to *Sclerotium rolfsii*.

Seed inoculation of these three isolates, resulted in a significantly higher pod yield than the control, in pots, during rainy and post-rainy seasons. The contents of nitrogen and phosphorus in soil, shoot and kernel were also enhanced significantly in treatments inoculated with these rhizobacterial isolates in pots during both the seasons. The isolates, viz. PGPR1, PGPR2 and PGPR4, significantly enhanced pod yield (23–26%, 24–28% and 18–24%, respectively), haulm yield and nodule dry weight over the control in 3 years.

Seed bacterization with these three isolates, suppressed the soil-borne fungal diseases like collar rot of peanut caused by *A. niger* and PGPR4 also suppressed stem rot caused by *S. rolfsii* (Dey *et al.*, 2006).

Six strains of PGPR containing ACC-deaminase were studied for their growth-promoting activity in maize roots under gnotobiotic conditions in soil amended with N at a concentration of 175 kg ha^{-1} (1050 mg pot^{-1}). Significant increases in plant height, root weight and total biomass were observed in response to inoculation. The three most efficient strains were tested in the field for their effectiveness in the presence and absence of N fertilizer. Results revealed that the inoculum performed relatively better in the absence of N-fertilizer application. *Pseudomonas fluorescens* biotype G (N_3) was the most effective strain both in the presence and absence of N fertilizer. Results may imply that even in the presence of optimum levels of nitrogenous fertilizers, inoculation with rhizobacteria containing ACC-deaminase activity could be effective to improve the growth and yield of inoculated plants (Shaharoona *et al.*, 2006).

A study on co-inoculation dose of plant growth promoting rhizobacteria (PGPR) strains, *Serratia proteamaculans* 1-102 and *S. liquefaciens* 2-68, with *Bradyrhizobium japonicum* on soybean (*Glycine max* (L.) Merr.) growth, nodulation and nitrogen fixation under controlled root zone temperatures (RZTs; 25, 20 and 15 °C) in soil-less media revealed that the effective dose range for both PGPR strains is 10^7–10^9 and their optimal co-inoculation dose was 1×10^8 cells per seedling. This was not affected by RZT. The co-inoculation of PGPR at their optimal dose increased nodule number, plant dry weight and fixed nitrogen. The calculated dynamic parameters revealed that PGPR co-inoculation shortened the nodule initiation time and increased the nodulation rate (Bai *et al.*, 2002).

7.0 APPLICATION OF MICROBES IN SPICE CROPS

Beneficial microbes namely mycorrhiza, PGPR and *Trichoderma* have been used in spice crops for growth enhancement and management of diseases and plant parasitic nematodes.

7.1 Use of Mycorrhiza

Black pepper (*Piper nigrum* L.) is mycotrophic and responds well to mycorrhizal inoculation (Manjunath and Bagyaraj 1992, George and Ghai 1988, Bopaiah and Khader 1989). An effort was made to collect and assemble a large number of isolates from black pepper growing areas and from the Western Ghats where the black pepper has originated and still surviving in the wild. These isolates were maintained on *Zea mays* and several experiments were conducted to isolate efficient strains and also to test their effect on disease suppression.

7.1.1 Short listing of efficient isolates

Single spore isolates of the collections were cultured on *Chloris gayana* (Rhodes grass) in a mixture of soil and perlite at 1:1. The inoculum consisting of potting mixture and root bits are collected after six weeks of growth (Anandaraj and Sarma 1998). The percentage of rooting and production of roots were higher in cuttings treated with Mycorrhiza. This property of enhanced rooting was used for selection of isolates. Among 60 isolates tested 7 isolates were short listed as efficient isolates The short listed isolates were tested on different varieties of black pepper and seven strains belonging to *Glomus fasciculatum* and *Gigaspora gigantea* have been short listed and two strains were identified as most efficient. These are used regularly in the nursery mixture while raising black pepper cuttings (Anandaraj and Sarma 1994).

7.1.2 Effect of Mycrorrhiza on root pathogen of black pepper

In black pepper slow decline is caused by the combined damage of feeder roots by *Phytophthora capsici, Radopholus similis* and *Meloidogyne incognita* and these pathogens are often carried inadvertently from the nursery to the main field (Anandaraj 2000). The effect of various species of mycorrhiza when tested for the suppression of *M. incognita* in black pepper revealed that the protection offered by mycorrhiza was comparable to that offered by chemicals (Anandaraj et al 1991.) In another experiment with mycorrhiza on suppression of root pathogen of black pepper, it was found that mycorrhiza inoculated plants had significantly higher biomass than non- Mycorrhizal plants. The expression of symptoms due to feeder roots expressed as foliar index and root rot index due to the by pathogens was low in mycorrhizal plants. The populations of the inoculated nematodes in mycorrhizal plants were significantly lower both in roots and soil (Anandaraj *et al* 1996). The mechanism appears to be both competitions for infection court and due to altered physiology of roots (Garcia- Garrido *et al* 1993.) In case of *P. capsici* and *M. incognita* the preferred site of infection is the zone of elongation. As the site is already occupied by mycorrhizal fungi, site elimination appears to be one of the basic mechanisms for the suppression root rot (Hussey and Rancodri 1982). In case of *R. similis* the enhanced root mass offsets the damages caused by the pathogen. Based on these studies application of mycorrhizal inoculum in the nursery mixture has been recommended.

7.1.3 Mycorrhizal inoculation with other biofertilizers

Black pepper is propagated vegetatively through cuttings. These cuttings are produced through out the year in multiplication sheds. When mycorrhiza was inoculated along with *P. fluorescens* the effect was synergetic *P. fluorescens* also enhances growth and suppress root pathogens. The production of healthy and robust cuttings was achieved when a combination of Mycorrhiza with *Pseudomonas fluorescens* (IISR-6) was used (Anandaraj and Sarma 2003, Thankamani *et al* 2005). The height and number of nodes of plants recorded were analysed statistically after 4 months of planting and bacterial isolates were short-listed based on efficiency in promoting growth of black pepper. Bacteria treated plants showed significantly increased growth with respect to the control. The fresh weight of roots and shoots of 6 months old bacteria treated pepper cuttings were recorded. *Pseudomonas* spp. treated plants showed a maximum of up to 200 % increase in biomass when compared to untreated check. The fluorescent pseudomonad strains, IISR-11, IISR-6, IISR-43, IISR-23, IISR-51, IISR-13 and IISR-8 imparted enhanced growth in treated black pepper in terms of shoot and root biomass and also in terms of height of the plant (Diby et al 2002, Saju et al 2002,Thankamani et al 2003). The efficient strains of Mycorrhiza *G. fasciculatum* was tested on black pepper along with *Azospirillum* and *Bacillus megaterium* for growth and of nutrient uptake in a pot culture study. After six months, the combination of *Azospirillum* and *Phosphobacteria* along with Mycorrhiza recorded maximum growth, increased biomass and nutrient contents (Kandiannan *et al* 2000).

7.2 Evaluation of biocontrol agents

Black pepper specific bacteria and fungi were isolated from Silent valley biosphere reserve and were short-listed based on its antagonistic properties on *P. capsici in vitro.* The morphological characterization of the bacterial strains was carried out. Fluorescent pseudomonads showed a maximum of 82% and *Trichoderma* a maximum of 47% inhibition of *P. capsici* over control. These isolates were also tested for the production of volatile and non- volatile metabolite production and also for the production of HCN. Bacterial strain IISR-310 was found to produce maximum level of HCN. The percentage inhibition of growth of *P. capsici by* the inhibitory volatiles produced by the isolates IISR-25 and IISR-36 were 39.7. In the green house assay performed, the bacterial strain, IISR-51 promoted growth of the black pepper up to 57.15% and among the fungal isolates IISR-24 performed best (29.7 % increase in growth of the black pepper cuttings (Lisha *et al* 2002)

7.2.1 Efficiency of rhizobacteria in protecting black pepper from its three root pathogens viz. R. similis, M. incognita and P. capsici.

An experiment on the efficiency of *Bacillus* (7 strains), fluorescent pseudomonads (3 strains) and two strains of bacteria isolated from silent valley in protecting black pepper from three pathogens viz. *R.similis, M. incognita* and *P. capsici* were conducted. The root rot was indexed as 1 to 5 (No root rot was indexed as 0, 0-25 % root rot as 1, 25-50 % as 3, 50-75% as 4 and 75-100% as 5). The fluorescent pseudomonads protected black pepper significantly from all the pathogens (IISR 2001, IISR 2002).

7.2.2 Rejuvenative capacity of fluorescent pseudomonads in black pepper

The efficient five rhizobacteria, alone and in combinations with themselves and in combination with a systemic fungicide metalaxyl (Ridomil - Mancozeb RMZ 72 wp) were tested for the rejuvenation of black pepper cuttings infected with *P. capsici*. It was found that the strains IISR-8, IISR-51 and IISR-11 could effectively rejuvenate the black pepper cuttings when treated alone and also in combination. However, the percent rejuvenation of the infected cuttings increased when the bacterial treatments were with metalaxyl (Diby *et al* 2005d).

7.2.3 Effect of PGPR on rejuvanation of slow decline in the field

The bacterial isolates which were short listed based on *in vitro* and green house experiments both for suppression of *P. capsici* and nematodes were tested on a standing crop of black pepper showing slow decline symptoms at IISR experimental farm Peruvannamuzhi. The evaluation on the disease severity was made based on foliar yellowing. There were 34.7% of plants were showing yellowing before the application of PGPR. After 60 days of application of PGPR, it was found that all the bacterial treatments significantly reduced the foliar yellowing. The intensity of yellowing was graded based on visual observations. There were three grades (up to 25%, up to 50% and greater than 50% yellowing). In all the PGPR treated vines the grades have come down and the vines produced new foliage. The recovery of vines was total in vines treated with IISR-853, followed by IISR-51. Whereas the untreated vines deteriorated in whereas, the untreated vines deteriorated in health as the intensity of yellowing has increased. Among the strain IISR-853 and IISR-51 were found to be most effective in reducing the disease (Sabir, 2002).

7.3 Trichoderma for disease management in spice crops

Trichoderma harzianum isolated from rhizosphere of black pepper, ginger and cardamom have been evaluated for their efficiency and efficient isolate for each crop were identified and used against the pathogens of respective crops. The experiments on black pepper and ginger were performed in the green house and that of cardamom was carried out in already existing field. Out of the 22 different treatments, 3 treatments were found to be effective in suppressing root rot (*Phytophthora capsici*) disease in black pepper, soft rot (*Pythium aphanidermatum*) in ginger and clump rot (*Pythium vexans*) in cardamom (Diby *et al* 2001). The maximum disease suppression obtained by the treatment combination, *T. harzianum* isolate, IISR - 1369 and *P. fluorescens* strain, IISR–11 in black pepper was 63 and for cardamom, it was 36% over control. The same treatment could impart 66.2 % survival of ginger tillers after challenge inoculation with the pathogen. The combination of *T. harzianum* isolate, IISR-1369 and *P. fluorescens* strain, IISR –11 could improve the vigour of the plant both in black pepper and ginger. The same treatment combination imparted maximum yield in ginger and cardamom. When these biocontrol agents were applied in combination there was synergistic effect both for growth promotion and disease suppression (Diby *et al* 2001, Minimol 2002, IISR, 2002). The integrated disease management strategies involved for foot rot management includes biocontrol agents (Rajan *et al* 2002). Mass multiplication of *Trichoderma* using commonly available organic matter in the farm has been standardized (Prakash *et al* 1999, Saju *et al* 2002). The compatibility of chemical components used in IPM of black pepper with biocontrol agents also has been studied and at the recommended levels chemicals are compatible with *T. harzianum* (Anandaraj, 2000; Stephen et al, 2000).

7.4 Rhizobacteria for enhancing productivity of seed spices

All India Coordinated Research Project on Spices (AICRPS) has developed several technologies involving organics alone or in combination with inorganic inputs specific to varied agro-climatic regions of the country for spice production. The results indicated that an integrated approach involving use of inorganic source of nitrogen along with organic manures and biofertilizers has resulted in higher production in black pepper, ginger and turmeric. In cardamom, the use of phosphate solubilizing bacteria along with Phosphorus application resulted in higher yields. In seed spices such as cumin, fennel and fenugreek also higher production was obtained with the combined application of inorganic and organic nutrition along with biofertilizers.

8.0. CONCLUSIONS

Microorganisms have evolved to survive in unique niches in the ecosystem and have several mechanisms to adapt to varied climatic and soil conditions. They also evolved mechanisms to mobilize nutrients needed for the plants and adapted to the rhizosphere niche. Understanding the complicated interactions among the microbes, plants and environment holds the key to deploy consortia of compatible micro organisms for optimum use of the available resources for the sustained productivity of crop plants.

REFERENCES

Alexandre, A. P. F., Cirano J. U., Marcelo, V. S., Edivaldo, X. F. F. and Carlos, A. O. R. (2002) Involvement of G proteins and cAMP in the production of chitinolytic enzymes by *Trichoderma harzianum. Brazilian J. Microbiol.* 33: 169-173.

Allen, M. F., Moore, T. S. and Christenson, M. (1980). Phytohormone Changes in *Bouteloua gracilis* infected by vesicular-arbuscular mycorrhizae Can. J. Bot. 58:371-374.

Altomare, C., Norvell, W. A., Bjorkman, T. and Harman, G.E. (1999) Solubilization of phosphates and micronutrients by the plant growth-promoting and biocontrol fungus *Trichoderma harzianum* Rifai 1295-22. Appl. Environ. Microbiol. 65:2926-2933.

Anandaraj, M. (2000) Diseases of black pepper. In: *Black pepper Piper nigrum* P.N. Ravindran. (ed). Medicinal and Aromatic plants-Industrial profiles Harwood Academic publishers. PP 239-267.

Anandaraj, M., Ramana K. V. and Sarma, Y. R. (1991) Interaction between vesicular arbuscular mycorrhizal fungi and *Meloidogyne incognita* in black pepper. In: Bagyaraj D J & Manjunath A (Eds.) *Mycorrhizal Symbiosis* and *Plant Growth* (pp.110-112). Mycorrhiza Network Asia and University of Agricultural Sciences, Bangalore.

Anandaraj, M., Ramana, K. V. and Sarma, Y. R. (1996) Suppressive effects of VAM on root pathogens of black pepper - a component of Western Ghat forest ecosystem. In K.S.S. Nair, J.K. Sharma, and R. V. Varma, (eds.), *IUFRO Symposium on impact of diseases and insect pests in tropical forests*, Kerala Forest Research Institute, Peechi, India. pp 232-238.

Anandaraj, M. and Sarma, Y.R. (1994) Effect of vesicular arbuscular Mycorrhiza on rooting of black pepper. *J. Spices Arom. Crops* 3: 39-42.

Anandaraj, M. and Sarma, Y. R. (1998) Vesicular Arbuscular Mycorrhizae for black pepper nurseries, *Spice India* July 1998: 11-14.

Anandaraj, M. and Sarma, Y. R. (2003) The potential of PGPR in disease management in spice crops In: *Oral presentations, 6th International Workshop on Plant Growth Promoting* Rhizobacteria, (eds) M. S. Reddy, M. Anandaraj, S. J. Eapen, Y.R. Sarma & J. W. Kloepper. Indian Institute of Spices Research, Calicut PP. 8-12.

Andrews, J. H. and Harris, R. E. (2000) The ecology and biogeography of microorganisms on plant surfaces. *Ann. Rev. Phytopathol.*38:145-180.

Bai, Y., Pan, B., Charles, T. C. and Smith, D. L. (2002) Co-inoculation dose and root zone temperature for plant growth promoting rhizobacteria on soybean [*Glycine max* (L.) Merr] grown in soil-less media. *Soil Biol. Biochem.* 34: 1953-1957.

Baker, K. F. and Cook, R. J. (1974) Biological control of plant pathogens. Freeman , W.H Fransisco. p 433.

Bakker, P. A. H. M., Raaijmakers, J. M. and Schippers, B. (1993) Role of iron in the suppression of plant pathogens by fluorescent pseudomonads. In: *Iron chelation in plants and soil microorganisms*. L. L. Barton and B. C. Hemmings, (eds) pp 269-281. Academic Press, London/New York.

Barbieri, P., Zanelli, T., Galle, E. and Zanetti, G. (1993) Effect on wheat root development of innoculation with an *Azospirillum brasilense* mutant with altered indole-3-acetic acid production. *Res. Microbiol.* 144: 69–75.

Bashan, Y. and Holguin, G. (1997) Proposal for the division of plant growth-promoting rhizobacteria into two classifications: biocontrol-PGPG (plant growth-promoting bacteria) and PGPB. *Soil Biol. Biochem.* 30:1225–1228.

Bashan, Y. and Levanony, H. (1991) Alterations in membrane potential and in proton efflux in plant roots induced by *Azospirillum brasilense*. *Plant Soil* 137: 99-103.

Barea, J. M. and Azcon-Aguillar, C. (1982) Production of plant growth regulating substances by the vesicular arbuscular fungus *Glomus mosseae*. *Appl. Envl.Micro.* 43:810-813.

Bastian, F., Cohen A, Piccoli, P., Luna, V., Baraldi, R., Bottini, R. (1998) Production of indole-3-acetic acid and gibberellins A_1 and A_3 by *Acetobacter diazotrophicus* and *Herbaspirillum seropedicae* in chemically-defined culture media. *Plant Growth Reg.* 24: 7–11.

Benhamou, N. and Chet, I. (1993) Hyphal interactions between *Trichoderma harzianum* and *Rhizoctonia solani:* Ultrastructure and gold cytochemistry of the microparasitic process. *Phytopathology* 83: 1062-1071.

Benitez, T., Rincon, A. M., Limon, M. C. and Codon, A. C. (2004) Biocontrol mechanisms of *Trichoderma* strains. *Intl. Microbiol.* 249-260.

Billington, D. C., Golding, B. T. and primrose, S. B. (1979) Biosynthesis of ethylene from methionine. *Biochem, J,* 182; 827-836.

Bloemberg, G. V. and Lugtenberg, B. J. J. (2001) Molecular basis of plant growth promotion and biocontrol by rhizobacteria, *Current Opinion in Plant Biology*, 4, 343-350.

Bloemberg, G. V., O'Toole, G. A. Lutenberg, B. J. J. and Kolter. (1997) Green fluorescent protein as a marker for *Pseudomonas* spp. *Appl. Environ. Microbial.* 63: 4553-4551.

Bolan, N. S. (1991) A crtical review of the role of mycorrhizal fungi in the uptake of phosphorus by plants. *Plant Soil,* 134; 189-208.

Bopaiah, B. M. and Abdul Khader, K. B. (1989) Effect of biofertilizers on growth of black pepper *(Piper nigrum)*. *Indian J. Agric. Sci.* 59: 682-683.

Bottini, R., Fulchieri, M., Pearce, D. and Pharis, R. P. (1989) identification of gibberellins, A_1 A_3 and iso-A_3 in cultures of *Azospirillum lipoferum. Plant Physiol.* 90: 45-47.

Brown, M. E. (1974). Seed and root bacterization. *Ann. Rev. Phytopathol.* 12: 181-197.

Brunner, K., Peterbauer, C. K., Mach, R. L., Lorito, M., Zeilinger, S. and Kubicek, C. P. (2003) The Nag1 N-acetylglucosaminidase of *Trichoderma atroviride* is essential for chitinase induction by chitin and of major relevance to biocontrol. *Curr. Genet.* 43: 289-295.

Brunner, K., Zeilinger, S., Ciliento, R., Woo, S. L., Lorito, M., Kubicek, C. P. and Mach, R. L. (2005) Genetic improvement of a fungal biocontrol agent to enhance both antagonism and induction of plant systemic disease resistance. *Appl. Environ. Microbiol.* 71:3959-3965.

Carrol, H. Y., Moenne-Loccoz, Dowling, D. N. and O'Gara, F. (1995) Mutational disruption of the biosynthesis genes coding for the anti-fungal metabolite 2,4-diacetylphloroglucinol does not influence the ecological fitness of *Pesudomonas fluorescens* F113 in the rhizosphere of sugar beets. *Appl. Environ. Microbiol.* 61:3002-3007.

Chernin, L. and Chet, I. (2002) Microbial enzymes in the biocontrol of plant pathogens and pests. In:Burns, RG, Dick RP (Eds.) *Enzymes in the Environment - Activity, Ecology and Applications.* Marcel Dekker, New York, pp 171-226.

Chet, I. and Inbar, J. (1994) Biological control of fungal pathogens. *Appl. Biochem. Biotechnol.* 48:37-43.

Cooper, K. M. (1984) Physiology of VA Mycorrhizal associations In: *VA Mycorrhiza* Powell, C, L. and Bagyaraj, D. J.155-186.CRC Press, Boco Raton.

Cox, G. and Tinker, P.B. (1976) Translocation and transfer of nutrients in vesicular-arbuscular mycorrhizas. I1.The arbuscle and phosphorus transfer:a quantitative ultrastructural study. *New Phytol.* 77:371-378.

Cunningham, J. E. and Kuiack, C. (1992) Production of citric and Oxalic acids and solubilization of calcium phosphates by *Penicillium bilaii. Appl. Environ.microbiol.* 58: 1451-1458.

Davies, F. T., Porter, J. R. and Linderman, R. G. (1993) Drought resistance of mycorrhizal pepper plants: Independent of leaf phosphorus concentration, response in gas exchange and water relations. *Physiol. Plant.* 87:45-53.

Diby Paul, Kumar, A., Anandaraj, M., and Sarma, Y. R. (2001) Studies on the suppressive action of fluorescent pseudomonads on *Phytophthora capsici,* the foot rot pathogen of black pepper. *Indian Phytopath.* 54(4):515 Abs.

Diby Paul, Srinivasan, V., Anandaraj, M. and Sarma, Y. R. (2003) *Pseudomonas fluorescens* mediated nutrient flux in the black pepper rhizosphere microcosm and enhanced plant growth. In: Reddy, M. S., Anandaraj, M., Eapen, S. J., Sarma, Y. R. and Kloepper,

J. W. (eds) *Oral presentations. 6th International PGPR workshop*, 5-10 October 2003, Indian Institute of Spices Research, Calicut, India, pp 18-24.

Diby, P., Anandaraj, M., Kumar, A. and Sarma, Y. R. (2005a) Antagonistic mechanisms of fluorescent pseudomonads against *Phytophthora capsici* in black pepper (*Piper nigrum.* L). *J. Spices Aromatic Crops*, Vol.14 (2): 122-129.

Diby, P., Saju, K. A., Jisha, P. J., Sarma, Y. R., Kumar, A. and Anandaraj, M. (2005b) Mycolytic enzymes produced by *Pseudomonas fluorescens* and *Trichoderma* species against *Phytophthora capsici*, foot rot pathogen of black pepper. *Ann. of Microbiol.* Vol. 55(2): 45-49.

Diby, P., Sarma, Y. R., Srinivasan, V. and Anandaraj, M. (2005c) *Pseudomonas fluorescens* mediated vigour in black pepper (*Piper nigrum.* L) under green house cultivation. *Ann. Microbiol.* Vol. 55(3): 171-174.

Diby Paul, Jisha, P. J., Sarma, Y. R. and Anandaraj, M. (2005d) Rhizospheric *Pseudomonas fluorescens* as rejuvenating and root proliferating agents in Black Pepper. J. Biol. Control. 19: 173-178.

Dey, R., Pal, K. K., Bhatt, D. M. and Chauhan, S. M. (2006) Growth promotion and yield enhancement of peanut (*Arachis hypogaea* L.) by application of plant growth-promoting rhizobacteria. *Microbiol. Res.*, 159: 371-394.

Dobbelaere, S., Croonenborgh, A., Thys, A., Ptacek, D., Vanderlcyden, J., Dutto, P., labendara-Gonzalez, C. Caballero-mellado, J. Agirre, J. F. Kapulnik, Y., Berner, S., Burdman, S., Kadouri, D., Sarig, S. and okan, Y. (2001) Response of agronomically important crops to inoculation with *Azospirillum. Aust. J. Plant Physiol.* 28:871-879.

Dori, S. Z., Solel, Z., Kashman, Y. and Barash, I. (1990) Characterization of hydroxamatesiderophores and siderophore mediated iron uptake in *Gaeumannmyces graminis var. tritici. Physiol. Plant Pathol.* 37:97-106.

Dubeikovesky, A. N., Modokhova, E. A., Kocheskov, V. V., Polikarpova, F. P. and Boronin, A. M. (1993) Growth promotion of black currant softwood cuttings by recombinant strain *Pseudomonas fluorescens* BSP53a synthesizing an increased amount of indole-3-acetic acid. *Soil Biol. Biochem.* 25:1211–1221.

Elmerich, C. (1984) Molecular biology and ecology of diazotrophs associated with non-leguminous plants. *Bio. Tech.* 2: 967-978.

Esitken, A., Pirlak, L., Turan, M. and Sahin, F. (2006) Effects of floral and foliar application of plant growth promoting rhizobacteria (PGPR) on yield, growth and nutrition of sweet cherry. *Sci. Horti.*, 110: 324-327.

Fukoda, H., Takahashi, M., Fujii, T. and ogawa, T. (1989) Ethylene production from L-methionine by *Cryptococcus albidus. J. Ferment. Bioeng.* 67: 173-175.

Garbeva, P., van Veen, J. A. and van Elsas, J. D. (2004) Microbial diversity in soil:selection of microbial populations by plant and soil type and implications for disease suppressiveness. *Ann. Rev. Phytopathol.* 42:243-270.

Garcia de Salamone, I. E., Hynes, R. K. and Nelson, L. M. (2001) Cytokinin production by plant growth promoting rhizobacteria and selected mutants. *Can. J. Microbiol.* 47: 404–411.

Gardner, I. C, Clelland, D. M. and Scott, A. (1984) Mycorrhizal improvement in non-leguminous nitrogen-fixing associations with particular reference to *Hippophae rhamnoides L, Plant Soil* 78:189–199.

Garcia-Garrido, J. M., Toro, N. and Ocampo, J. A. (1993) Presence of specific polypeptides in onion roots colonized by *Glomus mosseae. Mycorrhiza,* 2: 175-177.

Garmendia, I., Goicoechea, N., and Aguirreolea, J. (2004) Effectiveness of three Glomus species in protecting pepper (*Capsicum annuum* L.) against verticillium wilt. *Biol. Control* 31, 296-305.

George V Thomas, and Ghai, S. K. (1988) Relative efficiency of different VA Mycorrhizal fungi on black pepper (*Piper nigrum* L.). In: *Mycorrhiza Round Table. Proc. Workshop held in New Delhi,* 13-15 March 1987. Manuscript Report 201e (pp.421-430). The International Development Research Centre and Jawaharlal Nehru University, Delhi.

Glick, B. R. (1995a) The enhancement of plant growth by free-living bacteria, *Can. J. Microbiol.* 41, 109–117.

Glick, B. R. (1995b) Metabolic load and heterologous gene expression. *Biotechnol. Adv.* 13: 247-261.

Glick, B. R., Jacobson, C. B., Schwarze, M. M. K. and Pasternak, J.J. (1994) 1-Aminocyclopropane-l-carboxylic acid deaminase mutants of the plant growth promoting rhizobacterium *Pseudomonas putida* GR12-2 do not stimulate canola root elongation. *Can. J. Microbiol.* 40: 911-915.

Gray, E. J. and Smith, D. L. (2005) Intracellular and extracellular PGPR: commonalities and distinctions in the plant–bacterium signaling processes. *Soil Biol. Biochem.* 37, 395-412.

Gutierrez Munero, F. J, Ramos-Solano, B., Probanza, A., Mehouachi , J., Tadeo, F. R. and Talon, M. (2001) The plant-growth-promoting rhizobacteria *Bacillus pumilus* and *Bacillus licheniformis* produce high amounts of physiologically active gibberellins. *Physiol. Plant* 111: 206–211.

Hackett, B. P. (1985) Hormonal regulation of leaf senescence in Rumex: ethylene and cytokinins. *Dissertation Abstracts International B Sciences & Engineering* 46: 1018B

Hagen, G. (1990) The control of gene expression by auxin In: P.J. Davies, Editors, *Plant Hormones and their Role in Plant Growth and Development,* Kluwer Academic Publishers, Dordrecht, The Netherlands, pp. 149–163.

Halvorson, H. O., Kenyan, A. and Kornberg, H. L. (1990) Utilization of calcium phosphate for microbial growth at alkaline pH. *Soil Biol. Biochem.* 22:887-890.

Harman, G. E., Charles, R., Howell, Ada Viterbo, Ilan Chet, and Matteo Lorito, (2004) *Trichoderma* species-Opportunistic, avirulent plant symbionts. *Nat. Rev. Microbiol.* 2 ,43-56.

Haas, D. and Keel, C. (2003) Regulation of antibiotic production in root colonizing *Pseudomonas* Spp. and relevance for biological control of plant disease. Ann. Rev. Phytopathol. 41: 117- 153.

Hjelford, L. and Tronsmo, A. (1998) *Trichoderma and Gliocladium* in biological control : an overview. *In: Trichoderma and Gliocladium* Vol. 2 (eds) G.E. Harman and C.P. Kubicek Taylor and Francis, U.K.

Holland, M. A. (1997) Occam's razor applied to hormonology (are cytokinins produced by plants?). *Plant Physiol.* 115: 865–868.

Hussey, R. S. and Roncodri, R. W. (1982) Vesicular- arbuscular Mycorrhizal fungi may limit nematode activity and improve plant growth. *Plant Disease*, 66: 9-14.

Howell, C. R. (2003) Mechanisms employed by *Trichoderma* species in the biological control of plant diseases: The history and evolution of current concepts. *Plant Dis.*, 87:4-10.

Hubbel, D. H., Tien, T. M., Gaskins, M. H. and Lee, J. (1979) Physiological interactions in the Azospirillum-grass root associations. *In: Associative N2 fixation*, P.B. Vose and A.P. Ruschel (eds) pp1-6, CRC Press, Boca Raton, FL.

Illmer, P. and Schinner, F. Solubilization of inorganic calcium phosphates- solubilization mechanisms. *Soil Biol.Biochem.* 27:257-263.

Inbar, J. and Chet, I. (1995) The role of recognition in the induction of specific chitinases during mycoparasitism by *Trichoderma harzianum. Microbiol.* 141:2823-2829.

Indian Institute of Spices Research (IISR) 2001. *Annual Report 2000* 155 pp.

Indian Institute of Spices Research (IISR) 2002. *Annual Report 2001-02.* 131. pp.

Indian Institute of Spices Research (IISR) (2003) *Annual Report 2002-03.* 145 pp.

Kandiannan, K., Sivaraman, K, Anandaraj, M. and Krishnamurthy, K. S. (2000) Growth and nutrient content of black pepper (*Piper nigrum* L.) cuttings as influenced by inoculation with biofertilizers. *J. Spices Aromatic Crops* 9 (2): 145-147.

Jacobsen, I., Gazey, C. and Abbot, L. K. (2001) Phosphate transport by communities of arbuscular mycorrhizal fungi in contact soil cores. *New Phytol.* 149: 95-103.

Kaya, C.Higgs, D., Kimak, H. and Tas, I. (2003) Mycorrhizal colonization improves fruit yield and water use efficiency in watermelon (*Citrullus lanatus* Thunb.) grown under well-watered and water-stressed conditions. *Plant Soil* 253: 287-292.

Jalali, B. L. (1991) Relevance of VA-mycorrhizal systemin the management of root-borne diseases. Second Asian Conference on Mycorrhiza. *BIOTRAP* special publication 42: p 46.

Karagiannidis, N., Bletsos, F. and Stavropoulos, N. (2002) Effect of Verticillium wilt (*Verticillium dahliae* Kleb.) and mycorrhiza (*Glomus mosseae*) on root colonization, growth and nutrient uptake in tomato and eggplant seedlings. *Sci. Hort.* 94:145-156.

Khalid, A., Arshad, M. and. Zahir, Z. A. (2001) Factors affecting auxin biosynthesis by wheat and rice rhizobacteria. *Pak. J. Soil. Sci.* 21:1-10.

Kloepper, J. W. (1996) Biological control agents vary in specificity for host, pathogen control, ecological habitat and environmental conditions. *Bio. Sci.* 46; 406-409.

Kloepper, J. W., Leong, J., Teintze, M. and Schroth, M. N. (1980) Enhanced plant growth by siderophores produced by plant growth-promoting rhizobacteria, *Nature* 286: 885–886.

Kloepper, J. W., Lifshitz, R. and Zablotowicz, R. M. (1989) Free-living bacterial inocula for enhancing crop productivity. *Trends Biotech.* 7:39-44.

Kloepper, J. W. and Schroth, M. N. (1978) Plant growth-promoting rhizobacteria on radishes. *Proc. of the 4th Int. Conf. on Plant Pathogenic Bacter* Vol. 2, Station de Pathologie Végétale et Phytobactériologie, INRA, Angers, France, pp. 879–882.

Kobayashi, M., Suzuki, T., Fujita, T., Masuda, M. and Shimizu, S. (1995) Occurrence of enzymes involved in biosynthesis of indole-3-acetic acid from indole-3-acetonitrile in plant-associated bacteria, *Agrobacterium and Rhizobium*, *Proc.Natl.Acad.Sci.* 92: 714–718.

Kohler J, Caravaca F, Carrasco L, Roldán A (2007) Interactions between a plant growth-promoting rhizobacterium, an AM fungus and a phosphate-solubilising fungus in the rhizosphere of *Lactuca sativa*. *Appl. Soil Ecol.* 35: 480-487.

Krishna, K. R. and Bagyaraj, D. J. (1983). Interaction between *Glomus fasciculatus* and *Sclerotium rolfsii* in peanut (*Arachis hypogea.*) *Can. J. Bot.* 61:2349-2351.

Kubicek, C. P., Mach, R. L., Peterbauer, C. K. and Lorito, M. (2001) *Trichoderma*: from genes to biocontrol. *J. Plant Pathol.* 83:11-23.

Lisha, K. P., Anandaraj, M., Diby Paul, Jisha, P. J. and Sarma, Y. R. (2002) Evaluation of biocontrol agents obtained from Silent valley biosphere reserve against *Phytophthora capsici* the foot rot pathogen in black pepper (*Piper nigrum* L.). Indian Phytopathol. 55: 3, pp.373 (Abs).

Linderman, R. G. (1988) Mycorrhizal interactions with the rhizosphere microflora;the mycorrhizosphere effect. *Phytopathology* 78: 366-370.

Lorito, M. (1998) Chitinolytic enzymes and their genes. In: *Trichoderma* and *Gliocladium Vol.2. Enzymes biological control and commercial application.* (eds) G. E. Harman and P. C. Kubicek, Taylor and Francis, U.K.

Lorito, M., Harman, G. E., Hayes, C. K., Broadway, R. M., Tronsmo, A., Woo, S. L. and Di Pietro, A. (1993) Chitinolytic enzymes produced by *Trichoderma harzianum*: antifungal activity of purified endochitinase and chitobiosidase. *Phytopathology* 83:302-307.

Lorito, M., Peterbauer, C. K., Hayes, C. K., Woo, S. L. and Harman, G. E. (1994b) Synergistic combination of cell wall degrading enzymes and different antifungal compounds enhances inhibition of spore germination. *Microbiology* 140:623-629.

Lorito, M., Woo, S. L., Garcia, I., Colucci, G., Harman, G. E., Pintor-Toro, J. A., Filippone, E., Muccifora, S., Lawerence, C. B., Zoina, A., Tuzun, S. and Scala, F. (1998) Genes from mycoparasitic fungi as a source for improving plant resistance to fungal pathogens. *Proc Natl. Acad. Sci.* (USA) 95: 7860-7865.

Lugtenberg, B. J. J., Dekkers, L. and Bloemberg, G. V. (2001) Molecular determinants of rhizosphere colonization by *Pseudomonas. Ann. Rev. Phytopathol.* 39: 461-490.

Mahmoud, S. A. Z, Ramadan, E. M. Thabet, F, M. and Khater, T. (1984) Production of plant growth promoting substances by rhizosphere microorganisms. *Zentrabl. Mikrobiol.* 139: 227-232.

Manjunath, A. and Bagyaraj, D. J. (1982) Vesicular arbuscular mycorhhizas in three plantation crops and cultivars of field bean. *Curr. Sci.* 51: 707-709.

Manjunath, A. and Habte, M. (1988) Develpoment of vesicular-arbuscular mycorrhizal infection and the uptake of immobile nutrients in Leucaena leucocephala. *Plant Soil* 106:97-103.

Mansouri, S. and Bunch, A. W. (1989) Bacterial ethylene synthesis from 2-oxo-4-thiobutyric acid and from methionine. *J. Gen. Microbiol.* 135: 2819-2827.

Manulis, S., Kashman, Y. and Barash, I. (1987) Identification of siderophores and siderophore mediated uptake of iron in Stemphylium botryosum. *Phytochemistry* 26: 1317-1320.

Martinez-Romero, E. and Wang, E. T. (2000) *Sesbania herbacea–Rhizobium huautlense* nodulation in flooded soils and comparative characterization of *S. herbacea* nodulating rhizobia in different environments. *Microb. Ecol.* 41: 25–32.

Mathivanan, N., Kabilan, V. and Murugesan, K. (1998) Purification, characterization and antifungal activity of chitinase from *Fusarium chlamydosporum*, a mycoparasite to groundnut rust, *Puccinia arachidis. Can. J. Microbiol.* 44:646-651.

McSpadden, G. B. and Fravel, D. (2002) Biological control of plant pathogens: Research, commercialization and application in the USA. *Plant health progress Doi.* 10. 1094

Meyer, J. M. and Abdallah, M. A. (1978) The fluorescent pigment of *Pseudomonas fluorescens:* biosynthesis, purification and physicochemical properties. *J. Gen. Microbiol.* 107:319-328.

Minimol, (2002) Effect of plant growth promoting rhizobacteria on growth of black pepper(*Piper nigrum* L.) and disease suppression of *Phytophthora* root rot. M.Sc., Thesis, University of Madras. 60pp.

Mosse, B, Stribley, D. P. and Letacon, F. (1981) Ecology of mycorrhizae and mycorrhizal fungi. *Adv. Microbiol. Ecol.* 5:137-210.

Nickel, A., Pelz, O., Hahn, D., Saurer, M., Siegwolf, R. and Zeyer, J. (2001) Effect of inoculation and leaf litter amendment on establishment of nodule-forming *Frankia* populations in soil. *Appl. Environ. Microbiol.* 67:2603–2609.

Nieto, K. F. and Frankenberger, W. T. Jr. (1989) Biosynthesis of cytokininsis by *Azotobacter chroococcum. Soil Biol. Biochem.* 21; 967- 972.

Nordlund, D. A. (1996) Biological control, integrated pest management and conceptual models. *Biol. News Infmn.* 17; 33N-44N.

Pazout, J., Wurst, M. and vancura, V. (1981) Effect of aeration on ethylene production by soil bacteria and soil samples cultivated in a closed system. *Plant Soil* 62: 431-437.

Ping, L., Boland, W. (2004) Signals from the underground: bacterial volatiles promote growth in *Arabidopsis. Trends in Plant Science*, 9, 263-266.

Lambrecht, M., Okan, Y., vandeBroock, A. and Vanderleyden, J. (2000) Indole-3-caetic acid: a reciprocal signaling molecule in bacteria-plant interactions. *Trends Microbiol.* 8: 298-300.

Lucangeli, C. and Bottini, R. (1997) Effects of *Azospirillum* spp.on endogenous gibberellin content and growth of maize (*Zea mays* L.) treated with unicanazole. *Symbiosis* 23:63-71.

Persello-Cartieaux, F., Nussaume, L. and Robaglia, C. (2003) Tales from the underground: molecular plant-rhizobacteria interactions. *Plant, Cell and Environment* 26 (2), 189–199.

Powell, C. L. and Bagyaraj, D. J. (eds). (1984) *VA Mycorrhiza.* CRC Press, Bocca Raton.

Prakash, M. G., Vinaya Gopal, K., Anandaraj, M. and Sarma, Y. R. (1999) Evaluation of substrates for mass multiplication of fungal biocontrol agents. *J. Spices Aromatic Crops* 8: 207-210.

Prithiviraj, B., Zhou, X., Souleimanov, A., Kahn, W. and Smith, D. L. (2003) A host-specific bacteria-to-plant signal molecule (Nod factor) enhances germination and early growth of diverse crop plants, *Planta* 216: 437–445.

Rajan, P. P. Sarma, Y. R. and Anandaraj, M. (2002) Management of foot rot disease of black pepper with *Trichoderma* spp. *Indian Phytopathol.* 55 (3):34-38.

Ramos, H. C., Hoffmann, T., Nedjari, H., Presecan-Siedel, E., Dreesen, O., Glaser, P. and Jahn, D. (2000) Fermentative metabolism of *Bacillus subtilis*: physiology and regulation of gene expression. *J. Bacteriol.* 182: 3072–3080.

Reddell, P. and Spain, A. V (1991) Transmission of infective *Frankia* (actinomycetales) propagules in casts of the endogeic earthworm *Pontoscolex corethrurus* (Oligochaeta: Glossoscolecidae), *123*: 775–778.

Rodríguez, H. and Fraga, R. (1999) Phosphate solubilizing bacteria and their role in plant growth promotion, *Biotechnology Adv.*, 17,319-339.

Rosbrook, P. A. and Reddell, P. (1995) Isolation of Frankia from root nodules of three species of *Casuarina, Soil Biol. Biochem.* 27: 427–429.

Srinivasan, M., Peterson, D. J. and Holl, F. B. (1996) Influence of IAA producing *Bacillus* isolates on the nodulation of *Phaseolus vulgaris by Rhizobium etli.* Can. J. Microbiol. 42: 1006–1014.

Rekha, P. D, Lai, W. A., Arun, A. B. and Young, C. C. (2007) Effect of free and encapsulated *Pseudomonas putida* CC-FR2-4 and *Bacillus subtilis* CC-pg104 on plant growth under gnotobiotic conditions. *Bio. Res. Tech.* 98: 447-451.

Rojas, N. S, Li, C. Y., Perry, D. A. and Ganio, L. M. (2001) *Frankia* and nodulation of red alder and snowbrush grown on soils from Douglas-fir forests in the H.J. Andrews experimental forest of Oregon, *Appl. Soil Ecol.*17: 141–149.

Ryu, C. M., Farag, M. A., Hu, C. H., Reddy, M. S., Kloepper, J. W., Pare, P. W. 2004. Bacterial Volatiles Induce Systemic Resistance in *Arabidopsis Plant Physiol.,* 134: 1017 - 1026.

Sabir, R. K. (2002) Effect of plant growth promoting rhizobacteria on slow decline disease of black pepper (*Piper nigrum* L.) M. Sc Project report, Calicut University 61pp.

Safir, G. R. (ed) (1987) *Ecophysiology of VA Mycorrhizal plants*, CRC Press, Boco Raton.

Saju, K. A., Anandaraj, M. and Sarma, Y. R. (2002) On farm production of *Trichoderma harzianum* using organic matter. *Indian Phytopathol.* 55(3): 277-281.

Shaharoona, B., Arshad, M., Zahir, Z. A. and Khalid, A. (2006) Performance of *Pseudomonas* spp. containing ACC-deaminase for improving growth and yield of maize (*Zea mays* L.) in the presence of nitrogenous fertilizer. *Soil Biol. Biochem.* 38: 2971-2975.

Sieverding, E. (1991). *Vesicular-Arbuscular mycorrhiza management in tropical agrosystems.* GTZ, Eschborn, Germany.

Sivan, A. and Harman, G. E. (1991). Improved rhizosphere competence in a protoplast fusion progeny of *Trichoderma harzianum. J. Gen. Microbiol.*137: 23-29.

Steenhoudt, O. and Vanderleyden, J. (2000) *Azospirillum*, a free-living nitrogen-fixing bacterium closely associated with grasses: genetic, biochemical and ecological aspects. *FEMS Microbiol. Rev.* 24: 487–506.

Stenlid, G. (1982) Cytokinins as inhibitors of root growth. *Physiol. Plant* 56: 500–506.

Sturz, A. V., Christie, B. R. and Nowak, J. (2000) Bacterial endophytes: Potential role in developing sustainable systems of crop production, *Crit. Rev. Plant Sci.* 19: 1–30.

Stephen, R. J., Anandaraj, M. and Sarma, Y. R. (2000) Compatibility of chlorpyriphos with *Trichoderma harzianum* (Rifai) applied for integrated disease management in black pepper (*Piper nigrum* L.) *J. Spices Aromatic Crops* 9(2): 111-115

Sturz, A. V, Christie, B. R. and Nowak, J. (2000) Bacterial endophytes: Potential role in developing sustainable systems of crop production, *Crit. Rev. Plant Sci.* 19: 1–30.

Strzelczyk, E., Kampert, M. and Li, C. Y. (1994) Cytokinin like substances aqnd ethylene production by *Azospirillum* in media with different carbon sources. *Microbiol. Res.* 149: 55-60.

Swanson, B. T., Wilkins, H. F. and Kennedy, B. W. (1979) Factors affecting ethylene production by some plant pathogenic bacteria. *Plant Soil* 51- 19-26.

Tang, W. H. (1994) Yield-increasing bacteria (YIB) and biocontrol of sheath blight of rice. In: *Improving plant productivity with rhizosphere bacteria*, M.H. Ryder, P.M. Stephens and G.D. Bowen (eds.), pp. 267-278, CSIRO, Adelaide.

Taylor, T. N. (1990) Fungal associations in terrestrial *pale* system. *Trends Ecol. and Evol.* 5:21-25.

Tello, J. C., Vares, F. Notario, A. and Lacasa, A. (1987) Mycorrhizae: a potential against plant diseases. ITEA (*Informacion Tecnica Economica Agraria*) 18:40-64.

Thankamani, C. K., Anandaraj, M., Diby Paul, and Kandiannan, K. (2003) Compatability of *Pseudomonas fluorescens*, VAM (*Glomus fasciculatum*), *Trichoderma harzianum* and pesticides on establishment and growth of black pepper under nursery condition. In: Reddy, M.S., Anandaraj, M., Eapen, S. J., Sarma, Y.R. and Kloepper, J. W (eds) *Proc. 6th International PGPR workshop*, 5-10 October 2003 , Indian Institute of Spices Research, Calicut, India, pp 82-88.

Thankamani, C. K., Sreekala, K. and Anandaraj, M. (2005) Effect of *Pseudomonas fluorescens* (IISR-6) and *Trichoderma harzianum* (P-26) on growth of black pepper (*Piper nigrum* L.) in the nursery. *J. Spices Aromatic Crops* 14:112-116.

Thanuja, T. V., Hegde, R. V. and Sreenivasan, M. N. (2002) Induction of rooting and root growth in black pepper cuttings (*Piper nigrum* L.) with the inoculation of arbuscular mycorrhizae. *Sci. Hort.* 92:339-346.

Tien, T. M., Gaskins, M. H. and Hubbell, D. H. (1979) Plant growth substances produced by *Azospirillum brasilense* and their effect on the growth of pearl millet (*Pennisetum americanum* L.). *Appl. Environ. Microbiol.* 37: 1016–1023.

Troxler, J. Azelvandre, P. Zala, M, Defago, G. and Haas, D. (1997) Conjugative transfer of chromosomal genes between fluorescent pseudomonads in the rhizosphre of wheat. *Appl. Environ. Microbiol.* 63: 213-219.

Velazhahan, R., Samiyappan, R. and Vidhyasekaran, P. (1999) Relationship between antagonistic activities of *Pseudomonas fluorescens* isolates against *Rhizoctonia solani* and their production of lytic enzymes. *J. Panat Dis. Prot.* 106: 244-250.

Viswanathan, R. and Samiyappan, R. (1999) Induction of systemic resistance by plant growth promoting rhizobacteria against red rot disease caused by *Colletotrichum falcatum* Went. in sugarcane. *Proc. Sugar Technol. Assn. India.* 61:24-39.

Walling, L. L. (2000) The myriad of plants responses to herbivores. *J. Plant Growth Regul.* 19:195–216.

Weidhase, R. A., Lehmann, J., Kramell, H., Sembdner, G. and Parthier, B. (1987) Degradation of ribulose-1,5-bisphosphate carboxylase and chlorophyll in senescing barley leaf segments triggered by jasmonic acid methyl ester, and counteraction by cytokinin, *Physiol. Plant* 69:161–166.

Wicklow, D. T., Joshi, B. K., Gamble, W. R., Gloer, J. B. and Dowd, P. F. (1998) Antifungal Metabolites (Monorden, Monocillin IV, and Cerebrosides) from *Humicola fuscoatra* Traaen NRRL 22980, a Mycoparasite of *Aspergillus flavus, Sclerotium*, Appl. and Environ. Microbiol., 64, 4482-4484.

Wilhite, S. E., Lumbsden, R. D. and Straney, D.C. (2001) Peptide synthetase gene in *Trichoderma virens. Appl. Environ. Microbiol.* 67:5055-5062.

Winklemann, G. (1992) Structure and functions of fungal siderophores containing hydroxamate and complex one type iron binding ligands. *Mycol. Res.* 96:529-534.

Yanni, Y. G., Rizk, R. Y., Corich, V., Squartini, A., Ninke, K., Philip-Hollingsworth, S., Orgambide, G., de Bruijn, F., Stoltzfus, J., Buckley, D., Schmidt, T.M., Mateos, P. F., Ladha, J. K. and Dazzo, F.B. (1997) Natural endophytic association between *Rhizobium leguminosarum* bv. trifoli and rice roots and assessment of its potential to promote rice growth. Plant Soil 194; 99-114.

Zahir, A. Z., Arshad, M. and Khalid, A. (1998) Promoting crop yields by *Azotobacter* inoculation. *J. Agri. Res.* 36; 369-380.

Zahir, A. Z., Arshad, M., Frankenburger, J. T. (2004) Plant growth promoting rhizobacteria: Applications and perspectives in Agri. Adv. Agron. 81:97-165.

Zakharova, E., Shcherbakov, A., Brudnik, V., Skripko, N., Bulkhin Sh, N. and Ignatov, V. (1999) Biosynthesis of indole-3-acetic acid in *Azospirillum brasilense*: insights from quantum chemistry, *Eur. J. Biochem.* 259: 572–576.

Zhang, R. and Smith, D. L. (1997) Application of genistein to inocula and soil to overcome low spring soil temperature inhibition of soybean nodulation and nitrogen fixation, *Plant Soil* 192: 141–151.

4

Biological Control of Insect Pests

S. Devasahayam

1.0 INTRODUCTION

Infestation by insect pests is a major factor responsible for the low productivity of spice crops in India. Spice crops such as black pepper (*Piper nigrum* L.), cardamom (*Elettaria cardamomum* Maton), ginger (*Zingiber officinale* Roscoe) and turmeric (*Curcuma longa* L.) are infested by many species of insect pests, some of which cause severe crop losses. Adoption of integrated pest management (IPM) schedules in which biological control forms a major component is more relevant in these crops since excessive use of insecticides could lead to pesticide residues in the produce and other ecological hazards. Spices are high-value and export-oriented in nature and pesticide residues are the most important non-tariff barriers on trade in these produces. There has been a renewed interest during the past decade in developing environment-friendly crop management schedules in sustainable agriculture including spice crops, reflecting the increasing concern over pesticide misuse.

2.0 BIOLOGICAL CONTROL-GENERAL PRINCIPLES

Biological control is the regulation of plant and animal populations in nature by another living organism (natural enemy), which keeps them in a state of balance in an ecosystem. DeBach (1964) defined biological control as 'the action of parasites, predators and pathogens in maintaining another organism's density at a lower average than would occur in their absence'. The natural enemies are conserved, multiplied by artificial means and disseminated by man by his own efforts to bring down the population of insect pests.

Biological control of insect pests is undertaken in three principal ways namely, collection of natural enemies from the field where they are naturally found and releasing them in areas where they are not found; rearing of natural enemies in the laboratory and releasing them in the field wherever needed; and importation of natural enemies from other countries and rearing and releasing them in the field.

Before initiating a programme on biological control on a particular crop or area, a survey should be made for natural enemies already existing in the area which would help in avoiding importation of natural enemies that are already existing. If it is necessary to import natural enemies, it should be done from countries with similar climatic conditions so that the imported predator / parasitoid can successfully establish in the new area. A through knowledge of the species of the host to be controlled and that of the predator / parasitoid to be imported is essential and all aspects of the biology and ecology of the host and natural enemy should be studied. Generally biological control has been most successful in ecologically insular areas and homopteran insects like aphids, mealybugs and scale insects, followed by coleopterans and lepidopterans provide the majority of successful instances of biological control.

2.1 Use of Predators and Parasitoids

A predator is comparatively larger than its host (prey) and consumes a large number of hosts during its development. Predatory insects feed on all stages of the host, namely, egg, larval, pupal and adult stages. Some predators are predaceous only as larvae while others are predaceous as larvae and adults. Predators either have biting and chewing mouth parts and chew and feed on insects, or have piercing and sucking mouth parts and pierce and suck fluids from their prey. Predatory species of insects occur in most orders with the maximum in Coleoptera. A parasitoid is invariably parasitic only in its immature stages and develops within a single host and kills it during its course of development. The adult is generally free-living feeding on nectar and honey dew. Many parasitoids are host-specific developing on only one or a few closely related species. Parasitoids are recorded from five insect orders with the maximum occurring in Diptera and Hymenoptera.

A potential predator or parasitoid should have certain qualities for it to be successful as a biocontrol agent. It should quickly adapt itself to the environmental conditions in the introduced locality and should be able to survive in all habitats inhabited by the pest; it should not affect other beneficial fauna in the introduced locality; it should have a high female:male ratio and fecundity; it should have a short life-cycle synchronizing with that of the host and multiply faster than the host; it should be amenable for easy mass culturing in the laboratory; it should possess high host-searching capacity; and, it should be able to disperse quickly in the introduced habitat.

Table 1. Natural enemies of insect pests of spice crops

Crop / Pest	Predator	Parasitoid	Nematode	Pathogen
Black pepper				
Pollu beetle	2	-	1	-
Scale insects	17	4	-	-
Top shoot borer	-	7	-	-
Leaf gall thrips	7	-	-	-
Cardamom				
Cardamom thrips	3	-	-	1
Shoot and capsule borer	-	20	-	-
Root grub	-	-	2	2
Cardamom aphid	6	-	-	2
Ginger and Turmeric				
Shoot borer	6	17	1	1
Rhizome scale	2	2	-	-

2.2 Use of Microbial Pathogens

The pathogenic organisms causing infectious diseases in insects are primarily viruses, bacteria, fungi and protozoans. Viruses have been isolated and described from over 800 species of insects. Viruses are generally very specific affecting a single genus or a species. Viruses are effective only when they are consumed by the insects. Certain viruses have inclusion bodies which may be polyhedral or granular and the polyhedral virus may be called nuclear or cytoplasmic according to its site of multiplication. Nuclear polyhedrosis viruses are more common in Lepidoptera, Hymenoptera and Diptera and granular inclusion viruses in Lepidoptera. Insect viruses are particularly suited to low input local production units and though labour intensive, does not require high-cost capital investment.

The bacterial pathogens that affect insects are broadly classified as spore formers and non-spore formers. There are both obligate and facultative spore formers. About 100 species of entomopathogenic bacteria have been described among which *Bacillus thuringiensis* is well known. Bacterial pathogens must be consumed by the insect to be effective. Various commercial formulations of the bacteria are now available that are widely used for the management of various insect pests in the world. The most widely used biopesticides are those based on *B. thuringiensis*. Wider use of *B. thuringiesis* in developing countries not only depends on effective control and environmental safety aspects but also on reduced cost effective procedures.

Fungi form the largest group of insect pathogens and more than 500 species have been reported from different insect hosts among which Lepidoptera, Coleoptera, Homoptera, Hymenoptera and Diptera are more susceptible. In most of the cases the adults are more commonly affected than the larvae. Unlike viruses and bacteria, fungi need not be consumed to be effective. The fungal conidia germinate on the insect cuticle and produce specialized structures that penetrate the cuticle and enter into the insect's body. Entomopathogenic fungi have been exploited on a reasonably large scale in many developing countries in predominantly local production units utilizing agricultural byproducts.

Protozoan pathogens belonging to the Class Sporozoa are more important containing several pathogens infecting Lepidoptera, Coleoptera, Orthoptera, Hemiptera and Diptera. Though most of the protozoans can kill their insect hosts, they generally cause debilitating effects on their hosts. However, the role of protozoa as microbial agents in insect pest management is limited due to their difficulty in mass multiplication and field application.

2.3 Use of Entomopathogenic Nematodes

Nematodes affecting insects belong to the families Heterorhabditidae, Mermithidae and Steinernematidae, and include species that are parasitic on insects and either kill or sterilize them, or alter their development. Nematode species belonging to the families Heterorhabditidae and Steinernematidae are highly virulent because they have a symbiotic relationship with bacteria forming a disease complex in the infected insects which is lethal to their host. The nematode serves as a vector for the bacteria, which are released into the blood stream when the nematode enters the insect body. The nematodes feed on these bacteria as they multiply and cause septicemia in the insect hosts. This peculiar characteristic makes them more suitable for biological control of insects than any other nematode group. Entomopathogenic nematodes are known to be effective against over 400 species of insects, mostly coleopterans and also dipteran and lepidopteran larvae.

2.4 Use of Agronomic Practices

Various agronomic practices or cultural methods not only help in reducing the population of insect pests in the field but also help in conservation of natural enemies (van den Bosch & Telford 1973; Pickett & Bugg 1998).

2.4.1 Cultural control

The control of insect pests through adoption of farm practices in such a way that insect pest populations are reduced in the field is called cultural control. These practices are generally adopted in such a way so as to affect the life history or host adaptation of an insect. This method of control is generally more economical in homestead farms where the cost of labour is low. Though these methods do not control pest populations completely, they are important in reducing them and making them more amenable for other methods of control.

2.4.2 Crop rotation

Crop rotation involves planting of various crops (non-host and host crops) in a rotation, which helps to break the continuity of host material being available for feeding and multiplication of pest species.

2.4.3 Trap cropping

Trap cropping is planting of a susceptible or preferred crops near a major crop along the border or within the field to act as a 'trap' to attract the pest to this crop. The pest populations which are thus 'trapped' are collected and mechanically killed or by treating with insecticides.

2.4.4 Tillage operations

Many insect pests spend their larval and pupal stages in soil. Deep ploughing will bring these insects to the surface to be exposed to desiccation and predators. Ploughing also helps in removal of weeds which may otherwise serve as host plants for insect pests. Earthing-up with soil so as to cover roots and tubers also helps in reducing the level of pest infestation in these crops.

2.4.5 Crop sanitation

Pruning of dried twigs or branches will eliminate borers and scale insects in many crops. Pruning dried leaf sheaths helps in elimination of aphids and thrips on cardamom. Removal of infested plants, crop remnants and fallen infested fruits helps in reducing pest populations in the field.

2.4.6 Time of sowing / harvesting

Early or late sowing / transplantation of crops can be practiced to avoid egg laying by certain pests. Timely harvest is also important in managing pest species which oviposit on mature and ripe fruits.

3.0 CONSERVATION OF NATURAL ENEMIES

The agronomic practices that help in conservation of natural enemies involve altering the habitat or environment to augment or enhance the effectiveness of a natural enemy. The efficacy of natural enemies depends to a large extent, upon the degree of stability and general favourability of environmental conditions. Habitat modification includes:

3.1.1 Construction of artificial structures

Construction of artificial structures can be resorted to increase the population of many predaceous insects. Use of 'nesting shelters' for wasps and predaceous ants around agro-ecosystems is known to reduce the population of insect pests.

3.1.2 Provision of supplementary food

Adult natural enemies need nectar and pollen as a source of nourishment for their growth and development. In nature these requirements are provided by a variety of plants that flower during one season or the other. However, in agro-ecosystems many of these plants are considered as weeds and are removed. Inter planting of suitable crops provides food and shelter for natural enemies. Many adult parasitoids and predators also benefit from the protection provided by refuges such as cover crops and weeds in the borders of the fields. Provision of adequate mulches is known to enhance the populations of many species of predatory spiders.

3.1.3 Provision of alternate hosts

Alternate host plants generally reduce conditions of asynchrony between preferred hosts and their parasites and predators. In many cases this has been achieved by intercropping host plants of the pest thus preventing total destruction of the habitat during harvest. Intercrops provide shade and alternate food and higher humidity and lower temperatures and reduce wind speed thus favouring the multiplication of natural enemies.

3.1.4 Modification of agricultural practices

Clean cultivation (elimination of weeds) can result in the elimination of alternate hosts and supplementary sources of nourishment of natural enemies. Burning of harvest residues can destroy natural enemies which develop on hosts remaining in the residues. These can be overcome by allowing weeds to grow to a certain extent in the borders and also by using harvest residues as mulch.

4.0 BIOLOGICAL CONTROL IN SPICE CROPS

The information available on management of major insect pests of spice crops through biological means and the scope and challenges in its adoption in integrated management of insect pests are highlighted in this chapter. The information available on other eco-friendly methods of control such as cultural methods and natural products for the management of insect pests is also included since these methods are compatible with biocontrol agents and are to be utilized in an integrated manner.

4.1 Black Pepper

4.1.1 Pollu beetle (Longitarsus nigripennis Mots.)

The *pollu* beetle (*Longitarsus nigripennis* Mots.) (Coleoptera, Chrysomelidae) is the most destructive insect pest of black pepper in India. The pest infestation is higher in plains and midlands in areas below 300 m MSL and 10%-32% of berries were reported to be damaged by the pest in Kerala (Premkumar & Nair 1988). The adult feeds on tender shoots, leaves and spikes resulting in black patches on shoots and spikes and small irregular holes on tender leaves. The damage caused by the grub is more serious and it bores into tender berries and feeds on the internal tissues. The infested berries turn black and crumble when pressed (the term *pollu* denotes the hollow nature of the infested berries in Malayalam). The grub also bores into tender spikes resulting in necrotic patches on them. The pest infestation is more severe in shaded areas in the plantation and on vines trailed on standards (support trees) producing heavy shade like jack, mango, etc. (Devasahayam 2000a).

Natural enemies

Very few natural enemies have been recorded on *pollu* beetle in the field. These include entomophagous nematode (Mermithidae), spider (Araneae) and *Oecophylla smaragdina* Fabr. (Formicidae). However the level of parasitization/predation by these natural enemies (Table 2) is quite low and they may not be of much significance in the control of the pest in the field (Devasahayam & Koya 1994a; Devasahayam 1996). Evaluation of *Beauvaria bassiana* (Bals.) Vuill, *B. brongniartii* and *Metarhizium anisopliae* (Metsch) Sorokin (Hyphomycetes) from other sources indicated their limited scope for use as biopesticides for the management of the pest (IISR 1996).

Plant products

Evaluation of neem products in laboratory bioassays against p*ollu* beetle indicated that Repelin, Neemazal-F and neem seed kernel extract were more promising (Devasahayam & Anandaraj 1997). An integrated strategy involving regulation of shade in the plantation and spraying endosulfan or quinalphos (0.05% each) during July followed by 3–4 sprays (at 21 day intervals) of neem-based product such as Neemgold 0.6% during August-October, or spraying Neemgold 0.6% during July–October (4–5 sprays at 21 day intervals) has been suggested for controlling the pest infestation (Devasahayam & Koya 1999).

Various plant extracts have been evaluated in the laboratory for their antifeedant activity against *pollu* beetle. Among them root bark extracts of *Uvaria narum* Blume and *U. hookeri* King (Annonaceae) (Babu *et al.* 1996), leaf extracts of *Strychnos nux-vomica* L. (Loganiaceae), *Chromolaena odorata* L. (Asteraceae) and seed extract of custard apple (*Annona squamosa* L.) (Annonaceae) (Devasahayam & Leela 1997) possessed significant antifeedant activity against the pest indicating their potential in utilizing them in IPM schedules.

4.1.2 Scale insects (Lepidosaphes piperis Green and Aspidiotus destructor Sign.)

Among the various species of scale insects recorded on black pepper in India, mussel scale *Lepidosaphes piperis* Green and coconut scale *Aspidiotus destructor* Sign. (Homoptera, Diaspididae) are more common and widely distributed at higher altitudes (above 600 m MSL) in Kerala, Karnataka and Tamil Nadu (Koya *et al.* 1996). The mussel scale encrusts stems of younger vines, lateral branches, mature leaves and berries resulting in chlorotic patches, yellowing and drying of and lateral branches, and mortality of young vines. *L. piperis* also infests older cuttings in the nursery in the plains. The coconut scale infests leaves resulting in chlorotic patches and very rarely stems of lateral branches and berries (Koya *et al.* 1996).

Natural enemies

Twenty-one species of natural enemies have been recorded on scale insects in the field among which *Chilocorus circumdatus* (Gyllen.) and *C. nigrita* (Fabr.) were identified to be potential biocontrol agents of *A. destructor* and *L. piperis*, respectively (Devasahayam 1996; Selvakumaran *et al.* 1996). Techniques were standardized for mass rearing of *C. circumdatus* and *C. nigrita* on *Aonidiella orientalis* (Newst.) (host scale) raised on pumpkins (*Cucurbita moscahata*) in the laboratory. Evaluation of *C. circumdatus* and *C. nigrita* in the field at Wyanad (Kerala) for the management of *A. destructor* and *L. piperis* indicated that release of these predators was promising in reducing the populations of scale insects (Devasahayam 1998).

Plant products

Plant and organic products such as neem oil 0.3%, Neemgold 0.3% and fish oil rosin 3% were promising in controlling both the species of scale insects and were also safe to the coccinellid predators indicating their potential use in IPM schedules (Devasahayam 1998). Spraying of Neemgold was also found to reduce coconut scale infestations on black pepper in the field (Selvan *et al.* 1996).

4.1.3 Top shoot borer (Cydia hemidoxa Meyr.)

The top shoot borer (*Cydia hemidoxa* Meyr.) (Lepidoptera, Tortricidae) is a serious insect pest of black pepper in younger plantations being widely distributed in all the areas in Kerala. The larva of top shoot borer infests tender terminal shoots especially those that adhere to the standards. The earlier instar larva scrapes and feeds on the epidermis of tender terminal shoots and sometimes on tender leaves. The later instar larva bores into tender shoots and feeds on the internal contents resulting in decay and drying of infested shoots. Repeated infestation of tender terminal shoots affects the growth and establishment of newly planted vines (Devasahayam & Koya 1993a; Devasahayam 2000a).

Natural enemies

A few natural enemies have been recorded on larva of top shoot borer among which *Hexamermis* sp. (Mermithidae) and *Apanteles cypris* Nixon

(Braconidae) are more common and were identified as potential biocontrol agents (Devasahayam & Koya 1994a; Devasahayam 1996). The mermithid nematode was more common during the monsoon period (June to September) at Peruvannamuzhi (Kerala) with a peak parasitism of 66.6%-76.7% during July-August (Devasahayam & Koya 1994b). The hymenopterous parasitoid was more common during the post monsoon period (September-November) with a peak parasitism of 20% during October (Devasahayam & Koya 1993b).

4.1.4 Leaf gall thrips (Liothrips karnyi Bagn.)

The leaf gall thrips (*Liothrips karnyi* Bagn.) (Thysanoptera, Phlaeothripidae) is a widespread insect pest of black pepper especially on younger vines at higher altitudes and also on rooted cuttings in nurseries in the plains in Kerala, Karnataka and Tamil Nadu. Leaf gall thrips infest tender leaves causing the leaf margins to curl downwards and inwards resulting in the formation of marginal, tubular, hypophyllous galls. As the infested leaves grow they become crinkled, malformed and reduced in size; the leaf margins also turn chlorotic later. The growth of younger vines and rooted cuttings in the nursery and development of spikes are affected in severe cases of infestation (Devasahayam 1992).

Natural enemies

A few predators have been recorded on leaf gall thrips in the field among which *Montandoniola moraguesi* (Puton) (Anthocoridae) and *Androthrips flavipes* Schmutz (Phlaeothripidae) are more common and widely distributed; both the predators were identified as potential biocontrol agents of the pest. *M. moraguesi* predates on all stages of leaf gall thrips whereas, *A. flavipes* generally prefers eggs and juvenile stages of leaf gall thrips. The predatory potential, life cycle and seasonal population of these predators have been studied (Devasahayam 1992).

4.2 Cardamom

4.2.1 Cardamom thrips (Sciothrips cardamomi Ramk.)

The cardamom thrips (*Sciothrips cardamomi* Ramk.) (Thysanoptera, Thripidae) is the most destructive pest of cardamom occurring in all cardamom areas. The adults and larvae lacerate the tissues of leaves, shoots, panicles, flowers and immature capsules and feed on the exuding sap. The pest infestation results in shedding of flowers and immature capsules and the formation of corky, scab-like encrustation on the capsules. The affected capsules are malformed, shriveled and without seeds. The pest is reported to cause 42%-47% yield loss to the crop in various areas (Varadarasan 1995; Gopakumar & Chandrasekhar 2002).

Table 2. Potential biocontrol agents of major insect pests of spice crops

Crop / Pest	Biocontrol agent
Black pepper	
Pollu beetle (*Longitarsus nigripennis*)	Nil
Scale insects (*Lepidosaphes piperis* and *Aspidiotus destructor*)	*Chilocorus nigrita*, *C. circumdatus*, *Aphytis* sp.
Top shoot borer (*Cydia hemidoxa*)	*Apanteles cypris*, *Hexamermis* sp.
Leaf gall thrips (*Liothrips karnyi*)	*Montandoniola moraguesi*, *Androthrips flavipes*
Cardamom	
Cardamom thrips (*Sciothrips cardamomi*)	Nil
Shoot and capsule borer (*Conogethes punctiferalis*)	*Xanthopimpla australis*, *Eriborus trochanteratus*, *Friona* sp., *Agrypon* sp.
Root grub (*Basilepta fulvicorne*)	*Metarrhizium anisopliae*, *Beauveria bassiana*, *Heterorhabditis* sp., *Steinernema* sp.
Cardamom aphid (*Pentalonia nigronervosa* f. *caladii*)	*Beauveria bassiana*, *Pochonia chlamydosporia*
Ginger and Turmeric	
Shoot borer (*Conogethes punctiferalis*)	*Apanteles taragammae*, *Hexamermis* sp.
Rhizome scale (*Aspidiella hartii*)	*Physcus comperei*, *Adelencyrtus moderatus*

Natural enemies

Very few natural enemies have been recorded on cardamom thrips. *Chrysopa* sp. (Chrysopidae) and an unidentified anthocorid bug (Anthocoridae) have been observed to feed on larvae and adults in the field; an unidentified fungus has also been recorded on larvae in the field (Varadarasan 1995).

Cultural methods

Cultural methods such as shade regulation, removal of alternate host plants and removal of dried leaf sheaths thrice a year during early monsoon, mid-monsoon and late monsoon reduces the population of cardamom thrips in the field (Venugopal *et al.* 2006).

4.2.2 *Shoot and capsule borer (Conogethes punctiferalis Guen.)*

The shoot and capsule borer (*Conogethes punctiferalis* Guen.) (Lepidoptera, Pyralidae) is a serious pest of cardamom in the nursery and main field. The early stages of larvae bore into unopened leaf buds, panicles and immature capsules. The late stages of larvae bore into pseudostem and feed on the central inner portion resulting in drying of the terminal leaf and producing the characteristic 'dead heart' symptom of pest attack. Yield losses of 70%–80% have been recorded in severely infested plantations (Varadarasan 1995; Gopakumar & Chandrasekhar 2002).

Natural enemies

Twenty species of hymenopterous parasitoids have been recorded on the shoot and capsule borer (Varadarasan 1995). Among them, *Eriborus trochanteratus* (Morley), *Xanthopimpla australis* Kr., *Friona* sp. and *Agrypon* sp. (Ichneumonidae) have been identified as potential biocontrol agents. Parasitization by *E. trochanteratus* and *X. australis* ranged between 10.5%–11.1% at Appangala (Karnataka) (CPCRI 1985). *Friona* sp. and *Agrypon* sp. parasitized 8%–10% and 16%–18% of larvae/pupae, respectively, in the field at Idukki (Kerala) (Varadarasan 1995).

Cultural methods

Cultural methods such as removal of alternate host plants and freshly infested tillers and collection and destruction of adult moths that are generally found on the lower surface of cardamom leaves reduces the population of shoot and capsule borer in the field (Venugopal *et al.* 2006).

4.2.3 *Root grub (Basilepta fulvicorne Jacoby)*

The root grub (*Basilepta fulvicorne* Jacoby) (Coleoptera, Chrysomelidae) is a serious pest of cardamom, being distributed in all cardamom areas. The larvae feed on roots and rhizomes and in severe cases of infestation, the entire root system is eaten away. The infested plants turn yellow and become stunted; severely infested plants, especially seedlings, succumb to the pest attack. Seedlings damaged by root grubs are subsequently infected by secondary pathogens resulting in rotting. Crop losses of 29–66% have been reported in various cardamom areas (Varadarasan 1995; Gopakumar & Chandrasekhar 2002).

Natural enemies

The natural enemies recorded on the root grub include the green muscardine fungus *Metarrhizium anisopliae* (Metsch) Sorokin (Hyphomycetes) and the entomogeneous nematode *Heterorhabditis* sp. (Heterorhabditidae) infesting grubs, and the white muscardine fungus *Beauvaria bassiana* (Bals.) Vuill. (Hyphomycetes) infesting adults in the field. These pathogens are highly pathogenic to the host under laboratory conditions and also in the field indicating their potential as biocontrol agents (Varadarasan & Naidu 1993; Vardarasan *et al.* 1993; Vardarasan 1995; Varadarasan & Sivasubramonian 1996).

Cultural methods

Cultural methods such as shifting of nursery sites and collection and killing of adult beetles during peak periods of emergence during April-May and September-October reduces the population of the pest in the field. Removal of dried leaf sheaths during early and late monsoon periods adopted for the management of cardamom thrips also helps in prevention of egg laying by adults of root grub the field (Venugopal *et al.* 2006).

4.2.4 Cardamom aphid (Pentalonia nigronervosa f. caladii van der Goot)

Colonies of cardamom aphid (*Pentalonia nigronervosa* f. *caladii* van der Goot) (Homoptera, Aphididae) occur within leaf sheaths of pseudostem feeding on plant sap. Though the direct damage caused by them is not very serious, they are vectors of *katte* and *kokke kandu* diseases of cardamom.

Natural enemies

The natural enemies recorded on cardamom aphid include predatory chrysopids, coleopterans and dipterans and entomopathogenic fungi (Devasahayam 1996). Laboratory bioassays indicated that entomopathogens such as *B. bassiana* and *Pochonia chlamydosporia* (Hyphomycetes) were promising for the management of the pest (Mathew *et al.* 1997).

Plant products

Laboratory bioassays of plant extracts and commercial neem products indicated that seed extract of *Annona squamosa* L. (Annonaceae), leaf extracts of *Lawsonia inermis* L. (Lythraceae) and *Leucas aspera* Sprengel (Lamiaceae), rhizome extract of *Acorus calamus* L. (Acoraceae) and Bioneem and Margolin significantly affected the settling and colonization behaviour and multiplication of cardamom aphid. Trials laid out in the field also indicated that neem products such as Margolin and Bioneem (1.6% each) and extracts of *A. squamosa* and *A. calamus* (5% each) were promising for the management of the pest (Venugopal 1999).

Cultural methods

Removal of alternate hosts in the vicinity of plantations and removal of dried leaf sheaths thrice a year during early monsoon, mid-monsoon and late monsoon reduces the population of cardamom aphids in the field (Venugopal 1999).

4.3 Ginger and Turmeric

4.3.1 Shoot borer (Conogethes punctiferalis Guen.)

The shoot borer (*Conogethes punctiferalis* Guen.) (Lepidoptera, Pyralidae) is the most serious insect pest of ginger and turmeric being widely distributed in India. The larvae bore into pseudostems and feed on the central growing shoot resulting in yellowing and drying of infested shoots. The presence of a bore-hole in the pseudostem through which frass is extruded and the withered central shoot is a characteristic symptom of pest infestation. In ginger significant reduction in yield occurs when more than 50% of the pseudostems in a clump are infested by the pest (Koya *et al.* 1986, Devasahayam and Koya, 2005; 2007).

Natural enemies

Seventeen species of natural enemies including parasitoids and predators have been recorded on shoot borer in the field. At Peruvannamuzhi (Kerala) among the various natural enemies, *Hexamermis* sp. (Mermithidae) and *Apanteles taragammae* (Braconidae) were more common and were identified as a potential biocontrol agents of the pest (Devasahayam 1996). Commercial formulations of *Bacillus thuringiensis* such as Dipel 0.3% were effective in controlling the pest infestation on ginger and turmeric in the field when sprayed at 21 day intervals during July – October (Devasahayam 2000b; 2002).

4.3.2 Rhizome scale (Aspidiella hartii Ckll.)

The rhizome scale (*Aspidiella hartii* Ckll.) (Homoptera, Diaspididae) infests rhizomes of ginger and turmeric both in the field (especially during the later stages of the crop) and in storage. The pest infestation is generally seen in the form of encrustation on the rhizomes during the later stages of the crop and severely infested plants wither and dry. In storage, the pest infestation results in shriveling of buds and rhizomes affecting their sprouting (Devasahayam & Koya 2005; 2007).

Natural enemies

A few hymenopterous parasitoids and general predators have been recorded as natural enemies of rhizome scale among which *Physcus comperei* Hayat (Aphelenidae) and *Adelencyrtus moderatus* Howard (Encyrtidae) were more common and identified as a potential biocontrol agents of the pest (Devasahayam 1996).

Plant products

Discarding of severely infested rhizomes and storage in dried leaves of *Strychnos nux-vomica* L. (Loganiaceae) and saw dust in 1:1 proportion was effective in reducing the damage caused by rhizome scale to ginger during storage (IISR 2003).

5.0 STRATEGIES FOR THE FUTURE

There is tremendous potential for inclusion of biological control as a component in integrated pest management schedules against many major insect pests of spice crops. A number of natural enemies have been recorded on most of the major insect pests of these crops and potential biocontrol agents have also been identified (Tables 1 & 2). Techniques are also available for the mass culture of most of these natural enemies. However, information on their efficacy in the field and strategies for their release are not available for many of the species.

The perennial nature / long duration of most of the spice crops and the availability of other crops / plant species along with various insect species in the ecosystem on which the natural enemies can multiply and survive are ideal factors for the use of biocontrol strategies for the management of insect pests of spice crops. Improvement of habitat and alternate food sources by cultural manipulations and selective use of safer insecticides and plant products for the control of insect pests would help in conservation of natural enemies.

The ecosystem in which spice crops are grown is a dynamic system in which many other crops, plants and indigenous trees that are used as shade and support trees are also grown. The role of plant volatiles of various crops and plant species and kairomones of insect pests and their effect on the behaviour of natural enemies in relation to their searching and feeding potential needs to be investigated. A wide variation in the susceptibility of black pepper, cardamom, ginger and turmeric cultivars to major insect pests has also been observed. The tritrophic relationships between the resistant crop plants, insect pests and their natural enemies also needs to be investigated for increasing the effectiveness of natural enemies.

Systematic efforts have not been made to document the microbial pathogens of insect pests of spice crops in the field. The humid conditions under which many of the spice crops are grown is an ideal factor for the use of microbial pathogens in the management of insect pests of spices. Information on bioecological aspects of insect-microbe interactions especially variations in epizootic potential of various strains and sporulation and dispersal of the pathogen in the ecosystem are to be generated. The feasibility of multiplication of these pathogens in various cheap substrates including agricultural wastes such as coir pith, coffee husk and tea wastes that are commonly available in plantation crop ecosystems need to be studied.

REFERENCES

Babu, P. B. S., Padmaja, V. and Hishan, A. (1996) Isodesacetyluvaricin-insect antifeedant against *Longitarsus nigripennis* Mots. *Indian J. Exptl. Biol.* 34: 377–379.

(CPCRI) Central Plantation Crops Research Institute (1985) *Annual report* for 1983. Kasaragod: Central Plantation Crops Research Institute.

DeBach, P. (ed.) (1964) *Biological control of insect pests and weeds*. London: Chapman and Hall.

Devasahayam, S. (1992) *Bioecology and control of leaf gall thrips* (*Liothrips karnyi* Bagnall) *infesting black pepper*. Final report of research project. Calicut: National Research Centre for Spices.

Devasahayam, S. (1996) *Biological control of insect pests of spices*. In: Anandaraj, M. and Peter, K. V. (eds.) Biological control in spices. Calicut: Indian Institute of Spices Research. pp. 33-45.

Devasahayam, S. (1998) *Biological control of scale insects infesting black pepper*. Final report of ad-hoc research scheme. Calicut: Indian Institute of Spices Research.

Devasahayam, S. (2000a) *Insect pests of black pepper*. In: Ravindran, P. N. (ed.) Black pepper *Piper nigrum*. Amsterdam: Harwood Academic Publishers. pp. 309-334.

Devasahayam, S. (2000b) Evaluation of biopesticides for the management of shoot borer (*Conogethes punctiferalis* Guen.) on ginger (*Zingiber officinale* Rosc.). In: Ramana, K. V., Eapen, S. J., Babu, K. N., Krishnamurthy, K. S. and Kumar, A. (eds.) *Spices and aromatic plants, challenges in the new millennium*. Calicut: Indian Society for Spices, pp. 276–277.

Devasahayam, S. (2002) Evaluation of biopesticides for the management of shoot borer (*Conogethes punctiferalis* Guen.) on turmeric (*Curcuma longa* L.). In: Rethinam, P., Khan, H. H., Reddy, V. M., Mandal, P. K. and Suresh, K. (eds.). *Plantation crops research and development in the new millennium*. Kochi; Coconut Development Board. pp. 489–490.

Devasahayam, S. and Anandaraj, M. (1997) Black pepper (*Piper nigrum* L.) research. In : Narwal, S. S., Tauro, P. and Bisla, S. S. (eds.). *Neem in sustainable agriculture.* Jodhpur: Scientific Publishers, pp. 117-122.

Devasahayam, S. and Koya, K. M. A. (1993a) Effect of top shoot borer (Cydia hemidoxa) infestation on young vines of black pepper (*Piper nigrum*). *Indian J. Agric. Sci.* 63: 762–763.

Devasahayam, S. and Koya, K. M. A. (1993b) Seasonal incidence of hymenopterous parasites of top shoot borer *Cydia hemidoxa* (Lepidoptrea: Tortricidae) on black pepper. *J. Entomol. Res.* 17: 205-208.

Devasahayam, S. and Koya, K. M. A. (1994a) Natural enemies of major insect pests of black pepper (*Piper nigrum* L.) in India. *J. Spices Aromatic Crops* 3: 50–55.

Devasahayam, S. and Koya, K. M. A. (1994b) Seasonal incidence of *Hexamermis* sp. (Dor., Mermithidae) parasitising larvae of top shoot borer *Cydia hemidoxa* Meyr. (Lep., Tortricidae) on black pepper. *J. Appl. Ent.* 117: 31-34.

Devasahayam, S. and Koya, K. M. A. (1999) Integrated management of insect pests of spices. *Indian J. Arecanut Spices Med. Plants* 1: 19–23.

Devasahayam, S. and Koya, K. M. A. (2005) Insect pests of ginger. In : Ravindran, P. N. and Babu, K. N. (eds.) *Ginger. The genus Zingiber.* Washington: CRC Press. pp. 367–389.

Devasahayam, S. and Koya, K. M. A. (2007) Insect pests of turmeric. In : Ravindran, P. N., Babu, K. N. and Sivaraman, K. (eds.) *Turmeric. The genus Curcuma* . Washington: CRC Press.

Devasahayam, S. and Leela, N. K. (1997) Evaluation of plant products for antifeedant activity against pollu beetle (*Longitarsus nigripennis* Motschulsky), a major pest of black pepper. In: Reddy, P. P., Kumar, N. K. K. and Verghese, A. (eds.) *Advances in IPM for horticultural crops: environmental implications and thrusts.* Bangalore: Association for Advancement of Pest Management in Horticultural Ecosystems. pp. 172–174.

Gopakumar, B. and Chandrasekhar, R. (2002) Insect pests of cardamom. In: Ravindran P N and Madhusoodanan, K. J. (eds.) *Cardamom. The genus Elettaria.* London: Taylor and Francis pp. 180–206.

(IISR) Indian Institute of Spices Research (1996) *Annual report* for 1995–96. Calicut: Indian Institute of Spices Research.

(IISR) Indian Institute of Spices Research (2003) *Annual report* for 1902–03. Calicut: Indian Institute of Spices Research.

Koya, K. M. A., Devasahayam, S., Selvakumaran, S. and Kallil, M. (1996) Distribution and damage caused by scale insects and mealybugs associated with black pepper (*Piper nigrum* Linnaeus) in *India. J. Entomol. Res.* 20: 129–136.

Koya, K. M. A., Balakrishnan, R., Devasahayam, S. and Banerjee, S. K. (1986) A sequential sampling strategy for the control of shoot borer (*Dichocrocis punctiferalis* Guen.) in ginger (*Zingiber officinale* Rosc.) in India. *Trop. Pest. Manage.* 32: 343–346.

Mathew, M. J., Saju, K. A. and Venugopal, M. N. (1997) Efficacy of entomogenous fungi on biological suppression of *Pentalonia nigronervosa f. caladii* van der Goot of cardamom (*Elettaria cardamomum* Maton). *J. Spices Aromatic Crops* 7: 43–46.

Pickett, C.H. & Bugg, R.L. (1998) *Enhancing biological control*: Habitat management to promote natural enemies of agricultural pests. Berkeley: University of California Press.

Premkumar, K. and Nair, M. R. G. K. (1988) Effect of some planting conditions on infestation of black pepper by *Longitarsus nigripennis* Mots. *Indian Cocoa Arecanut Spices J.* 10: 83–84.

Selvakumaran, S., Kallil, M. and Devasahayam, S. (1996) Natural enemies of two major species of scale insects infesting black pepper (*Piper nigrum* L.) in India. *Pest Manage. Hort. Ecosyst.* 2: 79–83.

Selvan, M. T., Gopakumar, B., Chandrasekhar, S. S., Singh, J., Kumaresan, D. and Varadarasan, S. (1996) Survey on the occurrence of coccids infesting black pepper in higher altitudes of Idukki District, Kerala and adaptive trial for their management. *Pestology* 20 (1): 19–23.

van den Bosch, R. and Telford, A. D. (1973) Environmental modification and biological control. In: *Biological control of insect pests and weeds.* DeBach, P and Schlinger, E.I. (eds.) London : Chapman and Hall Ltd.1973.

Varadarasan, S. (1995) Biological control of insect pests of cardamom. In: Ananthakrishnan, T. N. (ed.). *Biological control of social, forest and plantation crops insects.* New Delhi: Oxford and IBH Publishing Company Private Limited. pp. 109–119.

Varadarasan, S. and Naidu, R. (1993) *Studies on root grubs.* Consolidated report on research project financed by NABARD (1989–93). Myladumpara: Indian Cardamom Research Institute.

Varadarasan, S. and Sivasubramonian, T. (1996) Studies on natural occurrence of bioagents and their bioassay on cardamom root grub, *Basilepta fulvicorne* Jacoby (Coleoptera: Chrysomelidae). *J. Plantn. Crops* 24 (Suppl.): 286–290.

Varadarasan, S., Sivasubramonian, T., Manimegalai, R. and Naidu, R. (1993) Integrated management of cardamom root grub *Basilepta fulvicorne* Jacoby (Eumolpinae: Chrysomelidae: Coleoptera). *J. Plantn. Crops* 21 (Suppl.): 191–194.

Venugopal, M. N. (1999) *Characterisation, early detection and management of 'kokke kandu' disease of cardamom*. Final report of ad-hoc research scheme. Calicut: Indian Institute of Spices Research.

Venugopal, M. N., Ankegowda, S. J. and Prasath, D. (2006) Cardamom. In : Parthasarathy, V. A. and Rajeev, P. (eds.) *Major spices-production and processing*. Calicut: Indian Institute of Spices Research. pp. 62–114.

5

Biological Control of Fungal Diseases

R. Suseela Bhai

1.0 INTRODUCTION

"Biological control is the reduction of the amount of inoculum or disease producing activity of a pathogen or parasite in its active or dormant state, by one or more organisms, accomplished naturally or through manipulation of the environment, host or antagonist or by mass introduction of one or more antagonists " (Baker & Cook 1974). The biological world is a vast interacting network of living organisms in a state of dynamic equilibrium, reflecting changes in their physical environment and their relations to each other. Organisms exhibit many type of interactions in the process of achieving biological balance. An organism may exert no effect on another or it may stimulate or inhibit the growth or development of the associate, it may stimulate or inhibit the formation of resting spores of the associate or it may enforce dormancy or cause lysis of the associate (Cooke & Baker, 1983).

2.0 BIOLOGICAL CONTROL OF FUNGAL PATHOGENS

Biocontrol of plant pathogens by mycoparasites was borne by the work of Weindling (1932) when he discovered that *Trichoderma lignorum* could parasitize a number of soil borne fungi in culture and he suggested controlling certain pathogenic fungi by augmenting the soil with large quantities of mycoparasites. Biological control is promising in an environment and economic sense as it offers long lasting, safe and cost effective alternatives to soil applied chemicals. In depth knowledge of the biology of the pathogen, impact of inoculum levels and the distribution of inoculum on disease is an important prerequisite for biological control because it is used to decide when and where to apply candidate treatments and also to determine population thresholds where damage caused by the pathogen is significant in terms of crop loss (Merriman & Russell, 1990).

Controlling fungal pathogens in a crop is a cumbersome process. Cultural techniques like adequate drainage, sanitation and clean stock have been used successfully to reduce the disease incidence. The greatest potential for the development of control may be in the integration of cultural treatments that create an environment that is physically, chemically and microbiologically unfavorable to the pathogen and that increases host resistance (Shea & Broadbent, 1983).

There are a number of fungi that parasitizes plant pathogens (Barnett & Bender, 1973, Boosalis & Mankau, 1965). *Trichoderma* and *Pseudomonas fluorescens* represent two common biocontrol agents used world wide for the control of many diseases including soil born and phylloplane diseases. Others mainly reported are *Bacillus* sp, *Pseudomonas* sp. *Aspergillus, Coniothyrium minitans, Laetesaria arvalis, Pythium nunn, Talaromyces flavus* and *Sporidesmium sclerotivorum etc.* (Adams 1990).

The first successful application of *Trichoderma* to control *P. cactorum* on apple was reported by Valdebenito-Sanhueza (1987). Gees and Coffey (1989) have reported promising activity of *Myrothecium roridum* for the control of *P. cinnamomi* or *Persea indica* in green house trails using natural or artificial inoculum. Inoculation of established trees of *Citrus clementina* with agar disc of *Myrothecium roridum* controlled previously existing infections (Tuset *et al.* 1990). *Penicillium funiculosum* and *Pythium nunn* controlled *Phytophthora* root rots of azalea and orange *in vitro* (Fang & Tsao 1995 a and b). Kim C H (1989 a and b) reported biological control activity of antagonists such as *Pseudomonas cepacia* and *T. harzianum* in reducing *Phytophthora* blight of red pepper caused by *P. capsici*. The above are some of the examples where biological control is applied successfully to reduce the disease incidence caused by soil borne fungi.

Many microbial biological control agents (BCA) that are effective against plant diseases are available commercially. These BCA's target plant pathogenic fungi such as *Botrytis cinerea, Fusarium oxysporum, Pythium* spp., *Phytophthora* spp. and powdery mildews. These pathogens cause significant losses in crop yield and quality if not managed properly. Among the Rhizobacteria, Pseudomonas *fluorescens* Migula and *Bacillus* sp. are effective candidates for biocontrol owing to their rhizosphere competence (Ahmad & Baker, 1987). These organisms are reported to be very useful as bio control agents against several soil borne pathogens such as *Sclerotium sclerotiorum, Fusarium oxysporum* Schl. *Rhizoctonia solani* Kuhn., *Macrophomina phaseolina* Tassi, *Pythium vexans* de Bary and *Ralstonia solanacearum* Yabuuchi etc.(Rama Singh *et al.* 2003).

Disease management in organic farming

Disease management in organic farming basically starts from the selection of disease free planting materials. Prior to the initial planting or raising the nursery, the soil sterilization is to be done to kill the soil borne pathogens. The soil sterilization can be done by exploiting solar energy (soil solarization). Further the soil may be fortified with proven microbial agents to prevent the access of infectious pathogens. Enriching the soil with potential biocontrol agents lead to increased competition and antagonism in the rhizosphere. The presence of beneficial root colonizing bacteria such as *Pseudomonas fluorescence, Bacillus subtilis* etc. were identified as major contributors in the control of root diseases besides vesicular Arbuscular Mycorrhizae.

Direct disease control measures are only the last resort of defence. The use biological controls of fungal disease are possible within the organic system.

All the above mentioned antagonists / biocontrol agents can be used in the organic system without any restriction; however the number of products available is limited due to constraints of registration of such products and lack of economic mass production.

2.1 Suppressive soil

Suppressive soil is yet another indirect form of biological control. Here the disease development is suppressed even though the pathogen is available in the presence of a susceptible host. Factors involved in suppressive soils are fungistasis, competitive saprophytic ability and other disease pathogen interactions. Suppressive soils are also called antagonistic, biologically buffered, competitive declining, fungistatic, long living, low *Pythium* and resistant which is in contrast to conducive, receptive, sterile, non-competitive, short life, high *Pythium* and susceptible soils. But mere absence of a disease is not a justification in conclude that a suppressive mechanism is involved even if a susceptible host is present. (Schneider, 1982; Cook & Baker, 1983).

Two types of disease suppressive soils were identified depending on whether micro environmental or microbial factors are responsible for reduced diseases. The disease suppressiveness of a particular soil has frequently been recognized after cultural or cropping modifications. Heat, chemical or irradiation sterilization of soil, crop sequence, tillage etc, are some of the factors where the critical association of biological activity or microbial balance with disease suppression in the soil exists.

Some of the mechanisms involved in soil suppressiveness are ammensalism, parasitism, predation competition and all of them may operate together or individually to reduce inoculum. Suppressiveness can be induced by practices such as monoculture, transfer of suppressive soil to conducive soil and by adding supplemental antagonists. Effect of organic and inorganic amendments also influences the suppressiveness. Addition of fertilizers and micronutrients also may affect suppressiveness either directly or indirectly by alteration in the soil microflora. The evidence for biological factors in suppression is that the suppressive factors can be transferred to conducive soils.

Management of suppressive soil involves the incorporation of large quantities of organic material to maintain the level of organic matter. The organic material include cow dung, chicken manure, cereal straw, crop residue and other readily available substance etc. The soil managed by these systems is known as Ashburner system (Schneider, 1982).

The potential for disease control by natural antagonism probably exists in many soils. Generally suppressiveness is related to the total microbiological activity of the soil. Management of the microbiota is accomplished by cultural practices that increase the total microbiological activity or maintain the microbiological balance in favor of the antagonists. The suppressive agents can be isolated, mass cultured and introduced into the soil for biological control.

Soil suppressive to some of the most important diseases caused by soil borne plant pathogens have been described from different parts of the world (Schneider 1982; Cook and Baker 1983). The most known are the soil suppressive to *Fusarium* wilts. These soils limit the incidence or severity of wilts in banana, carnation, cucumber, cotton, flax, muskmelon and tomato. This suppressiveness is described as specific to *Fusarium* wilts. It is expressed against all formae specialis of *F. oxysporum*, but is ineffective against other soil borne pathogens and non-vascular *Fusarium* species. Many other groups of microorganisms such as actinomycetes, *Arthrobacter*, *Clostridium*, *Bacillus* and *Trichoderma* have also been proposed to be involved in suppression of *Fusarium* wilts. But experimental evidence indicated that fluorescent pseudomonades are involved in the suppression of *Fusarium* wilts in these suppressive soils (Alabouvette 1990).

Phytophthora species has been considered to possess little competitive saprophytic ability (Garrett 1950). Hence promotion of an antagonistic microflora was the classic approach to biocontrol diseases caused by *Phytophthora* species. Organic matter can affect the vigor of the host by improving the structure, moisture and nutrient holding capacity of the soil, but its principle effect is to provide a more complex antagonistic microflora and fauna (Broadbent & Baker 1975). High levels of organic matter in soil have been associated with some soils that are suppressive to *P. cinnamomi* (Broadbent & Baker 1974 a).

The efficacy of fluorescent pseudomonas against *Pythium* sp. have been proved by many workers (Ongena *et al.* 1999; Vogt & Buchenauer, 1997) and similar results have been obtained in rice (Sulochana *et al.* 2003), wheat (Mukerjee & Rai, 2000) and chillies (Srinivasa & Krishnaraj, 1992). In chillies, simultaneous inoculation resulted in increased shoot dry mass, fruit yield; shoot phosphorus content and also micronutrient content.

Use of bioprotective and Plant Growth Promoting Rhizobacteria (PGPR) and Vesicular Arbuscular Mycorrhizae (VAM) are some among the effective methods which can be exploited for disease control. Rhizobacteria are ideal for use as biocontrol agents as they can provide the frontline defence for plant root against the attack by various pathogens (Pal & Jalali, 1998).

2.2 Mycoparasites

Mycoparasites can also be exploited for biological control of fungal pathogens. They are fungi which parasites other fungi especially pathogens. They can be easily isolated from any field inoculum using a baiting technique called the *precolonised plate method.* They can be used against soil-borne pathogens and also against above-ground pathogens (Foley & Deacon, 1985; Krauss *et al.* 1998).

2.2.1 Procedure for isolation of Mycoparasites

First, agar plates are uniformly precolonised by the pathogen. Then field inoculum is collected and in the laboratory, field inoculum is shredded into small pieces to increase the surface area. These pieces are allowed to air-dry overnight to avoid bacterial contamination. Then, they are carefully placed on the surface of the pathogen mycelium. There should be good contact to the mycelium but none to the underlying agar. If the plates are not completely precolonised, it is necessary to excise noncolonised parts of the agar. Very few fungi have the ability to invade precolonised substrate. The ones that do are either mycoparasites or exceptionally strong competitors and have potential for biocontrol. They are easily isolated onto fresh agar plates because they suppress the pathogen host. Pure cultures can usually be obtained with the second subculture without a need for antibiotics in the medium.

2.3 Mechanisms of Biocontrol

Any disease control in which antagonists are involved is biological control. It is the use of natural or modified organisms, genes, or gene products to reduce the effect of detrimental organisms. Biocontrol agents are also called as antagonists and antagonism is the generalized mechanism used to reduce the survival or disease causing activities of the pathogen. The term antagonism includes a number of mechanisms that result in the suppression of pathogenic organisms in the soil rhizosphere or infection areas. The mechanisms are ammensalism (antibiosis and lysis), competition, parasitism and predation all of which may act together or independently to reduce inoculum (Baker & Cook, 1974; Malajczuk, 1983)

i) *Ammensalism* includes antibiosis and lysis. It is the production of antibiotics or toxic metabolic by-products either volatile or non-volatile that cause disintegration of living cells. Antagonists can produce powerful growth inhibitory compounds called antibiotics that are detrimental to a wide array of microorganisms. They may penetrate a cell and inhibit its activity by chemical toxicity. The amount of antibiotic produced by an antagonist depends on its population density and also on environmental conditions. Two factors limit the utilization of antibiotics. First the antibiotics must continuously be present on the surface. Antibiotics are not generally stable on field conditions for long periods or may be bound on plant surfaces. Hence antagonist populations must continuously be present in sufficiently high densities to produce antibiotics. Second, the pathogen may build up resistance against the antibiotics. Lysis is the destruction, disintegration, dissolution or decomposition of biological materials.

ii) *Competition* is the injurious effect of one organism on another because of the utilization or removal of some resources of the environment (Clark, 1965). It is the struggle for nutrients such as high energy carbohydrates and nitrogen, certain growth factors, oxygen and space. Two or more organisms trying to utilize the same food or mineral source, or occupy the same niche or infection site. The successful competitor excludes the others by way of faster growth or reproductive rate.

iii) *Parasitism and Predation:* involve active contact among microorganisms resulting in either the hyphal degradation or the mycophagy of the whole propagules (Alexander, 1971). Predation is the feeding of one organism on another organism.

iv) *Mycoparasitism:* is the fungus- fungus relationship. Fungal cell wall degrading enzymes such as chitinases and b- glucanases are commonly produced by hyperparasites. Parasitism depends on the close contact between antagonist and the host. It is often exploited for biocontrol of biotrophic pathogens. Mycelial development of necrotrophic pathogens can also be controlled by mycoparasites. Four stages can be identified in mycoparasitism namely 1) Chemotrophic growth 2) Recognition of the pathogen 3) Attachment and 4) Degradation of host wall. The degradation of host wall by production of lytic enzymes such as chitinases and glucan 1, 3 glucosidases. Other mechanisms of biocontrol include i) Hypovirulence ii) Induced Systemic Resistance and iii) Increased Growth Response.

3.0 IMPORTANT BIOCONTROL AGENTS

Principal fungi used as biological control agents against soil borne diseases include species of *Gliocladium* and *Trichoderma*. They are used in potting mixtures or used as solid matrix in seed priming treatments. They are effective against damping off diseases of ornamentals and vegetables caused by species of *Pythium, Phytophthora, Botrytis, Sclerotium* etc. Other fungi that can be used as biocontrol agents include *Sporidesmium sclerotivorum, Coniothyrum minitans, Talaromyces flavus and* species *Paecilomyces sp, Penicillium, Aspergillus, Verticillium* etc. Non-pathogenic strains of *Fusarium* were also used as biocontrol agents against *Fusarium oxysporum* causing vascular wilts. Some of these strains have been isolated from the vascular tissue of the host plants that remained healthy while nearby plants had been killed by the wilt inducing strain of the fungus (Agrios, 2005). Species of *Pythium* such as *P. nunn* and *P. oligndrum were* also used as biocontrol agents against pathogenic *Pythium*. Bacterial antagonists include species *Agrobacterium, Bacillus* and fluorescent pseudomonads. They are being used for the biological control of many crop diseases including horticultural crops as well as plantation and spice crops (Table 1 & Table 2).

3.1 Advantages of Pseudomonas fluorescens as bacterial antagonists

The major group of bacterial antagonists used in biological control of plant pathogens is fluorescent pseudomonas species. They produce several types of metabolites having antifungal activities. The lipopolysaccharides present in their outer membrane induces systemic resistance. They produce salicylic acid at low Fe availability which in turn is also involved in the induction of resistance. Species of fluorescent pseudomonas also produce siderophores called pseudobactines or pyoverdin which are very efficient competitors for iron and that com1petition for iron is one of the mechanisms responsible for soil suppressiveness. Reducing iron availability by addition of ethylene diamine di-o-hydroxy phenyl acetic acid (EDDHA) resulted in an increased level of soil suppressiveness and on the contrary, soil suppressive ness was decreased by addition of Fe-ethylene diamine tetra acetic acid (Fe EDTA) providing iron available for the pathogenic *Fusarium* sp. The pyoverdin produced by the fluorescent pseudomonas species was able to reduce the rate of chlamydospore germination and growth of germ tubes arising from chlamydospores in *Fusarium* sp.

3.2 Advantages of Trichoderma as fungal antagonist

Trichoderma and *Gliocladium* species have many advantages as biocontrol agents. They have fast growth and have an array of inducible polysaccharides - degrading enzymes that are very good sources of various toxic metabolites. *Trichoderma* isolates possess three antagonistic mechanisms *viz*.: competition, antibiosis and mycoparasitism. Antibiosis occurs when the production of toxic metabolites or antibiotics produced by *Trichoderma* has a direct effect on the pathogen. Even though many antagonist are able to produce antibiotics or toxins in pure culture there is little proof of a major effect of such compounds in biocontrol under field condition with *Trichoderma*. Similarly in *Trichoderma* also, all the four mechanisms could be distinguished.

This fungi can be propagated on a wide variety of carbon sources, so it is easy to find a reasonably cheap substrate for biomass production of the biocontrol agent. They proliferate abundantly in various natural soils when added as young mycelia in intimate contact with food base, proliferation and further establishment depend on inoculum age and how it is added in relation to the food base (Papavizas, 1985). Another advantage is the wide range of environmental conditions tolerated by *Trichoderma* sp. and isolates by which it is possible to select isolates suited for most of the environmental conditions under which the plant pathogens can cause diseases. For e.g. some isolates can grow at low temperature others at high temperature. Isolates also vary in their tolerance to various chemicals and many show remarkable resistance to fungicides, either inherently or through mutation or adaptation. It is therefore possible to select fungicide tolerant or resistant biocontrol agents for use in integrated control. This great adaptability among *Trichoderma* isolates provide us with a large arsenal from which to choose in the design of biocontrol programme.

3.3 Vesicular Arbuscular Mycorrhizae in biological control

The vesicular Arbuscular Mycorrhizae inherently associated with roots of certain plants are also considered as another group of biocontrol agents. The roots of most plants are colonized by a group of soil borne fungi which

Table 1. List of antagonist used against major soil borne diseases

Pathogen	Crop/disease	Antagonists	Reference
P. boehmeriae	Cotton	P.olegandrum, *Trichoderma* sp.	Ma & Shen, 1993
P. drechsleri	Cucumber	*Trichoderma* sp.	Xue *et al.*1994
P. capsici	Black pepper, Cucumber, Red pepper	*Aeromonas hydrophilla, serratia marscence Chromobacterium violaceum, Pseudomonas cepacia, T. harzianum Enterobacter agglomerans*	Anandaraj & S J Eapen 2003 Yeom *et al.*1995 Park *et al.*995 Kim 1989b Park & Kim 1989
P. cinnamomi	Avocado, Pine apple	*T. harzianum, T. hamatum, Paecilomyces lilacinus, A.candidus, Bacetoformans, B. megaterium, Myrothecium roridum, Micromonospora , streptomyces violaceus, chaetomium globosum,* VAM	You *et al.*1996 Mc Leod *et al.*1995, Duvenhage & Kotze, 1993 Guillemin *et al.*1994
P. cactorum	Apple, Gerbera	*T. virens, T. Viride, B. subtilis, E. aerogenes, T. harzianum*	Orlikowski, 1995, Duskova, 1992, Pasini *et al.*1991 Roiger & Jeffers 1991,
P. fragariae	Raspberry	Glucanolytic actinomycetes	Valois *et al.*1996
P. nicotianae var. *parasitica*	Coorg mandarin, Tomato, *Casuarinia equisitifolia*	*T. harzianum , P. fluorescens, P. putida , Penicillium funiculosum, G.mossae,*	Trotta *et. al., 1996* Sawant *et al.*1995 Yang *et al.,1994* *Pena et al.1994*
P. meadii	Cardamom	*T.harzianum , T. Viride, Latesaria arvalis, Pseudomonas fluorescens*	Suseela Bhai &Y. R Sarma 2003 Suseela Bhai *et al.*1993
P. sojae	Alfalfa	*B. cereus* UW 85	Handelsman *et al.*1990
Rhizoctonia solani	Black-scurf of Potato Root rot of Chick pea Root rot of Cotton	*Verticillium biguttatum* *T.harzianum , T. Viride* *Pseudomonas fluorescens*	Van den Boogert *et. al.,* 1990 Arora , 1999 Prasad & Rangeswaran, 2001 Laha & Verma 1998

Contd...

Table 1. Contd...

F. udum *F. oxysporu m f.sp* *F. oxysporum f.spricini* *F. oxysporum f.sp cubense*	Pigeon pea wilt Castor wilt Panama wilt of Banana	*T.harzianum , T. Viride,T. koningii* *Pseudomonas fluorescens*	Sumitha & Gaikwad, 1995 Raoof & HariNarayanan 2001 Raguchander *et al.*2001
Thielaviopsis basicola	Black root rot of Potato	*Pseudomonas fluorescens*	Defago *et al.*1990
Sclerotium rolfsii	Collar rot in Chick pea	*T.harzianum , T. Viride*	Prasad & Rangeswaran, 2001
Pythium aphanidermatum	Rhizome rot of ginger	*T.harzianum , T. Virens*	Usman *et al.*1996
Gliocladium roseum,	Guava wilt	*Penicillium citrinum, Aspergillus niger(AN17) Trichoderma sp.*	Om Prakash *et al.*2002
Colletotrichum falcatum	Red rot of sugar cane	*Pseudomonas fluorescensVPT4 ARRIGand EPI and P. putida KKMI*	Viswanathan & Samiyappan 2000

Source: Suseela Bhai (1998)

Table 2. List of biocontrol products used against soil borne diseases

S.No	Bioagents	Trade name/formulation	Diseases/pathogens controlled	Crops	Manufacturers	Application
1	*Agrobacterium radiobacter* Strain 84	**Galltrol** (Pure culture grown on agar)	*Agrobacterium tumefaciens*	Fruit, nut, and ornamental nursery stock	AgBioChem, Inc., 3 Fleetwood Ct., Orinda, CA 94563, USA	Bacterial mass from one plate transferred to one gallon of non-chlorinated water; suspension applied to seeds, seedlings, cuttings, roots, stems, and as soil drench
2	*Agrobacterium radiobacter* K1026	**Nogall** (Pure culture in pre-sterilized, neutralized, finely divided peat)	*Agrobacterium tumefaciens*	Fruit and nut trees, caneberries, roses, and other ornamental nursery stock	Bio-Care Technology Pty. Ltd., Somersby, Australia/New BioProducts, Inc., 4272 Pintail Place, Corvallis, OR 97330	Culture diluted in nonchlorinated water then used as for seed, root, or cutting dips or sprays
3	*Ampelomyces quisqualis* isolate M-10	**AQ10 Biofungicide** (water-dispersible granule)	Powdery mildew	Apples, cucurbits, grapes, ornamentals, strawberries, and tomatoes	Ecogen, Inc., 2005 Cabot Blvd. West, Langhorne,	Spray
4	*Bacillus licheniformis* SB3086	**EcoGuard** liquid spore concentrate	Dollar spot, low and moderate disease pressure	Many	Novozymes Biologicals, Inc., 5400 Corporate Circle, Salem, VA	Standard spray equipment
5	*Bacillus subtilis*GB03, *other B. subtilis, B. lichenformis, B. megaterium*	**Companion** (Liquid)	*Rhizoctonia, Pythium, Fusarium,* and *Phytophthora*	Greenhouse and nursery	Growth Products, PO Box 1259, Westmoreland Ave., White Plains, NY 10602 USA;	Drench at time of seeding and transplanting or as a spray for turf
6	*Bacillus pumillus* GB34	**YiedShield** (dry powder)	Soil borne fungal pathogens causing root diseases.	Soybean.	Gustafson, Inc., 1400 Preston Rd, Plano TX 75093 USA;	Added to a slurry mix for seed treatment; hopper box treatment
7	*Bacillus subtilis* GB03	**Kodiak** (several formulations - dry powder; usually applied with chemical fungicides)	*Rhizoctonia solani, Fusarium* spp., *Alternaria* spp., and *Aspergillus* spp. that attack roots	Cotton, legumes	Gustafson, Inc., 1400 Preston Rd, Plano TX 75093 USA	Added to a slurry mix for seed treatment; hopper box treatment
8	*Bacillus subtilis* MBI600 (rhizobia also in formulation)	**HiStick N/T**	*Fusarium* spp., *Rhizoctonia* spp, *Aspergillus*	Soybean, alfalfa, dry/snap beans, peanuts	Becker Underwood, 801 Dayton Ave., PO BOX 667, Ames, IA 50010	Slurry, damp and dry inoculation of seed

Contd...

Table 2. Contd...

9	*Bacillus subtilis* FZB24	**Rhizo-Plus** (Water dispersible granule)	*Rhizoctonia solani*, *Fusarium* spp., *Alternaria* spp., *Sclerotinia*, *Verticillium*, *Streptomyces scabies*	field (potatoes, corn), vegetables, and ornamental plants	KFZB Biotechnik GmbH, Glienicker Weg 185, D-12489 Berlin, Germany	As suspension for seed treatment, soil drench, dip, and addition to nutrient solutions
10	*Bacillus subtilis* MBI600	**Sutilex** (Powder)	*Fusarium* spp., *Rhizoctonia* spp., and *Pythium* spp. that cause seed and root rots	Field, ornamental, and vegetable crops	Becker Underwood, 801 Dayton Ave., PO BOX 667, Ames, IA 50010	Drench, spray or through irrigation system
11	*Bacillus subtilis* QWT713	**Serenade** (Wettable powder)	Powdery mildew, downy mildew, Cercospora leaf spot, early blight, late blight, brown rot, fire blight, and others	Cucurbits, grapes, hops, vegetables, peanuts, pome fruits, stone fruits, and others	AgraQuest, Inc., 1530 Drew Avenue, Davis, CA 95616 USA	Spray
12	*Burkholderia cepacia* type Wisconsin	**Deny** (Peat-based dried biomass from solid fermentation; aqueous suspension)	*Rhizoctonia, Pythium, Fusarium*, and disease caused by lesion, spiral, lance, and sting nematodes	Alfalfa, barley, beans, clover, cotton, peas, grain sorghum, vegetable crops, and wheat	Stine Microbial Products, 6613 Haskins, Shawnee, KS 66216;	Applied to seeds with a sticking agent in planter box (aqueous suspension formulation is for use in drip irrigation or as a seedling drench)
13	*Burkholderia cepacia*	**Intercept**	*Rhizoctonia solani, Fusarium* spp., *Pythium* spp.	Maize, vegetables, cotton	Soil Technologies Corp., 2102 185th St., Fairfield, IA 52556, USA	
14	*Coniothyrium minitans*	**Contans WG/ Intercept WG** (water dispersible granule)	*Sclerotinia sclerotiorum* and *S. minor*	Many	Prophyta Biologischer Pflanzenschutz GmbH, Inselstrasse 12, D-23999 Malchow/Poel, Germany	Spray
15	*Coniothyrium minitans*	**KONI** (granule)	*Sclerotinia sclerotiorum* and *S. minor*	Cucumber, lettuce, capsicum, tomato, and ornamental flowers in greenhouse production	BIOVED, Ltd., Ady Endre u. 10, 2310 Szigetszentmiklos, Hungary	Granules incorporated into soil or soilless mix
16	*Candida oleophila* I-182	**Aspire** (wettable powder)	*Botrytis* spp., *Penicillium* spp.	Citrus, pome fruit	Ecogen, Inc	Post harvest application to fruit as drench, drip or spray

Contd...

Table 2. Contd...

17	*Erwinia amylovora* HrpN harpin protein	**Messenger** (powder)	Many.	Field, ornamental, and vegetable crops	Bioscience Corporation,11816 North Creek Parkway North, Bothell, WA 98011 -8205, 1-800-635-6866	Drench or spray
18	*Fusarium oxysporum non-pathogenic)*	**Fusaclean** (spores, microgranule)	*Fusarium oxysporum*	Asparagus, basil, carnation, cyclamen, gerbera, tomato	Natural Plant Protection, Route d'Artix B.P. 80,64150, Nogueres, France;	In drip to rock wool; incorporate in potting mix; in row
19	*Fusarium oxysporum* (nonpathogenic)	**Biofox C** (Dust or alginate granule)	*Fusarium oxysporum, Fusarium moniliforme*	Basil, carnation, cyclamen, tomato	S.I.A.P.A., Via Vitorio Veneto 1 Galliera, 40010, Bologna, Italy	Seed treatment or soil incorporation
20	*Gliocladium virens* GL-21	**SoilGard** (granules)	Damping-off and root rot pathogens especially *Rhizoctonia solani* and *Pythium* spp	Ornamental and food crop plants grown in greenhouses, nurseries, homes, and interiors capes	Certis, Inc., 9145 Guilford Road, Suite 175, Columbia, MD 21046 USA	Granules are incorporated in soil or soilless growing media prior to seeding
21	*Gliocladium catenulatum*	**Primastop** (Powder)	Soil borne pathogens that cause seed, root, and stem rot, and wilt disease	Ornamental, vegetable, and tree crops	Kemira Agro Oy, Porkkalankatu 3, P.O. Box 330, 00101 Helsinki, Finland;	drench, spray or through irrigation system
22	*Myrothecium verrucaria (*fermentation product of killed *Myrothecium verrucaria)*	**DiTera** (Powder and liquid suspension from fermentation)	Parasitic nematodes	Cole crops, grape, ornamentals, turf, trees	Valent Biosciences, Inc., 1333 N. California Boulevard #600, Walnut Creek, CA 94596, USA	applied to soil and incorporated by various methods
23	*Paecilomyces lilacinus*	**Paecil** (also known as **Bioact**) (Dry spore concentrate)	various nematode spp.	Banana, Tomatoes, sugar cane, pineapple, citrus, wheat, potatoes, and others	Technological Innovation Corporation Pty. Ltd., Innovation House, 124 Gymnasium Dr., Macquarie University, Sydney NSW, 2109 Australia;	Seedling or soil drench
24	*Phlebia gigantea*	**Rotstop** (Spores in inert powder)	*Heterobasidium annosum*	Trees	Kemira Agro Oy, Porkkalankatu 3, P.O. Box 330, 00101 Helsinki, Finland	Spray, chain saw oil

Contd...

Table 2. Contd...

25	*Streptomyces lydicus*	**Actinovate*** (water-dispersible granule)	Soil borne disease	Greenhouse and nursery crops, turf	Natural Industries, Inc. 6223 Theall, Houston, TX.77066.	Drench
26	*Pseudomonas aureofaciens*	**BioJect** (Spot-Less liquid)	Dollar spot, Anthracnose, *Pythium aphanidermatum*, Michrochium patch (pink snow mold)	Turf and others	Eco Soil Systems, Inc. 10740 Thornmint Rd., San Diego, CA	Over-head irrigation; can only be used with the BioJect Automatic Fermentation System
27	*Pseudomonas syringae*	**Bio-save 10LP, 110** (Lyophilized product, frozen cell concentrated pellets)	*Botrytis cinerea, Penicillium* spp., *Mucor pyroformis, Geotrichum candidum*	Pome fruit, citrus, cherries, and potatoes	EcoScience Corp., 153 Sable Palm Dr., Longwood, FL 32779	Pellets added to water to produce liquid suspension, post harvest application to fruit as drench, dip or spray
28	*Pseudomonas fluorescens* A506	**Blight Ban A506** (Wettable powder)	Frost damage, *Erwinia amylovora*, and russet-inducing bacteria	Almond, apple, apricot, blueberry, cherry, peach, pear, potato, strawberry, tomato	NuFarm Inc., 1-708-754-3330	Bloom time spray of the flower and fruit
29	*Pseudomonas chlororaphis* strain	**Cedomon**	Leaf stripe, net blotch, *Fusarium* sp., spot blotch, leaf spot, and others	Barley and oats; potential for other cereals	BioAgri AB, P.O. Box 914	Seed dressing
30	*Pythium oligandrum*	**Polyversum /Polygandron** (Wettable powder)	*Pythium* spp., *Fusarium* spp., *Botrytis* spp., *Phytophthora* spp., *Aphanomyces* spp., *Alternaria* spp., *Tilletia caries, Pseudocercosporella herpotrichoides, Gaeumannomyces graminis, Rhizoctonia solani, Sclerotium cepivorum*	Vegetables (tomatoes, potatoes, pepper, cucumbers, Brassicaceae vegetables), fruits (grapes, strawberries, citrus), legumes, cereals, canola, forest nurseries and ornamental plants	Biopreparaty Ltd., Tylisovska 1, 160 00, Prague 6, Czech Republic;	Root and stem drench, spray
31	*Streptomyces griseoviridis* strain K61	**Mycostop** (Powder)	*Fusarium* spp., *Alternaria brassicola, Phomopsis* spp., *Botrytis* spp., *Pythium* spp., and *Phytophthora* spp. that cause seed, root, and stem rot, and wilt disease	Field, ornamental, and vegetable crops	Kemira Agro Oy, Porkkalankatu 3, P.O. Box 330, 00101 Helsinki, Finland;	Drench, spray or through irrigation system

Contd...

Table 2. Contd...

32	*Trichoderma* products	**Binab** (wettable powder and pellets)	pathogenic fungi that cause wilt, take-all, root rot, and internal decay of wood products and decay in tree wounds	flowers, fruit, ornamentals, turf, and vegetables	Binab, Box 161, 546 22 Karlsbor, Sweden,	Spray, mixing with potting substrate, mixing with water and painting on tree wounds, Inserting pellets in holes drilled in wood
33	*Trichoderma* spp.	**Bio-Fungus-Supresivit** (Granular, wettable powder, sticks and crumbles)	*Sclerotinia, Phytophthora, Rhizoctonia solani, Pythium* spp., *Fusarium, Verticillium*	Flowers, strawberries, trees, vegetables	BioPlant, Helsingforsgade 27B, DK-8200, Arhus N, Denmark, 45	Applied after fumigation; incorporated in soil; sprayed or injected
34	*Trichoderma harzianum*	**Root Pro/ RootProtato** (fungal spores mixed with peat and other organic material)	*Rhizoctonia solani, Pythium spp., Fusarium spp.,* and *Sclerotium rolfsii*	All susceptible flower and vegetable plants, potatoes	Efal Agri, POB 8213, Netanya, Israel.	Agent is mixed into growing media at time of seeding or transplanting
35	*Trichoderma harzianum* Rifai strain KRL-AG2 (T-22)	**RootShield / PlantShield/ T-22 Planter box** (Granules or Wettable powder)	*Pythium* spp., *Rhizoctonia solani, Fusarium* spp.	Trees, shrubs, transplants, all ornamentals, cabbage, tomato, cucumber	Bioworks, Inc., 122 North Genesee St., Geneva, NY 14456 USA	Granules mixed with soil or potting medium; powder mixed with water and added as a soil drench
36	*Trichoderma harzianum*	**Trichodex** (Wettable powder)	Primarily *Botrytis cinerea*, also *Colletotrichum* spp., *Fulvia fulva, Monilia laxa, Plasmopara viticola, Pseudoperonospora cubensis, Rhizopus stolonifer, Sclerotinia sclerotiorum*	Cucumber, grape, nectarine, soybean, strawberry, sunflower, tomato	Makhteshim Chemical Works, Ltd., P.O. Box 60, Beer Sheva, Israel	Spray
37	*Trichoderma harzianum* and *T. viride*	**Trichopel**	*Armillaria, Botryosphaeria, Chondrosternum, Fusarium, Nectria, Phytophthora, Pythium, Rhizoctonia*		Agrimm Technologies, Ltd., P.O. Box 13-245, Christchurch, New Zealand	

Contd...

Table 2. Contd...

38	*Trichoderma viride*	**Trieco** (Powder)	*Rhizoctonia spp., Pythium spp., Fusarium spp.,* root rot, seedling rot, collar rot, red rot, damping-off, Fusarium wilt	Mustard, masoor, oilseeds, soybean, cotton, chilies, chickpeas, green arhar dal, tobacco, cardamom, tumeric, moong, udad, chawli, tea, coffee, rubber, ginger, tomato, sugar cane, citrus, grapes, sunflower, cereals, vegetables, and others	Ecosense Labs Pvt. Ltd., 54 Yogendra Bhavan, J.B. Nagar, Andheri (E), Mumbai 400 059 India;	dry or wet seed, tuber, or set dressing or soil drench, spread/broadcast over field

Source: http://www.cabi-commodities.org/Acc/ACCrc/ACCrcWM.htm

form a specific symbiotic relationship called mycorrhiza means fungus root. They are of three group's *viz.* ecto mycorrhizae, endomycorrhizae (Vesicular arbuscular mycorrhizae-VAM) and ectendo mycorrhizae. They increase the uptake of phosphorus and other nutrients by plants and also affect susceptibility of plants to disease. The management of these symbionts is another way to achieve biological control of soil borne plant pathogens. Experiments with mycorrhizae indicated that roots with mycorrhizae are less damaged by pathogenic nematodes and fungi than non-mycorrhizal roots. AM can reduce the severity of diseases caused by soil- borne pathogens (Linderman, 1994). The mechanism of suppression may be due to morphological, physiological and biological alterations in the host. VAM colonization induces remarkable changes in root morphology, as well as in the meristamatic and nuclear activities of root cells (Atkinson *et al.* 1994) which in turn affect the rhizosphere interactions and pathogen- infection development. Thickening of the cell walls through lignification and production of other polysaccharides in mycorrhizal plants also prevent penetration and growth of pathogens. Mycorrhizal plants harbor higher populations of microorganisms in the rhizosphere thus making it difficult for the pathogen to compete and gain access to the root (Secilia & Bagyaraj, 1987). Addition of suitable mycorrhizal inoculum to the nursery bed would result in healthier seedlings which would not only be able to withstand transplant shock better but also challenge infection by root pathogens and grow better when planted in the field. Being obligate symbionts, these fungi can be multiplied only in roots of plants such as sorghum. The harvested roots along with root zone soil act as the source of inoculum for biological protection or growth promotion in plants.

4.0 QUALITY REQUIREMENTS FOR BIOCONTROL AGENTS

Biocontrol agents must satisfy several requirements before being developed and marketed. This can be considered in four general categories viz. economic, environmental, technical and regulatory (Otis, 1993). The product must be technically feasible, highly consistent performers and must offer some advantage in performance and cost. Bioagents should be indigenous microbes so that there should not be any danger of introducing potentially damaging organisms. Organism should have saprophytic abilities so that they can maintain population during the absence of target pathogens or they should have resting structures that ensure survival. Products should be available to conventional and available technology.

Knowledge of mechanisms involved in biological control is important for estimating and predicting its reliability and selection of better strains. The choice of an antagonist with characteristic mechanisms depends on the stage of the life cycle of the pathogen the antagonist is aimed at. Allowable interaction times and niche characteristics determine the suitability of certain modes-at action during different developmental stages of the pathogen. The scientific procedure involved in developing biocontrol agents involves the following steps (Otis 1993).

i) Select potential biological agents by mass isolation of organisms from either target pathogen or environment or cultivating a single known organism.

ii) Screen agents to further determine that they have properties requisite for acceptable biological control agents such as production of propagules, longevity and compatibility with pesticides and other chemicals.

iii) Characterize agents to identify those features that make particular organism candidates for development

iv) Maintain a required level of quality control.

v) Develop a suitable delivery system to apply the agents.

vi) The procedure must be tested under field conditions.

Isolation of effective biocontrol strains

As the geographical origin of many plant diseases is not known, it is not really practical to search for antagonists in the centre of origin. A rich source of antagonist is an area continuous in monoculture where the disease has declined to a low level and not in places where no pathogen exists (Baker & Cook, 1974). The following points should be taken care of while isolating antagonists.

i) Antagonists should only be isolated from healthy tissue in areas where a high disease pressure exists which guarantee that natural biocontrol is in effect.

ii) Fields free of disease should also be avoided because the required high inoculum pressure may not exist.

iii) The antagonists should always be isolated from the same habitat in which they are desired to work in order to guarantee good adaptation to that habitat and better survival and physiological activity required for biocontrol Krauss *et al.* (1998).

Native antagonists can be exploited for biocontrol by two methods

(a) Augmentation of existing populations to a level where they suppress the disease. It requires either repeated application or co-application with a suitable substrate to maintain sufficiently high population levels of the antagonist for effective and consistent biocontrol.

(b) Application at an early stage of crop development before pathogen populations build up and long before hyperparasites would respond to that built-up.

Mass multiplication of Biocontrol agents

Large quantity of the bioagents is necessary to achieve biocontrol when applied directly to soil in the field. If it were applied to a diseased crop after harvest, the mycoparasite could use the crop refuse and simultaneously parasitize the pathogen propagules and lead to reduction in inoculum density of the pathogen. After application to several consecutive crops, the soil could very well become suppressive to one or more pathogens of one or more crops. The bioagents selected for mass multiplication should have the following qualities:-

i) It should survive and grow in the rhizosphere or in the hyphosphere, near the pathogen resting structures, or in the soil mass.

ii) It should easily grow in organic matter and crop refuse.

iii) It should produce even when growing slowly on simple available substrates, a broad spectrum highly toxic antibiotic that is effective at low concentrations and that is not sorbed or degraded in the soil.

iv) The antibiotic produced by one antagonist should not be inhibitory to other associated antagonists.

v) The antibiotic produced should not harm the host.

vi) It should be adaptable to large scale commercial production and handling.

ix) It should have fast germinating spores.

x) It should be more adaptable to the environmental extremes

xi) It should have wider adaptability to changing environmental conditions.

xii) Nutritional requirement of the antagonists should not be narrower than those of the pathogen.

A bioagent satisfying all these qualities can be subjected for mass multiplication and field application. Development of a stable, cost effective and easy-to-apply biocontrol formulation is critical for the advancement of biological control of plant pathogens with introduced antagonists. The formulation may be a liquid or semisolid fermented product (Lisansky 1985).

Application methods of Biocontrol agents

There are four different techniques for the application of *Trichoderma* spp. as biocontrol agents. They are i) Broadcast-Preparation can be broadcasted on the surface and incorporated into infested soil ii) Furrow application-Preparation can be put into infested soil in the planting furrow iii) Root zone application-Preparation loaded into the root zone by mixing before transplanting seedlings in the infested field and iv) Seed coating- Preparation can be coated on seeds using an adhesive. It was established that except broadcasting all the other three methods are economical especially the method that introduce the antagonist with the planting material.

The biocontrol agents are generally most effective when applied to the plants prior to the onset of infection by the pathogen (preventative) or in some cases, just after the first appearance of symptoms. They are seldom effective when the disease is well-established in an environment. In most instances, multiple applications of the BCA are required, either as a root drench to control root-infecting pathogens, or as a foliar spray to control foliar pathogens.

Agronomic Practices for dealing with biocontrol agents in disease management

There are two traditional approaches for the biological control of soil borne plant pathogens in the field. They are crop rotations and the addition of organic amendments to soil. The former allow time for resident antagonists to sanitize the soil and the latter stimulate resident antagonists. Soil fumigation is a substitute for long crop rotation as it sanitizes the soil from all resident microbes. Tillage has also been important to biological control in some agro ecosystems. It is a means of accelerating the displacement of certain pathogens by resident antagonists in crop residues through exposure of new sites for colonization on the residue fragments and through intensification of microbial activity that accompanies soil disturbance (Cook, 1980).

The other applied agronomic techniques for dealing with soil borne pathogens (Baker & Cook, 1974) are

a) Altering pH of the soil to one favorable to the antagonists and / or unfavorable to the pathogen. The use of inorganic fertilizers and altering the soil pH with lime or sulfur helps to achieve biological control. For *eg.* it was reported that lowering the soil pH with sulfur prevented sporangial formation by *P. cinnamomi* and favored *Trichoderma* population which in turn controlled root rot of pine apple (Pegg, 1977). The altered rhizosphere pH affects the susceptibility of the plant by reducing the availability of essential micronutrients.

b) Tillage methods modify the soil structure, crumb size or aeration. Tillage may affect the activity of soil borne pathogens and its interaction with antagonists. It may influence the physical environment for biological activity or the location of the crop residue occupied by the pathogen or the location of the pathogen as such. Most fungal pathogens, plant parasitic nematodes, pathogenic bacteria and antagonists reside or complete their life cycle in the soil and when the soil is tilled, the ecology of the organisms will also undergo a drastic change. Thorough tillage is generally more advantageous than reduced tillage in disease control.

c) Selection of planting date is also a means to achieve biocontrol. It will expose the principle infection areas to the pathogen at a time when the environment is more favorable to the antagonists or the host.

d) Organic amendments favor biological protection of the plant. The biological control achieved with organic amendments results partly by enhanced competition among the soil micro organisms for nitrogen, carbon or both. Applying organic amendments reduce the available nitrogen at the infection site. Fertilizer placement should be outside the infection court. The potential for disease control by use of proper form of nitrogen was also reported (Huber *et al.* 1968). It was found that nitrate nitrogen favored the disease whereas ammonium nitrogen suppresses it (eg. take- all of wheat). This in turn was found to be due to the change in pH. Ammonium nitrogen lowers the rhizosphere pH whereas nitrate nitrogen enhances it (Bruehl, 1975).

e) Employ irrigation to maintain a soil water potential favorable to antagonists at the infection site. But it should ensure water available to the host. Irrigation affects biological control of plant pathogens through effects on the physical environment of the soil and the plant surface. It also enhances the resistance of plants to pathogens by alleviating predisposing stresses.

f) Flooding the soil during hotter period provides highly effective biological control of soil borne pathogens. It helps to destroy the inoculum, which is accomplished by the anaerobes that decay the metabolically inactive propagules and produce toxic metabolites inhibitory to the propagules. The disadvantage of the method is that the root infecting pathogens introduced into the soil after flooding become more destructive because of the biological vacuum created from the indiscriminate destruction of soil organisms .But it may be effective if used in combination with crop rotation.

g) Trap and inhibitory plants can be identified and promoted to reduce injury from nematodes and some fungal pathogens.

h) Selective treatment can be given in the soil to reduce competitors of the antagonists, but provide as many antagonists as possible and maintaining a favorable environment for the antagonists.

i) Manipulating the environment to produce maximal resistance to the host.

j) A method of culture should be selected that favors the antagonists at the soil depth where infection by the pathogen occurs.

5.0 BIOLOGICAL CONTROL OF IMPORTANT DISEASES OF SPICES

5.1 Foot rot disease of Black pepper

Phytophthora foot rot, previously known as "Quick wilt" is caused by *Phytophthora capsici,* a soil borne fungus and slow decline caused by feeder root damage either by *Phytophthora capsici* or plant pathogenic nematodes namely *Radopholus similis* and *Meloidogyne incognita* or their combinations; *Phytophthora* foot rot is serious and devastating disease causing large scale death of black pepper vines in all pepper growing areas of the country.

About US $ 4.5-7.5 million loss per annum has been reported at global level due to foot rot in black pepper (De waard, 1979). Non-adoption of phytosanitory measures, non-availability of disease free planting material, and lack of disease resistant varieties contributed substantially to the increase in disease incidence. Earlier, the disease infection was a sporadic occurrence, but in early nineties, it reached epidemic form and created untold miseries to lakh of farmers, particularly small and marginal farmers in Kerala state.

An array of experiments has been done at Indian Institute of Spices Research Calicut, in order to establish the biocontrol efficacy of these organisms under field conditions to reduce the damage caused by *Phytophthora capsici* in causing foot rot disease (Saju *et al.* 2003). Several antagonists like *T. harzianum, T. virens and VAM* fungi were isolated and evaluated for the disease suppression (Sarma *et al.* 1996, Rajan, 1999, Rajan *et al.* 2002). T. *harzianum* isolate IISR 1369 along with either of the two isolates of P. *fluorescens* viz. IISR 41 and IISR 6 are found highly growth promotive as well as disease suppressive and compatible combination for the management of foot rot disease. *T. harzianum* multiplied in a suitable carrier media having a CFU of 10^8/g has to be applied around the plant base @ 50g/vine at the onset of monsoon (May-June) and a second application during August- September. Similarly *P. fluorescens* @ 50g/vine having a CFU of 10^8/g of formulation may also be applied along with *T. harzianum*. Addition of organic mulches and oil cakes in the plant basins improve the texture of the soil and enhance the growth of antagonistic organisms

5.2 Rot diseases of cardamom

Capsule rot ('Azhukal') disease of cardamom caused by Phytophthora *meadii* Mc Rae of *A2* mating type is a serious threat to cardamom in the high ranges of Kerala. An Ecofriendly and economic disease management strategy for the capsule rot by the use of introduced antagonists such as *Trichoderma* species was studied in detail as an important component of integrated disease management system which fits in well into the organic farming system. (Suseela Bhai, 1998, Suseela Bhai *et al.* 1993, 1998)

Though many species of *Trichoderma* were isolated and identified, all of them were not equally effective or antagonistic to Phytophthora of cardamom. They varied in their capacity to produce antibiotics (volatile and non volatile), hyperparasitism and predation. Some of them showed volatile inhibitory effect (antibiosis) along with predation and parasitism. Among the isolates tested, *T. harzianum* and T. *viride* having fast growth of >90mm in 48 hrs were found to be inhibitory to *Phytophthora* species of cardamom under *in vitro* conditions. They over grew the *Phytophthora* culture and parasitized the hyphae.

In vitro inhibition due to volatile was tested for 25 isolates of *Trichoderma*. Among the isolates, T. *harzianum* and *T. hamatum* were found to be more inhibitory than others with an inhibition rate of 44%. The growth inhibition due to volatiles varies with species (1.65-44.31%). All the species do not have the volatile production.

Rhizome rot caused by *Pythium vexans* and *R. solani* is also a serious soil borne infection in cardamom. An integrated disease management strategy comprising of plant sanitation, soil amendments fungicidal applications and the use of bioagents was developed to control the disease (Dhanapal *et al.* 1997).

Seedling rot of cardamom

Seedling rot of cardamom caused by *Pythium vexans is* a serious disease problem in cardamom nurseries. Attempts were made to control the disease using biological methods by exploiting species of *Trichoderma. T.viride and T. harzianum* as soil drenching as well as seed dressing prior to sowing of seeds and as soil application in nursery beds at the time of transplanting did not alter the germination percentage but reduced the disease incidence considerably in conducive soil. Soil drenching of *Trichoderma* species one week prior to transplanting of seedlings was effective in reducing the seedling mortality compared to *Trichoderma* application after transplanting or as seedling dip (Suseela Bhai & Joseph Thomas, 1998).

Effect of soil amendments on the proliferation stability of T. harzianum

The augmentation of the soil with organic matter to promote biocontrol activity against soil borne pathogens has been established both under green house and field conditions (Baker & Cook, 1974). In cardamom, farm yard manure, poultry manure, coffee husk and neem cake were used as soil amendment with and with out *Trichoderma* to evaluate their efficacy in controlling *Phytophthora* under *in vitro* conditions. These were selected because of their availability and low cost. Addition of organic plant residues were found to be effective at high moisture levels i.e., at 75% and field capacity in suppressing the *Phytophthora* population. With farm yard manure and poultry manure amendments microbial population was very low when compared to plant residues. But addition of *Trichoderma* along with these organic matter strongly influenced the proliferation stability of *Trichoderma* as well as suppression of *Phytophthora.* The proliferation capacity varied with the organic matter. Addition of coffee husk and neem cake enhanced the population of other soil microflora such as *Aspergillus, Penicillium, Rhizopus, Mucor* etc. and there is a decrease in the population of *Trichoderma* which indicated relatively poor competitive saprophytic ability of *Trichoderma* species compared to above fungal flora in the soil. Though poultry manure supported the growth of *Trichoderma,* it did not in turn found to be suppressive to *Phytophthora* population at all moisture levels. Organic matter can affect the vigour of the host by improving the structure and moisture and nutrient holding capacity of the soil but its principal effect is to provide a more complex antagonistic soil microflora and fauna (Suseela Bhai, 1998).

Under sterile conditions *Phytophthora* did not survive but is suitable for the proliferation of *Trichoderma.* Under sterile conditions most of the organic manures and decomposed plant residues supported the growth of *Trichoderma* species and under natural non-sterile conditions the competitive saprophytic ability rules and the media which supported the growth of *Trichoderma* under sterile conditions would be less conspicuous under *in vitro* conditions. Most of the fungicide used even at low concentration were found to be inhibitory to *Trichoderma*. These observations indicated that at 50% Field capacity, the application of organic manures and NPK fertilizers facilitate the build-up of *Phytophthora* population. This might be due to the increase N_2 availability in the soil. Based on these observations, it was found that FYM was a suitable soil amendment at all moisture levels in enhancing the population level of *Trichoderma* as well as reducing the population build up of *Phytophthora* in combination with *Trichoderma*. Though coffee husk was found to be a suitable carrier medium for mass multiplication as mentioned elsewhere in the results, as a soil amendment it was not found very effective in maintaining the stability of *Trichoderma*. Mere increase in *Trichoderma* population never means that there is a proportional decrease in *Phytophthora* population. Plant residues were found to be poor in maintaining the *Trichoderma* population except under sterile conditions.

Decomposed coffee husk and FYM in a 1:1 ratio showed the highest population (15 x 10^8 cfu/g in one month followed by fresh coffee husk and FYM (13 x10^8 cfu/g). Spent tea waste and FYM alone (9.15 x10^8) where also found to be good in the multiplication of *T. harzianum* under sterile conditions. Of the various carrier media tested coffee husk FYM 1:1 was found as a suitable growth cum storage medium for the field application of *T. harzianum* (Suseela Bhai *et al.* 1994).

5.3 Soft rot disease of Ginger

Ginger is affected by a number of fungal, bacterial and viral diseases among which soft rot caused by *Pythium* spp., bacterial wilt by *Ralstonia solanacearum* Yabuuchi and yellows by *Fusarium* sp. cause heavy crop losses in many ginger growing regions. The disease being both seed and soil borne, the strategy of disease management involve measures that effectively suppress the pathogen through cultural, biological and chemical control methods (Dake & Edison, 1988; Sarma, 1994).

Nine *Trichoderma* species and fluorescent pseudomonads isolated from rhizosphere and rhizomes were tested for their *in vitro a*ntagonism and hyperparasitism to the fungus among which *T. harzianum* was most effective seed treatment with biocontrol inoculum in combination with starch 0.2% was also effective. Field trails carried out in *Pythium* sick soils with seed and soil treatment with *T. harzianum* mixed neem cake @ one kg/bed reduced the incidence of the disease and increased the yield. A combination of biocontrol agent with soil solarization is more effective in suppression of the disease and for obtaining higher yields. Fluorescent pseudomonas with *T. harzianum* is also effective against the pathogens under pot culture studies (Sarma *et al.* 2001).

Twenty rhizobacterial cultures were tested both under *in vitro* and *in vivo* conditions to evaluate their efficacy in inhibiting *Pythium myriotylum* as well as reducing the soft rot disease incidence in ginger. The tested cultures included *Pseudomonas fluorescens, Enterobacter agglomerans* and *Bacillus* sp. Among the 20 isolates selected, two isolates (IISR 859 and 853) have nematicidal action on *Radopholus similis* and *Meloidogyne incognita,* the major nematode parasites of ginger (Anandaraj & Sarma, 2003; Beena *et al.* 2003). Under in *vitro* conditions 16 of the tested rhizobacterial cultures inhibited *P. myriotylum* to 65-85%. In the preliminary i*n vitro* screening experiment, significant reduction in growth of *P. myriotylum* was obtained with 17 rhizobacterial isolates and among the 20 isolates tested, the percentage inhibition ranged from 0-85%. Except IISR 6, 152 and 914, all the other isolates recorded more than 50% inhibition. Maximum inhibition (84.7%) over control was achieved in IISR 909 (*Bacillus polymixa).* The *P. fluorescens* isolates (IISR 13, 51, 859), *Bacillus* isolates (IISR 150, 151, 907, 913, and 915) and an isolate of *B. lentus* (IISR 906) were found to be at par giving 70-75% inhibition. Isolate IISR 914 (*Bacillus* sp) was found ineffective under the same conditions (Table 3).

Table 3. In *vitro* inhibition of *Pythium myriotylum* by rhizobacterial cultures

Isolate No.	Radial growth of *P.myriotylum* (mm)1	*In vitro* Inhibition (%)
IISR 6	26.4	38.8 b
IISR 13	11.0	74.1 hi
IISR 51	11.3	73.4 ghi
IISR 853	12.5	70.5 fgh
IISR 859	12.8	69.9 efg
IISR 906	10.5	75.3 i
IISR 907	10.8	74.6 i
IISR 909	6.5	84.7 j
IISR 910	25.5	40.0 b

Contd...

Table 3. Contd...

IISR 912	13.8	67.5 def
IISR 913	11.0	74.1 hi
IISR 914	42.5	0.0 a
IISR 915	11.3	73.4 ghi
IISR 147	14.8	65.2 d
IISR 149	14.3	66.4 de
IISR 150	11.5	72.9 ghi
IISR 148	12.0	71.8 ghi
IISR 151	10.5	75.8 i
IISR 152	22.5	47.1 c
IISR 153	12.0	71.8 ghi
Control	42.5	0.0 a

Evaluation of these cultures in comparison with metalaxyl-mancozeb 0.2% revealed that the cultures IISR 51, 859, 906, 915 and 151 were superior to fungicide treatment in reducing the disease incidence to less than 20% (Table-4) (Suseela Bhai & Kumar, 2005).

Table 4. *Efficacy of rhizobacterial cultures against P. myriotylum*

Treatments	Disease incidence (%)	Disease reduction (%)
IISR 6	32.2 abcd	52.0
IISR 13	26.8 abc	60.1
IISR 51	4.2 a	93.7
IISR 853	33.3 abcd	50.4
IISR 859	14.2 ab	78.8
IISR 906	3.3 a	95.1
IISR 907	45.7 abcd	31.9
IISR 909	75.0 d	0.0
IISR 910	55.6 bcd	17.1
IISR 912	41.7 abcd	37.9
IISR 913	20.0 ab	70.2

Contd...

Table 4. Contd...

IISR 914	16.4 ab	75.6
IISR 915	14.8 ab	77.9
IISR 147	33.6 abcd	49.9
IISR 149	41.9 abcd	37.6
IISR 150	36.2 abcd	46.1
IISR 148	33.1 abcd	50.7
IISR 151	6.7 a	90.0
IISR 152	34.2 abcd	49.0
IISR 153	21.4 abc	68.1
Metalaxyl- mancozeb	36.7 abcd	45.3
P.myriotylum alone	67.1 cd	0.0
Absolute control (No pathogen)	0.0 a	100.0

The predominant mycorrhiza associated with ginger was found to be *Glomus* species. The rhizobacterial cultures and the native VAM were used to study the combined effect of both rhizobacteria and VAM on disease incidence. The experiment revealed the efficacy of rhizobacteria especially IISR 51, 915, 859 and 906 in improving the ability of native population of VAM fungi to colonize ginger plant roots. The combined inoculation enhanced the root development and reduced root rot. This is evidenced from the enhanced root length and biomass development (Table 5 & Table 6).

Table 5. Intensity of VAM colonization & spore load in rhizobacterial treatments in Ginger

Treatments	***Intensity of root colonization**	***Spore load/cc**
IISR 6	+++	32.5 cdef
IISR 13	+	16.0 fg
IISR 51	+++	61.5 a
IISR 853	-	21.0 efg
IISR 859	++	34.0 cde
IISR 906	+	32.0 cdef
IISR 907	++	27.0 abc
IISR 909	++	12.5 bcd
IISR 910	+++	49.0 abc

Contd...

Table 5. Contd...

IISR 912	++	55.0 ab
IISR 913	++	53.5 ab
IISR 914	++	61.0 a
IISR 915	++	60.0 a
IISR 147	+	16.5 fg
IISR 149	+++	41.0 bcd
IISR 150	+	17.5 efg
IISR 148	++	10.0 g
IISR 151	++	31.0 def
IISR 152	++	55.0 ab
IISR 153	+	18.0 efg
Metalaxyl - Mancozeb	+	11.0 g
Pythium alone	+	31.0 def
Absolute control	+	11.0 g

Intensity of VAM colonization in roots: + Low ++ Medium, +++ High

Table 6. Effect of PGPR on growth Promotion and disease suppression in Ginger

Treatments	**Root length (cm)**	**Biomass (g)**
T_1 IISR 51+ VAM+*P.myriotylum*	20.83 C	6.0 A
T_2 IISR 151+ VAM+*P.myriotylum*	21.0 C	4.9 B
T_3 IISR 859+ VAM+*P.myriotylum*	23.7 B	2.2 D
T_4 IISR 906+ VAM+*P.myriotylum*	25.0 A	3.0 C
T_5 IISR 913+ VAM+*P.myriotylum*	23.8 D	1.5 E
T_6 IISR 915+ VAM+*P.myriotylum*	20.0 B	6.0 A
T_7 *P.myriotylum*	3.9 F	0.2 F
T_8 Absolute Control	14.2 E	1.2 E

Direct suppression by rhizobacteria could be attributed to the production of iron sequestering siderophores which inhibit the growth of *Pythium* spp, by reducing the availability of iron (Whipps & Lumsden, 1991) or through antibiosis and HCN production (Diby Paul & Sarma, 2003; Sikora *et al.* 1990). Indirectly, rhizobacteria enhance the VAM colonization which in turn improve the plant's resistance through enhanced root development and also by way of nematode suppression. More over one of the pseudomonas isolate (IISR 859) found effective in suppressing the disease was antagonistic to both *Meloidogyne incognita* and *Radopholus similis* (Anandaraj & Sarma, 2003 & Beena *et al.* 2003).

Table 7. Biopesticide registered with central insecticide board for commercial purpose

vide notification No. G.S.R. 224 (E) dated 26.03.99		
S.No	**Common name**	**Chemical/commercial name**
588.	*Bacillus subtilis*	*Bacillus subtilis cohn*
589.	*Gliocladium Species*	*Gliocladium Spp*
590.	*Pseudomonas Spp*	*Pseudomonas spp*
591.	*Trichoderma Species*	*Trichoderma spp.*
592.	*Beauveria bassiana*	*Beauveria bassiana*
593.	*Metarrhizium anisopliae*	*Metarrhizium anisopliae var. acridum*
594.	*Nomuraea rileyi*	*Nomuraea rileyi*
595.	*Verticillium lecanii*	*Verticilium lecanii*
596.	Grannulosis Viruses (GV)	Grannulosis Viruses (GV)
597.	Nuclear Polyhedrosis Viruses (NPV)	Neodiprion sertifer NPV and Neodiprion lecontei NPV
vide notification No. G.S.R. 378 (E) dated June 24th, 2004		
675.	*Verticilium chlamydosporium*	*Verticilium chlamydosporium*
677.	*Streptomycin griseoviridis*	*Streptomyces griseoviridis*
678.	*Streptomyces lydicus*	*Streptomyces lydicus*
679.	*Ampelomyces quisqualis*	*Ampelomyces quisqualis*
680.	*Candida oleophila*	*Candida oleophila*
681.	*Fusarium oxysporum (non pathogenic)*	*Fusarium oxysporum (non pathogenic)*
682.	*Burkholderia cepacia*	*Burkholderia cepacia*
683.	*Coniotyrium minitans*	*Coniotyrium minitans*
684.	*Agrobactarium radiobacter strain 84*	*Agrobactarium radiobacter strain 84*
685.	*Agrobactarium tumefaciens*	*Agrobactarium tumefaciens*
686.	*Pythium oligandrum*	*Pythium oligandrum*
687.	*Erwinia amylovora (Hairpin protein)*	*Erwinia amylovora (Hairpin protein)*
688.	*Phlebia gigantean*	*Phlebia gigantean*
689.	*Plaecilomyces lilacinus*	*Plaecilomyces lilacinus*
690.	*Penicillium islanidicum (for groundnut)*	*Penicillium islanidicum (for groundnut)*
691.	*Alcaligenes spp.*	*Alcaligenes spp.*
692.	*Chaetomium globosum*	*Chaetomium globosum*
693.	*Aspergillus niger-strain AN27*	*Aspergillus niger-strain AN27*
694.	*VAM-Vesicular arvescular mycorrhijal (Fungus)*	*VAM-Vesicular arvescular mycorrhijal (Fungus)*

Source: http://cibrc.nic.in/insecticides_rules.htm

6.0 REGISTRATION OF BIOCONTROL AGENTS

The Central Insecticide Board constituted under the Insecticides Act, 1968, Govt of India regulated the import, manufacture, sale, transport, and use pesticides. It has approved inclusion of the following antagonistic organisms (Table-7) to the schedule of insecticides Act to regulate their manufacture and use in the country, vide Notification No. G.S.R. 224 (E) dated 26.03.99, G.S.R.621 (e) dated 27-10-99 AND G.S.R. 69 (E) dated 5-11-2001) (Pawar, 2003). To ensure registration of the BCA, appropriate information on efficacy, mode of action, persistence, non-target effects, formulation, delivery method, and mammalian toxicity is required to meet the approval criteria of the Pest Management Regulatory Agency (http://pbi-ibp.nrc-cnrc.gc.ca/en/bulletin/2004issue3/page4.htm). At present around 30 microbial pesticides have been registered under insecticides Act. Important ones are *Gliocladium* sp, *Trichoderma* sp, *and Pseudomonas* sp. *Paecilomyces* sp. *Verticillium* sp. *etc.* And many others which are mostly used against soil borne plant pathogens of a variety of horticultural and plantation crops Quality biopesticide is of paramount importance for protecting the crops from possible damages by pests and diseases. The use of microbial pesticides reduces the load of toxic pesticide residues in the agricultural food commodities and in the environment. They are eco-friendly and economical but care has to be taken to see that the quality of the bio product is maintained to promote IPM.

7.0 CONCLUSION

Several fungal antagonists have potential to become biocontrol agents for disease management. Among these, species of *Trichoderma, Pochonia, and Paecilomyces* are very promising which can be grown in both semi solid or liquid fermentation and applied either alone or in combination with low doses of fungicides to eventually establish the fungus in the soil. If it were applied to diseased crop after harvest the mycoparasite could use the crop residues and simultaneously parasitize the pathogen propagules, this would lead with time a reduction of inoculum density of the pathogen. After application to several consecutive crops, the soil could very well become suppressive to one or more pathogens of one or more crops. But one of the biggest obstacles to practical biocontrol is the largest quantity of the agent necessary to achieve biocontrol when applied directly to soil in the fields.

In spite of all these experimental evidence the farmers are not using some of the antagonists to control plant diseases because of the following reasons

i) Few organisms have been reported to provide economic disease control either because of the high rates of application needed for the minute level of disease control obtained.

ii) Lack of ecological knowledge of biocontrol agent.

iii) Lack of practical or realistic approach to the study of biocontrol

iv) Lack of concern for the spatial relationship of the target pest and its control agent and carryover of the fungicide mentality to the field of biocontrol.

7.1 Future strategies

To minimize the practical difficulties of widespread utilization of biopesticides, it is necessary to

- Provide basic training in all relevant areas of biocontrol agent's development - from concept through to complete product that includes collection, isolation and characterization of the pathogen, production and storage, formulation and application, registration and commercialization.
- Relevant training and financial assistance, combined with access to relevant specialist facilities and equipment, advice, support and information will significantly empower the beneficiaries and scientist who are engaged in biocontrol of pests and diseases and ensure the development of biocontrol agents
- Toxicological and ecotoxicological tests with the biopesticide are required for registration with the relevant regulatory authorities.
- Appropriate technical assistance is provided to carry out relevant toxicological and ecotoxicological studies, to meet the international standards for registration and to provide support to regulatory authorities in line with FAO and OECD harmonization programmes.
- To safeguard IPR and the transfer of the technology and subsequent product support.
- Create awareness of the opportunities offered by biocontrol agents with international and national agencies, decision-makers, extension agencies, entrepreneurs, potential commercial investors and farmers.

REFERENCES

Adams, P. B. (1990) The potential of mycoparasites for biocontrol of plant diseases. *Ann.Rev. Phytopathol*. 28: 59-72.

Ahmad, and Baker. (1987) Rhizosphere competence of *Trichoderma harzianum*. *Phytopathol*. 77: 182-189.

Alexander, M. (1971) Microbial ecology. John Wiley & Sons, New York 511p

Anandaraj, M. and Sarma, Y. R. (2003) The potential of PGPR's in disease management of spice crops. In: Reddy M.S, Anandaraj M, Eapen S J, Sarma Y R & J W Kloepper (Eds.) Oral Presentations, *International workshop on Plant Growth Promoting Rhizobacteria*, 5-10th October 2003, Calicut (p 27) Indian Institute of Spices Research. Calicut.

Anandaraj, M. and Eapen, S. J. (2003) Achievements in biological control of diseases of spice crops with antagonistic organisms at Indian institute of Spices Research, Calicut. *In: Current Status of Biological Control of plant Diseases using Antagonistic organisms in India* (Eds.) B Ramanujam and R J Rabindra. Project Directorate of Biological control, Bangalore.

Arora, R. K. (1999) Evaluation of bioagents for control of soil and tuber diseases of Potato. *Indian Phytopathol.* 52:310.

Atkinson, D., Berta, G. and Hooker, J. E. (1994) Impact of mycorrhizal colonization on root longevity and the formation of growth regulators. *In: Impact of arbuscular mycorrhizas on sustainable agriculture and natural ecosystems.* Birkhauser, Basel, pp. 89-99.

Baker, K. F. and Cook, R. J. (1974) *Biological control of Plant pathogens.* SanFrancisco: Freeman.433pp.

Barnett, H. L. and Binder, F. L. (1973) The fungal host-parasite relationship. *Annu. Rev. Phytopathol.* 11: 273- 292.

Beena, B., Eapen, S. J. and Ramana, K. V. (2003) Native rhizobacteria for the biological suppression of *Radopholus similis* infecting black pepper (*Piper nigrum* L.). In(Eds.): Reddy M.S, Anandaraj M, Eapen S J, Sarma Y R & J W Kloepper Abstracts and Short papers, *International workshop on Plant Growth Promoting Rhizobacteria.*, 5-10th October 2003, Calicut (p 12) Indian Institute of Spices Research. Calicut.

Bhai, R. S., Thomas, J. and Naidu, R. (1994) Evaluation of carrier media for field application of Trichoderma spp. in cardamom growing soils. *J. Plant. Crops* 22 (1). 50-52.

Bonnie, H. Ownley, and Mark T. Windham, (2003) Biological control of plant pathogens. *In: Plant Pathology Concepts and Laboratory exercises* (Eds. Robert N. Trigiano, Mark T. Windham & Alan S. Windham). CRC Press, New York.

Boosalis, M. G. and Mankau, R. (1965) Parasitism and Predation of soil microorganisms. In Baker K F and Snyder W C (Eds.) *Ecology of soil borne plant pathogens-Prelude to biological control.* University of California Press. Berkley, California p. 374-391.

Broadbent, P. and Baker, K. F. (1974) Association of bacteria with sporangium formation and breakdown of sporangia in *Phytophthroa* spp. *Austral. Journ. Agric. Res.* 25:139-145.

Broadbent, P. and Baker, K. F. (1975) Soil suppressive to *Phytophthora* root rot in eastern Australia. In: (G W. Bruchl Ed.) *Biology and Control of Soil-borne Plant Pathogens.* American. Phytopathological Society, St. Paul. Minnesota p 152-157.

Bruehl, G. W., ed. (1975) *Biology and control of soil borne plant pathogens.* American Phytopathological Society, St. Paul. Minnesota p.216.

Claude Alabouvette, Bob Schippers, Lemanceau, P. and Peter A. H. M. Bakker, (1997) Biological control of Fusarium wilts: Toward Development of Commercial products. In: Greg J Boland &L .David Kuykeendall Eds.) *Plant Microbe Interactions and Biological Control.* Marcel Dekker, Inc.

Clark, F. E. (1965) The concept competition in microbial ecology. *In Ecology of soil borne plant pathogens-Prelude to biological control* (Baker K F and Snyder W. C. Eds.). University of California Press. Berkley, California p. 339-345.

Cook, R. J. (1980) Fusarium foot rot wheat and its control in the Pacific Northwest. *Plant Disease* 64: 1061-1066.

Cook, R. J. and Baker, K. F. (1983) *In: Nature and Practice of Biological control of Plant Pathogens,* American Phytopathological Society, St. Paul. Minnesota. p539

Dake, G. N. and Edison, S. (1988) Survey for disease incidence in major ginger growing areas of Kerala during 1984 and 1985. *J. Plantn. Crops* 16: 55-57.

Defago, G., Berling, C. H., Burger, U., Haas, D., Kahr, G., Keel, C., Voisard, C., Wirthner, P. and Wuthrich, B. (1990) Suppression of black root rot of tobacco and other root diseases by strain of *Pseudomonas fluorescens;* Potential applications and Mechanisms. In: (D. Hornby Ed.) *Biological Control of Soil Borne Plant Pathogens* CAB International, Wallingford, UK.

De Waard, P. W. F. (1979) Evaluation of the results of research on eradication of *Phytophthora* foot rot of Black pepper (*Piper nigrum* L.) pp.1-47, Circulated during the first meeting of the pepper community permanent panel on Techno Economic Studies, 31 Jan - 4th Feb 1979. IPC, Cochin .India.

Dhanapal, K., Joseph Thomas, and Suseela Bhai, R. (1997) Use of bioagent and botanicals In the management of rot disease of small cardamom. Paper presented in the 49th Annual meeting and national Symposium on "*Emerging Issues in Plant Pathology*". (Ed.) Dharam Vir Singh. February 15-17, Rani DurgavathiViswavidhyalaya, Jabalpur.

Diby Paul, Kumar, A., Anandaraj, M. and Sarma, Y. R. (2001) Studies on the suppressive action of Fluorescent pseudomonads on *Phytophthora capsici. Indian Phytopathol.* 54: 515.

Diby Paul, and Sarma, Y. R. (2003) Induction of defence related enzymes with Plant Growth Promoting Rhizobacteria (PGPRS) in black pepper (*Piper nigrum*) and *P. capsici* pathosystem In: Reddy M.S, Anandaraj M, Eapen S J, Sarma Y R & J W Kloepper (Eds) Abstracts and Short papers, *International workshop on Plant Growth Promoting Rhizobacteria.*, 5-10th October 2003, Calicut (p 473) Indian Institute of Spices Research. Calicut.

Duskova, E. (1992) Efficacy of the biological fungicide Supersivit based on T. harzianum against *P. cryptogea. Zahradnicivi* 19 (3): 223-230.

Duvenhage, J. A. Kotze, J. M. (1993) *Biocontrol of root of avocado seedlings.* Year Book. South African Avocado growers Association 16: 70-72.

Fang, J. G. and Tsao, P. H. (1995a) Evaluation of *Pythium nunn* as a potential biocontrol agent against *Phytophthora* root rot of azalea and sweet orange. *Phytopathol.* 85: 29-36.

Fang, J. G. and Tsao, P. H. (1995b) Efficacy of *Penicillium funiculosum* as a biocontrol agent against *Phytophthora* root rot of azalea and citrus. *Phytopathol.* 85 (8): 871-878.

Foley, M. F. and Deacon, J. W. (1985) Isolation of *Pythium oligandrum* and other necrotrophic mycoparasites from soil. *Transactions of the British Mycological Society* 85,631-639.

Garret, S. D. (1950) Ecology of root inhabiting fungi. *Biol. Rev.* 25:220-254.

Garret, S. D. (1965) Towards biological control of soil borne plant pathogens. *In*: Baker K F and W C Snyder (Eds.) *Ecology of soil-borne Plant Pathogens.* University of California Press. Los Angels p 4-17, 425-435.

Gees, R. and Coffey, M. D. (1989) Evaluation of a strain of *Myrothecium roridum* as a potential biocontrol agent against *P. cinnamomi. Phytopathol.* 79:1079-1084.

Guillemin, J. P., Gianinazzi, S., Gianinazzi-Pearson, V. and Marchal, J. (1994) Contribution of arbuscular mycorrhizas to biological protection of micro propagated pine apple (*Ananas comosus* (L) Merr. against *P. cinnamomi Rands. Agriculture Science in Finland* 3(3): 241-251.

Huber, D. M., Painter, C. G., McKay, H. C. and Peterson, D. L. (1968) Effects of nitrogen fertilization on take all of winter wheat. *Phytopathol.* 58:1470-1472.

Ilan Chet, (1990) Biological Control of soil bone Plant pathogens with Fungal Antagonists in Combination with soil Treatments. In: D.Hornby (Ed.) *Biological Control of Soil Borne Plant Pathogens.* CAB International, Wallingford, UK.

Joseph Thomas, Suseela Bhai, R., Vijayan, A. K. and Naidu, R. (1993) Evaluation of antagonists and their efficacy in managing rot disease of small cardamom. *J. Biocontrol* 7: 29-36.

Joseph Thomas, Suseela Bhai, R., Vijayan, A. K. and Susan George, (1994) *Trichoderma* spp in vermicomposting systems- A new approach for bio control of rot diseases of small cardamom. Paper presented in "*National Symposium on frontiers in Plant Science Research*". February 13-24, 1994.

Joseph Thomas, and Suseela Bhai, R. (1998) Trichoderma, a potential bioagents: In *Integrated Management of Rot Diseases of Small Cardamom.* ICPP Abstr. 5

Joshi, L. K. and Sharma, N. D. (1982) Disease of Ginger and Turmeric. In: Nair, M K, Premkumar T, Ravindran P N and Sarma Y R (Eds.) *Proc. Nat. Seminar on ginger and Turmeric.* Calicut pp 104-117. CPCRI, Kasaragod.

Kenneth F Baker, and James Cook, R. (1973) *Biological control of plant pathogens* W.H. Freeman & Co, San Francisco.

Kim, C. H. (1989a) Soil borne diseases. Phytophthora blight and other diseases of red pepperin. Korea Xtension Bulletin-ASPAC, Food and Fertilizer Technology Center, 302, 10-17.

Kim, C. H. (1989b) Biological Control of Phytophthora blight of Red pepper in Korea. Korean J. Plant Pathol. 5(1): 100-106.

Krauss, U., Bidwell, R. and Ince, J. (1998) Isolation and preliminary evaluation of mycoparasites as biocontrol agents against crown rot of banana. Biological Control -Theory and Application 13, 111-119.

Krauss, U. and E. Bustamante, http://www.cabi-commodities.org/Acc/ACCrc/ACCrcWM.htm.

Laha, G. S. and Verma, J. P. (1998) Role of fluorescent pseudomonads in the management of sclerotial wilt of cotton. *Indian Phytopathol.* 51: 275-278.

Linderman, R. G. (1994) Role of VAM fungi in bio-control. *In: Mycorrhiza and Plant Health.* ASP. St. Paul, pp. 1-26.

Lisansky, S. G. (1985) Production and commercialization of pathogens. In: Hussey, N W and Scopes. N (Eds.) *Biological Pest Control.* Blandford Press, Poole, UK, pp.210-218.

Ma, P. and Shen, C. Y. (1993) .Potential for biological control of *Phytophthora* boll rot of cotton by *Trichoderma and Pythium. Chinese J. Biological Control* 9 (3) 122-125

McLeod, A., Labuschagne, N. and Kotze, J. M. (1995) Evaluation of *Trichoderma* for biological control of avocado root rot in bark medium artificially infested with *P. cinnamomi.* Year Book. *South African Avocado growers Association* 18: 32-37

Malajczuk, N. (1983) Microbial antagonism to *Phytophthora.* In: D.C. Erwin, S. Bartnicki-Garcia and P.H. Tsao (Eds.) *Phytophthora, Its Biology Taxonomy, Ecology and Pathology,* p. 197-218.

Merriman, P. and Russell, K. (1990) Screening strategies for biological control. In: D. Hornby (Ed.) *Biological Control of Soil bore Plant Pathogens.* p 427-436.

Mitchell, R. M. and Alexander, M. (1962) Microbiological processes associated with the use of chitin for biological control. *Proc. Soil. Sci. Soc. Amer.* 26, 556-558.

Mukherjee, P. K. and Rai, P. K. (2000) Effect of VAM and Phosphate solubilizing bacteria on growth, yield and phosphorus uptake by wheat(*Triticum aestivum*) and Chick pea (*Cicer arietinum*). *Indian J. Agron.* 45: 602-607.

Okumoto, S. and Bustamante, E. (1993) Efecto de enmiendas sobre bacterias antagonicas a *Alternaria solani* en tomate. *Manejo Integrado de Plagas* (C.R.) 28:1-6.

Om Prakash, Misra, A. K. and Shukla, S. K. (2002) Penicillium citrinum, a potent biocontrol agent against wilt disease of guava. *Asian Congress of mycology and plant pathology.* Dept. of Applied Botany, Seed Pathology and Biotchnology, Univ. of Mysore, October 1-4, 2002: 118.

Ongena, M., Daayf, F., Jacques, P., Thonart, P., Benhamou, N., Paulitz, T. C., Cornelis, P., Kedam, N. and Belanger, R. R. (1999) Protection of Cucumber against Pythium root rot by fluorescent pseudomonads: Predominant role of induced resistance over siderophores and antibiosis. *Plant Pathol.* 48: 66-76.

Orlikowski, L. B. and Schmidle, A. (1995) On the biological control of *P. cactorum* with *T. viride. Nachrrichtenblatt des. Deutschen Pflanznschutzdienstes,* 37 (5): 78-79.

Otis C Maloy, (1993) *In: Plant Disease Control. Principles and Practice,* John Wiley & Sons INC p.235-250.

Papavizas, G.C. (1985) Trichoderma and Gliocladium: Biology, Ecology, and Potential for bio control. *Annu. Rev. Phytopathol.* 23:23-54.

Pal, V. and Jalali, (1998) Rhizosphere bacteria for biological control of plant diseases. *Indian J. Microbiol.* 38: 187-204.

Park, J. H. and Kim, H. K. (1989) Biological control of *Phytophthora* Crown and root rot Green house pepper with *Trichoderma harzianum* and *Enterobacter agglomerans* by improved Method of application. *Korean J. Plant Pathol.* 5 (1): 1-12.

Parke Scurkee, Lee Hyo Yeon, Kim Kichung, (1995) Antagonistic effect of Chitinolytic bacteria on soil borne plant pathogens *Korean J. Plant Pathol.* 11 (1): 47-52.

Pasini, C., Cassini, M. D., Aquila, F. and Garibaldi, A. (1991) Soil suppressive to *Phytophthora cryptogea* in Haly. Preliminary results. *Bulletin SROP* 14 (5): 178-180.

Pawar, A. D. (2003) Registration requirements and quality parameters for antagonistic organisms. In: (B Ramanujam and R J Rabindra Eds.).*Current Status of Biological Control of Plant Diseases using Antagonistic organisms in India,* Project Directorate of Biological control, Bangalore, 176-178.

Pegg, K. G. (1977) Soil application of elemental sulphur as a control of *Phytophthora cinnamomi* root and heart rot of Pine apple. *Aust. J. Exp. Agric. Anim. Husb.* 17: 859-865.

Phillips, J. M. and Hayman, D. S. (1970) Improved procedures for clearing roots and staining parasitic and vesicular arbuscular mycorrhizal fungi for rapid assessment of infection. *Trans. Brit. Mycol. Soc.* 55: 158-161.

Prasad, R. D. and Rangeswaran, R. (2001) Biological control of root and collar rot of chick pea caused by *Sclerotium rolfsii. Annals of plant protection Sciences.* 9: 297-303.

Raguchander, T., Shanmugam, V. and Samiyappan, R. (2001) Biological control of panama wilt disease of banana. *Madras Agri.* J. 87: 320-321.

Rajan, P. P. (1999) Approaches towards Integrated management of Phytophthora infections in black pepper Ph D thesis, University of Calicut. Calicut.

Rajan, P. P., Anandaraj, M. and Sarma, Y. R. (2002) Management of foot rot disease of black Pepper with *Trichoderma* spp. *Indian Phytopathol.* 55: 34-38.

Rama, S., Singh, Jaspal Kaur, Astha, and G. D. Munshi, (2003) Efficacy of bacterial antagonists for the suppression of various soil borne diseases of vegetable and enhanced plant growth In: Reddy M.S, Anandaraj M, Eapen S J, Sarma Y R & J W Kloepper (Eds.) Abstracts and Short papers, *International workshop on Plant Growth Promoting Rhizobacteria.* 5-10th October 2003, Calicut (Pp 212) Indian Institute of Spices Research. Calicut.

Raoof, M. A. and Hari Narayanan, A. B. (2001) Development of improved biocontrol agent against castor wilt pathogen, *F. oxysporum* f. sp. *ricini* Nanda & Prasad. In: *Biopesticide Conference–Biopesticides Emerging Trends* (BET 2001) February, 7-9 2001, Chandgarh, India

Roiger, D. J. and Jeffers, S. N. (1991) Evaluation of *Trichoderma* sp for biological control of Phytophthora crown and root rot of apple seedlings. *Phytopathol.* 81(8): 910-917

Saju, K. A., Anandaraj, M. and Sarma, Y. R. (2003) Evaluation of *Trichoderma* spp and *Pseudomonas fluorescens* for suppression of *Phytophthora capsici* infecting Black Pepper. Abstracts and short papers: In: Reddy M.S, Anandaraj M, Eapen S J, Sarma Y R & J W Kloepper (Eds.) Abstracts and Short papers, *International workshop on Plant Growth Promoting Rhizobacteria.*, 5-10th October 2003, Calicut (p 52), Indian Institute of Spices Research. Calicut.

Sarma, Y. R. (1994) Rhizome rots disease of ginger and turmeric. In: Chadha K L and Rethinam P (Eds.) Plantation and spices crops, *Advances in Horticulture,* 10 (2): 1113-1138.

Sarma, Y. R., Anandaraj, M. and Venugopal, M. N. (1996) Biological control of diseases of spices. In: M. Anandaraj and K V Peter (Eds.), *Biological Control in Spices.* Indian Institute of Spices Research, Kerala, India. P l-19.

Sarma, Y. R., Ramana, K. V., Devasahayam, S. and Rema, J. (2001) *The Saga of Spices Research,* Indian Institute of Spices Research (ICAR), Calicut, Kerala. p 29.

Sawant, I. S., Sawant, S. D. and Nayana, K. A. (1995) Biological control of *Phytophthora* root rot of Coorg mandarin (*Citrus reticulate*) by *Trichoderma* species grown on coffee waste. *Indian J. Agric. Sci.* 65(11): 842-846.

Schneider, R. W. (1982) *Suppressive soils and Plant Disease.* American Phytopathology Society, St.Paul, Minnesota.

Secilia, J. and Bagyaraj, D. J. (1987) Bacteria and Actinomycetes associated with pot cultures of vesicular arbuscular mycorrhizas. *Canadian J. Microbiol.* 33: 1069-1073.

Sikora, R. A., Bodenstein, F. and Nicolay, R. (1990) Effect of Sugar beet seed treatment with rhizosphere bacteria on infection by fungi of the genus Pythium .II. Studies on the mode of action. *J. Phytopathol.* 129:121-132.

Shea, S. R. and Broadbent, P. (1983) Developments in cultural and biological control of *Phytophthora* Diseases. In: D.C.Erwin, S.Bartnicki-Garcia and P.H.Tsao (Eds.), In *Phytophthora Its Biology, Taxonomy, Ecology and Pathology.* p. 335-350.

Srinivas, M. N. and Krishnaraj, P. U. (1992) Synergistic interaction between VA Mycorrhizal fungi and Phosphate solubilizing bacteria in Chilli (*Capsicum annuum*). *Zentralblatt fur Microbiologie* 147:126-130.

Sumitha, R. and Gaikwada, S. J. (1995) Checking *Fusarium* wilt of pigeon pea by biological means. *J. Soils and Crops* 5: 163-165.

Sulochana, K. K., Sivaprasad, P., Soumya, V. I. and Kamala Nayar, (2003) Biological control of sheath blight of rice using fluorescent pseudomonas and AMF in the field. In: Reddy M.S, Anandaraj M, Eapen S J, Sarma Y R & J W Kloepper (Eds.) Abstracts and Short papers, *International workshop on Plant Growth Promoting Rhizobacteria*, 5-10th October 2003, Calicut (p 331) Indian Institute of Spices Research. Calicut.

Suseela Bhai, R. (1998) Studies on Capsule rot (Azhukal) disease of cardamom. Ph.D Thesis, University of Calicut, Kerala, India.

Suseela Bhai, R. and Sarma, Y. R. (2003) In vitro effect of Fluorescent Pseudomonas on capsule rot infection by *Phytophthora meadii* in Cardamom (*Elettaria cardamomum* Maton). In: Reddy M.S, Anandaraj M, Eapen S J, Sarma Y R & J W Kloepper (Eds.) Abstracts and Short papers, *International workshop on Plant Growth Promoting Rhizobacteria.*, 5-10th October 2003, Calicut (p 74) Indian Institute of Spices Research. Calicut.

Suseela Bhai, R., Joseph Thomas, and Naidu, R. (1993) Biological control of Azhukal disease of cardamom caused by *Phytophthora meadii* Mc Rae. *J. Plantn. Crops* 21(suppln.): 134-139.

Suseela Bhai, R., Joseph Thomas, and Naidu, R. (1994) Evaluation of carrier media for field application of *Trichoderma* sp. in cardamom soils. *J. Plantn. Crops* 22: 50-52

Suseela Bhai, R. and Joseph Thomas, (1998) Trichoderma spp. An effective bioagent for the control of seedling rot disease in cardamom nurseries. *J. Spices and Aromatic Crops.* 7 (2): 85-88.

Suseela Bhai, R. and Kumar, A. (2005) Screening of Rhizobacteria for growth promotion and pathogen suppression in vanilla (*Vanilla planifolia* Andrews.). *Indian Phytopathol.* (In press).

Tuset, J. J., Hinarejos, C. and Garcia, J. (1990) Phytophthora foot rot control in citrus with *Myrothecium roridum. Bulletin OEPP* 20 (1) 169-176.

Trotta, A., Varese, G. C., Gnavi, E., Fusconi, A., Sampo, S. and Berta, G. (1996) Interactions between the soil borne root pathogens of *Phytophthora nicotianae* var.*parasitica* and the vesicular arbuscular fungus *Glomus mosseae* in tomato plants. *Plant Soil.* 185 (2): 199-209.

Usman, N. M., Balakrishnan, P. and Sarma, Y. R. (1996) Biocontrol of Rhizome rot of Ginger. *J. Plantn. Crops (suppl.)* 24: 184-191.

Valois, D., Fayad, K., Baradubiye, T., Garon, M., Dery, C., Brzenski, R. and Beaulieu, C. (1996) Glucanolytic actinomycetes antagonistic to *Phytophthora fragariae* var. *rubi* the causal agent of raspberry root rot. *Applied and Environmental Microbiol.* 62 (5): 1630-1635.

Van den Boogert, P. H. J. F., Jager, G. and Velvis, H. (1990) *Verticillium biguttatum* an important mycoparasite for the control of *R. solani* in Potato. In: D. Hornby (Ed.) *Biological Control of Soil-borne Plant Pathogens.* CAB International. p77-93.

Viswanathan, R. and Samiyappan, R. (2000) Efficacy of *Pseudomonas* spp. Strains against soil borne and sett borne inoculum of *Colletotrichum falcatum* causing red rot disease of sugar cane. *Sugar Tech.* 2: 26-29.

Vogt, W. and Buchenauer, H. (1997) Enhancement of biological control of combination of antagonistic fluorescent pseudomonas strains and resistance inducers against damping off and powdery mildew in cucumber. *Zeitschrift Fur Pflanzenkrankheiten und Pflanzenschutz* 104: 272-280

Weindling, R. (1932). *Trichoderma lignorum* as parasite of other soil fungi. *Phytopathol.* 22: 837-845.

Whipps, J. M. and Lumsden, R. D. (1991) Biological control of *Pythium* species. *Biocontrol Science and Technology* 1: 75-90.

Xue, B. D., Li, J., Chenn, Y. X. (1994) Studies on antagonism of *Trichoderma* sp. against six pathogenic fungi and biological control. *J. Nanjing Agric. Univ.* 18(1): 31-36.

Yang, C. H., Menge, J. A. and Cooksey, D. A. (1994) Mutations affecting hyphal colonization and Pyoverdine production in Pseudomonas antagonistic towards *P. parasitica, Applied and Environ. Microbiol.* 60 (2): 473-481.

Yeom, J., Park, V. (1995) Enhancement of plant growth and suppression of damping off of cucumber by low temperature growing Pseudomonas isolates. *Korean J. Plant Pathol.* 11 (3): 252-257.

You, M. P., Sivasithamparam, K. and Kurtboke, D. I. (1996) Actinomycetes in organic mulch used in avocado plantations and their ability to suppress *Phytophthora cinnamomi. Biology and fertility of Soils* 22 (3): 237-242.

6

Management of Bacterial Diseases

A. Kumar

1.0 INTRODUCTION

The bacteria in the genus *Ralstonia, Xanthomonas, Agrobacterium, Xylella, Pantoea, Leifsonia, Curtobacterium, Enterobactor, Pseudomonas, Corynebacterium, Erwinia* cause the most important bacterial diseases of crop plants. Few of them are directly responsible for post harvests losses during storage or transit and many of them are known to cause disease in the field. These diseases are economically important due their potential to cause crop failure in crops including spice crops. Among the bacteria genus *Ralstonia* is very important owing to its broad host range across the plant kingdom. Traditionally bacterial diseases of crops are prevented by the principle of 'Disease Avoidance' rather than disease control in the field. Their fast rate of multiplication, selective host and rapid spread in the field makes the disease control one of the challenges in the production system. The organic strategies are very well fit in the disease avoidance in the field. The preemptive strategies works exceedingly well for management of bacterial disease in any crop. Few genus such as *Ralstonia, Enterobacter* and *Xanthomonas* have been frequented in spice crops.

Ralstonia solancearum induced bacterial wilt of ginger inflicts economic losses to small and marginalized farmers in many ginger-growing countries who depend on this crop for their livelihood (Kumar and Sarma, 2003). The pathogen assumes significance owing to its wide host range, its genetic variability, the complexity of its epidemiology and multiple modes of transmission. When ginger becomes infected with bacterial wilt it is doubly devastating, because the surviving rhizome carry the bacterial inoculum to next season or newer lications besides loosing present crop. So controlling bacterial wilt in ginger could have a large impact. Various control measures has been suggested to combat the disease with limited success. The general strategies for management of bacterial wilt are: selection of healthy rhizome material from a disease free area; selection of field with no previous history of bacterial wilt; pre-plant treatment of rhizomes by application of heat or biological control agents (if any); strict phytosanitation in the field including restrictions on movement of farm workers and irrigation water across the field; clean cultivation and minimum tillage; crop rotation with non-host plants like paddy, maize, cole crops etc; insect pest and nematode control in the field; soil amendments including biological control agents. Some of these control methods in order of their priority are discussed in great detail below. Organic method for ginger production can aim the disease control strategies in two ways. One is targeted on the planting seed rhizomes, the key input in the production system and the other one is the field or soil where the material is to be planted.

2.0 STRATEGIES FOR CONTROL OF BACTERIAL DISEASE

2.1 Exploitation of host resistance

Bacterial wilt disease has been reported in 44 families of plants including economically important plant species in the families *Solanaceae, Leguminosae, Musaceae,* and *Zingiberaceae*. Among the *Zingiberaceae* members, ginger is one of the important hosts of bacterial wilt pathogen *Ralstonia solanacearum*. Disease management strategies are not very effective owing to the complex epidemiology where both seed and soil borne inoculum play an important role in disease occurrence. Selection of healthy seed rhizomes and pre-plant seed rhizome chemical and heat treatments are prerequisite for prevention of the disease. Though effective, the rhizome treatment methodologies are known to reduce 'only the primary inoculum' of the pathogen latently carried in the planting material or the rhizomes. Therefore, the practical bacterial wilt management option in any crop plants is, undoubtedly, through use of resistant cultivars of the crops.

The use of cultivars resistant to bacterial wilt is the most effective mechanism to manage the spread of bacterial wilt under organic agriculture. Resistant varieties have been reported in *Solanaceae* members. Resistant cultivars/varieties are available or reported in hosts such as tomato, potato, brinjal and chilli belonging to the family, *Solanaceae*. During last five years, Indian Institute of Spices Research, Calicut has evaluated over 650 germplasm accessions of ginger for bacterial wilt resistance. None of these accessions could resist the infection by *Ralstonia solanacearum*. In the recent years the search for resistance has been extended to other genus in the family, *Zingiberaceae*. Among the *Zingiberaceae* members such as *Curcuma amada, C. longa, C. zedoria, C. aromatica, Kaempferia galanga, Elettaria cardamomum, Zingiber zerumbet* and *Z. officinale* assayed for their reaction to ginger strain of *Ralstonia solanacearum* biovar 3, only *C. amada* Roxb, the Indian Mango Ginger (Rhizomes have an aroma of green mango) was found to resist infection when the plant was repeatedly soil inoculated with the suspension of bacterial wilt pathogen at 10^8 cells ml^{-1}. None of the plants inoculated with the pathogen-expressed wilt till the end of the crop period. Besides, the *Curcuma amada* plant yielded 200-275g of rhizome that was comparable to that of uninoculated controls. The pathogen, *R. solanacearum* could be detected in soil, root and rhizome surface of mango ginger in PCR mediated assay using *R.solanacearum* specific primers. The survival of *R. solanacearum* in soils planted with *C.amada* was confirmed by bioassay where the ginger plants transplanted in soil in the vicinity of surviving *C. amada* was found to wilt within 5-7 days. This clearly indicated that the pathogen was unable to infect the plant when inoculated in soil. This, could be due to non-recognition of the plant by *R. solanacearum*. However, the *C. amada* succumbed to wilt when the pathogen was directly delivered into the pseudostem through pinpricking. This further indicated that the plant *per se* is not antagonistic to the pathogen in soil. The *C. amada* plants hold promise for developing bacterial wilt resistant plants of ginger if the exact mechanism of resistance is understood.

2.2 Cultural methods for management of bacterial wilt of ginger

Prevention is better than cure applies well to the management of bacterial diseases; the bacterial wilt is no exception. Cultural practices play a pivotal role in the prevention and management of plant diseases. The benefits of cultural control begin with the establishment of a growing environment that favors the crop over the pathogen. Reducing plant stress through environmental modification promotes good plant health and aids in reducing damage from some plant diseases. Sanitation practices aimed at excluding, reducing, or eliminating pathogen populations are critical for management of infectious plant diseases. In order to reduce dispersal of soil borne pathogens between fields, stakes and farm equipment should be decontaminated before moving from one field to the next. Reduction of pathogen survival from one season to another may be achieved by destruction of volunteer plants and crop rotation. Avoid movement of soil from one site to another to reduce the risk of moving pathogens. Crop rotation is a very important practice, especially for soil borne disease control. For many soil borne diseases, at least a 3-year-rotation using a non-host crop will reduce pathogen populations below threshold. The bacterial wilt infected potato fields in Peru is crop rotated with maize (Elphinstone and Aley, 1993). In India, rotation with finger millet or maize reduces wilt of eggplants and tomato (Sohi *et al*, 1981).

2.2.1 Phytosanitation

Good sanitation practices play an important role in the control of the bacterial wilt of ginger. Sanitation is essential in the ginger field to protect against vascular pathogens such as *Ralstonia solanacearum*. As the crop experiences heavy rainfall during the active tillering state this may not be always possible in an open field. However, care should be taken while designing the field for ginger cultivation so that the runoff water flows into a trench, which can be diverted to an area where there is no crop. Weed control programs are in place within and around field. It has been observed that the pathogen is contained within the wilted plant as long as the plant is intact and green. The pathogen is released into the soil when the pseudostem breaks away from the rhizomes and the plant lodges on the ground. The pathogen could not be tracked in the rhizosphere soil or on the roots of the plants freshly wilted in PCR based assay whereas the pathogen could be tracked in the rhizosphere soil and soil away from the rhizosphere soil of the completely collapsed plants (Kumar unpublished data) This clearly emphasized the suitable timing of the rouging operation in the diseased garden. The affected plant must be pulled out from the soil as early as possible. The best time for the removal of the affected plant is when the plant expresses the early wilting symptoms. Any rouging operation after the collapse of the plant on the field will be counter-productive as it is dangerous to meddle with severely contaminated field. In order to execute this, the inspection of the field should be done early in the morning so that the bacterial wilted plant can be distinguished from the other normally wilted plants, which recovers towards the end of the day.

Under controlled production system the sanitation procedure can be as follows. Limited access to greenhouses is accomplished through a single entry with double doors. Production greenhouses have concrete or rock floors, raised benches and drip irrigation (Klopmeyer, 2002). There is no water flow between plants. Water quality is ensured through different treatments and routine testing (Anonymous 2002a, 2002b). Wash basins with soap and water located at the entryway help prevent the movement of organisms in and out of the greenhouse structure. Workers are clothed with lab coats or aprons and wear latex gloves. Gloves are changed or disinfected between every plant (Klopmeyer, 2002). Knives and other tools are disinfested between plants either with quaternary ammonium, sodium hypochlorite or flame sterilization (Anonymous, 2002b; Klopmeyer, 2002). Footbaths at greenhouse entries are used to prevent organisms from being tracked in and out of the greenhouse. Soilless, sterilized or solarized growth media/soil can be used. Greenhouses are cleaned at the end of the season with hypochlorite or quaternary ammonium, and in some regions, with a formaldehyde drench (Anonymous, 2002b; Klopmeyer, 2002). The plant material is culture indexed over 1-3 years under temperatures that favor expression of *R. solanacearum.* This ensures that stock is pathogen free which can be certified. Throughout the production process, greenhouses are scouted and ELISA or PCR assays routinely run to ensure clean stock (Klopmeyer, 2002).

Excessive handling of plants such as in thinning, pruning and tying may be involved in spread of pathogens, particularly bacteria. It is advisable to handle plants in the field when plants are driest. Because some pathogens can only enter the host through wounds, situations that promote plant injury should be avoided. During pruning process and harvest, workers should periodically clean their hands and tools with a disinfectant, such as isopropyl alcohol. In pot culture experiments soil amendments were effective in controlling bacterial wilt (Chang and Hsu, 1988).

2.2.2 Selection of healthy rhizome material from disease free area

The use of seed rhizomes procured from previously disease-affected areas or the known endemic areas (eg. Wayanad in Kerala, East and North Districts of Sikkim in India) as planting material (mainly out of ignorance) invariably results in severe disease when such material is planted in virgin soil or fallowed soil or even soil which has been rotated with non-host crops. This experience emphasizes the importance of healthy or pathogen free seed rhizomes in order to avoid disease outbreaks. The basic principles of disease control option by DISEASE AVOIDANCE works very well in this case. In the absence of effective chemical and biological control methods, the best possible approach would be planting of "*pathogen or disease free rhizomes*" in "*pathogen free soil or field*" in order to avoid or prevent the occurrence of bacterial wilt epidemics (Pordesimo and Raymundo 1963; Supriadi et al 2000). Rhizome borne nature of *Ralstonia solanacearum* has been proved using serological and molecular approaches (Anila 2003; Supriadi *et al.* 1995, Kumar unpublished). In spite of the availability of excellent pathogen detection technologies to detect the pathogen in rhizomes, soil and irrigation water, their use in indexing of planting material is negligible among ginger farmers. Though very sensitive and selective these techniques, for detection of *R. solanacearum* in ginger rhizomes, are not readily adapted to processing of large volumes of planting material; they are almost impractical under the instructured farming conditions of developing nations in Asia and they have not been adopted. The only method that has been practised is selection of seed rhizomes from disease free fields by visual inspection. This unscientific method of planting material selection often results in risk of triggering severe epidemics of bacterial wilt disease in India and other Southeast Asian nations. Moreover, pathogen free rhizomes are not readily available to all farmers owing to scarcity of seed material during peak season of planting especially in crops like ginger, which require one tonne of seed rhizome per acre of land. In several occasions the quality of the planting material is compromised to great extent. Testing of such planting material for freedom from phytopathogenic bacteria is an important, although not exclusive, method for control of bacterial diseases of plants. Ideally, pathogen-free or, if possible, pathogen-/disease-resistant planting material is desirable.

2.2.3 Production of wilt free rhizome for planting material

In a recent publication by Hawaiian group, a greenhouse production system has been described based on the non-circulating hydroponic method" of Kratky (1998) where the initial planting material is obtained from *in vitro* culture of ginger which is reported to be pathogen free. (Hepperly *et al.* 2004). The greenhouse production and management system using rhizomes produced from tissue-cultured plants was developed. Such as system has got the following multitude of advantages over the open field production, (Hepperly *et al.* 2004).

- The possible contamination of rhizome and soil is negligible in the controlled production system. Therefore a "clean start" is ensured by using clean seed rhizomes planted in a *Ralstonia* -free greenhouse using a *Ralstonia* free commercial growing medium.
- Homogeneity of planting unit is possible thus qualifies for DUS testing as the seed rhizome pieces are of high quality due to its selection from second-generation plants of tissue-culture origin, which allows for elimination of the off type rhizomes that may be produced from first-generation tissue-cultured plants.
- Stringent control over growing conditions is assured as the green house area is strictly monitored and protected from weather throughout the growing season, thereby reducing the chances of accidental introduction of the pathogen.
- As the individual production bag is isolated from others the phytosanitary practices can be done without flaw allowing for quick removal from the area of a plant suspected of being contaminated.
- *Ralstonia*-free seed-pieces can be regenerated year after year.
- The green house facility and production system can be monitored, cleaned and disinfected for each growing season, thus eliminating the need for a new pathogen free field.

Though advantageous, from the point of view of healthy seed rhizome production, some requirements of the controlled production system that may be considered disadvantages such as Initial infrastructure investment that is high for greenhouse and its accessories, a reliable source of irrigation water and its quality and strict sanitation practices that is needed to prevent introduction of diseases. For most bacterial pathogens, resistance is not available in cultivated hosts, and production of pathogen-free planting material requires strict certification schemes. These include (i) indexing, with subsequent removal of infected/contaminated material from the production chain; (ii) meristem and other tissue culture production systems; (iii) thermo- or chemotherapy; (iv) plant or seed surface disinfection for epiphytic bacterial pathogens; (v) avoidance or decontamination of contaminated production factors such as substrate, soil or irrigation water. Methodologies such as NCM-ELISA, PCR for indexing ginger for bacterial wilt pathogen is available (Kumar *et al* 2002, Kumar and Anandraj 2006 & Anila 2003).

These methods cannot guarantee cent per cent freedom from the pathogen or disease during crop multiplication from certified planting material, because of factors such as sampling error, experimental error and selectivity of the assay, test sensitivities, limitations of therapies (e.g. heat injury or insufficient penetration), insufficient hygiene or decontamination during planting and multiplication of clean propagating material, and post harvest manipulations during marketing. The most efficient control of bacterial diseases can be expected through a combination of the use of healthy/tested planting material and good cultivation practice, including strict crop and storage hygiene. Proper sanitation is paramount in ginger cultivation. The field should be free from contamination from flowing water and external soil. Entry should be limited to soil free equipment, machinery and personnel.

2.2.4. Selection of field with no previous history of bacterial wilt

Sites should be avoided which have a history of banana production as the nematodes, which infect banana, will infect the rhizomes and make them unsaleable. It has been observed that a soil with no history of bacterial wilt often results in healthy crops of ginger if the seed rhizomes are healthy and free from the pathogen. Traditionally ginger is grown in a fallowed soil or uncultivated forest soil in many parts of India and other countries. This kind of long crop rotation often results in a healthy crop of ginger. Another alternative is to plant them underneath perennial trees or in social amenity forests with regulated shade (Supriadi *et al* 2000). French (1994a,b) has emphasized that pathogen free soil and the use of certified seed contribute most to avoidance of brown rot of potato and it is likely that the same is true for the avoidance of bacterial wilt of ginger. Soil can also be indexed for the presence of the *Ralstonia solanacearum* by polymerase chain reaction (PCR) based assays. Techniques are available to detect the pathogen in soil using ELISA and PCR (Priou 2001; Priou *et al* 1999; Priou *et al* 2002; Kumar *et al* 2002, Kumar and Anandaraj 2006). The PCR-based method for detection of the bacterium in soil has been based on universal primers specific for *R. solanacearum* (Kumar and Anandaraj 2006).

3.0 SOIL DISINFECTION

Increasing constraints of public opinion and regulations regarding chemical pollution of the soil, irrigation water and environment by fungicides, herbicides pesticides and the potential danger associated with their continuous use have focused attention on alternative methods soil treatment, such as soil solarization. The application of chemical fumigants (methyl bromide) to soil for plant disease control has a long history dating back to 1869, and for soil steaming, back to 1893 (Baker and Cook 1974). However, the near-sterilization of agricultural soils by soil steaming or chemical fumigation is now viewed with concern since these procedures destroy many of the beneficial and essential biotic and abiotic elements of soil, including biological control of plant pathogens, which are necessary for plant health (Baker and Cook. 1974, Dawson et al. 1965). In this sense, soil solarization is an important advancement because it targets mesophilic organisms, which include most plant pathogens, without destroying the beneficial mycorrhizal fungi and the growth-promoting *Bacillus spp.* (Stapleton and DeVay, 1984; 1986); it contributes significantly to the principles of sustainable agriculture (Paul and Robertson, 1989).

3.1 Methods of soil disinfection

Disinfestations of soil for soil borne plant pathogens have been primarily based on methods of soil fumigation and soil steaming (Baker and Cook. 1974, Dawson et al 1965). However, the report of Katan *et al* in 1976 on the use of solar heating for disinfesting soil of pathogens and pests, now known as soil solarization, attracted the attention of many agriculturalists and is now widely used (Katan *et al* 1987). In contrast to soil steaming or chemical fumigation of soil, which are broad spectrum in their killing action of soil borne organisms (Baker and Cook. 1974, Dawson *et al* 1965), soil solarization is non-chemical, is selective in its action, and improves the nutrient status of soil (Katan, 1987, Stapleton and DeVay, 1986). Temperatures associated with soil steaming usually exceed the maximum temperatures (60° to 61.5°C) that can be tolerated by eukaryotic organisms and some non-spore-forming bacteria, although for some bacteria, life-limiting temperatures are over 100°C (Brock, 1978).

3.1.1 Solarization

Solarization of field prior to planting has been widely used to control soil borne pathogens and pests in various crops including potato, ginger, onion, carrot, peanut etc, without any damage to the environment as occurs when methyl bromide is used for shallow-rooted, short season crops (Katan 1981; Katan and Devay 1991; Stapleton 1994). Solarization is a hydrothermal process dependent on moisture of the sample for maximum heat transfer. The soil is covered with transparent plastic sheets to trap solar radiation in order to raise the temperature sufficient enough to suppress or eliminate soil borne pathogens and pests (Katan 1981; Katan and Devay 1991; Kumar *et al* 2005). Polyethylene is a suitable cover because it transmits the germicidal component of sunlight besides heating up the soil.

Soil temperature is a fundamental factor in the solarization effect. Maximum temperatures at a soil depth of 10 cm were 8°C higher and 20° C higher in a vinyl-mulched field in a closed plastic house, when compared with those in the mulched open field and bare open field, respectively (Kodama and Fukui, 1979). In the greenhouse treatment, the temperature range of 40 to 45^0 C was obtainable at least for several hours a day at the soil depth of 20 cm (Kodama and Fukui, 1979, 1982), within which most of soil borne pathogens are inhibited. No appreciable difference was found in soil temperatures between sheeting film of vinyl chloride and polyethylene (Sakamoto *et al.* 1986). Used film was also applicable for covering material of the soil (Sakamoto *et al* 1986). Ridges on the soil were believed to raise soil temperatures, since they received much solar radiation (Sakamoto *et al.* 1986).

In vitro laboratory experiments (Horiuchi et al 1982, Kodama and Fukui, 1979) proved that the disease-suppressing effect of heat treatment was higher in moist soils than in dry soils, both of which were infested with soil borne pathogens. It is generally accepted that wet heat causes much more serious damage on microorganisms than dry heat. In addition, higher water contents in soil would give rise to soil reduction or soil anaerobiosis, which is enhanced by airtight and high temperature conditions in solarization. Such conditions in soil possibly affect the viability and stability of mesophilic pathogenic propagules. Organic materials further accelerate both disease suppression and soil reduction, when soils were heat-treated in laboratory experiments (Horiuchi *et al.* 1982, Kodama and Fukui, 1979, 1982). Similarly, organic amendments employed in solarization may have caused an increased reductive condition in the soil, although those materials were used to improve soil fertility and soil physical properties (Kodama and Fukui, 1979). In the practical scale of solarization, however, no clear relationship has been found between soil reduction and the disease-suppressing effect (Kodama and Fukui, 1979).

Solarization can be effective against a broad spectrum of soil borne diseases caused by fungi, nematodes, bacteria etc. Effectiveness of this method is directly linked to climatic factors (Katan and Devay 1991). Solarization causes complex biological, physical and chemical changes in soil which improve plant growth, quality, and yield for up to several years (Stapleton 1994). Solarization has proven to be an effective disease management option for tomato, pepper, and eggplant production in northern parts of Florida and California (Gamliel and Stapleton 1993). The success of solarization is based on the fact that most plant pathogens and pests are mesophiles, which do not produce heat-resistant spores, and they cannot survive for long periods at high temperature. The greater the temperature less is the time needed to attain a lethal effect. Solarization studies have shown that: solarization reduces or eliminates pathogens and pests prior to planting; crop yields can be significantly increased, and; the beneficial effect of solarization can extend through several growing seasons. (Afek *et al.* 1991). Soil solarization has been studied as an option for soil borne disease control with potential for bacterial wilt control. The method should be used during the hot season and depends on weather conditions. It is important to evaluate this method in other seasons, associated with other control methods. In infested areas, only methyl bromide application reduced bacterial wilt incidence, but methyl bromide application and chicken litter amendment combined with soil solarization allowed higher potato tuber production. Solarization is positively evaluated in improving soil fertility and removing accumulated salts, owing to decomposition of organic amendments and temporary submerging of the soil, respectively.

3.1.2 Effects of Solarization on Soil borne Bacteria

Limited studies have been made on the survival of plant pathogenic bacteria in solarized soil; however, *Agrobacterium* species are highly sensitive to solarization and populations are reduced up to 72 percent (Stapleton and DeVay, 1982). Similarly, species of *Rhizobium* are heat sensitive and their populations are greatly reduced during soil solarization (Abdel-Rahim et al 1988, Katan, 1987). Since none of the plant pathogenic bacteria form spores, their decline in solarized soil is expected to parallel that of green fluorescent pseudomonads (Stapleton and DeVay, 1982). During soil solarization, populations of Oxidase negative fluorescent pseudomonads and gram-positive bacteria, including *Bacillus* species and populations of Actinomycetes has been reduced over 50 percent compared with non-solarized soil (Stapleton and DeVay, 1984). Interestingly *Pseudomonas* species quickly recolonize the soil after solarization and their populations reach high levels (Gamliel *et al* 1987).

Like the pseudomonads, some strains of *Bacillus* species are rhizosphere competent and either through aggressive growth or the production of antibiotics, they appear to be major contributors to the disease suppressiveness of soils following solarization (Stapleton and DeVay, 1984, Katan, 1987). The change in populations of *Bacillus* species during solarization is of significance as the percentage of colonies in solarized soil, which exhibited antibiosis to *Geotrichum candidum*, was increased nearly 20-fold when compared with non-solarized soil (Stapleton and DeVay, 1984). These bacteria are among those, which are rhizosphere competent and are PGPRs (Katan, 1987, Stapleton and DeVay, 1984). Although initial populations of *Bacillus* species are greatly reduced, they are spore formers and are a major component of the soil microflora. Though initially reduces *Bacillus* species are the predominant Gram-positive bacteria that survive soil solarization (Stapleton and DeVay 1982). Recolonization of solarized soils includes saprophytic bacteria, which have less stringent nutritional requirements than plant pathogens (Misaghi and Grogan 1969). Soil solarization is a useful alternative in many places for the disinfestation of soil for the control of numerous and diverse fungi and bacteria and various crop pests. Depending on several variables, soil solarization has met with great success in some situations and failure in others for managing plant diseases. With greater understanding of the components of soil solarization and their interaction, the uses of this alternative for chemical or steam treatments for disinfesting soil should increase. Anith *et al.* (2000) documented a synergistic interaction between *Pseudomonas fluorescens* and soil solarization that decreased *Ralstonia solanacearum* bacterial wilt and increased the yield of ginger.

3.1.3 Limitations of soil solarization

Though effect in a place where the ambient temperature is above 30°C the solarization has got few limitations. Soil solarization is not always sufficiently acceptable for farm management. One limitation is the inevitable problem due to meteorological restriction. The effectiveness of solarization is greatly influenced by the weather (sunshine hours) during the treating period and locality of the field. Solarization requires high soil moisture content, and the treatment is believed to be less effective in a field with poor water retentivity. The effect is relatively low for deep-rooted crops and for pathogens inhabiting deeper soil layers.

Growers regard solarization as too laborious in balance with its beneficial effects. They eventually have to employ other measures of disease control, when solarization is not fully successful. Growers prefer short-term soil disinfestation such as fumigation. Also, solarization is not attractive when diseases are satisfactorily controlled by other means, such as solanaceous rootstocks resistant to bacterial wilt. Moreover, some growers are averse to solarization due to the difficulty in regulating nutrient level in soil, and consequent plant growth. Other problems involve obvious damage due to the treatment on the material and structure of plastic houses, such as cover film and iron pipe frames.

3.2 Soil disinfection by biofumigation

One of the alternative management options for soil borne pathogens in horticulture is soil treatment by Biofumigation. Biofumigation refers to the use of plants containing biologically active compounds as rotation crops or green manures or soil amendments to suppress soil-borne pests and diseases in agricultural production systems. Research focuses on plants in the *Brassica group* because they can produce compounds similar to the commercial fumigant methane sodium that will help avoid potentially unsustainable total reliance on a single synthetic compound for management of soil-borne pests and diseases. *Brassicaceae* plants have tremendous potential to be used in crop rotations as a natural pesticide source, thereby decreasing the dependence on synthetic chemicals. Various research groups around the world have been exploring the prospective use of these plants to control insects, nematodes, weeds, and fungal pathogens.

Biofumigation refers to the suppression of soil-borne pests and pathogens by biocidal compounds, principally isothiocyanates (ITCs) released when glucosinolates (GSLs) in the tissues of *Brassica* plants are hydrolyzed in soil. Biofumigation potential of brassica in cereal farming systems was investigated by considering the profile of GSL in the roots, and the suppression of the Take-all fungus (*Gaeumannomyces graminis*) by the *Brassica* roots in soil. The results showed that canola roots contain predominantly the aromatic GSL, 2-phenylethyl GSL, which yields an ITC upon hydrolysis, which is highly toxic to cereal fungal pathogens.

Brassica species and other members of Cruciferae contain significant quantities of the thioglucoside compounds known as glucosinolates (GSLs) in their tissues. GSLs are hydrolysed by the myrosinase enzyme (present endogenously in *Brassica* tissues) to release a range of hydrolysis products including oxazolidinethiones, nitriles, thiocyanates and various forms of volatile isothiocyanates (ITCs). These hydrolysis products, in particular the ITCs, are known to have broad biocidal activity (Brown and Morra, 1997). In *Brassica* oilseed rotation crops such as canola (*B. napus*) or Indian mustard (*B. juncea*) crops are grown to maturity for seed production and shoot biomass, if incorporated, is generally mature and low in GSL concentration (Kirkegaard, *et al* 1996). An experimental result has been published on the antibacterial effect of volatile from mustard leaf discs. In an innovative test the antibacterial effect of mustard volatile is studied in a small test tube. The experiment was run by removing the tubes from the freezer and adding an appropriate amount of soil on top of the leaf (to achieve 5% tissue w/w but no mixing), followed by a suspension of bacterial inoculum in water. The tubes were then sealed and left for three days before the soil was plated to count surviving bacteria. Only volatile compounds released from the thawing leaf discs contacted the soil during the 3-day experiments, maximizing the ITC impacts and minimizing other nonvolatile mechanisms. The data clearly indicated that mustards were generally the most toxic brassica to the bacterial wilt. In a field experiment conducted with five different *Brassica* green manures including two mustards, two rapes, a radish and white mustard along with weedy fallow and soybean control together revealed the positive effect of mustard on the bacterial wilt incidence. One month after incorporation, tomatoes were transplanted onto the beds and the incidence and severity of wilt and yield of tomatoes was recorded during the growing season (Baptista *et al.* 2006a). The results showed that all of the brassica were more effective than the weedy fallow or soybean control, but that the mustard was clearly superior to other species in reducing wilt incidence and increasing yield of tomatoes. Positive effect of biofumigation on potato yield due to bacterial wilt control has been reported (Baptista *et al.* 2006b).

4.0 DISINFECTION OF PLANTING MATERIAL

4.1 Pre-plant rhizome treatment by heat

Pre plant disinfection of seed rhizomes is an important approach to control of bacterial wilt of ginger; as noted earlier contaminated planting material is one of the primary inoculum sources for field infection. A good starting material is an insurance against crop failure due to wilt. Since no chemical and biological control approaches are available for control of bacterial wilt a possible and viable alternative is heat treatment to inactivate or kill pathogenic bacteria, fungi and nematodes right on the rhizomes (Janse and Wenneker 2002). This kind of ecofriendly *in situ* decontamination would be advantageous, as it does not leave any toxic compounds on the produce or the environment. Pathogens are killed either directly by heat or weakened by sub lethal heat to the extent that they are unable to cause economical damage to the crop. This kind of rhizome treatment can very well fit into the organic production system of ginger.

Among the various means of inducing heat in the rhizomes, hot water (Tsang and Shintaku 1998), hot air, water vapor, solar energy rays and microwaves (Kumar *et al* 2005, Yathon *et al.* 1990) have been reported. Of this hot air, hot water and water vapor have been used for many years as effective treatments for planting material to rid them of various pathogenic microorganisms including nematodes. Heat treatment is known to improve the post harvest quality of several horticultural crops (Klein and Lurie 1992). (Colbran and Davis 1969). These approaches are discussed in length below.

4.1.1 Heat induction by hot air

Use of unsaturated heated air is a potential treatment for disinfection of ginger seed pieces. Heat inactivation using hot air has had wide application in control of post harvest disease and insect pests in fruit and vegetables (Couey 1989). Exposure of ginger seed rhizomes to hot air at 75 per cent RH until their center temperature reaches 49°C (112 F) for 30 minutes or 50°C (122° F) for 30 minutes results in minimal injury to the host without adverse effect either on germination or subsequent growth. Rhizomes harvested from plants grown from seed pieces inoculated with *R. solanacearum* and subsequently heat treated with hot air at 75 per cent RH until their central temperature reached 49°C for 30 minutes were free of the bacterial wilt pathogen. The bacterial wilt pathogen was destroyed in ginger seed pieces treated with hot air at 75 % RH and this method is recommended for disinfection. Heat treatment with 75% RH heated air may prove to be a convenient method to disinfect ginger seed rhizomes of the bacterial wilt pathogen. (Tsang and Shintaku 1998). Heat treatment also serves to release the seed piece from dormancy. To be effective heat treatment must be long enough to penetrate the seed piece to full depth, but not so prolonged as to be injurious to the host. Selection of similar sized seed pieces is an aid in maintaining uniform thermal gradients in a batch. There are many examples of exploitation of heat to kill pathogens without affecting the viability of the planting units (Kuniyasu 1983; Shiomi 1992; van der Hulst and de Munk 1992; Dhanvantari and Brown 1993; Tsang and Shintaku 1998). In ginger planting of seed pieces after exposure to 50°C for 30 minutes often resulted in healthy plantlets in Hawaii (Tsang and Shintaku 1998).

4.1.2 Heat induction by hot water treatment

Most insects on or in flowers, foliage and roots, including ants, foliar and root aphids, armored scales, soft scales, foliar and root mealybugs and whiteflies are killed at 49°C (120° F) from 5 to 12 minutes (Hara *et al.* 1993, 1994, 1996, Hata *et al.* 1994). In addition, hot water treatment at 49° to 50° C for 10 mm. is highly effective against burrowing, citrus, lesion and root knot nematodes (Baines *et al.* 1949, Birchfield 1954, Lear & Lider 1959, Kaplan 1982). Hot water dip is more effective against many cryptic insect and mite pests because heat penetrates plant tissue while insecticidal dips are only contact in action (Tsang *et al.* 1995). Soaking of ginger seed in hot water at 50°C for 10 minutes (Nishina *et al* 1992; Trujillo 1963) is the usual pre-plant preparation in Hawaii. Shorter exposure times give insufficient heat penetration and longer soaking periods result in heat injury to the seed piece and growth of stunted crops (Nishina *et al.*1992).

4.1.3 Heat induction by rhizome solarization

Disinfection of rhizomes with solar radiation, a method called rhizome solarization, has been developed for bacterial wilt management (Kumar *et al* 2005). This is one of the most ecofriendly and energy efficient methods available for rhizome treatment. Rhizome temperatures of 40°C and 50°C were recorded after one and two hours of solarization from 9:00 am to 11:00 am on a bright sunny day (January to May in India) (Prasheena 2003). Plants emerging from solarized rhizomes often escape the disease due to *in situ* killing of the pathogen in the seed rhizome or in the vascular tissue itself (Kumar *et al* 2005). Serological evidence for elimination of *R. solanacearum* from ginger a rhizome is reported (Prasheena 2003). When rhizomes are exposed to solar radiation the rhizome temperature rises especially in the vascular region. Incidentally the thermal inactivation point for *R. solanacearum* is 46-50°C at 30 minutes of continuous exposure *in vitro* (Kumar *et al* 2005). However, data obtained from studies *in vitro* with bacterial suspensions are not comparable to the situation where the bacterium is well protected in vascular tissues of ginger rhizomes. To achieve the requisite temperature inside the vascular tissue or in a site where the pathogen is located, 2 hours of rhizome solarization is sufficient. A temperature of over 46-49°C was recorded in almost all the locations where the trial was conducted. The consistency with which the rhizome temperature increases in the rhizome confirms the effectiveness of rhizome solarization for heat induction in rhizomes (Prasheena 2003). However, one of the major sources of variability in heat build up *vis-à-vis* the fate of pathogen, *R. solanacearum* is variation in the size and shape of the rhizome. As the size increased the heat build up was also increased. Larger rhizomes recorded 1-3°C higher temperatures than the smaller rhizomes. The variation in the heat build up in the rhizome could be due to fact that the larger seed rhizome has a larger surface area to trap the sunlight which in turn results in higher temperature in rhizome (Prasheena 2003). Relationship between rhizome size and heat buildup has been recorded (Kumar 2005, Tsang and Shintaku 1998).

Rhizomes collected from ginger plants emerged from solarized infected rhizomes tested negative for *R. solanacearum* in NCM-ELISA (Kumar *et al* 2005). This result corroborates the finding of Tsang and Shintaku (1998) that the bacterial wilt pathogen was killed due to heat exposure as assayed by PCR using primers specific for *R. solanacearum*. This could be due to heat killing of the microbial cells on ginger rhizomes including *R. solanacearum*, as the rhizome temperature recorded after two hours of rhizome solarization was 50^{0}C. The thermal death point of *R. solanacearum* has been worked out to be 46^{0}-48° C for 30 minutes of continuous exposure. The effect of rhizome solarization on microbial populations has been reported (Prasheena 2003). Tsang and Shintaku (1998) reported that *R. solanacearum* was eliminated from ginger rhizomes when the rhizome was exposed to heat for 30 minutes at 50°C. Negative results obtained in post enrichment DAS-ELISA for *R. solanacearum* in solarized rhizomes confirm that the bacterium is unable in solarized rhizomes (Anila 2003; Prasheena 2003). The assay clearly indicates that rhizome solarization is capable of disinfecting the rhizomes infected by *R. solanacearum* either artificially or naturally. The temperature generated inside the rhizome may have decreased numbers of viable bacteria in the rhizome. The technology for disinfection is scaled up to accommodate large quantity of seed rhizomes of ginger (Kumar, 2005).

5.0 BIOLOGICAL CONTROL

The Biocontrol agents have the ability to compete with the other members of soil micro flora and also to produce antibiotics or induce defuse response in the host. Examples of commercially available biocontrol agents include the fungi *Trichoderma harzianum* and *Gliocladium virens*, an actinomycete *Streptomyces griseoviridis*, and a bacterium *Bacillus subtilis*. Bacteriophages (phages) have been found as an effective biocontrol agent for the management of bacterial spot on tomato. Bacterial antagonists and avirulent strains of *R. solanacearum* are effective in the control of wilt in ground nut (He, 1990). Positive results were achieved with the antagonistic bacteria *Bacillus polymyxa* and *Pseudomonas fluorescens* (Kempe and Sequeira, 1983, Ciampi-Panno *et al*: 1989) that controlled bacterial blight on potato in laboratory experiments in the Philippines (Aspiras and Cruz, 1985). Experiments in Chile were successful using *P. fluorescens* to control *R. solanacearum in* potato in field and laboratory trials. Avirulent mutants of the bacterium have also been used in some studies (Ciampi-Panno *et al*, 1989, Gallardo and Panno 1989, Hartman and Elphinstone, 1994).

6.0. SUMMARY AND CONCLUSION

Fundamentally bacterial diseases are controlled by the principle of Disease Avoidance. The foregoing discussion was about the management of bacterial wilt disease of ginger using the principles of non-chemical methods or popularly called ecofriendly methods. The chapter dealt with the various options available for preventing bacterial disease by cultural methods, soil treatment by solarization, and biofumigation, rhizome treatment by hot water treatment, solarization and hot air treatment, and biocontrol method by introduced microorganisms. Discussions were devoted to phytosanitation, the important aspects of bacterial disease control in any crops. Undoubtedly, the biofumigation for bacterial disease control would potentially benefit the crop productivity along with the improvement of soil characteristics. More and more choice crop can be identified by thorough screening of plant species suitable for biofumigation. For rhizome treatment by heat, mechanization can be brought in to suit bulk requirement as well as to control the temperature buildup very precisely. Rhizome treatment containers can be fabricated for this purpose. The various 'non chemical' options available and documented for bacterial wilt management further underscores the need for streamlining the organic strategies for bacterial wilt control.

REFERENCES

Abdel-Rahim, M. F., Satour, M. M., Mickail, K. Y., El-Eraki, S. A., Grinstein, A., Chen, Y. and Katan, J. (1988) Effectiveness of soil solarization in furrow-irrigated soils. *Plant Dis.* 72:143-146.

Afek, U., Menge, J. A. and Johnson, E. L. V. (1991) Interaction among mycorrhizae, soil solarization, metalaxyl, and plants in the field. *Plant Dis.* 75: 665-672

Anila, G. (2003) Phenotypic characterization of bacterial wilt pathogen Ralstonia solancearum and its detection in soil and rhizomes using Double Antibody Sandwich (DAS) ELISA. *MSc Thesis* submitted to Bharathiar University. 49pp

Anith, K. N., Manomohandas, T. P. and Jayarajan, M. (2000) Integration of soil solarization and biological control with a fluorescent Pseudomonas sp. for controlling bacterial wilt *Ralstonia solanacearum* (E. F. Smith) Yabuuchi et al. of ginger. *J. Biological Control* 14:25-29.

Anonymous (2002a) *Ralstonia solanacearum* race 3 biovar 2. Memo to State and Territory agricultural Regulatory Officials. DA-2002-04, February 27, 2002. p. 1-9.

Anonymous. (2002b) *Clean Stock Production for Geraniums.* Society of American Florists. Geranium Industry-USDA Meeting, April 2, Riverdale, MD.

Aspiras, R. B., Cruz, A. R. de la (1985) Potential biological control of bacterial wilt in tomato and potato with Bacillus polymyxa FU6 and Pseudomonas fluorescens. In: *Bacterial wilt disease in Asia and the South Pacific. Proceedings of an International Workshop* held at PCARRD, Los Banos, Philippines, 8-10 October 1985 (Ed. by Persley, G.J.), pp. 89-92. ACIAR Proceedings 13.

Baines, R. C., Klotz, L. J., Clarke, O. F. and DeWolfe, T. A. (1949) Hot water treatment of orange tress for eradication of citrus nematode. *Calif. Citrograph* 34: 482-484

Baker, K. F. and Cook, R. J. (1974) *Biological Control of Plant Pathogens.* Freeman, San Francisco. 433 pp

Baptista Mírian J, Souza Ronessa B de, Pereira Welington, Lopes Carlos Alberto, Carrijo Osmar A. (2006a) Effect of soil solarization and biofumigation on tomato bacterial wilt incidence. *Hortic. Bras.* 24(2): 161-165

Baptista, Mírian Josefina, Lopes, Carlos Alberto, Souza, Ronessa B De (2006b) Effect of Soil Solarization and Biofumigation During Autumn on Bacterial Wilt Incidence and Potato Yield. *Hortic. Bras.* 24 (1): 99-102. http://www.scielo.br/Scielo.php

Birchfield, W. (1954) The hot water treatment of nematode infested nursery stock. 67: 94-96.

Brock, T. D. (1978) *Thermophillic Microorganisms and Life at High Temperatures.* Springer-Verlag, New York. 465 pp

Brown P. D. and Morra M. J. (1997) *Advan. Agron.* 61, 167-231.

Chang, M. L. and Hsu, S. T. (1988) *Suppression of bacterial wilt of tomato by soil amendments,* Plant Protection Bulletin, Taiwan 30: 349-359.

Ciampi - Panno, L., Fernandez, C., Bustamante, P., Andrade, N., Ojeda, S and, Contreras, A. (1989) Biological control of bacterial wilt of potatoes caused by *Pseudomonas solanacearum*. *American Potato J.* 66, 315-332.

Colbran, R. C. and Davis, J. .J. (1969) Studies of hot water treatment and soil fumigation for control of root knot in ginger. *Queens land J. Agric. Animal Sci.*, 26:439-445

Couey, M. H. (1989) Heat treatment for control of post harvest and insect pests of fruits. *Horticultural Sci.*, 24: 194-202

Dhanvantari, B. N., Brown, R.J. (1993) Improved seed treatments for the control of bacterial canker of tomato. Canadian *J. Plant Pathol.* 15: 201–5.

Dawson, J. R., Johnson, R. A., Adams, P. and Last, F. T. (1965) Influence of steam/air mixtures, when used for heating soil, on biological and chemical properties that affect seedling growth. *Ann. Appl. Biol.* 56:24351.

Elphinstone, J. G. and Aley, P. (1993) Integrated control of bacterial wilt of potato in the warm tropics of Peru. *In*: *Proceeding of an international bacterial wilt symposium*. Hartman, G.L. and Hayward, A.C (eds). Kaohsiung, Taiwan, ROC, 29-30 October 1992, ACIAR Proceeding 45, p276-283, ACIAR, Canberra.

French, E. R. (1994a) Integrated control of bacterial wilt of potatoes. *CIP circular* 20(2): 8-11

French, E. R. (1994b) Strategies for integrated control of bacterial wilt of potatoes. *In*: *Bacterial wilt and its causatic agent, Pseudomonas solancearum*, Hayward, A.C and Hartman, G.L. (eds.) CAB international, Wallingford, Oxon, UK, pp 199-207

Gallardo, P. B., Panno, L. C. (1989) Biological control of bacterial wilt of potato induced by *Pseudomonas solanacearum*. *Revista de Microbiologia* 20, 18-26.

Gamliel, A. and Stapleton, J. .J. (1993) Effect of chicken compost or ammonium phosphate and solarization on pathogen control, rhizosphere microorganisms, and lettuce growth. *Plant disease*, 77: 886-891.

Gamliel, A., Hadar, E. and Katan, J. (1987) Microbial phenomena related to increased growth response in solarized soils and to monoculture systems. p. 72. *In*: *Proc. 7th Congress of the Mediter*. Phytopathol. Union (Spain).

Hara, A. H., Hata, T. V., Hu, B. K. S., Kaneko, R. T. and Tenbrink, V. L. (1994) Hot-water immersion of cape jasmine cuttings for disinfestation of green scale (Homoptera: Coccidae). *J. Econ. Entomol.* 87: 1569-1573.

Hara, A. H., Hata, T. Y., Hu, B. K. S. and Tenbrink., V. L. (1993) Hot water immersion as a potential quarantine treatment against *Pseudaulacaspis cockerellii* (Homoptera: Diaspididae). *J. Econ. Entomol.* 86: 1167-1170.

Hara, A. H., Hata, T. Y., Tenbrink, V. L., Hu, B. K. S. and Kaneko, R. T. (1996) Postharvest heat treatment of red ginger flowers as a possible alternative to chemical insecticidal dip. *Postharvest Biol. Tech.* 7: 137-144.

Hartman, G. L., Elphinstone, J. G. (1994) Advances in the control of Pseudomonas solanacearum race 1 in major food crops. *In: Bacterial wilt: the disease and its causative agent, Pseudomonas solanacearum* (Ed. by Hayward, A.C., Hartman, G.L.), pp. 157-177. CAB International, Wallingford, UK.

Hata, T. Y., Hara, A. H., Nagao, M. A. and Hu, B. K. S. (1994) Hot-water disinfestation treatment and Indole-3-butyric acid stimulate rooting and shoot development of tropical ornamental cuttings. *Hort Tech.* 4: 159-162.

He, L. Y. (1990) Control of bacterial wilt of groundnut in China with emphasis on cultural and biological properties. In; Middleton, K. J., Hayward, A. C. 1990.*Bacterial Wilt of Groundnuts.* Proc. ACIAR/ICRISAT collaborative res. plan. meet., Genting Highlands, Malaysia,18-19 March. ACIAR Proc. No. 31 pp. 22-25.

Hepperly, P., Zee, F., Kai, R., Arakawa C., Meisner, M., Kratky, B., Hamamoto, K. and Sato, D. (2004). Producing bacterial wilt free ginger in Green house culture. Soil and Crop Management 8: 1-6 cooperative Extension Service College of Tropical Agriculture and Human Resources University of Hawaii, Manoa.

Horiuchi, S., Hori, M., Takashi, S. and Shimizu, K. (1982) Factors responsible for the development of clubroot-suppressing effect in soil solarization. *Bull. Chugoku Nat. Agric. Exp. Sta.* 20: 2548.

Janse, J. D. and M. Wenneker (2002) Possibilities of avoidance and control of bacterial plant diseases when using pathogen-tested (certified) or-treated planting material. *Plant Pathology* 51 (5), 523–536.

Kaplan, D. T. (1982) Eradication of Pratylenchus coffeae 'Carizo' ctrange seedling. *Proc. Fla. State Hart. Soc.* 95: 70-72.

Katan, J. (1981) Solar heating (solarization) of soil for control of soil borne pests. *Annual Review of Phytopathol.* 19: 211-36.

Katan, J. (1987) Soil solarization. p. 77-105. *In: Innovative Approaches to Plant Disease Control.* 1. Chet, Ed. John Wiley & Sons, New York

Katan, J. and DeVay, J. E. (1991) *Soil solarization.* CRC Press Inc. Boca Raton, Ann Arbor, Boston, London. p. 267.

Katan, J., Greenberger, A., Alon, H. and Grinstein, A. (1976) Solar heating by polyethylene mulching for the control of diseases caused by soilborne pathogens. *Phytopathol.* 66:683-688.

Katan, J., Grinstein, A., Greenberger, A., Yarden, O. and DeVay, J. E. (1987) The first decade (1976-1986) of soil solarization (solar heating): a chronological bibliography. *Phytoparasitica* 15:229-255.

Kempe, J., Sequeira, L. (1983) Biological control of bacterial wilt of potatoes: attempts to induce resistance by treating tubers with bacteria. *Plant Dis.* 67:499-503.

Kirkegaard, J. A., Wong, P. T. W. and Desmarchelier, J. M. (1996) *Plant Pathol.* 45, 593-603.

Klein, J. D. and Lurie, S. (1992) Heat treatment for improved postharvest quality of horticultural crops. *HortTechnology* 2: 316-320.

Klopmeyer, M. (2002) *Geranium Industry-USDA Meeting*, April 2. Riverdale, MD.

Kodama, T. and Fukui, T. (1979) Solar heating sterilization in the closed vinyl house against soil-borne diseases. I. The movements of soil temperature and determination of thermal lethal conditions for some soil-borne pathogens. *Bull Nara Pref. Agric. Exp. Sta.* 10: 71-82.

Kodama, T. and Fukui, T. (1982) Solar heating in closed plastic house for control of soilborne diseases. V. Application for control of *Fusarium* wilt of strawberry. *Ann. Phytopath. Soc. Japan* 48:570-577. (Jap., Eng. sum.).

Kratky, B. A. (1998) Experimental non-circulating hydroponics methods for growing edible ginger. *Proc. National Agricultural Plastic Congress* 27:133–137.

Kumar, A 2005. Final Report Development of rhizome disinfection technology for management of bacterial wilt in ginger IFS, Sweden p. 97.

Kumar, A. and Sarma, Y. R. (2003) Characterization of Ralstonia solanacearum causing bacterial wilt of ginger. *Indian Phytopathology*, 57:12-17

Kumar, A., Anandaraj, M. and Sarma, Y.R. (2005) Rhizome solarization and microwave treatment: a ecofriendly methods for disinfecting ginger rhizomes *In: Bacterial wilt and Ralstonia solanacearum species complex* (Eds. P.Prior, C.Allen and A.C. Hayward) American Phytopathological Society Press pp: 185-196

Kumar, A. and Anandaraj, M. (2006) Method for isolation of soil DNA and PCR based detection of ginger wilt pathogen, *Ralstonia solanacearum Indian Phytopathol.* 59(2) : 154-160

Kumar, A., Sarma, Y. R. and Priou, S. (2002) Detection of bacterial wilt pathogen, *Ralstonia solanacearum* in ginger rhizomes using post-enrichment NCM ELISA, *J. Spices and Aromatic crops*, 11: 35-40.

Kuniyasu K. (1983) The present situation of seed borne diseases and seed disinfection in Japan. *Japan Pesticide Information* 43, 13–22.

Lear, B. and Lider, L. A. (1959) Eradication of root—knot nematodes from grapevine rootings by hot water. *Plant Dis. Reptr.* 43: 314-317.

Misaghi, I. and Grogan, R. G. (1969) Nutritional and biochemical comparisons of plant-pathogenic and saprophytic fluorescent *pseudomonads. Phytopathol.* 59:1436-1450

Nakahara, L. M. (1982) Notes and Exhibitions. *Proc. Hawaiian Entomol. Soc.* 24: 8.

Nishina, M. S., Sato, D. M., Nishijima, W. T. and Manu, R. F. L. (1992) *Ginger root production in Hawaii.* Commodity Fact Sheet Gin-3 (A). Manoa. Hawaii cooperative Extension service, University of Hawaii at Manoa.

Paul, E. A. and Robertson, G. P. (1989) Ecology and the Agricultural Sciences: A false dichotomy? *Ecology* 70:1594-1597

Pordesimo, A. N. and Raymundo, S. A. (1963) Rhizome rot of ginger and its control. *Coffee Cocoa*, 5:240.

Prasheena, E. (2003) Effect of rhizome solarization on microbial population of ginger with special reference to survival of bacterial wilt pathogen Ralstonia solanacearum. *MSc Thesis* submitted to Bharathiar University, Coimbatore, India, 44 p

Priou, S. (2001). Instruction Manual for DAS-ELISA Kit for the detection of *Ralstonia solanacearum* in soil, International Potato Centre (CIP), Lima, Peru.

Priou, S., Gutarra, L., Fernandez, H. and Alley, P. (1999) Sensitive detection of *Ralstonia solanacearum* in latently infected potato tubers and soil by post enrichment ELISA-CIP Programme Report 1997-98 International Potato Centre, Lima, Peru, p 111-121.

Priou, S., Gutarra, L., Rivera, M. and Alley, P. (2002) Sensitive detection of R. solanacearum (Races 3 and 1) in soil by post enrichment DAS-ELISA. In: *3rd International Bacterial Wilt Symposium,* White River, South Africa, February 4-8, 2002, p1-8.

Sakamoto, I., Jinnoh, K., Aino, M., Yoshikura, J. and Shiwaku, K. (1986) Control of soilborne diseases of Chinese cabbage by solar heating with plastic-film mulching in the out-door field. 1. On effective soil temperature for sterilization. *Bull. Hyogo Pref. Agric. Center* 34:63-68. (Jap.).

Shiomi T, (1992) Black rot of cabbage seeds and its disinfection under a hot air treatment. *Japan Agricultural Research Quarterly* 26, 13–8.

Sohi, H. S., Rao, M. V. B., Rawal, R. D. and Kishun, R. (1981) Effect of crop rotation on bacterial wilt of tomato and eggplants. *Indian Journal of Agricultural Science,* 8:572-573.

Stapleton, J. J. and DeVay, J. E. (1982) Effect of soil solarization on populations of selected soilborne microorganisms and growth of deciduous fruit tree seedlings. *Phytopathol.* 72:323-326.

Stapleton, J. J. and DeVay, J. E. (1984) Thermal components of soil solarization as related to changes in soil and root microflora and increased growth response. *Phytopathol.* 74:255-259.

Stapleton, J. J. and DeVay, J. E. (1986) Soil solarization: a non-chemical approach for management of plant pathogens and pests. *Crop Protection* 5: 190-198.

Stapleton, J. .J. (1994) Solarization as a framework for alternative soil disinfestations strategies in the interior valleys of California. In: Proceeding of the 1994 Annual International Research Conference on Methyl Bromide Alternatives and Emission Reduction. *Agricultural and Forest Meteorology,* 81:299-323

Supriadi (2000) Bacterial wilt disease of medicinal plants caused by *Ralstonia solanacearum* and strategy to control the disease (Indonesian) Original Penyakit.layu bacteri (R.so) pada tumuhan obat dan strategi Penan ggulan gannya source. *Jurnal Penelitian and Penyembangan.* Pertanian, 19:17-26

Supriadi, Elphinstone, J. G. and Hartati S. Y (1995) Detection of latent infection of *Pseudomonas solanacearum* in ginger rhizomes and weeds by Indirect ELISA. *J. Spice and Medicinal Crops,* 3: 5-10.

Trujillo, E. E. (1963) Fusarium Yellows and rhizome rot of common ginger. *Phytopathology,* 53:1370-1

Tsang, M. C. C., Hara, A. H., Hata, T. V., Hu, B. K. S., Kaneko, R. T. and Tenbrink, V. L. (1995) Hotwater immersion unit for disinfestation of tropical floral commodities. *Applied Engineering in Agriculture* 11: 397-402.

Tsang, M. M. C. and Shintaku, U.M. (1998) Applied Engineering in Agriculture. *American society of Agricultural engineer.* 14 (2): 159-163

Van der Hulst, C. T. C., de Munk, W. J., (1992) Non-parasitic disorders in hyacinths caused by a high-temperature treatment to control yellow disease (*Xanthomonas campestris* pv. *hyacinthi*), and the analysis of gaseous emanations from the bulbs. *Acta Horticulturae,* 325: 149–55.

Yathom, S., Padova, R., Tal, S. and Ross, I. (1990) Effects of gamma radiation on the immature-stages of *Liromyza trifolii. Phytoparasitica* 18: 117-124.

7

Management of Plant Parasitic Nematodes

Santhosh J. Eapen

1.0 INTRODUCTION

Plant parasitic nematodes are among the major soil-borne crop pests that cause significant economic losses to a wide variety of crops. They cause an estimated annual crop-yield loss of US$78 billion worldwide which warrants for effective control strategies (Sasser and Freckman, 1987). Nematodes have the greatest impact on crop productivity when they attack the roots of seedlings or young plantlets (Ploeg, 2001). Nematode feeding also creates open wounds that provide entry to a wide variety of plant-pathogenic fungi and bacteria. These microbial infections are often more economically damaging than the direct effects of nematode feeding.

Plant-parasitic nematodes—the majority of which are root feeders, completing their life cycles in the root zone—are found in association with most plants. About 66 species of plant parasitic nematodes belonging to 27 genera are reported on ten major spices (Koshy *et al.*, 2005). A list of economically important ones and the symptoms they cause on some important spice crops are given in Table 1. The most predominant and widely distributed among these are root knot nematodes (*Meloidogyne* spp.). Tremendous diversity is observed in the nematode fauna present in the rhizosphere of tropical spices like black pepper, ginger and turmeric.

2.0 NEMATODE CONTROL

Among all the options for plant-parasitic nematode control, chemical control is widely used, due to its non-selective nature in controlling nematodes.

Chemical nematicides have been the primary management tool for over fifty years. Chemicals are effective but the application rates needed to kill nematodes in soil are quite high and they are highly expensive. However, the use of these chemicals is increasingly restricted because inappropriate application may

Table 1. Economically important nematode problems of major spices

Crop	Nematode species	Symptoms
Black pepper	*Radopholus similis*	Yellowing of leaves, defoliation and dieback and 'slow decline' leading to death of vines Lesions, necrosis and rotting of roots, especially of feeder roots
	Meloidogyne spp.	Yellowing of leaves (inter-veinal chlorosis), unthrifty growth and gradual decline of vines Presence of moderate to large galls on roots, root rotting
Cardamom	*Meloidogyne* spp.	Poor germination of seeds in nurseries, stunted growth, reduced tillering, narrowing, yellowing and drying of leaf tips and margins Severe galling of roots in seedlings, galls not prominent in mature plants but unusual branching of roots is common
Ginger	*Meloidogyne* spp.	Chlorosis and marginal necrosis of leaves; stunted and patchy growth, poor tillering; early maturing of plants Galling and rotting of roots and underground rhizomes
	Radopholus similis	Stunting, reduced vigour and tillering; chorotic leaves with scorched tips; infested plants mature and dry out faster Small, shallow, water soaked lesions on rhizomes
	Pratylenchus spp.	Yellowing of leaves and dry rot symptoms on rhizomes Dark brown necrotic lesions within rhizomes
Turmeric	*Meloidogyne* spp.	Stunted growth; yellowing, drying of leaf tips and margins, poor tillering; wilting and pre-mature death of plants Heavy galling and rotting in roots; infested rhizomes have dull colour
	Radopholus similis	Infested plants dry faster Shallow, water-soaked brown areas on rhizomes; extensive rotting in roots
	Pratylenchus spp.	Rhizomes less turgid and wrinkled with dark brown necrotic lesions and dry rot symptoms

have negative environmental impacts and risks to human health. The non-target effects of nematicides are too great that have an impact on the environment and the whole soil community. The chemical nematicides are now being *reappraised* in respect of environmental hazard, high costs, limited availability in many developing countries or their diminished effectiveness following repeated applications. In recent years, continuing environmental problems associated with the use of nematicides have introduced a sense of urgency into the search for alternative methods of nematode management (Kerry, 2000).

Nematode control is essentially *prevention*, because once a plant is parasitized it is impossible to kill the nematode without also destroying the host (Guerena, 2006). This warrants a sustainable approach to nematode control integrating several tools and strategies, including cover crops, crop rotation, soil solarization, least toxic pesticides, biological control and plant varieties resistant to nematode damage (Barker and Koenning, 1998).

2.1 Preventive measures

Preventing nematodes from entering uninfested areas is important. Use of certified planting material, use of soil-less growing media in greenhouses/ nurseries, good agricultural practices and phytosanitation can reduce nematode spread to a greater extent. Sanitation covers a wide range of cultural practices such as weed control, crop-residue destruction, use of selective crops as mixed or companion crops, use of nematode-free planting materials etc. Though not very powerful, these tactics, when used along with other techniques, can bring down the nematode incidence significantly.

Excellent reduction in populations of plant parasitic nematodes can be achieved through use of various heat treatments like steam, hot water and sunlight. Soil solarization, a method of pasteurization, can effectively suppress most species of nematode wherever summers are predictably sunny and warm. It has been successfully employed in many crops, as a pre-sowing treatment to bring down the initial nematode population in a field. Solarization is more detrimental to plant parasitic than saprophytic nematodes (Ostrec and Grubisic, 2003). Biofumigation by incorporating brassica or cabbage residues under plastic mulches increases the fumigation-like effect (Gamliel and Stapleton, 1993; Chellami *et al.*, 1997). Fallowing and flooding are also being used to manage nematode problems in some crops. But all these techniques are employed only under special conditions and are not reliable for economic control of nematodes when used in isolation.

2.2 Resistant crops

The use of resistant or tolerant cultivars is an economically attractive and effective means of nematode management. Resistant cultivar is more effective against sedentary endoparasitic species such as root-knot and cyst nematodes. Most of these are narrow resistance governed by dominant R genes that trigger hypersensitive reactions in host plants (Dangl and Jones, 2001). Nematode-resistant rootstocks are used in perennial crops like apple, citrus and grapes to ensure protection against these unseen pests. Resistant varieties have been developed for only a few crops and are not widely used in practice (Starr *et al.*, 2002). These must be used judiciously to reduce the selection pressure for races that are able to break down the nematode resistance. Present-day research aims at introgression of polygenic resistance (gene pyramiding) for more reliable results. Application of chemicals like DL-b-amino-n-butyric acid (BABA) acid and salicyclic acid triggered genetic resistance response and suppressed root knot nematodes (Oka *et al.*, 1999; Vasyukova *et al.*, 2003). Transgenic crop resistance to nematodes and other pests is being developed for numerous crops by various companies worldwide. But the use of genetically modified organisms is not accepted in organic production systems.

2.3 Cropping practices

Crop rotation or planting crops in special sequences is an old but effective management practice to reduce nematode populations in the soil. A general rule of thumb is to rotate crops not related to each other or broad leaf to a grass. Researchers have observed that sesame, soybean and brassicas have a nematode-suppressive effect that benefits the following crop in a rotation. Crop rotation is effective, provided crops that are resistant or non-hosts for the target nematode are rotated with susceptible crops. The usefulness of crop rotation depends on the longevity of survival stages of the nematode (van der Putten *et al.*, 2006). Rotations like these not only help prevent nematode populations from reaching economic levels, they also help control other plant diseases and insect pests. Intercropping, on the other hand, not only controls plant parasitic nematodes but also a range of other pests and diseases. The mechanisms involved are release of nematicidal compounds by some component crops and limited availability of food resources for host-specific nematodes. Allelochemicals such as polythienyls, glucosinolates, cyanogenic glycosides, alkaloids, lipids, terpenoids, steroids, triterpenoids, and phenolics, produced by certain cover crops like sudan grass, marigolds exhibit nematode suppressive characteristics. But both these methods do not fit the intensive agricultural practices but are good options in subsistence agriculture.

2.4 Botanical nematicides

Extracts or essential oils of certain plants are able to kill or repel pests, disrupt their life cycle, or discourage them from feeding. A classical example is neem and its derivatives like neem cake that are toxic to plant parasitic nematodes and is not as detrimental to beneficial free-living soil organisms (Riga and Lazarovits, 2001; Abbasi *et al.*, 2005). Higher plants have yielded a broad spectrum of active compounds (Table 2). The development of nematicidal phytochemicals has not yet reached maturity but their agricultural utilization, although currently uneconomic in many situations, offers tremendous potential.

2.5 Soil amendments

There is substantial evidence that the addition of organic matter in the form of compost or manure will decrease nematode pest populations and associated damage to crops (Stirling, 1991; Akhtar and Alam, 1993; Oka and Yermiyahu, 2002; Walker, 2004). Some sources of organic matter known to be nematode-suppressive include oilcakes, sawdust, sugarcane bagasse, bone meal, horn meal, manures, compost, and certain green manures. Reduced nematode damage from increased organic matter in soil is likely a combination of improved soil structure and fertility, alteration of the level of plant resistance, release of nemato-toxins, or increased populations of fungal and bacterial parasites and other nematode-antagonistic agents. Most nematode species can be significantly reduced by tilling in chitinous materials such as crushed shells of crustaceans (Hallmann *et al.*, 1999; Guerena, 2006). To be effective against nematodes, the organic amendments should possess low C:N ratios and have a high protein or amine-type content (Rodriguez-Kabana *et al.*, 1987). Moreover, they should be applied at high rates (>1 ton ha^{-1}) in order to have significant effect on nematodes (van der Putten *et al.*, 2006).

3.0 BIOLOGICAL CONTROL

Suppression of plant-parasitic nematodes with nematode predators, parasites or disease agents is a desirable alternative to chemicals. This emerging area is called biological control. In a nematological perspective, biological control is the reduction of nematode populations, which is accomplished through the action of living organisms other than nematode-resistant host plants, which occur naturally or through the manipulation of the environment or the introduction of antagonists. The mechanisms involved in biological control of nematodes are mainly hyper-parasitism, predation, competition and antibiosis.

Table 2. A list of phytochemicals, their source and mode of action on nematodes

Herb	Family	Nematicidal compound	Mode of action	Reference
Marigold (*Tagetes* spp.)	Asteracea	Polythienyls	Causes mortality of nematodes; inhibits host finding	Uhlenbroek and Bijloo (1958)
Mustard (*Brassica* and *Sinapsis* spp.)	Brassicaceae	Isothiocyanates Glucosinolates	Inhibits egg hatching by reacting with the sulfhydryl groups of proteins	Brown and Morra (1997)
Sudan grass (*Sorghum sudanense*)	Poaceae	Glycosides	On hydrolysis releases cyanide that is toxic to nematodes	Widmer and Abawi (2000)
Black-eyed Susan (*Rudbeckia hirta*), *Carthamus tinctorius, Helenium*sp., *Cirsium japonicum, Solidago canadensis* etc.	Asteracea	Polyacetylenes	Broad spectrum nematicidal activity	Kogiso *et al.* (1976); Kawazu*et al.* (1980); Gommers (1981)
Calabar bean (*Physostigma venenosum*), *Crotalaria spectabilis,* Madagascar periwinkle (*Catharanthus roseus*) etc.	Fabaceae and other families	Alkaloids like physostigmine, Monocrotaline, Serpentine etc.	Acetylcholinesterase inhibitor that affects nematode movements, hatching of eggs etc.	Bijloo (1965); Fassuliotis and Skucas (1969); Chandravadana *et al.* (1994) etc.
Basil (*Ocimum basilicum*), tulsi (*O. sanctum*), peppermint (*Mentha piperatum*), bottle brush (*Callistemon lanceolatus*), clove (*Eugenia caryophyllata*), *Lantana camara,* neem (*Azadirachta indica*) etc.	Lamiaceae, Myrtaceae, Geraniaceae etc.	Terpenoids, diterpenoids, triterpenoids, sesquiterpenoids etc.	Direct toxicity, hatching stimulants etc.	Malik *et al.* (1987); Oka *et al.* (2000); Shakil *et al.* (2004) and several others
Knema hookeriana, betel vine (*Piper betle*), *Rhyncholacis penicillata* etc.	Myristicaceae, Piperaceae, Podostemaceae etc.	Phenolics such as lignans, chromenes, tannins, flavanoids etc.	Different mode of actions	Evans *et al.* (1984); Mahajan *et al.* (1992); Alen *et al.*, (2000) etc.
Asparagus *(Asparagus officinalis)*	Liliaceae	Asparagusic acid	Inhibition of hatching and mortality	Takasugi *et al.* (1975)
Garlic (*Allium sativum*)	Liliaceae	Allicin	Inhibition of hatching	Gupta and Sharma (1993)
Jackbean (*Canavalla ensiformis*)	Fabaceae	Concanavalin A	Blocks nematode's chemoreceptors	Marban-Mendoza *et al.* (1987)

Hundreds of organisms, which parasitize or prey on nematodes, are reported and they belong to diverse taxa including nematode trapping or endoparasitic fungi, predatory nematodes, arthropods (e.g. mites and collembola), bacterial parasites, and predatory protozoa. A number of such organisms have been identified from spice based cropping systems and further evaluated in greenhouse and field trials for managing nematode pests of spices (Table 3). Deploying and managing biocontrol agents will likely become increasingly important components of integrated pest management programs and sustainable agricultural systems.

3.1 Fungi

Interest in biological control of nematodes was elicited as soon as plant parasitic nematodes were recognized as pests. Fungi belonging to widely divergent orders and families, from Phycomycetes to Basidomycetes, were reported as antagonists of nematodes (Mankau, 1980). Nematophagous fungi are those fungi with the capacity to capture parasitise or paralyse nematodes at all stages of their life cycles. They are divided into groups depending on their mode of infection: nematode-trapping, endoparasitic, egg and female-parasitic and toxin-producing fungi (Jansson *et al.*, 1997).

The existence of fungi that trap and prey on nematodes was noticed in 19th century. Almost after one decade, the use of predacious or nematode-trapping fungi to control root-knot nematodes of pineapple was first attempted in Hawaii (Linford *et al.*, 1938), which stimulated interest among several workers in nematode-trapping fungi. Several reviews were published accounting apparently, over 100 species of these fungi that attack nematodes and the major genera were *Arthrobotrys, Dactylaria, Dactylella,* and *Monacrosporium* (Duddington, 1954; Christie, 1960; Sayre, 1971; Barron, 1977; Mankau, 1980; Gray, 1987). Majority of nematode-trapping species were reported to be fairly ubiquitous, but their population levels in soil were quite low, as they exist in soil as mycelium rather than as spores (Gray, 1987). Nematode-trapping fungi appear to have little host specificity and can also live at varying degrees of saprophytism in soil (Persmark *et al.*, 1995).

Another group was fungal endoparasites of vermiform nematodes belonging to Chitridiomycete (*Catenaria anguillulae*), Oomycetes (*Myzocytium humicola*), Zygomycete (*Meristacrum asterospermum*), Deuteromycete (*Harposporium anguillulae*), and Basidiomycete (*Nematoctonus* sp.) (Jatala, 1986). Two obligate parasites of females of cyst nematodes viz. *C. auxiliaries* (Tribe, 1979) and *Nematophthora gynophila* (Kerry and Crump, 1980) were isolated. Several isolates of *Hirsutella rhossiliensis* were screened and most of them showed parasitization of soybean cyst nematode (Liu and Chen, 2001).

Table 3. Potential microbial agents for biological control of plant parasitic nematodes infesting spices.

Crop	Nematode	Biocontrol organism	Result obtained	Reference
Black pepper	Root knot nematodes (*Meloidogyne* spp.)	*Pochonia chlamydosporia*	Reduction in nematode population and >70 per cent increase in yield	Eapen (2003)
		Pseudomonas fluorescens	Growth promotion from 26.7 to 55.6% and nematode suppression in greenhouse trials and in nurseries	Eapen *et al.* (1997); Beena *et al.* (2001)
	Burrowing nematode (*Radopholus similis*)	*Pseudomonas fluorescens*	More than 80% suppression of nematodes in greenhouse evaluation	Beena *et al.* (2003)
		Curtobacterium luteum	More than 70% suppression in greenhouse trails	Eapen (2007)
		Bacillus megaterium	-do-	Eapen (2007)
Cardamom	Root knot nematodes (*Meloidogyne* spp.)	*Trichoderma harzianum*	Reduction in nematode population and higher number of quality seedlings	Eapen (2003); Eapen *et al.* (2000)
		Paecilomyces lilacinus	-do-	Eapen (2003)
		Pochonia chlamydosporia	-do-	Eapen (2003)
Ginger	Root knot nematodes (*Meloidogyne* spp.)	*Pochonia chlamydosporia*	Significant reduction in nematodes and increase in yield under field conditions	Ramana *et al.* (2002)
Turmeric	Root knot nematodes (*Meloidogyne* spp.)	*Fusarium oxysporum*	Significant reduction in nematodes and increase in yield under field conditions	Ramana *et al.* (2002)

The endoparasites are obligate parasites that do not form trapping organs, but use their spores that are either motile or non-motile, to infect the nematode hosts. Many of these fungi spend their vegetative lives inside infected nematodes, and in the soil, they mainly exist as spores. They generally have a more restricted host range than nematode-trapping fungi (Jansson and Lopez-Llorca, 2001). Eggs and females that are sedentary stages in the life cycle of nematodes are also vulnerable to the infection by several fungi. Some of them produce toxins which are nematicidal or nematistatic in nature (Dong and Zhang, 2006). Nematophagous fungi have long been considered promising biological agents for control of plant-parasitic nematodes (Jansson *et al.* 1985; Duponnois *et al.* 1995; Sorbo *et al.* 2003).

A large number of species have been recorded as facultative parasites of root-knot nematode females and eggs (Stirling and Mankau, 1978; Godoy *et al.*, 1982; Morgan-Jones and Rodriguez-Kabana, 1988; Stirling, 1991). The first report on the natural occurrence of *Pochonia chlamydosporia* (formerly *Verticillium chlamydosporium*) was from England (Kerry, 1974). Natural parasitism of *Meloidogyne* females and eggs by this fungus has been reported from several crops cultivated in tropical and subtropical soils across the world. Many species of *Verticillium* have been recorded from eggs of *Meloidogyne* (Godoy *et al.*, 1982; Morgan Jones and Rodriguez-Kabana, 1988). The genus *Paecilomyces*, a close relative of *Penicillium*, was first observed to be associated with nematode eggs in 1976 (Lysek, 1976). Later *P. lilacinus* was found parasitizing eggs of *M. incognita* and *M. arenari* in several countries. Two other species, *P. marquandii* from 'chinampa' soils (Marban-Mendoza *et al.*, 1992) and *P. fumosoroseus* from Cuba (Cuadra *et al.*, 2000), were equally suppressive of root-knot nematodes. *Dactyle1lla oviparasitica* has been the first parasite of *Meloidogyne* eggs to be described (Stirling and Mankau, 1978). Some isolates of *Trichoderma* spp., widely used as antagonists of many fungal diseases in several crops, were found to have good potential as a nematophagous fungus (Saifullah and Thomas, 1996; Spiegel and Chet, 1998; Sharon *et al.*, 2001; Eapen *et al.*, 2005). *Cylindrocarpon destructans* was frequently found in association with nematode eggs (Freitas *et al.*, 1995). A hyphomycete similar to the genus *Scytalidium* was isolated from black-coloured egg masses of *M. javanica* (Oka *et al.*, 1997). *Beauveria bassiana*, a known insect pathogen, parasitized eggs of *M. incognita* (Saikia, 1998). *Phoma glomerata* and *Curvularia* species were also reported to be parasitic on juveniles of *M. javanica* (Varaprasad *et al.*, 1998). *Aspergillus niger, A. terreus, Cladosporium oxysporum, Fusarium oxysporum* and *F. dimerum* (*Microdochium dimerum*) were observed inhibitory to nematodes (Goswami *et al.*, 1998). Among these, *C. oxysporum* parasitized root-knot nematode larvae and eggs.

A survey for fungi associated with root-knot nematode egg masses in India showed the highest frequency of occurrence of *F. oxysporum* followed by *F. solani, F. pallidoroseum* and *F. moniliforme* (Goswami and Uma, 1996).

Many plants harbor endosymbiotic microorganisms within their tissues and some of them viz. *F. oxysporum* (Hallman and Sikora, 1996), *Lecanicillium lecanii* (Monfort *et al.*, 2005), *Melanconium betulinum* (Schwarz *et al.*, 2004) and *Phomopsis phaseoli* (Schwarz *et al.*, 2004) are nematophagous. Antagonism of endophytic fungi is by making the plant hosts more resistant and putting the nematodes at a disadvantage. Among endophytic fungi, mycorrhizal fungi constitute an important component of the agricultural and forestry ecosystems. Arbuscular mycorrhizal fungi (AMF) have a protective role and mechanisms like improved plant nutrition, competition with pathogens for photosynthates and infection sites, local elicitation of plant defense mechanisms etc. are reported. *Glomus* species are the most diverse of the arbuscular mycorrhizal fungi and are found in many soils all over the world (Talavera *et al.*, 2002). AMF would be more effective against endoparasitic nematodes than ectoparasites (van der Putten *et al.*, 2006).

Commercially available fungal formulations include *T. harzianum, Hirsutella rhossiliensis, H. minnesotensis, Pochonia chlamydosporia, A. dactyloides,* and *P. lilacinus. P. lilacinus*, a common soil hyphomycete with a cosmopolitan distribution (Dube and Smart, 1987), is the one in which the largest number of formulated biological nematicides has been reported.

3.2 Actinomycetes

Actinomycetes like *Streptomyces* spp. are antagonistic towards plant parasitic nematodes (Dicklow *et al.*, 1993). The natural and semi-synthetic derivatives of avermectins, produced by them, are potent anthelmintic and insecticidal compounds. One such compound, Abamectin, has been commercialized to be used as a prophylactic treatment for the control of *M. incognita* (El-Nagdi and Youssef, 2004).

3.3 Bacteria

Bacteria antagonistic towards plant-parasitic nematodes include rhizobacteria, endophytic bacteria, rhizobia and bacterial symbiont of entomopathogenic nematodes. But species belonging to *Pasteuria* are the only obligate parasites of nematodes (Poinar and Hansen, 1986; Sayre and Starr, 1988). This was first described as a protozoan namely *Duboscqia penetrans* (Thorne, 1940) and was subsequently related to bacteria through electron microscopic studies and was named as *Bacillus penetrans* (Mankau, 1975).

Because of the resemblance of *B. penetrans* with *Pasteuria ramose*, a parasite of water fleas, it was renamed as *P. penetrans* (Sayre and Starr, 1985). Soon it became apparent that the bacterium constituted an assemblage of numerous pathotypes and those that parasitized primarily root-knot nematodes were called *P. penetrans*. Members of the *P. penetrans* group were found to infest 323 nematode species in about 116 genera from 80 countries (Sturhan, 1988; Chen and Dickson, 1998). However, the inability to mass-produce this fastidious bacterium *in vitro* used to be the barrier to making *P. penetrans* a marketable product.

Several root-colonizing bacteria were reported to inhibit the development of plant parasitic nematodes (Zavaleta-Mejia and van Gundy, 1982; Becker *et al.*, 1988). These included several coryneform and Gram-negative bacteria like *Serratia* (Zavaleta-Mejia and vanGundy, 1989), *Pseudomonas* spp. (Oostendorp and Sikora, 1989; Spiegel *et al.*, 1991; Kloepper *et al.*, 1992; Kluepfel *et al.*, 1993), *Bacillus* spp. (Becker *et al.*, 1988), *Enterobacter cloacae* (Duponnois *et al.*, 1999), *Agrobacterium radiobacter* (Jonathan *et al.*, 2000), and many others from egg masses of *M. incognita* (El-Sherif *et al.*, 1995).

Second stage juveniles of *M. javanica* were disorientated in the presence of the bacterial symbiont, *P. oryzihabitans* of the entomopathogenic nematode, *Steinernema abbasi* (Samaliev *et al.*, 2000). The control mechanisms include suppression of root-gall development (Chen *et al.* 2000; Jonathan *et al.* 2000), nematode reproduction (Al-Rehiayani *et al.*, 1999; Duponnois *et al.*, 1999; Hackenberg *et al.*, 2000; Jonathan *et al.*, 2000), egg hatching (Cronin *et al.*, 1997; Neipp and Becker, 1999; Insunza *et al.*, 2002; El-Nagdi and Youssef, 2004) or juvenile survival (Neipp and Becker, 1999; El-Nagdi and Youssef, 2004), nematode killing (Huang *et al.*, 2005), toxin production (e.g., Jacq and Fortuner, 1979; Ali *et al.*, 2002; Jagdale and Grewal, 2002) or endoparasitism (Davies *et al.*, 2001; Talavera *et al.*, 2002).

Several endophytic bacteria that exist in a variety of tissue types within numerous plant species, also have been found antagonistic towards nematodes like root lesion nematodes (Sturz and Kimpinski, 2004). Apart from their nitrogen fixing capabilities, some rhizobia also benefit economically important crops by protecting them from various soil-borne plant pathogens including root-knot nematodes (Siddiqui and Shaukat, 2002). Further, two genera of symbiotic bacteria, *Photorhabdus* (Hu *et al.*, 1999) and *Xenorhabdus* (Grewal *et al.*, 1999), have been reported antagonistic towards plant-parasitic nematodes. Commercial formulations of *P. penetrans*, *Bacillus thuringiensis* (available in insecticidal for mulations) and *Burkholderia cepacia* are currently available.

One limiting factor of biological control is the inconsistent performance of microbial biocontrol agents under field conditions, especially in tropical and subtropical countries, due to a range of complex, integrated, intrinsic (microbial) and external (host and environmental) factors. Lack of adequate understanding of these has hampered the development of effective biocontrol agents (Kerry, 1987). For fine tuning microbial control of nematodes, the five party interactions naturally occurring among 'host plant-nematode-soil-microbial biocontrol agent-environment' have to be studied and anlysed (Dong and Zhang, 2006). The consistency and efficacy of biological control can be increased by the proper handling of the above intrinsic and external factors and by the careful integration of different control measures. For example, crop rotation (Roberts *et al.*, 1981; Chen and Reese, 1999) and organic amendment (Jaffee *et al.*, 1994) can also affect performance of biocontrol agents (Woodward *et al.*, 2005). Nematode-trapping or closely related endoparasitic fungi have been reported being stimulated by organic amendments (Cooke, 1962; Eren and Pramer, 1978; van den Boogert *et al.*, 1994). In another experiment, soil application of *Paecilomyces lilacinus* with oil cake of neem (*Azadirachta indica*) showed synergistic effect with respect to nematode management and improvement in plant growth parameters (Anver *et al.*, 2001). By combining two or more microbial enemies the success of biological control can be improved (Meyer and Roberts, 2002).

4.0 MANAGING SOIL BIOLOGY AND SOIL HEALTH

The ultimate goal of plant-parasitic nematode control is to improve soil health and gain economic benefits. Soil health is the capacity of a soil to function within ecosystem boundaries to sustain biological productivity, maintain environmental quality, and promote plant and animal health (Doran *et al.*, 1996). A healthy soil is a biologically or chemically buffering and equilibrium system, possessing high biological diversity and providing resilience to disturbance (short-term) or stress (long-term) by pests or pathogens and most importantly, improving plant health. The basis of sustainable nematode control is the maintenance of a healthy soil food-web which can be achieved with routine application of organic matter. Organic amendments lead to improved soil structure and fertility, alteration of the level of plant resistance, release of nemato-toxins, or increased populations of indigenous fungal and bacterial parasites and other nematode-antagonistic agents (Akhtar and Malik, 2000). Natural enemy communities tend to build up relatively slowly over a period of 3-4 years under perennial crops or crops grown in monocultures (van der Putten *et al.*, 2006). Higher organic matter content increases soil's water-holding capacity, and supports thriving communities of the decomposers and predators. Nematodes regulate mineralization processes in the soil and are important participants in this underground energy-transfer system. Evidence suggests that between 30 and 50 percent of the nitrogen present in crop plants was made available by the activity of bacteria-consuming nematodes (Ingham, 1996; Ferris *et al.*, 2004). More diversity in nematode communities is observed in relatively undisturbed soil. Higher biodiversity and thus supposedly a higher soil health is observed in organic soil due to the lower plow depth and especially the use of the organic amendments and the absence of artificial fertilizer (van Diepeningen *et al.*, 2006). Agricultural intensification disrupts the biological equilibrium in the soil and the stable food-web that results in build up of pests and plant parasites (Giller *et al.*, 1997). Therefore, the presence of large populations of plant parasitic nematodes is an indication of poor ecosystem health (Berkelmans *et al.*, 2003). Therefore, it is important to actively manage soil biology using minimum-tillage practices, compost, animal manures, green manures, cover crops, and crop rotations.

5.0 CONCLUSION

Complex agro-ecosystems involving mixed cropping, rotations etc. are common in developing countries of the tropics. Biodiversity of plant-parasitic nematodes in tropical agro-ecosystems is greater than in temperate regions. Control methods targeting specific nematodes are then not very effective in reducing nematode damage in such locations. Such complex agro-ecosystems are more similar to natural ecosystems that may enhance competitive interactions, thereby preventing outbreaks of individual plant parasitic nematode species (Brinkman *et al.*, 2005). Such systems are more resilient and therefore, by monitoring the soil diversity in such systems more sustainable ways of nematode management can be identified.

REFERENCES

Abbasi, P.A., Riga, E., Conn, K.L. and Lazarovits, G. (2005) Effect of neem cake soil amendment on reduction of damping-off severity and population densities of plant-parasitic nematodes and soil-borne plant pathogens. *Canadian J. Plant Pathol.*, 27: 38-45.

Akhtar, M. and Alam, M. M. (1993) Utilization of waste materials in nematode control: A review. *Bioresource Technology*, 45: 1–7.

Akhtar, M. and Malik, A. (2000) Roles of organic soil amendments and soil organisms in the biological control of plant-parasitic nematodes: A review. *Bioresource Technology*, 74(1): 35-47.

Alen, Y., Nakajima, S., Nitoda, T., Baba, N., Kanzaki, H. and Kawazu, K. (2000). Two antinematodal phenolics from *Knema hookeriana*, a Sumatran rainforest plant. *Z. Naturforsch.*, 55c: 300–303.

Al-Rehiyani, S., Hafez, S.L., Thornton, M. and Sunderaraj, P. (1999) Effect of *Pratylenchus neglectus, Bacillus megaterium,* and oil radish or rapeseed green manure on reproductive potential of *Meloidogyen chitwoodi* on potato. *Nematropica,* 29: 37-49.

Ali, N.I., Siddiqui, I.A., Shaukat, S.S. and Zaki, M.J. (2002) Nematicidal activity of some strains of *Pseudomonas* spp. *Soil Biology and Biochemistry,* 34:1051–1058.

Anver, S., Khan, A.A. and Alam, M.M. (2001) Integrated management of root-knot and reniform nematodes with neem cake and biocontrol fungus *Paecilomyces lilacinus* on chickpea and pigeonpea. *Arch Phytopath Pflanz,* 34: 255–264.

Barker, K.R. and Koenning, S.R. (1998) Developing sustainable systems for nematode management. *Annual Review of Phytopathology,* 36: 165–205.

Barron, G.L. (1977) *The Nematode Destroying Fungi.* Guelph, Ontario: Canadian Biological Publications.

Becker, J. O., Zavaleta-Mejia, E., Colbert, S.F., Schroth, M.N., Weinhold, A.R., Hancock, J.G. and van Gundy, S.D. (1988) Effect of rhizobacteria on root knot nematodes and gall formation. *Phytopathology,* 78: 1466–1469.

Beena, B., Santhosh J. Eapen and Ramana, K. V. (2001) Antagonistic rhizobacteria of root knot nematode infesting black pepper (*Piper nigrum* L.). *National Congress on Centenary of Nematology in India-Appraisal and Future Plans,* 5-7 December 2001. Indian Agricultural Research Institute, New Delhi. p. 143.

Beena, B., Santhosh J. Eapen and Ramana, K.V. (2003) Native rhizobacteria for the biological suppression of *Radopholus similis* infesting black pepper (*Piper nigrum* L.). 6th International PGPR Workshop, 5-10 October 2003, Indian Institute of Spices Research, Calicut. pp. 12.

Berkelmans, R., Ferris, H., Tenuta, M. and van Bruggen, A.H.C. (2003). Effects of long-term crop management on nematode trophic levels other than plant feeders disappear after 1 year of disruptive soil management. *Applied Soil Ecology,* 23: 223–235.

Bijloo, J.D. (1965) The "Pisum" test: a simple method for the screening of substances on their therapeutic nematicidal activity. *Nematologica,* 11: 643–644.

Brown, P.D. and Morra, M.J. (1997) Control of soil-borne plant pests using glucosinolate-containing plants. *Advances in Agron.,* 61: 167–231.

Chandravadana, M.V., Nidiry, E.S.J., Khan, R.M. and Rao, M.S. (1994) Nematicidal activity of serpentine against *Meloidogyne incognita. Fundamental and Applied Nematology,* 17: 185–192.

Chellami, D.O., Olson S.M. , Mitchell D. J., Secker I., and McSorley R. (1997) Adaptation of soil solarization to the integrated management of soil borne pests of tomato under humid conditions. *Phytopathology,* 87: 250–258.

Chen, J., Abawi, G.S. and Zuckerman, B.M. (2000) Efficacy of *Bacillus thuringiensis, Paecilomyces marquandii*, and *Streptomyces costaricanus* with and without organic amendments against *Meloidogyne hapla* infecting lettuce. *J. Nematology,* 32: 70–77.

Chen, S.Y. and Reese, C.D. (1999) Parasitism of the nematode *Heterodera glycines* by the fungus *Hirsutella rhossiliensis* as influenced by crop sequence. *J. Nematology,* 31: 437–444.

Chen, Z.X. and Dickson, D.W. (1998) Review of *Pasteuria penetrans*: biology, ecology and biological control potential. *J. Nematology*, 30: 313-340.

Christie, J.R. (1960) "Biological control - predacious nematodes". In: Sasser, J.N. and Jenkins, W.R. (eds.) *Nematology: Fundamentals and Recent Advances with Emphasis on Plant Parasitic and Soil Forms.* Chapel Hill: University North Carolina Press. Pp. 466-468.

Cooke, R.C. (1962) Behavior of nematode-trapping fungi during decomposition of organic matter in soil. *Transactions of British Mycological Society*, 45: 314–320.

Cronin, D., Moenne-Loccoz, Y., Dunne, C. and O'Gara, F. (1997) Inhibition of egg hatch of the potato cyst nematode *Globodera rostochiensis* by chitinase producing bacteria. *European J. Plant Pathol.*, 103: 433-440.

Cuadra, R., Castaneda, R. and Rodriguez, N. (2000) Pathogenicity of a Cuban isolate of *Paecilomyces fumosoroseus* on *Meloidogyne incognita. Revista de Proteccion Vegetal*, 15: 114-117

Dangl, J.L. and Jones, J.D.G. (2001) Plant pathogens and integrated defence response to infection. *Nature*, 411: 826-833

Davies, K.G., Fargette, M., Balla, G., Daudi, A., Duponnois, R., Gowen, S.R., Mateille, T., Phillips, M.S., Sawadogo, A., Trivino, C., Vouyoukalou, E.and Trudgill, D.L. (2001) Cuticle heterogeneity as exhibited by *Pasteuria* spore attachment is not linked to the phylogeny of parthenogenetic root-knot nematodes (*Meloidogyne* spp.). *Parasitology,* 122: 111–120

Dicklow, M.B., Acosta, N. and Zuckerman, B.M. (1993) A novel *Streptomyces* species for controlling plant-parasitic nematodes. *J. Chemical Ecology,* 19: 159-173

Dong, L. Q. and Zhang, K. Q. (2006) Microbial control of plant-parasitic nematodes: A five-party interaction. *Plant Soil*, 288 (1-2): 31-45

Doran, J.W., Sarrantonio, M. and Liebig, M.A. (1996) Soil health and sustainability. *Advances in Agronomy*, 56: 1–54

Dube, B. and Smart, G.C. Jr. (1987) Biological control of *Meloidogyne incognita* by *Paecilomyces lilacinus* and *Pasteuria penetrans. J. Nematology,* 19: 222–227

Duddington, C.L. (1954) Nematode-destroying fungi in agricultural soils. *Nature,* 173: 500-501

Duponnois, R., Mateille, T. and Gueye, M. (1995) Biological characteristics and effects of two strains of *Arthrobotrys oligospora* from Senegal on *Meloidogyne* species parasitizing tomato plants. *Biocontrol Science and Technology*, 5: 517–525

Duponnois, R., Ba, A.M. and Mateille, T. (1999) Beneficial effects of *Enterobacter cloacae* and *Pseudomonas mendocina* for biocontrol of *Meloidogyne incognita* with the endospore-forming bacterium *Pasteuria penetrans. Nematology*, 1: 95-101

Eapen, S.J. (2003) *Biological Control of Plant Parasitic Nematodes of Spices*. Ph.D. Thesis, Calicut University, Kerala.

Eapen, S.J. (2007). Endophytic Bacteria for the Biological System Management of *Radopholus similis*, the Key Nematode Pest of Black Pepper (*Piper nigrum* L.) - Final Report of the DBT Ad hoc Project. *Indian Institute of Spices Research*, Calicut, India. 56 pp.

Eapen, S.J., Ramana, K.V. and Sarma, Y.R. (1997). Evaluation of *Pseudomonas fluorescens* isolates for control of *Meloidogyne incognita* in black pepper (*Piper nigrum* L.). *In Biotechnology of Spices, Medicinal and Aromatic Plants*. pp. 129-133. (Eds. S. Edison *et al.*) Indian Society for Spices, Calicut, India.

Eapen, S.J., Venugopal, M.N., Ramana, K.V. and Sarma, Y.R. (2000) *Trichoderma* spp. for the management of root knot nematodes and rhizome rot disease in cardamom nurseries. *In: Recent Advances in Plantation Crops Research*. pp. 382-386. (Eds.) N. Muraleedharan and R. Raj Kumar. Allied Publishers Limited, Mumbai, India.

Eapen, S.J., Beena, B. and Ramana, K.V. (2005) Tropical soil microflora of spice-based cropping systems as potential antagonists of root-knot nematodes. *J. Invertebrate Pathology*, 88: 218–225.

El-Nagdi, W.M.A. and Youssef, M.M.A. (2004) Soaking faba bean seed in some bio-agents as prophylactic treatment for controlling *Meloidogyne incognita* root-knot nematode infection. *J. Pesticide Science*, 77: 75–78.

El-Sherif, M.A., Ali, A.H. and Barakat, M.I. (1995) Suppressive bacteria associated with plant parasitic nematodes in Egyptian agriculture. Japanese *J. Nematology*, 24: 55-59.

Eren, J. and Pramer, D. (1978) Growth and activity of the nematode-trapping fungus Arthrobotrys conoides in soil. *In:* Microbial Ecology. Loutit, M.W. and Miles, J.A.R. (eds.) Berlin: Springer-Verlag. pp. 121–127.

Evans, P.H., Bowers, W.S. and Funk, E.J. (1984). Identification of fungicidal and nematicidal components in the leaves of *Piper betel* (Piperaceae). *J. Agric. Food Chem.*, 32: 1254–1256.

Fassuliotis, G. and Skucas, G.P. (1969) The effect of a pyrrolizidine alkaloid ester and plants containing pyrrolizidine on *Meloidogyne incognita acrita*. *J. Nematology*, 1: 287–288 (Abstr.).

Ferris, H., Venette, R.C. and Scow, K.M. (2004) Soil management to enhance bacterivore and fungivore nematode populations and their nitrogen mineralisation function. *Applied Soil Ecology*, 25: 19–35.

Freitas, L.G., Ferraz, S. and Muchovej, J.J. (1995) Effectiveness of different isolates of *Paecilomyces lilacinus* and an isolate of *Cylindrocarpon destructans* on the control of *Meloidogyne javanica*. *Nematropica*, 25: 109-115.

Gamliel, A. and Stapleton, J.J. (1993) Characterization of antifungal volatile compounds evolved from solarized soil amended with cabbage residues. *Phytopathol.*, 83: 899–905.

Giller, K.E., Beare, M.H., Lavelle, P., Izac, A.M.N. and Swift, M.J. (1997) Agricultural intensification, soil biodiversity and agroecosystem function. *Applied Soil Ecology*, 6: 3-16.

Godoy, G., Rodriguez-Kabana, R. and Morgan-Jones, G. (1982) Parasitism of eggs of *Heterodera glycines* and *Meloidogyne arenaria* by fungi isolated from cysts of *H. glycines*. *Nematropica*, 12: 111-119.

Gommers, F.J. (1981). Biochemical interactions between nematodes and plants and their relevance to control. *Helminthological Abstracts*, 50B: 9–24.

Goswami, B.K. and Uma Rao (1996) Fungi associated with the egg masses of *Meloidogyne incognita* from different agroclimatic regions in India. *Indian J. Nematology*, 26: 268-269.

Goswami, B.K., Uma Rao and Singh, S. (1998) Association of some deuteromycetous fungi with the egg masses of *Meloidogyne incognita* infecting vegetables. *Annals of Agricultural Research*, 19: 149-152.

Gray, N.F. (1987) Nematophagous fungi with particular reference to their ecology. *Biological Reviews*, 62: 245-304.

Grewal, P.S., Lewis, E.D. and Venkatachari, S. (1999) Allelopathy: a possible mechanism of suppression of plant-parasitic nematodes by entomopathogenic nematodes. *Nematology*, 1: 735–743.

Guerena, M. (2006) Nematodes: alternative controls [online]. Fayetteville, Arizona, USA: ATTRA. Available from http://www.attra.org/pest.html [Accessed 10 February 2007].

Gupta, R. and Sharma, N.K. (1993) A study of the nematicidal activity of allicin—an active principle in garlic, *Allium sativum* L., against root-knot nematode, *Meloidogyne incognita* (Kofoid and White, 1919) Chitwood, 1949. *Int. J. Pest. Manag.*, 39: 390–392.

Hackenberg, C., Muehlchen, A., Forge, T. and Vrain, T.C. (2000) *Pseudomonas chlororaphis* strain Sm3, bacterial antagonist of *Pratylenchus penetrans*. *J. Nematology*, 32: 183-189.

Hallmann, J. and Sikora, R.A. (1996) Toxicity of fungal endophyte secondary metabolites to plant parasitic nematodes and soil-borne plant pathogenic fungi. *European J. Plant Pathol*, 102:155–162.

Hallmann, J., Rodriguez-Kabana, R. and Kloepper, J.W. (1999) Chitin mediated changes in bacterial communities of the soil, rhizosphere and within roots of cotton in relation to nematode control. *Soil Biology and Biochemistry*, 31: 551-560.

Hu, K.J., Li, J.X. and Webster, J.M. (1999) Nematicidal metabolites produced by *Photorhabdus luminescens* (Enterobacteriaceae), bacterial symbiont of entomopathogenic nematodes. *Nematology*, 1: 457–469.

Huang, X.W., Tian, B.Y., Niu, Q.H., Yang, J.K., Zhang, L.M. and Zhang, K.Q. (2005) An extracellular protease from *Brevibacillus laterosporus* G4 without parasporal crystals can serve as a pathogenic factor in infection of nematodes. *Res Microbiol*, 156: 719–727.

Ingham, E. (1996) The soil food web: Its importance in ecosystem health [online]. Santa Barbara, California: Visible Light, Inc. Available from http://rain.org:80/~sals/ingham.html [Accessed on 12 February 2007].

Insunza, V., Alstrom, S. and Eriksson, K.B. (2002) Root bacteria from nematicidal plants and their biocontrol potential against trichodorid nematodes in potato. *Plant Soil*, 241: 271–278.

Jacq, V.A. and Fortuner, R. (1979) Biological control of rice nematodes using sulphate reducing bacteria. *Revue de Ne´matologie*, 2: 41–50.

Jaffee, B.A., Ferris, H., Stapleton, J.J., Norton, M.V.K. and Muldoon, A.E. (1994) Parasitism of nematodes by the fungus *Hirsutella rhossiliensis* as affected by certain organic amendments. *J. Nematology*, 26: 152–161.

Jagdale, G.B. and Grewal, P.S. (2002) Identification of alternatives for the management of foliar nematodes in floriculture. *Pest Management Science*, 58: 451-458.

Jansson, H.B. and Lopez-Llorca, L.V. (2001) "Biology of nematophagous fungi" *In*: Misra, J.K. and Horn, B.W. (eds.) Trichomycetes and Other Fungal Groups: Robert W. Lichtwardt Commemoration Volume. Enfield, NH: Science Publishers Inc. pp. 144–173.

Jansson, H.B., Jeyaprakash, A. and Zuckerman, B.M. (1985) Control of root-knot nematodes on tomato by the endoparasitic fungus *Meria coniospora*. *J. Nematology*, 17: 327–329.

Jansson, H.B., Tunlid, A. and Nordbring-Hertz, B. (1997) "Nematodes" *In*: Anke, T. (ed.) Fungal Biotechnology. Weinheim, Germany: Chapman and Hall. pp. 38–48.

Jatala, P. (1986) Biological control of plant-parasitic nematodes. *Annual Review of Phytopathology*, 24: 453-489.

Jonathan, E.I., Barker, K.R., Abdel-Alim. F.F., Vrain, T.C. and Dickson, D.W. (2000) Biological control of *Meloidogyne incognita* on tomato and banana with rhizobacteria, actinomycetes, and *Pasteuria penetrans*. *Nematropica*, 30: 231–240.

Kawazu, K., Nishii, Y. and Nakajima, S. (1980) Two nematicidal substances from roots of *Cirsium japonicum*. *Agric. Biol. Chem*, 44: 903–906.

Kerry, B.R. (1974) A fungus associated with young females of the cereal cyst nematode, *Heterodera avenae*. *Nematologica*, 20: 259-261.

Kerry, B. R. (1987) "Biological Control" *In*: Brown, R.H. and Kerry, B.R. (eds.) Principles and Practices of Nematode Control in Crops. Sydney: Academic Press. pp. 233-264.

Kerry, B. R. (2000) Rhizosphere interactions and the exploitation of microbial agents for the biological control of plant-parasitic nematodes. *Annual Review of Phytopathology*, 38: 423-441.

Kerry, B.R. and Crump, D.H. (1980) Two fungi parasitic on females of cyst-nematodes (*Heterodera* spp.). *Transactions of the British Mycological Society*, 74: 119-125.

Kloepper, J.W., Rodriguez-Kabana, R., Mc Inroy, J.A. and Young, R.W. (1992) Rhizosphere bacteria antagonistic to soybean cyst (*Heterodera glycines*) and root-knot (*Meloidogyne incognita*) nematodes: Identification of fatty acid analysis and frequency of biological control activity. *Plant Soil*, 139: 75-84.

Kluepfel, D.A., McInnis, T.M. and Zehr, E.I. (1993) Involvement of root-colonizing bacteria in peach orchard soils suppressive of the nematode, *Criconemella xenoplax*. *Phytopathol.*, 83: 1240-1245.

Kogiso, S., Wada, K. and Munakata, K. (1976) Isolation of nematicidal polyacetylenes from *Carthamus tinctorius* L. *Agric. Biol. Chem.*, 40: 2085–2089.

Koshy, P.K., Eapen, S.J. and Rakesh Pandey (2005) "Nematode Parasites of Spices, Condiments and Medicinal Plants" *In*: Luc, M., Sikora, R.S. and Bridge, J. (eds.) *Plant Parasitic Nematodes in Tropical and Subtropical Agriculture -2 Edition.*, Wallingford, U.K.: CAB International. pp. 751-791.

Linford, M.B., Yap, F. and Oliveira, J.M. (1938) Reduction of soil populations of the root-knot nematode during decomposition of organic matter. *Soil Sci.*, 45: 127-141

Liu, X.Z. and Chen, S.Y. (2001) Screening isolates of *Hirsutella* species for biocontrol of *Heterodera glycines*. *Biocontrol Science and Technology*, 11: 151-160

Lysek, H. (1976) Autodehelminthization of soil in lowland deciduous forests. *Universitatis Palackianae Olomucensis Facultatis Medicae*, 41: 73-106

Mahajan, R., Kaur, D.J. and Bajaj, K.L. (1992) Nematicidal activity of phenolic compounds against *Meloidogyne incognita*. *Nematol. Medit.*, 20:217–219.

Malik, M.S., Sangwan, N.K., Dhindsa, K.S., Verma, K.K. and Bhatti, D.S. (1987) Nematicidal efficacy of some monoterpenes and related derivatives. *Pesticides*, 21:30–32.

Mankau, R. (1975) *Bacillus penetrans* n. comb. causing a virulent disease of plant parasitic nematodes. *J. Invertebrate Pathology,* 26: 333-339.

Mankau, R. (1980) Biological control of nematode pests by natural enemies. *Annual Review of Phytopathology,* 18: 415-440.

Marban-Mendoza, N., Garcia-e, R., Dicklow, M.B. and Zuckerman, B.M. (1992) Studies on *Paecilomyces marquandii* from nematode suppressive chinampa soils. *J. Chemical Ecology,* 18: 775-783.

Marban-Mendoza, N., Jeyaprakash, A., Jansson, H-B., Damon, R.A. and Zuckerman, B.M. (1987) Control of root-knot nematodes on tomato by lectins. *J. Nematology,* 19:331–335.

Meyer, S.L.F. and Roberts, D.P. (2002) Combinations of biocontrol agents for management of plant parasitic nematodes and soil-borne plant pathogenic fungi. *J. Nematology,* 34: 1-8.

Monfort, E., Lopez-Llorca, L.V., Jansson, H.B., Salinas, J., Park, J.O. and Sivasithamparam, K. (2005) Colonisation of seminal roots of wheat and barley by egg-parasitic nematophagous fungi and their effects on *Gaeumannomyces graminis* var. *tritici* and development of root-rot. Soil Biology and Biochemistry, 37: 1229–1235.

Morgan-Jones, G. and Rodriguez-Kabana, R. (1988) "Fungi colonizing cysts and eggs" *In*: Poinar, G.O. and Jansson, H.B. (eds.) *Diseases of Nematodes* Vol. II. Boca Raton: CRC Press Inc. pp. 39-58.

Neipp, P.W. and Becker, J.O. (1999) Evaluation of biocontrol activity of rhizobacteria from *Beta vulgaris* against *Heterodera schachtii. J. Nematology,* 31: 54-61.

Oka, Y. and Yermiyahu, U. (2002) Suppressive effects of composts against the root-knot nematode *Meloidogyne javanica* on tomato. *Nematology,* 4: 891-898.

Oka, Y., Chet, I., Mor, M. and Spiegel, Y. (1997) A fungal parasite of *Meloidogyne javanica* eggs: evaluation of its use to control the root-knot nematode. *Biocontrol Science and Technology,* 7: 89-97.

Oka, Y., Cohen, Y. and Spiegel, Y. (1999) Local and systemic induced resistance to the root knot nematode in tomato by DL-b-amino-n-butyric acid. *Phytopathol.,* 89: 1138-1143.

Oka, Y., Nacar S. , Putieusky E. , Ravid U. , Zohara Y. , and Spiegal Y. (2000) Nematicidal activity of essential oils and their components against the root knot nematode. *Phytopathol.,* 90: 710–715.

Oostendorp, M. and Sikora, R.A. (1989) Seed treatment with antagonistic rhizobacteria for the suppression of *Heterodera schachtii* early root infection of sugar beet. *Revue* de *Nematologie,* 12: 77–83.

Ostrec, L.J. and Grubisic, D. (2003) Effects of soil solarization on nematodes in Croatia. *J. Pesticide Sci.* 76: 139-144.

Persmark, L., Marban-Mendoza, N. and Jansson, H.B. (1995) Nematophagous fungi from agricultural soils of Central America. *Nematropica*, 2: 117–124.

Ploeg, A. (2001) When nematodes attack is important. California Grower. October. pp. 12-13.

Poinar, G. O. and Hansen, E. L. (1986) Associations between nematodes and bacteria. *Helminthological Abstracts* Series B, 55: 61–81.

Ramana, K.V., Santhosh J. Eapen and Beena B. (2002) *Biological Control of Plant Parasitic Nematodes of Major Spice crops (Final Report).* Indian Institute of Spices Research, Calicut, India. 90 pp.

Riga, E. and Lazarovits, G. (2001) Development of an organic pesticide based on neem tree products., *Phytopathol.* 91: S141.

Roberts, P.A., Thomason, I.J. and McKinney, H.E. (1981) Influence of nonhosts, crucifers, and fungal parasites on field populations of *Heterodera schachtii. J. Nematology,* 13: 164–171.

Rodriguez-Kabana, R., Morgan-Jones, G. and Chet, I. (1987) Biological control of nematodes: Soil amendments and microbial antagonists. *Plant Soil,* 100: 237-247.

Saifullah and Thomas, B.J. (1996) Studies on the parasitism of *Globodera rostochiensis* by *Trichoderma harzianum* using low temperature scanning electron microscopy. *Afro-Asian J. Nematology,* 6: 117-122.

Saikia, M.K. (1998) Colonization of nematode eggs by *Beauveria bassiana* (Bals.) Vuill - a new report from north-east India. *Annals of AgriBio Research,* 3: 97-99.

Samaliev, H.Y., Andreoglou, F., Elawad, S., Hague, N. and Gowen, S. (2000) The nematicidal effects of the bacteria *Pseudomonas oryzihabitans* and *Xenorhabdus nematophilus* on the root-knot nematode *Meloidogyne javanica. Nematology,* 2: 507-514.

Sasser, J.N. and Freckman, D.W. (1987) "A world perspective on nematology: the role of the society" *In*: Veech, J.A. and Dickson, D.W. (eds.) Vistas on Nematology: *A Commemoration of the Twenty-fifth Anniversary of the Society of Nematologists.* Society of Nematologists, Lakeland, FL: Society of Nematologists. pp. 7–14.

Sayre, R.M. (1971) "Biotic influences in soil environment" *In*: Zuckerman, B.M., Mai, W.F. and Rohde, R.A. (eds.) *Plant-Parasitic Nematodes,* Vol. I. New York: Academic Press. pp 235-266.

Sayre, R.M. and Starr, M.P. (1985) *Pasteuria penetrans* (ex Thorne, 1940) nom. rev., comb. n., sp. n., a mycelial and endospore-forming bacterium parasitic in plant-parasitic nematodes. *Proceedings of the Helminthological Society of Washington,* 52: 149-165.

Sayre, R.M. and Starr, M.P. (1988) "Bacterial diseases and antagonism of nematodes" *In*: Poinar, G.O. and Jansson, H.B. (eds.) *Diseases of Nematodes* Vol. I. Boca Raton: CRC Press Inc. pp 69–101.

Schwarz, M., Kopcke, B., Weber, R.W.S., Sterner, O. and Anke, H. (2004) 3-Hydroxypropionic acid as a nematicidal principle in endophytic fungi. *Phytochemistry*, 65: 2239–2245.

Shakil, N. A., Prasad, D., Saxena, D. B. and Gupta, A. K (2004) Nematicidal activity of essential oils of *Artemisia annua* against root-knot and reniform nematodes. Annals of *Plant Protection Sci.*, 12: 403-408.

Sharon, E., Bar Eyal, M., Chet, I., Herrera Estrella, A., Kleifeld, O. and Spiegel, Y. (2001) Biological control of the root-knot nematode *Meloidogyne javanica* by *Trichoderma harzianum. Phytopathol.*, 91: 687-693.

Siddiqui, I.A. and Shaukat, S.S. (2002) Mixtures of plant disease suppressive bacteria enhance biological control of multiple tomato pathogens. Biology and Fertility of Soils, 36: 260-268.

Sorbo, G.D., Marziano, F. and D'Errico, F.P. (2003) Diffusion and effectiveness of the nematophagous fungus *Hirsutella rhossiliensis* in control of the cyst nematode *Heterodera daverti* under field conditions. *J. Plant Pathol.*, 85: 219–221.

Spiegel, Y. and Chet, I. (1998) Evaluation of *Trichoderma* spp. as a biocontrol agent against soil-borne fungi and plant-parasitic nematodes in Israel. *Integrated Pest Management Reviews*, 3: 169-175.

Spiegel, Y., Cohn, E., Galper, S., Sharon, E. and Chet, I. (1991) Evaluation of a newly isolated bacterium, *Pseudomonas chitinolytica* sp. nov. for controlling the root-knot nematode *Meloidogyne javanica. Biocontrol Science and Technology*, 1: 115-125.

Starr, J. L., Cook, R. and Bridge, J. (2002) *Resistance to Plant Parasitic Nematodes.* Wallingford, UK: CABI Publishing.

Stirling, G.R. (1991) *Biological Control of Plant Parasitic Nematodes.* Wallingford, UK: CABI Publishing.

Stirling, G.R. and Mankau, R. (1978) *Dactylella oviparasitica*, a new fungal parasite of *Meloidogyne* eggs. *Mycologia*, 70: 774-783.

Sturhan, D. (1988) New host and geographical records of nematode-parasitic bacteria of the *Pasteuria penetrans* group. *Nematologica*, 34: 350-356.

Sturz, A.V. and Kimpinski, J. (2004) Endoroot bacteria derived from marigolds (*Tagetes* spp.) can decrease soil population densities of root-lesion nematodes in the potato root zone. *Plant Soil*, 262: 241–249.

Takasugi, M., Yachida, Y., Anetai, M., Masamune, T. and Kegasawa, K. (1975) Identification of asparagusic acid as a nematicide occurring naturally in the roots of asparagus. *Chem. Lett.*, 1975: 43–44.

Talavera, M., Itou, K. and Mizukubo, T. (2002) Combined application of *Glomus* sp. and *Pasteuria penetrans* for reducing *Meloidogyne incognita* (Tylenchida: Meloidogynidae) populations and improving tomato growth. *Appl Entomol Zool*, 37: 61–67.

Thorne, G. (1940) *Duboscqia penetrans* n.sp. (Sporozoa, Microsporidia, Nosematidae), a parasite of the nematode *Pratylenchus pratensis* (de Man) Filipjev. *Proceedings of the Helminthological Society of Washington*, 7: 51-53.

Tribe, H.T. (1979) Extent of disease in populations of *Heterodera* with special reference to *H. schachtii*. *Annals of Applied Biology*, 92: 61-72.

Uhlenbroek, J.H. and Bijloo, J.D. (1958) Investigations on nematicides I. Isolation and structure of a nematicidal principle occurring in *Tagetes* roots. *Rec. Trav. Chim.Pays-Bas*, 77: 1004–1009.

van den Boogert, P.H.J.F., Velvis, H., Ettema, C.H. and Bouwman, L.A. (1994) The role of organic matter in the population dynamics of the endoparasitic nematophagous fungus *Drechmeria coniospora* in microcosms. *Nematologica*, 40: 249–257.

van der Putten, W.H., Cook, R., Costa, S., Davies, K.G. Fargette, M., Freitas, H. Hol, W.H.G., Kerry, B.R., Maher, N., Mateille, T., Moens, M., Pena, E. de la, Piskiewicz, A.M., Raeymaekers, A.D.W., Rodriguez-Echeverria, S. and van der Wurff, A.W.G. (2006) Nematode interactions in nature: Models for sustainable control of nematode pests of crop plants. *Advances in Agron.*, 89: 227-260.

van Diepeningen, A.D., de Vos, O.J., Korthals, G.W. and van Bruggen, A.H.C. (2006) Effects of organic versus conventional management on chemical and biological parameters in agricultural soils. *Applied Soil Ecology*, 31: 120–135.

Varaprasad, K.S., Bhat, B.N., Girish, A.G. and Lenne, J.M. (1998) Parasitism of *Meloidogyne javanica* juveniles by *Phoma glomerata* and *Curvularia* spp. *International Chickpea and Pigeonpea Newsletter*, No.5, pp. 42-43.

Vasyukova, N.I., Zinov'eva, S.V., Udalova, Zh.V., Panina, Ya.S., Ozeretskovskaya, O.L. and Sonin, M.D. (2003) The role of salicyclic acid in systemic resistance of tomato to nematodes. *Doklay Biol. Sci.*, 191: 343-345.

Walker, G. E. (2004) Effects of *Meloidogyne javanica* and organic amendments, inorganic fertilisers and nematicides on carrot growth and nematode abundance. *Nematologia Mediterranea*, 32: 181-188.

Widmer, T.L. and Abawi, G.S. (2000) Mechanism of suppression of *Meloidogyne hapla* and its damage by a green manure of Sudan grass. *Plant Disease*, 84: 562–568.

Woodward, J.E., Walker, N.R., Dillwith, J.W., Zhang, H.L. and Martin, D.L. (2005) The influence of fungicides on *Arthrobotrys oligospora* in simulated putting green soil. *Annals of Applied Biology*, 146: 115–121

Zavaleta-Mejia, E. and Gundy, S.D. van (1982) Effects of rhizobacteria on *Meloidogyne* infection. *J. Nematology*, 14: 475–476

Zavaleta-Mejia, E. and Gundy, S.D. van (1989) Volatile toxicity of *Serratia marcescens* Bizio and other bacteria on the root-knot nematode *Meloidogyne incognita* (Kofoid & White) *Chitwood. Revista Mexicana de Fitopatologia*, 7: 188-194.

8

Black Pepper

C. K. Thankamani

1.0 INTRODUCTION

Black pepper popularly known as 'King of spices' or 'Black gold' is one of the important and earliest known spices produced and exported from India. Its original home is the dense evergreen forests of Western Ghats in south India. The flavour and fragrance of this oriental spice had a magic spell in human civilization and culture since very ancient days. Expeditions in search of pepper led to the discovery of new trade routes. Wars have been fought and countries colonized to gain control over the supply of pepper and other spices. The course of Indian history would have been a different one had not the Portuguese sailor Vasco-de gamma took the historic journey from Lisbon in 1497 to Calicut on 20th May, 1498. The Greek word 'peperi', the Latin 'piper' and the English word 'pepper' all came from Sanskrit word 'pipali' which was the name for the long pepper. Black pepper is believed to have been introduced into Indonesia and the Malay Archipelago as early as the first century AD, by Arab traders. In Malaysia pepper was first grown in the north Malay Penisular as early as the Ist century AD and later on encouraged by the British into other parts of the world.

Pepper is a popular ingredient in different cusines, in the west as well as in oriental cooking. It is being used in traditional medicines in many parts of the world: Pepper is widely used in Ayurvedic treatment in India and preparations of jams in Indonesia.

Products that are certified and sold as "Organic" black pepper fetch a premium price compared to conventionally produced ones. The general guide lines for organic production describing nursery management, crop production and crop protection technologies including harvest and post harvest processing strategies, applicable to Black pepper are described here.

1.1 Area and production

Important black pepper producing countries in the world are India, Indonesia, Malaysia, Brazil, Sri Lanka (who are members of the international pepper community, IPC), Thailand, China, Vietnam, Ecuador, Madagascar, and Cambodia. Other countries where pepper is gown on small scale are Mexico, Brunei, Fiji, Samoa, Micronesia, Honduras, St. Lucia, Guatemala, Tanzania, Malawi, Zimbabwe, Benin, Kenya, Cameroon, Ethiopia, Uganda and Zambia. In India, black pepper cultivation is mainly confined to Kerala, Karnataka and Tamil Nadu, and to a certain extent in Andaman and Nicobar Islands, Goa, Pondicherry and north-eastern states (Table 1). Kerala alone contributes about 95% of total production of black pepper. However, the crop has great potential to be grown in non-traditional areas such as Andhra Pradesh, West Bengal and in entire north-eastern India. In Kerala, more than 80% of the pepper farms are small (0.2 ha) or medium (0.2 ha to 0.8 ha^{-1}) and black pepper is grown in every homestead and also trailed on arecanut, mango, jack, coconut, shade trees in coffee and tea plantations and on avenue trees.

Table 1. Area and Production of black pepper in India in major states

Average of five years (1999-2005)		
State	**Area (ha)**	**Production (MT)**
Goa	588.00	55.0
Karnataka	7378.30	2192.0
Kerala	181441.80	63012.0
Meghalaya	950.00	680.0
Tamil Nadu	4218.30	933.3
A & Nicobar	511.67	375.0
Pondicherry	10.00	10.01
Total	**195098.07**	**67257.3**

Source: Anon, 2007.

1.2 Composition and uses

1.2.1 Composition

Black pepper is used either as whole as dried berries or in powdered form, besides oil and oleoresin extraction from the berries. The aroma, flavour and medicinal property of black pepper is attributed to the volatile oils. Pepper oil contains a range of organic compounds, which belong to the group of terpenes and hydrocarbons. The major aroma compounds of black pepper are β- caryophyllene, sabinene, myrcene, limonene and linalool. The non-volatile part of black pepper which is extracted using organic solvents is called black pepper oleoresin. The major pungent alkaloid present in black pepper is piperine. In general, black pepper contains about 3-5% volatile oil, 8-16% oleoresin and 2-6% piperine. However, starch, is the predominant constituent ranging from 35% to 40% in black pepper.

1.2.2 Uses

Black pepper is the most widely used spice in the world and is used in the cuisines of both East and West. It is a popular ingredient in sauces, soups, curry powders and pickles. Black pepper is preferred in USA while mild-flavored white pepper is generally consumed in Europe. Considerable amount of black pepper is consumed in the preparation of processed meat of all kinds in many countries. The oleoresin fraction of black pepper has bactericidal and fungicidal properties. It is much employed in indigenous medicine systems as an aromatic stimulant for enhancing salivary and gastric secretions and also as a stomachic. Oil of black pepper, obtained by steam distillation of crushed black pepper, is used in the flavoring of sausages, canned meat, soups and beverages. Black pepper hulls, recovered after the preparation of white pepper, is used for flavoring tinned foods.

2.0 BOTANY AND VARIETIES

2.1 Botany

Black pepper is a perennial woody climber which can grow to 10 m or more depending on height of support trees. *P. nigrum* belongs to the family Piperaceae which has two types of branches namely, orthotropic and plagiotropic branches. Orthotropic branches are straight, upward growing, with monopodial growth habit and after climbing on a support they become woody with a thick bark and form the central axis of the vine column. The nodes are swollen, has a leaf, and each node may have 10-15 short adventitious clinging roots which adhere to the support while climbing. At the axil of each leaf of the orthotropic branches there is an axillary bud which develops into a plagiotropic branch. The plagiotropic branches are laterally growing with sympodial growth habit and produce flowers and fruits. This can be rooted and it will grow as a 'bush' and ideal for pot growing or in field as hedge. The plagiotropic branches do not produce aerial roots and the leaf size and shape are slightly different from that of the orthotropic branches (Devasahayam *et al.*, 2006). Black pepper is a shallow rooted plant with the root system within 30 -60 cm of the soil surface. On the West Coast of India, the vine flowers during May-June after the receipt of a few pre-monsoon rains. In localities where South West Monsoon rains are not considerable, the vine may flower with the North East monsoon rains. The flowers are very small and are borne on pendulous spikes, which have a catkin like appearance. The fruit is small, usually globular but sometimes slightly elongated or oval, dark or pale green turning to red or orange red when ripe.

One of the most interesting features of the cultivated black pepper is that it may be male, female or bisexual. However, purely male vines occur very rarely in nature. A majority of the cultivated black pepper vines are bisexual (Devasahayam *et al.*, 2006). The percentage of bisexual flowers ranges from 85 to 99 in the improved varieties. But in some cultivars like Cheriyakanniakadan, bisexual flowers are only 1.64% and 98.36% are female flowers. Mundi, Perambramunda and Perumkodi also have a higher percentage of female flowers.

Table 2. Black pepper diversity in Kerala

S. No	Cultivar	Remarks
A. Central and South Kerala		
1.	Aimpirian	High yielding, performance excellent in higher elevations, good in quality, but late maturing, vigorous vines.
2.	Arakulam munda	Moderate and regular bearer, medium in quality, well adapted
3.	Balankotta	Cultivar with large droopy leaves, moderate and irregular bearing.
4.	Chengannurkodi	Moderate yielder from South Kerala, medium in quality.
5.	Cheppakulamundi	Moderate yielder from Central Kerala, medium in quality
6.	Jeerakamundi	Cultivar with small leaves and short spikes, alternate bearing nature, small berries.

Contd...

Table 2. Contd...

7.	Karimunda	Most popular cultivar suitable for most of the black pepper growing areas, high yielder and medium in quality, shade tolerant.
8.	Kottanadan	A high yielding cultivar from South Kerala, drought tolerant type.
9.	Kuthiravally	A cultivar with long spikes, high yield and good quality.
10.	Kuttianikodi	A moderate yielder from Central Kerala with relatively long spikes and good spiking intensity.
11.	Malamundi	A moderate yielder, medium in quality.
12.	Narayakodi	Popular in South Kerala, moderate yielder with medium quality, not easily affected by foot rot.
13.	Neelamundi	A good yielder from Central Kerala, medium in quality, tolerant to *Phytophthora* infection.
14.	Nedumchola	A cultivar with small leaves and short spikes, moderate yielder.
15.	Neyyattinkaramundi	A cultivar from Central Kerala, medium in quality and yield.
16.	Perumkodi	A cultivar from Central Kerala, moderate in yield and quality.
17.	Poonjaranmunda	A cultivar originally from Central Kerala, sporadically found in gardens of North Kerala, moderately good in yield and quality.
18.	Thommankodi	A cultivar from Central Kerala, moderately good in yield and quality.
19.	Thulamundi	A Central Kerala cultivar, medium in yield and quality.
20.	Vattamundi	A moderate yielder from Central Kerala.
21.	Vellanamban	Moderate yielder, medium in quality.
B. North Kerala		
22.	Cheriyakaniakadan	Moderate and early bearing cultivar.
23.	Kalluvally	Good yielder, medium quality with high dry recovery, drought tolerant.
24.	Kottan	Moderate yielder, medium in quality

Contd...

Table 2. Contd...

25.	Manjamundi	Moderate yielder, medium in quality.
26.	Perambramunda	Moderate yielder, medium in quality.
27.	Vadakkan	Medium in quality and yield with relatively large berries.
28.	Valliyakaniyakadan	A cultivar with large leaves, medium in yield and quality.

Table 3. Black pepper diversity in Karnataka

S. No	Cultivar	Remarks
1	Bilimallegesara	Moderate yielder with light green spikes.
2	Kurimalai	Moderate yielder with medium quality.
3	Karimaratta	Moderate yielder with uniform bearing
4	Karimalligesara	Moderate yielder with dark green spikes
5	Malligesara	Common cultivar, good in yield.
6	Uddagara	Good in yield and medium in quality.

Source: Devasahayam *et al.*, 2006.

2.2 Varieties

2.2.1 Cultivar diversity

Cultivated black pepper, unlike wild *P. nigrum*, is bisexual. Clonal selection, hybridization and open pollinated progeny selection have been used for evolving new varieties of black pepper. Black pepper is a predominantly self pollinated (geitonogamy) perennial vine propagated by stem cuttings.

Cultivar diversity (Tables 2 & 3) is one of the principal features of diversity in black pepper. The cultivars are evolved directly from wild *P. nigrum*. Natural selection and conscious selection by human endeavor for various traits have created diversity in cultivars. Most of the vernacular names of black pepper cultivars indicate a specific feature of the vine such as colour or appearance of the vine, leaf shape, spike features or the place from which the vine originated initially. More than 100 cultivars are known today and 60-65 of them are being cultivated commonly in various regions.

2.2.2 Selection

Wide variability for yield and quality characters, occasionally even within a cultivar, is common in black pepper. Clonal selection in highly popular cultivars like Karimunda and Kottanadan have resulted in identifying superior varieties such as Sreekara, Subhakara, PLD-2, etc.

2.2.3 Open pollinated progeny selection

Three black pepper varieties, namely, Panniyur 2, Panniyur 5 and IISR Sakthi, an open pollinated progeny of Perambramunda tolerant to foot-rot disease have been released, so far, through open pollinated progeny selection. The varieties developed/released through the above strategies and their salient features are given in Table 4. Root knot nematode (*Meloidogyne incognita*) tolerant Pournami and IISR Thevam a *Phytophthora* foot rot tolerant varieties have been developed through germplasm selection.

3.0 CLIMATE AND SOIL

3.1 Climate

Black pepper is crop of humid tropics requiring 2000-3000 mm rainfall with a relative humidity of 60%-95% with a little variation in day length throughout the year. The black pepper growing tracts in India receives rainfall during monsoon seasons namely, South West monsoon (June to September) and North-East monsoon (October-November) and the pattern of rainfall is bimodal. The rainfall is received in 120 to 135 rainy days and the summer months are dry. The distribution of rainfall and drainage and moisture holding capacity of the soil are more important than the total quantum of rainfall. Once the flowering process is set in, there needs to be continuous though not heavy rainfall for proper fruit set. A dry spell for a few days during this critical period will result in spike shedding and reduction in yield (Pillai *et al.*1988). Though the crop comes up even in low rainfall areas, uniform distribution of rainfall is important for obtaining higher yield. Rainfall after stress induces profuse flowering (Pillai *et al.*1988). Black pepper does not tolerate excessive heat and dryness. Long spells of dry periods are harmful. The crop tolerates temperature range of 10^0-40^0C. Optimum soil temperature for root growth is 26^0-28^0C. (De Waard 1969; Wahid and Sitepu 1987) A hot and humid climate is good for growth and development even though it encourages the incidence of diseases. A relative humidity of 60-95 % is optimum for various stages of growth. Fifty percent shade boosted the growth of black pepper cuttings in the nursery (Senanayake and Kirthisinghae, 1983).

Table 4. Improved varieties/hybrids of black pepper and their salient features

Name	Pedigree	Released from	Yield (kg/ha)	Oleo-resin (%)	Pip-erine (%)	Essential oil (%)	Remarks
Panniyur-1	F_1 of Uthirankotta x Cheriyakaniyakadan	Pepper Research Station, Panniyur (Kerala Agriucultural University)	1242	11.8	5.3	3.5	Suited to all black pepper growing regions. Not suited to heavily shaded areas.
Panniyur-2	Open pollinated progeny selection of Balankotta	-do-	2570	10.9	6.6	3.4	Shade tolerant.
Panniyur-3	F_1 of Uthirankotta x Cheriyakaniyakadan	-do-	1953	12.7	5.2	3.1	Late maturing. Suited to all black pepper growing regions.
Panniyur-4	Clonal selection from Kuthiravally	-do-	1277	9.2	4.4	2.1	Performs well under a variety of conditions. Stable yielder.
Panniyur-5	Open pollinated progeny selection of Perumkodi	-do-	1098	12.3	5.5	3.8	Tolerant to nursery diseases and shade.
Panniyur-6	Clonal selection of Karimunda	-do-	2127	8.3	4.9	1.3	Suited to all black pepper tracts.
Panniyur-7	Open pollinated progeny of Kalluvally	-do-	1410	10.6	5.6	1.5	Suited to all black pepper tracts.
Subhakara	Clonal selection from Karimunda	Indian Institute of Spices Research, Calicut (IISR)	2352	12.4	3.4	6.0	Suited to all black pepper growing regions.
Sreekara	Clonal selection from Karimunda	-do-	2677	13.0	5.1	7.0	Suited to all black pepper growing regions.
Panchami	Germplasm selection	-do-	2828	12.5	4.7	3.4	Late maturing type suited to all black pepper growing areas.
Pournami	Germplasm selection	-do-	2333	13.8	4.1	3.4	Tolerant to root knot nematode (*Meloidogyne incognita*).
PLD-2	Clonal selection from Kottanadan	IISR, Calicut and CPCRI.	2475	15.5	3.3	3.5	Suited to Thiruvananthapuram and Kollam districts of Kerala.

Contd...

Table 4. Contd...

IISR Thevam	Clonal selection of Thevanmundi	Indian Institute of Spices Research, Calicut	2148	8.2	1.6	3.1	Tolerant to *Phytophthora* foot rot disease. Suited to high altitudes and plains.
IISR Malabar Excel	F_1 of Cholamundi x Panniyur-1	-do-	1440	13.5	3.0	3.2	Suited to high altitudes and rich in oleoresin.
IISR Girimunda	F_1 of Narayakodi x Neelamundi	-do-	2880	9.7	2.2	3.4	Suited to high altitudes.
IISR Sakthi	Open pollinated progeny of Perambramundi	-do	2253	10.2	3.3	3.7	Tolerant to *Phytophthora* foot rot disease.

Fig. 1: Some high yielding varieties of black pepper. (a) Panniyur 1 (b) Subhakara (c) IISR Thevam (*Source:* Director, IISR, Calicut)

3.2 Soil

Black pepper grows successfully between 20^0 North and 20^0 South of Equator and from sea level up to 1500 m above mean sea level. Black pepper plantations are established on a wide variety of soils, their texture varying from sandy loam to clayey loam. Moisture availability is a major factor that governs the suitability of establishing black pepper plantations. Well-drained deep loamy soil, rich in organic matter with good water holding capacity and pH ranging from 5 to 6 are adequate for growth of the plants. In soils having pH less than 5, lime is to be applied to neutralize aluminium toxicity. Well-drained soil with near neutral pH, high organic matter and high base saturation enhances productivity (Mathew *et al*, 1995) .The major black pepper growing soils in India can be broadly classified into four major orders namely, mollisols (forest loam), alfisols (red loam), oxisols (laterite), entisols (alluvium) and alfisols (red loam) (Sadanandan 1993). Soils of high yielding black pepper gardens are generally sandy to loam textured with near neutral pH, high in exchangeable bases, organic carbon and micronutrients, especially zinc. The soil conditions ideal for plant growth and development are adequate drainage, good water holding capacity of soil, and absence of rock or hard substratum within one m of the soil surface.

4.0 PLANTING MATERIAL

4.1 Mother plant selection

Black pepper can be propagated through seeds and by vegetative means. Owing to its heterozygous nature, seedlings will not breed to true to types. Hence, black pepper is propagated vegetatively mainly from cuttings. Traditional varieties adapted to agro ecological conditions in the respective areas are to be selected for organic cultivation to bring about sustainable yield.

4.2 Nursery management

The mother plants selected for production of planting materials should possess desirable attributes such as regular, high-yielding habit, free from pests and diseases, and preferably 5-12 years old. Organic planting materials are preferred for planting but in the non availability of the certified organic planting materials, conventional planting materials are allowed initially.

4.2.1 Conventional method

Mother plants possessing desirable attributes mentioned above and grown organically are selected during October-November. Runner shoots from these vines are kept coiled on wooden pegs fixed at the base of vines to prevent shoots from coming in contact with soil. The runner shoots are separated during February-March and 2-3 nodal cuttings of 20 cm length are made. These are planted either in nursery beds or polythene bags filled with potting mixture after trimming the leaves. Sufficient holes are to be provided at the base of polythene bags to ensure good drainage. Three node cuttings should be planted by keeping one node below the soil and kept under shade. Light irrigation is to be provided daily to maintain a humid and cool atmosphere around the cuttings. The cuttings will strike roots and become ready for planting in May-June when 4-5 leaves are produced. Direct field planting of cuttings results in poor establishment (Creech, 1955).

4.2.2 Rapid multiplication method

Rapid multiplication technique also called as Sri Lankan method utilizing bamboo splits was recommended for large scale production of planting materials (Bavappa and Gurusinghae 1978). A suitable area with good drainage is selected and leveled. Overhead shade is provided by using 50% shade net; coconut leaves can also be used for roofing. Trenches of 30 cm width, 45 cm depth and of convenient length are taken and filled with soil, sand and farmyard manure (1:1:1). Bamboos of 8-10 cm diameter are selected and cut into 1.25-1.50 m long pieces and split into halves keeping the septa intact. Coal tar is smeared to prolong the life of bamboo splits. The bamboo splits are arranged at an angle of 45° alternatively either side on straight wooden poles or strong supports fixed on small supports from ground and tied to each other with coir rope at the free end. Rooted cuttings are planted in the trench, one for each bamboo split.

As the cuttings start growing, the bamboo splits are filled with rooting mixture composed of farmyard manure, coir dust and sand in equal proportions. Each tender node is tied carefully to the bamboo using banana fibre, so that every node is in contact with the rooting medium. For rapid growth, daily irrigation through rose can is essential. Vermiwash may be applied (100 ml per bamboo) at monthly intervals. When the vines reach to the top of the bamboo, the tip is nipped off and the vine is crushed at the base of 3rd or 4th node from the ground, to activate the buds. After 7-10 days, the vine is cut at the crushed point and removed from the bamboo with the roots intact and with the adhering soil. The vine is cut into single nodded pieces and each cutting is planted in a polythene bag filled with fortified potting mixture of soil, sand and farmyard manure (2:1:1) or solarized soil enriched with biocontrol agents or vermicompost.

After planting in the bamboo, the first harvest of cuttings can be done after 3-3½ months and the subsequent harvest at every 2-2½ months. Each rooted vine can give about 10 cuttings in one harvest and about 40 cuttings will be obtained in a year. A shed of 6 m x 24 m would accommodate 600 bamboo splits. On an average 20,000 cuttings can be produced annually by this method. The method is thus advantageous for producing a large number of rooted cuttings within a short period, throughout the year. The cuttings are also robust due to the abundance of roots leading to more than 90% establishment in the field.

4.2.3 Serpentine method

This is the best propagation technique for black pepper. Three node cuttings planted in polythene bags are kept in a corner of the nursery. When the plant develops two leaves they are trailed horizontally in polythene bags containing solarized potting mixture fortified with VAM (100cc/kg potting mixture) and *Trichoderma* (1g/kg potting mixture) kept below each tender node. Each node should be pressed into the mixture with polythene bags with 'V' shaped midribs of coconut leaves. As new shoots arise these are to be trailed horizontally in polythene bags containing potting mixture. Once 20 nodes get rooted, the first 10 polythene bags with the rooted nodes should be separated by cutting at the inter nodes. The internodal stub should be pushed back into the potting mixture. These stubs also produce a secondary root system. Daily irrigation is to be given using a rose can. Application of vermiwash or *Pseudomonas* @ 25ml/plant encourages the growth of the cuttings. After 3 months these cuttings are ready for planting in the field. On an average 60 cuttings can be obtained in a year by this method from each mother cutting. Serpentine method is simple, cheap and quick and suited to small and marginal farmers. The recovery percentage is higher compared to rapid multiplication technique. Increased biomass production of black pepper plants was observed for the cuttings raised in media consists of coir pith. Compost and, granite powder in 2:1 proportion with *Azospirillum* and *Phosphobacteria* as nutrient source (Thankamani *et al,* 2006).

4.2.4 Soil solarization and potting mixture preparation

Soil solarization is a soil disinfection practice achieved by covering moist soil with transparent polythene film during the period of high temperature and intense solar radiation. For solarization, beds of 1 meter width, 20-30 cm height and convenient length may be prepared after removing the pebbles present on the soil surface. Incorporate adequate quantity of organic manure in the soil after digging and irrigate at the rate of 5l/m$^{2.}$ The beds may be covered with polythene sheet of 300 μ thicknesses. Seal the edge of the sheet with soil to keep it in position to maintain the required temperature. Soil, sand and farm yard manure 2:1:1 proportion can also be used for solarisation.

Adequate soil moisture is necessary during solarisation to increase the thermal sensitivity of the target organism, to improve heat conduction in the soil and to enable biological activity. Solarisation should be done during March to May under the Indian conditions when the solar radiation is most intense and the soil should be covered for 45 to 60 days.

4.3 Management of diseases and pests in the nursery

Various diseases, nematodes and insect pests occur in the nursery and affect the growth of cuttings.

4.3.1 *Phytophthora infection*

Phytophthora infections are noticed on leaves, stem and roots of cuttings in the nursery. Dark spots with fimbriate margins appear on the leaves, which on spread rapidly resulting in defoliation. The infections on the stem are seen as black lesions which result in blight. The symptom on the roots appear as rotting of the entire root system.

Spraying Bordeaux mixture 1% prevents the disease. The potting mixture may be sterilized through solarization. To the sterilized mixture, biocontrol agents such as VAM @ 100 cc/kg of soil (containing 500 infective propagules) and *Trichoderma harzianum* @ 1g/kg of soil (containing 10^8 cfu/g) may be added at the time of filling of nursery mixture in polythene bags. *Pseudomonas fluorescence* (IISR-6) may also be added to the potting mixture @ 1 g/ kg of soil (containing 10^8 cfu/g) to enhance growth of cuttings. If Bordeaux mixture is used, care must be taken to prevent dripping of the fungicide to the potting mixture. Hence, application of biocontrol agents may not be done simultaneously with Bordeaux mixture spray.

4.3.2 *Anthracnose*

The disease is caused by *Colletotrichum gloeosporioides* which infects the leaves causing yellowish brown to dark brown irregular leaf spots with a chlorotic halo. Spraying Bordeaux mixture 1% is effective against the disease.

4.3.3 *Leaf rot*

The disease is caused by *Rhizoctonia solani* and is often serious during April-May when warm humid conditions prevail. The fungus infects both leaves and stems. Grey sunken spots and mycelial threads appear on the leaves and the infected leaves are attached to one another with the mycelial threads. On stems, the infection occurs as dark brown lesions which spread both upwards and downwards. The new flushes subtending the points of infection gradually droop and dry up. A prophylactic spray with Bordeaux mixture 1% prevents the disease.

4.3.4 *Basal wilt*

The disease is mainly noticed in nurseries during June-September and is caused by *Sclerotium rolfsii*. Grey lesions appear on stems and leaves. On the leaves white mycelial are seen at the advancing edges of the lesions. The mycelial threads later girdle the stem resulting in drooping of leaves beyond the point of infection and in advanced stages the rooted cuttings dry up. Small whitish to cream coloured grain like sclerotia bodies appear on the mature lesions. The disease can be controlled, if noticed early, by adopting phytosanitory measures. The affected cuttings along with defoliated leaves should be removed and destroyed. All the other cuttings should be sprayed with Bordeaux mixture 1%.

4.3.5 *Viral infections*

Vein clearing, mosaic yellow specks, mottling and small leaf are the most obvious symptoms for identifying viral infections in the nursery. As virus is systematic in nature, primary spread occurs through planting material since black pepper is vegetatively propagated. When infected plants are used as source of planting material, the cuttings will also be infected. Hence, selection of virus-free healthy mother plants is very important. Secondary spread of the disease occurs through insect such as aphids and mealy bugs. Besides, regular inspection and removal of infected plants should also be done. Procurement of certified materials is the best method to contain virus.

4.3.6 *Nematode infestation*

Root-knot nematodes (*Meloidogyne incognita*) and the borrowing nematode (*Radopholus similis*) are the two important nematode species infesting rooted cuttings in the nursery. The damage caused to roots by nematodes result in poor growth, foliar yellowing and interveinal chlorosis of leaves. *R. similis* causes elongate dark brown necrotic lesions on roots at points of entry, which coalesce and result in root rotting. *M. incognita* causes elongate thick swelling on thick primary roots and typical galls or knots on secondary and fibrous roots. These galls gradually disintegrate leading to rotting of tissue.

The nursery mixture can be sterilized by solarization. The solarized nursery mixture may be fortified with biocontrol agents such as *Pochonia chlamydosporia* or *Trichoderma harzianum* @ 1- 2g/kg of soil, the product containing $10^{6.5}$ cfu fungus/g of substrate.

4.3.7 Leaf gall thrips (Liothrips karnyi Bagn*)*

Leaf gall thrips infest tender leaves causing the leaf margins to curl downwards and inwards resulting in the formation of marginal, tubular, hypophyllous galls. As the infested leaves grow they become crinkled, malformed and reduced in size and may also turn chlorotic later at the margins. A few predators have been recorded on leaf gall thrips in the field among which *Montandoniola moraguesi* (Anthocordiae) and *Andothrips flavipes* (Phlaeothripidae) are voracious feeders on all the stages of the pest.

4.3.8 Scale insects

The mussel scale (*Lepidosaphes saphes* Green) encrusts main stems and mature leaves of older cuttings resulting in chlorotic patches, yellowing and drying of leaves and mortality of cuttings. Spraying neem oil 0.3% or Neemgold 0.06% or fish oil rosin 3% twice at 21 day intervals is effective against the pest infestation. *Protopulvinaia longivlvata is observed on mature leaves of older cuttings.* The pest infestation results in yellowing and wilting of affected plants and the subsequent attack by sooty mould accelerate the deterioration of the infested plants.

4.3.9 Mealy bugs

Mealy bugs such as *Planococcus* spp., *Pseudococcus* spp. and *Ferrisia virgata* infest tender shoots and leaves and cause the plants to wilt. Colonies of root mealy bug (*Planococcus* sp.) occur at the basal portion of the stem under the soil and on roots causing yellowing, wilting and mortality of cuttings. The pest infestation is generally more serious on cuttings affected with *Phytophthora* and nematodes.

4.3.10 Gall midge

Maggots of the gall midge *Cecidomyia malabarensis* develop within tender leaf petioles, leaf veins and shoots of cuttings resulting in swelling of infested tissue.

5.0 PLANTING AND AFTER CARE

5.1 Conversion and isolation

The entire pepper plantation can be converted to organic production when pepper is grown as sole crop. When pepper is grown in a mixed cropping system, it is essential that all the crops in the field are also subjected to organic methods of production. For new planting, varieties that are resistant or tolerant to diseases, pests and nematode infection should be used. All crop residues and farm waste available on the farm is recycled so that soil fertility is maintained at higher level. Composting, including vermicomposting of organic residues may be carried out as far as possible. Vesicular arbuscular mycorhizae (VAM) and biofertilizers may be used (Anon, 2004). Weeding is to be limited to slashing as far as possible and burning should not be resorted to. Mulching should be practiced with slashed materials and ground should be covered with green manure crops. Prophylatic measures should be taken to prevent diseases. No chemical pesticides or fungicides may be applied. However, Bordeaux mixture for fungus disease control is generally allowed but certain countries restrict the quantity to 8 kg per hectare per annum.

An isolation belt must be maintained around an organic holding. In smallholder groups, where the pepper holdings are contiguous, the isolation belt need only be at the outer periphery of the entire group of holdings. Any produce grown on this isolation belt cannot be treated as organic. Precautions should be taken to avoid the entry of run off water and chemical drift from neighboring farms, if they are not organically cultivated. While converting an existing plantation to organic cultivation, a minimum conversion period of three years is required. The conversion period may be relaxed if the organic farm is being established on land where chemicals were not previously used, provided sufficient proof of history of the area is available.

5.2 Selection of site

The land that is proposed for planting black pepper should be cleared of weeds and undergrowth. In level and low lands proper drainage channels should be provided to prevent water stagnation during periods of high rainfall. Areas having 1%-3% slope are ideal for planting black pepper. Planting in slopes facing south should be avoided so that the vines are not subjected to the scorching effect of the southern sun during summer. In sloppy lands adequate soil and moisture conservation measures are to be adopted.

5.3 Standards

Black pepper vines require a support for its growth, development and yield. These supports are called standards. Providing ideal supports plays an important role in successful establishment of black pepper vines. Since the black pepper vine is productive for 15 years and more selection of standards assumes great significance. The standards used for trailing black pepper vines are of two types namely, living and non-living. The non-living standards include reinforced concrete posts, granite pillars and teak poles. Such non-living standards are more commonly used in Malaysia, Brazil and Indonesia (George, 1981). In Malaysia and parts of Indonesia, the preferred support material is the wooden pole of the Belian tree (*Eusideroxylon zwageri*, lauraceae) other wise called Borneo Iron wood, which is a high density, heavy, construction timber resistant to white ants.

The use of reinforced concrete (RCC) posts at closer spacing (2X1 m) accommodated 5000 plants per hectare and recorded higher yields (Reddy *et al.*, 1992). However, owing to the high capital investment and also due to their relative non-availability, they are not very popular in India. Moreover, some of the non-living standards like concrete poles get heated up during summer resulting in drying of clinging roots and, therefore, result in poor crop growth under exposed situations. The ideal living support should have the following traits (Sivaraman *et al.*, 1999).

a. It should establish easily and grow rapidly to provide shade and support for the establishment of the crop

b. It should tolerate regular and heavy pollarding/pruning

c. It should not compete for resources and should not harbor pests and pathogens

d. It should have a deep root system without being allelopathic in the black pepper rhizosphere

e. It should have rough bark and be strong enough to support the clinging roots of black pepper vine

f. It should provide useful products at the end of the pepper cropping cycle

A variety of trees are used as living standards for black pepper cultivation in India. In homesteads gardens in Kerala, black pepper is usually trained on arecanut and coconut and also on mango, jack, etc. When interplanted in cardamom and coffee plantations, especially in Kodagu and Chickmagalur districts in Karnataka and Idukki and Wyanad districts in Kerala, black pepper is trailed on various forest trees that provide shade, while it is cultivated both as a pure and mixed crop along with arecanut in the plains of Uttar and Dakshin Kannada districts in Karnataka. In Andhra Pradesh, which is a newly emerging black pepper growing region of India, the vines are trailed on coconut and oil palm. In north-eastern states especially Assam, most of the homestead gardens have arecanut palm where black pepper vines can be trailed.

However, such standards are not ideal to establish large plantations under organized cultivation as monocrop at a normal spacing of 2.5 x 2.5 m. Therefore, on a plantation scale, *Erythrina indica* is the common live standard planted for trailing black pepper especially in Kerala. *Erythrina* is prone for wasp *Quadristichus erythrinae and* nematode infestation which may kill *Erythrina*. Under organic management, it is better to avoid *Erythrina* as standard. Other common standards that can be used are *Garuga pinnata, Gliricidia sepium, Leucaena leucocephala, Ailanthus malabarica* and *Grevillia robusta*. Rajagopalan & Mammooty (1996) and Mathew *et al.*, (1996) reported that *Ailanthus malabarica* and *G. pinnata* are the best living standards in plains. Both coconut and arecanut palms are used for trailing black pepper when black pepper is used as a mixed crop.

For raising living standards, seedlings are preferred to root cuttings since seedlings have a tap root system and do not compete with black pepper for resources. Living standards should be planted 3-4 years in advance so as to attain sufficient height at the time of planting of black pepper. In case stems/ stem cuttings of *Gliricidia sepium* or *Garcinia pinnata* are used, they are to be cut to suitable lengths during March-April and stacked in shade. The stacked stems start sprouting in May. After the first rain in May-June the sprouted stems are planted at the edge of the pits dug for planting black pepper vines. The cuttings of standards are to be planted in narrow holes of 40 to 50 cm depth. The soil should be well pressed around the standards to avoid air pockets and keep the standards firm in the soil.

In case *G. sepium* is planted the base of the standard is to be observed periodically for infestation by sapling borer which may be controlled by removing the frass and plugging the opening with clay. Normally the height of standard is not restricted; however for easy harvesting the height may be restricted. When pepper plants are trailed on coconut or arecanut plants, pepper height may be restricted to 5 meters for easy harvest.

5.4 Spacing

Spacing of black pepper vines varies considerably and depends on the spacing at which the standards are planted. Under monocropping system the optimum spacing is 3 m x 3 m, which can accommodate 1100 standards/ ha whereas in slopy land 3 m x 2 m spacing is recommended. Kurien *et al.* (1985) recorded the highest competition between living standards and black pepper at closer spacing of 2 m x 2 m. When non-living standards are used, much closer spacing (1.5 m x 2.0 m) is recommended. Poor growth and productivity were observed on concrete poles (Wahid and Sitepu 1987). Dead standards are not used in India, though trial at IISR have shown that pepper trailed on dead standards gives higher yield (Menon *et al.*, 1992).

5.5 Planting

Pits of 50 cm x 50 cm x 50 cm at a distance of 30 cm away from the base of supporting tree are taken with the onset of monsoon. The pits are filled with a mixture of topsoil, farmyard manure @ 5 kg/pit and 150 g rock phosphate. In case of large pits alternate layers of coconut husk and the above mixtures can be filled to help the young plants to withstand hot summer (Pillai 1992). Neem cake @ 1 kg and *Trichoderma harzianum* @ 50 g may be mixed in the pit at the time of planting. Two rooted cuttings are planted individually in the pits on northern side of each standard. At least one node of the cutting should be planted below the soil for proper anchorage. Unrooted cuttings may be planted @ 4-5 cuttings per pit and number of nodes in this case may be 4-5 (Sadanandan, 2000).

5.6 Training

Training is an essential step for establishment of black pepper vines. As the cuttings grow, the shoots are to be tied to the standards regularly using suitable materials for anchorage. For the production of leader shoots from the nodes, the vines are to be lowered into the soil during the first year after planting. In this method, the leaves are removed from vines after they attain a height of 1 m and they are brought down and $3/4^{th}$ of the basal portion is buried around the standard and covered with good top soil. This induces a good root system and also helps in production of more leader shoots from nodes. Three to five leader shoots are enough to produce sufficient laterals and to form a full canopy around the trunk of the standard. Kurien and Nair (1988) showed that pruning of terminal shoot increased number of spikes produced and number of bearing laterals.

5.7 Shading

The young vines should be protected from hot sun during summer especially in open areas by providing shade. The young vines are to be covered with dry arecanut or coconut leaves or twigs of trees during summer which should be removed with the onset of rains. After establishment, the side branches of standards are pruned to enable the standard to grow erect. After complete establishment, periodical pruning is important to allow sufficient light penetration into the black pepper canopy. Normally pruning of live standards has to be carried out one week prior to fertilizer application. At the beginning of rainy season, heavy pruning is done, leaving only one or two twigs at the top of the trees, while at the end of the rainy season a moderate pruning is carried out, leaving three or four twigs at the top of support trees. Shade regulation by pruning branches of standards in black pepper gardens during March and July-August is an important cultural practice to allow sufficient light for crop growth and productivity and also to reduce the incidence of diseases. Pruning reduces leaf density on the live support and stimulates growth of new leaves, which reduces root activity and places the support tree in a weaker position to compete for light, water and nutrients, relative to the pepper vines (Zaubin and Manohara, 2004).

5.8 Mulching

Mulching increases infiltration of water, conserves moisture, regulates temperature, decreases evaporation, suppresses weed growth, enhances microbial activity and also improves soil fertility. Mulching around the basins of black pepper vines with organic materials especially green leaves @10 kg/vine to a radius of 1 m is required at the end of North-East monsoon (Sivaraman *et al.*, 1999). Application Mulching should be repeated once the applied mulch is decomposed. Mulching the basins with prunings of standards, enhanced soil organic matter content and moisture in the soil (Sadanandan, 2000). Application of *Chromolena odoratum* L. (Eupatorium) as mulch controlled yellow disease and improved the stand of black pepper in Cambodia (Litzenberger and Lip, 1961). Live mulch (cover crops) such as *Calapagonium mucanoids* and *Mimosa invisa* can also be grown to provide soil cover and to prevent soil erosion. These cover crops are to be cut back regularly from the base to prevent them from twining along with black pepper vines.

5.9 Irrigation

Moisture requirement of black pepper vines vary with stages of crop growth. Moisture stress is one of the most serious constraints affecting the productivity of black pepper. Blossoming, spiking, flowering, spike elongation and setting are the sensitive periods for black pepper. Following summer, the first 16 weeks could be considered as critical period for shoot growth, flower bud differentiation, spike emergence, flower opening and fertilization, berry formation and development (Pillai *et al.*,1988). Spike development ceases if there is a prolonged dry spell immediately after good summer showers. A high correlation between rainfall received during second half of May (responsible for spike formation) and June and setting of black pepper berries has been observed (Pillai *et al.*, 1988). Water requirement of pepper in a multiple cropping system was carried out using Bowen ratio-energy balance (BREB) method and vapour diffusion model for estimating the water consumption of multiple cropping systems with black pepper. In the multiple cropping system of pepper trailed on coconut and arecanut, evapo transpiration coefficient (ET_C) value of pepper ranged from 0.31-0.42 mm day $^{-1}$, while those of coconut and arecanut ranged from 3.25-3.60 mm day^{-1} and from 1.45-1.78 mm day^{-1} respectively. The crop coefficient of pepper in multiple cropping system ranged from 0.07-0.10 mm day^{-1}, while those of coconut and arecanut ranged from 0.74 -0.82 and from 0.33-0.38 mm day^{-1} (Satheesan *et al.*, 1997a). Irrigating black pepper vines @ 100 litres per vine (hose irrigation) once a week during summer is recommended (Satheesan *et al.*, 1997b). The water is to be applied in basins taken around the plants at a radius of 75 cm. In case drip irrigation is adopted, 7 litres of water per day through drip during October to May is recommended (IISR, 1998).

5.10 Weeding

Weeds are a major problem in black pepper plantations that are not maintained properly. The weeds growing in the basins remove considerable quantities of moisture and nutrients and should be removed. Weeding may be done only when necessary by slashing and the materials should be used for mulching. Cover crops like C. *mucunoides* and *M. invisa* can also be grown not only to prevent soil erosion during rainy season but also to control weeds. The borders of the farm and road edges should be raised with suitable leguminous cover crops to prevent soil erosion and weed growth.

5.11 Cropping System

Multiple cropping is the practice of growing two or more crops in the same field simultaneously. A systematically planted black pepper garden would provide adequate interspaces for cultivation of other crops especially during the pre-bearing period because of the negligible shade effect by the vines. The factors that assume significance in multiple cropping are probable competition for nutrients, moisture and sunlight between black pepper and other crops and its effect on black pepper yield. The goal of intercropping is to produce a higher yield on a given piece of land, by making use of the space that would otherwise be wasted with a single crop. However, careful planning is required before selection of crops, taking into account the soil, climate and crops and it important not have crops competing with each other for space, nutrients, water or sunlight.

Black pepper roots extend up to about 90 cm from the base and so ample space is available for multiple cropping without affecting black pepper. In black pepper gardens receiving well-distributed mean rainfall of not less than 1500 mm/year, the competition for soil moisture is less and in such areas multiple cropping is feasible. In any multiple cropping programme, competition for soil nutrients such as N, K, Ca and Mg is a potential danger and inadequate manuring will adversely affect the yield of black pepper vines. It will also increase the organic matter and nutrient status of soil, reduces soil erosion, bulk density, soil temperature and aluminum toxicity.

Multiple cropping in black pepper gardens is a routine practice in Kerala and parts of Karnataka. A variety of crops are grown along with black pepper, such as elephant foot yam, colacasia, ginger and turmeric and perennial fodder grass, vanilla and banana. At higher altitudes, black pepper is grown along with coffee, cardamom and tea. Cereals like upland paddy, pulses like red gram, vegetables, flowers, fodders and other annuls are intercropped with black pepper (Sivaraman *et al.*, 1999). Black pepper is also trailed on coconut and arecanut palms in most of the areas. In these cases the rooted black pepper plants are planted away from the base of palms and as and when the vines grow, they are trailed along the ground and then trailed on to the trunk of coconut or arecanut palms. The varieties Sreekara, Subhakara and Panniyur-5 perform well as intercrops in coconut and arecanut gardens. Multiple cropping with black pepper as one of the component crops also offers immense scope in north-eastern states. Mixed cropping of black pepper + rubber+ pineapple+ clove; black pepper + Cocoa + Hawaiin papaya + orange + Tahitian lemon + guava reduced the incidence of *Fusarium* disease in Brazil (Duarte and Albuquerque, 1988).

(a) (b)

Fig. 2: Black pepper mixed crop (a) with arecanut and coconut; (b) with coffee
(*Source:* Director, IISR, Calicut)

6.0 NUTRIENT MANAGEMENT

6.1 Nutrient requirement

Black pepper is a surface feeder and feeding roots are concentrated in 30 cm radius of the vine (Jayasree *et al.*, 1988). Nitrogen, phosphorus, potassium, magnesium and micro nutrients are the most important nutrients for growth, development and yield of black pepper and their influence depend on their ratios in the soil as well as in the plant. Studies on nutrient uptake by adult (8 years old) Panniyur I vines IISR, Calicut, indicated that it removes 292, 56 and 405 kg/ha of N, P and K, respectively, per year, while Karimunda removes 183, 49 and 376 kg/ha of N, P and K, respectively, per year (Sadanandan, 2000).

The availability of phosphorus depends on the dynamics of phosphorus fixation in the soil and uptake by the plant and in the process soil pH and organic matter content play a vital role. Bone meal and mussoorie rock phosphate are good phosphorous sources in acid soil for the black pepper varieties Panniyur-1 and Karimunda for obtaining higher yields. Black pepper vines require large quantities of potassium for growth and fruiting. A soil fertility survey carried out in major black pepper growing tracts of Kerala and Karnataka indicated that 10% of the gardens surveyed were low, 31% medium and 59% high in available potassium status. Among the sources, potassium sulphate has a positive influence on quality (oleoresin 13.5% and piperine 6.1%) of black pepper berries. The response to application of lime to black pepper is noticed mainly due to correction of soil acidity.

6.2 Organic manure

Ravikumar and Krishnamurthy (1983) found that the addition of farm wastes reduced bulk density, increased the pore space, water holding capacity, aggregate stability and available water. Addition of organic manures improved the physical properties like infiltration rate, total porosity and hydraulic conductivity of soil with hard pan (Loganathan, 1990). Several kinds of organic manures like farmyard manure, compost, poultry manure and organic cakes like neem and groundnut cakes are used by farmers in India. Farm yard manure supplies most of the plant nutrients required by the plants (Palaniappan and Natarajan, 1993) and application rate vary from 4-10 kg/plant in Black pepper (Anon, 2004). The organic ions released by the decomposition of organic matter complexes with Fe and Al, thereby blocking sites for P sorption and making more P available to the plant. The organic ions/acids also form chelates with the micronutrient cations Zn, Cu, thereby making them available to the crop. Addition of organic matter also promotes growth of beneficial microbes and reduces the population of pathogenic organisms leading to reduced disease incidence. Peat at five kg vine^{-1} serves as a good source of nutrient supplement to reduce the dosage of chemical fertilizers (Zulkifly, 1996). In India, green leaves, dry leaves, wood ash, cattle manure, oil cakes, neem cake and bone meal are the common organics used. In Malaysia, organic cakes, prawn, guano, fish meals etc are used in quantities of about 0.6-1.0 kg plant^{-1} in every two months for growing pepper. Experiments conducted with different organic composts at IISR, Calicut, showed that application of organics increased pH to neutrality, improved the fertility status of the soil by significantly increasing the availability of major, secondary and micronutrients and build up of microbial population. Growth and biomass production of black pepper plants were increased by the application of organic materials like leaves of *Erythrina indica, Garuga pinnata, Grevillea robusta,* but biomass production decreased at application of *Piper nigrum* and *Coffea arabica* leaves which was attributed to allelopathic effect of decaying material (Siva kumar and Wahid, 1994).

For nutritional management under organic farming, a judicious application of a combination of organic manures such as farmyard manure @ 5 kg/vine, neemcake @ 1 kg/vine and vermicompost @ 1 kg/vine per year can be made during May-June. Biofertilizers such as phosphobacteria and *Azotobactor* can also be applied @ 100 g/vine mixed with farmyard manure.

Optimum time of organic manuring is May-June. Compost made from green loppings, crop residues, grasses, cowdung, poultry droppings etc., fortified with wood ash and/or rock phosphate showed better response instead of FYM alone (Tiwari and Agarwal, 2005). It is desirable to apply lime at the rate of 500g/vine in April-May with receipt of pre monsoon showers in alternate years.

The experiments conducted with different organic composts on black pepper at IISR, Calicut, India showed that highest yield of 8.3 kg vine^{-1} with high benefit cost ratio of 4.15 was recorded with FYM application followed by application of vermicompost (7 kg vine^{-1}). Release of non exchangeable K to the crop uptake was minimum with the addition of FYM and composts as their addition significantly increased the available K content by 3.8 times over no organic application (Srinivasan *et al.,* 2005). For nutritional management under organic farming, a judicious application of a combination of organic manures such as farm yard manure 5 kg, neemcake 1 kg and vermicompost @ 1 kg per vine per year can be made during May-June.

6.3 Biofertilizers

Formulations containing living cells of beneficial micro organisms multiplied in suitable carrier are biofertilizers which can form low cost ecofriendly inputs for crop production. There are also indications that Black pepper plants can benefit from the use of biofertilizers. Bopaiah and Khader (1989) reported increased shoot and root weight of cuttings of pepper by dipping in a peat based culture slurry of *Azotobactor, Azospirillum* and *Glomus mosseae* (VAM). Kandiannan *et al.,* (2000) reported enhanced growth of pepper due to application of 10g *Azospirillum* into the growing medium under nursery and significant enhancement in plant height, leaf area, biomass and dry matter production and nutrient content of black pepper cuttings due to combined application of *Azospirillium*, phosphobacteria and VAM. Combined application of coir compost @ 1.25tha^{-1} and biofertilizer recorded highest BC ratio in black pepper compared to other treatment combinations. It was reported that significant uptake of nitrogen (N) and phosphorus (P) and enhanced vigour of black pepper when inoculated with *Pseudomonas fluorescens* strains, due to higher root proliferation and nutrient mobilisation in the rhizosphere. To take advantage of *Rhizobium,* leguminous cover crops should be grown in the pepper holdings as green manure crops (Anon, 2005). Similarly, application of *Azospirillum, Azotobacter* @15-20g/plant each and VAM @ 40 to 50g plant are recommended as nutritional source of black pepper (Anon, 2004).

7.0 DISEASES AND PEST MANAGEMENT

Organic farming should be carried out in such a way that it minimizes losses from pests, diseases and weeds. Emphasis is to be given on the use of a balanced nutritional progamme, use of crops and varieties which are tolerant or resistant and well adapted to the environment, fertile soils of high biological activity, crop rotations, green manures etc.

7.1 Diseases

The major diseases affecting black pepper vines in the field are *Phytophthora* foot rot and stunt diseases. Spike shedding due to anthracnose, pollu disease and phyllody are also seen in some areas.

7.1.1 Phytophthora foot rot

Phytophthora foot rot caused by *Phytophthora capsici* is the most serious disease affecting black pepper and occurs mainly during the south-west monsoon season. Crop loss surveys conducted in Kerala indicated that the disease caused 3.7% vine death in Kozhikode District leading to yield loss of 119.6 tonnes and 9.4% vine death leading to yield loss of 904.9 tonnes in Kannur District annually. In small holdings, the crop loss could also be total as entire plantations succumb to *Phytophthora* infections in a single season.

Symptoms

The pathogen *P. capsici* infects all parts of the vine and the expression of symptoms depends up on the site of infection and extent of damage. On tender leaves dark coloured spots appear with characteristic fimbriate advancing margins. When the conditions are ideal the lesion spreads on the entire lamina within a few hours resulting in defoliation. The succulent shoot tips of freshly emerging runner shoots trailing on the soil turn black when infected. Spike infection occurs on any part of the spike resulting in blackening and shedding of spikes. Infection on a branch results in dieback of affected branches. Infection on the runner shoot sometimes reaches the collar and cause sudden wilting followed by shedding of leaves and spikes The branches break up at nodes and the entire vine collapse within a month. Infection on collar through the roots of upper tier of roots closer to soil surface also leads to sudden wilting of vines. If the damage is confined to feeder roots, the expression of symptoms is delayed till the cessation of rain and the vine starts showing declining symptoms such as yellowing, wilting, defoliation and drying up of a part of the vine during October-November onwards. These vines may recover after the rains and survive more than two seasons till the root infection culminates in collar rot and death of the vine.

Epidemiology

Vines of all age groups are susceptible to the disease. Often the inoculum is introduced along with planting material from the nursery to the plantations. The aerial spread of the disease is mainly through rain splashes. *Phytophthora* foot rot occurs at random in plantations in the beginning and secondary spread is clustered around previously infected vines. The fungus survives in the form of thickened mycelium and chlamydospores for over 18 months and soil population build up is correlated with soil moisture. For the aerial phase of the disease there is a positive correlation between rainfall, number of rainy days and relative humidity but negative correlation with temperature and sunshine hours. While the aerial spread is rapid, spread of the soil phase of the disease in field is slow.

Management

A management strategy involving cultural practices, chemical and biological control and use of resistant variety has been developed for the management of the disease.

Cultural control: Infected plants serve as the foci of infection (Zadocks and Van den Bosch, 1994). Phytosanitory measures such as removal and destruction of affected vines to reduce the initial inoculum in the field plays a crucial role in disease management. Tillage operations are to be kept to the minimum to avoid soil disturbance and root damage. Water logged condition predispose plants to *Phytophthora* infections and disease incidence positively correlated with Moisture (Anandaraj, 1997). Adequate drainage should be provided to reduce water stagnation. Shade regulation by lopping of branches of support trees during the monsoon season ensures reduction in humidity build up within the canopy of black pepper vines and results in reduction in the severity of aerial infection. Application of oil cakes such as neem, soybean, groundnut and coconut enhanced growth of saprophytes (Kueh and Sim, 1992). In soils amended with organic matter the saprophytic activity is enhanced and *P.capsici* population drops to undetectable levels (Anandaraj, 1997).

(a)

(b)

Fig. 3: Symptoms due to Phytophthora capsici infection (a) on collar and roots; (b) aerial infection on leaves (*Source:* Director, IISR, Calicut)

Biological control

Bio control agents are natural enemies of pests like predators and parasitoids, fungal, bacterial and viral antagonists of crop pests and diseases, botanical extracts, pheromones and any naturally occurring, organic or live material which can effectively control all the above and restrict the damage to below economic Threshold levels. Bio control agents are preferred in the place of chemical pesticides, because they are hazardless eco friendly, effective and economical.

Application of antagonistic micro organisms such as *Trichoderma harzianum* and *Pseudomonas fluorescens* would help in reducing the intensity of the disease. *T. harzianum* is to be applied at the onset of the monsoon during May around the base of the vine @ 50 g /vine (containing 10^8 cfu/g of formulation). A second application of *T. harzianum* has to be done during August-September. *P. fluorescens* @ 50 g/vine (10^8 cfu /g of formulation) may also be applied along with *T. harzianum.* Incorporation of *VAM* alone or in combination with other beneficial microorganism like *Azotobacter and Azospirillum* enhance rooting and growth of pepper (Govindan and Chandy 1985; Bopaiah and Khader, 1989).*VAM* inoculation has been found to enhance rooting, growth and suppress root damage caused by *P. capsici, R. Similis* and *M. incognita* under artificial inoculation and under field conditions (Anandaraj and Sarma, 1994). Addition of organic mulches and oil cakes in the basins improves the texture of the soil and enhances the growth of antagonistic microorganisms. Aqueous extracts of *C. odorata* extract was found toxic to *P. capsici* (Anandaraj and Leela, 1996). Root exudates *of Allium* spp. are inhibitory to zoospores of *Phytophthora* (Manohara *et al.*, 1992). After the receipt of a few monsoon showers, all the vines are to be sprayed with Bordeaux mixture 1%. In case biocontrol agents are applied, care should be taken to prevent the entry of chemicals into the soil.

7.1.2 Stunt disease

Stunt disease caused by viruses is becoming increasingly important in recent years in black pepper growing areas especially at high altitudes because of their systemic nature, widespread occurrence and difficulty in control. This disease is known as mosaic disease in India (Prakasam *et al.,* 1990) and as 'little leaf' in Sri Lanka (Randombage and Bandara, 1984).

Symptoms

A wide range of symptoms are observed on stunt disease infected vines. The symptoms on leaves include vein clearing, yellow specks, distortion, reduction in size, mottling and mosaic, along with stunting of the whole plant. The infected vines produce short spikes with poor filling leading to yield reduction. In severe cases, the leaves become abnormally narrow and give a sickle shaped appearance. The internodes of vines become abnormally short leading to stunting of plants and the affected branches give a typical witches broom appearance in advanced stages. However, though external symptoms are good criteria for detection of the disease, sometimes, the disease is difficult to identify visually. Hence, ELISA and PCR based methods are the best methods for the detection of CMV and PYMV infections in plants.

Causal viruses and transmission

Two viruses are associated with stunt disease affected vines namely, *Piper yellow mottle virus* PYMV (Badnavirus) and *Cucumber mosaic virus* CMV (Cucumovirus) in India. The major means of spread of the viruses is through the use of infected stem cuttings. In general, CMV is known to have a very wide host range infecting several plant species and spread in nature through aphids in a non-persistent manner. On the other hand, badna viruses are known to have a narrow host range and spread in nature through mealy bugs.

Management

Infected planting materials are the major mode of disease spread since black pepper is vegetatively propagated. Adequate care should also be taken to plant healthy virus-free cuttings especially in new areas where the incidence of the disease is not observed in the field.

Production of virus-free planting material: The most viable method to control the disease would be identification and use of virus-free cuttings for planting. Once a good virus-free plant is identified, they must be maintained virus-free under insect-proof conditions. Cuttings or tissue culture raised plants obtained from these virus-free stocks are then grown up on a sufficiently large scale under insect-proof conditions.

Rouging of infected plants: Regular inspection and removal of infected plants and replanting with healthy plants should be resorted to in the field. Other potential weed and crop hosts, which might act as reservoir for the virus need to be removed. The removed plants may be burnt or buried deep in the soil.

Control of vectors: Because of the closed placing of seedlings in the nursery, chances of spread of mealy bug are more in the nursery and hence regular monitoring of the nursery for mealy bugs is important.

7.1.3 Spike shedding

Spike shedding is a major problem in high yielding varieties like Panniyur -1 when cultivated at higher elevations. The malady is caused by absence of bisexual flowers resulting in low fruit-set and infection by the fungus *Colletotrichum gloeosporioides*. The incidence is more pronounced when the emergence of spike is delayed due to delayed monsoon and flowering occurs during June-July. These spikes predominantly produce female flower instead of bisexual flowers.

Symptom

The symptoms include appearance of small dark colored spots surrounded by a halo on the leaves. In serious infestations the expansion of lamina is affected resulting in crinkled appearance. Infection on spikes results in spike shedding whereas infection on immature berries leads to formation of brownish split on the seeds due to unequal development.

Management

The disease can be managed by adopting an integrated, including irrigation of black pepper vines 4-5 times at an interval of 5-7 days @ 40-50 litres/plant commencing from the third week of March, followed by shade regulation of support trees to provide minimum 7,500-10,000 lux light under cloudy condition along with prophylactic spraying with Bordeaux mixture 1%.

7.1.4 Pollu disease

This disease is caused by *Colletotrichum gloeosporioides* and can be distinguished from the pollu (hollow berry) caused by the beetle by the presence of characteristic cracks on the infected berries. The disease appears towards the end of the monsoon.

Symptoms

The affected berries show brown sunken patches during early stages and their further development is affected. In later stages, the discolouration gradually increases and the berries show the characteristic cross splitting. Finally, the berries turn back and dry. The fungus also causes angular to irregular brownish lesions with a chlorotic halo on the leaves.

Management

The disease can be controlled by spraying Bordeaux moisture 1%.

7.1.5 Phyllody disease

This disease which is caused by phytoplasma belonging to aster yellows group. The disease is noticed in parts of Wayanad (Sarma *et al.*, 1988) and Kozhikode districts of Kerala.

Symptom

The affected vines exhibit varying stages of malformation of spikes. Some of the floral buds are transformed into narrow leaf like structures. Such malformed spikes show leafy structures instead of floral buds, exhibiting phyllody symptoms. In advanced stages, the leaves become small and chlorotic, and the inter nodes are also shortened. The effected fruiting laterals give a witches broom appearance. Severely affected vines become unproductive. In severely affected vines the entire spike is converted in to small branches which appear chlorotic and the vines decline rapidly. The infected vine becomes unproductive within 2 to 3 years.

Management

The infected vines are to be destroyed to prevent the further spread of the disease.

7.1.6 Nematodes

Among the plant parasitic nematodes infesting black pepper, the burrowing nematode and root knot nematode, causing slow decline are important. Among the insect pests, pollu beetle, mussel scale and coconut scale, top shoot borer and leaf gall thrips can be considered as major insect pests. Apart from these insects, scale insects, mealy bugs, gall midges and leaf feeding caterpillars also damage the crop in certain areas and could be considered as minor pests.

7.1.6.1 Slow decline

Slow decline occurs due to feeder root damage by burrowing nematode *Radopholus similis* and root knot nematode *Meloidogyne incognita* and also the fungal pathogen *P. capsici* either alone or in combination. Surveys conducted in major black pepper areas in Kerala and Karnataka indicated that *M incognita* and *R similis* were distributed in over 70% of gardens either alone or together.

Symptoms

The symptoms of the disease include yellowing of leaves, defoliation, die back and loss of vigour and productivity, leading to death of the vine. The diseased vines generally exhibit foliar yellowing from October onwards coinciding with depletion of soil moisture. With onset of south west monsoon during May-June, some of the affected vines recover and put forth fresh foliage. However, the symptoms reappear during the subsequent season after the cessation of the monsoon and succumb during the course of 2 to 3 years. The affected vines exhibit varying degrees of root degeneration due to infestation by plant parasitic nematodes. *R. similis* causes elongate dark brown necrotic lesion on roots at points of entry, which coalesce and result in root rotting. *M. incognita* causes elongate thick swelling on thick primary roots and typical galls or knots on secondary and fibrous roots. These galls gradually disintegrate leading to rotting of tissue. Black pepper vines of all ages are susceptible to the nematodes and they are passively distributed through planting material (rooted cuttings). Most of the nurseries harbour nematodes since they provide congenial conditions for their multiplication.

Management

An integrated management strategy is effective for the management of the disease.

Cultural control: Nematode and *P. capsici*-free rooted cuttings raised in solarized nursery mixture should be used for planting in the field. Severely affected vines which are beyond recovery should be removed from the field along with the entire root system and destroyed. Application of neemcake @1kg/plant during May-June is recommended.

Bio control agents like *Pochonia chlamydosporia* or *Trichoderma harzianum* can be applied @ 50 g/ vine (containing 10^8 cfu/g of formulation) twice a year (during April-May and September-October).

Resistant variety: In areas severally infested with root knot nematodes, cuttings of the resistance variety Pournami may be planted.

7.2 Pests

7.2.1 Pollu beetle

The pollu beetle (*Longitarsus nigripennis*) is the most destructive insect pest of black pepper and the pest infestation is higher in the plains and midlands below 300 m MSL in Kerala and Karnataka. The pest infestation is generally higher in shaded areas in the plantation. In endemic areas around 30% of the crop is lost due to the pest infestation.

Damage

The adult *pollu* beetle feeds on tender shoots, leaves and spikes of black pepper vines resulting in black patches on tender shoots and spikes and small irregular circular holes on tender leaves. Severely infested leaves and spikes often rot and drop due to invasion of secondary microorganisms. The damage caused by the larva (grub) is more serious and it bores into tender spikes and berries and feeds on the internal contents. The infested spikes develop necrotic patches and the berries turn black and crumble when pressed (the hollow nature of the infested berries is termed as pollu in Malayalam).

Management

Neem-based insecticide such as Neemgold 0.6% at 21 days interval from August-October is effective for controlling the pest infestation. The undersides of leaves are to be sprayed thoroughly where the adults are generally seen. Leaf extracts of *Chromolaena odorata* L., neem (*Azadirachta indica* A.juss) and nuxvomica (*Strychnos nux-vomica* L.) and neemoil, neem seed kernel extract and custard apple (*Annona squamosa* L.) seed extract possessed appreciable antifeedent activity against pollu beetle in laboratory bioassays (Devasahayam and Anandaraj, 1997).

7.2.2 Scale insects

Scale inset are major insect pest of black pepper at higher altitudes. Among the various species of scale insect recorded on the crop, the mussel scale (*Lepidosaphes piperis*) and coconut scale (*Aspidiotus destructor*) are the most common species and the pest infestation is higher at higher altitudes in Kerala, Karnataka and Tamil Nadu.

Damage

The mussel scale encrusts main stems, lateral branches, mature leaves and berries o black pepper vines resulting in chlorotic patches, yellowing and drying of leaves. Younger vines often succumb to the pest during the course of 1-2 years when the infestation occurs on the main stem. On older vines the infested lateral branches wilt and dry resulting in vacant spaces in the canopy. Infestations of mussel scales also are also observed in older cuttings in the nursery especially in the plains. Coconut scales infest mature leaves resulting in chlorotic patches on them and sometimes also infest barriers.

Fig. 4: Pollu infested black pepper berries (*Source:* Director, IISR, Calicut)

Management

Infestation of *A. destructor* is generally kept under check by various natural enemies observed in the field. Neemgold (a commercial formulation of neem) caused 58% reduction in scale population compared to 83-100 percent reduction.

7.2.3 Top shoot borer

The top shoot borer (*Cydia hemidoxa*) is a serious insect pest black of black pepper in younger plantations (1-3 years old) and is widely distributed in the plains and higher attitudes in Kerala.

Damage

The larvae of top shoot borer infest tender terminal shoots that climb on the standards. The earlier instar larvae scrape and feed on the epidermis of tender terminal shoots and sometimes on tender leaves, whereas the later instar larvae bore onto tender shoots and feed on the internal contents resulting in decay and drying of infested shoots. The growth of the vine and also the establishment of newly planted vines is affected due to repeated infestation of tender terminal shoots.

7.2.4 Leaf gall thrips

Leaf gall thrips (*Liothrips karnyi*) are serious insect pests of black pepper on younger vines at higher altitudes and also on rooted cuttings in nurseries in the plains, in Kerala, Karnataka and Tamil Nadu.

Damage

Leaf gall thrips infest tender leaves causing the leaf margins to curl downwards and inwards resulting in the formation of marginal, tubular, hypophyllous galls. As the infested leaves grow they become crinckled, malformed and reduced in size and may also turn chlorotic later at the margins.

7.2.5 Mealy bugs

Mealybugs such as *Planococcus* spp., *Pseudococcus* spp. and *Ferrisia virgata* infest tender shoots and leaves especially on rooted cuttings in the nursery and cause the plants to wilt. Colonies of root mealy bug (*Planococcus* sp.) are also observed at the basal portion of the stem under the soil and on roots causing yellowing, wilting and mortality of vines in the field and cuttings in the nursery especially at higher altitudes. The pest infestation is generally more serious on vines affected with *Phytophthora* and nematodes and hence management of these pathogens would reduce the incidence of pest in the field.

Fig. 5: Mealybug, *Planococcus* sp. (*Source:* Director, IISR, Calicut)

7.2.6 Scale insects

Marsipococcus marsupiale is commonly seen infesting black pepper leaves at higher altitudes though no visible symptom of damage is observed. *Protopulvinaia longivlvata* is observed on mature leaves of older cuttings in the nursery. The pest infestation result in yellowing and wilting of affected plants and the subsequent attack by sooty mould accelerates the deterioration of the infested plants. Spraying neem oil or neem gold 0.3% is effective for management of scale insects.

7.2.7 Leaf feeding caterpillars

The semilooper *Synegia* sp. is more common among the few species of leaf feeding caterpillars infesting black pepper. The pest damages tender shoots, leaves and spikes of younger vines especially during the monsoon season. Hand picking of larvae would help in reducing the pest infestation.

8.0 HARVEST AND POST HARVEST HANDLING

Black pepper takes 7-8 months after flowering to reach full maturity. The crop is harvested during December -January in plains and January-April in the high ranges of Western Ghats. It is important to harvest green pepper at proper stage of maturity in order to achieve a dried product of good colour and appearance. The spikes are nipped of by hand and collected in bags when one or two berries in the spike turn red or yellow. At the time of harvesting no chemical should be applied to ward off red ants. Normally, single pole bamboo ladder is used as a support for harvesting. If the berries are allowed to over ripe, there is heavy loss due to berry drop and damage by birds. Recent advances in product diversification have necessitated harvesting of the berries at different stages of maturity. The level of maturity required at harvest for processing into different pepper products is given in Table 5 (Govindarajan, 1979).

Table 5. Optimum maturity at harvest for pepper products

Product	Stage of maturity at harvest
Canned pepper	4-5 months
Dehydrated green pepper	10-15 days before maturity
Oleoresin and essential oil	15-20 days before maturity
Pepper powder	Fully mature with maximum starch
Black pepper	Fully mature and when 1-2 berries start turning from yellow to red in each spike
White pepper	Fully ripe

8.1 Post Harvest Management

There are several factors, which affect the quality of processed black pepper products. These include pests and micro-organisms which infest the product from the plantation, foreign matter and impurities from materials used in processing which contaminate the products, micro-organisms and dirt which are introduced through unhygienic practices of people who handle the produce, and loss of quality that results from short-comings in storage practices. Thus, the quality of the product depends on the quality of raw materials, the methods used in processing, the packaging methods and materials and marketing practices. Therefore, to ensure quality of the product, constant care should be maintained from the time of harvesting till the time it reaches the consumer.

8.1.1 Processing

Processing methods of black pepper should be based on mechanized, physical and biological processes. Care should be taken to maintain the vital quality of an organic ingredient through each step of its processing.

8.1.2 Threshing

The berries are separated from the spike usually by trampling. This operation is crude, tedious and unhygienic. Chances of extraneous, soil particles and filth contaminating the produce are also high. At few places especially in large plantations mechanical threshers are used. Mechanical threshers with capacities varying from 50 kg/h to 1200 kg/h are available which can thresh quickly and provide cleaner products are given in Table 6.

Table 6. Black pepper threshers developed by various agencies

Thresher	Capacity (kg/h)	Threshing mechanism	Max. efficiency (%)	Damage (%)
Gudalur Thresher	600	Wooden cylinder with aluminium angles	99.5	<0.3
KAU Thresher	50	Metal drum with rubber lining	98.0	<0.5
TNAU Thresher (Improved)	200	Metal drum with wooden rasp bar	99.6	Negligible
Malaysian Thresher	600	Metal drum lined with rubber blades	99.8	Negligible
Indonesian Thresher (Manual)	170-185	Pedal operated metal drum	94.13	3.78
Indonesian Thresher (Power)	425-450	Metal drum	92.21	6.13
Vivega Thresher (Manual)	50	Metal drum with angles	91.5	4.0
Vivega Thresher (Power)	200	Metal drum with angles	96.0	>4.0
Mohta Thresher	1200	Metal drum	Nearly 100	Negligible
Cafex	600	Metal drum	98.0	<0.5

Source: Dhas & Korikanthimath, 2003

8.1.3 Blanching

The quality of the black pepper can be improved by a simple treatment of dipping the mature berries taken in perforated vessel in boiling water for a minute before drying. This processing technique has several advantages:

a. Uniform coloured black pepper is obtained after drying.

b. Microbial load is reduced.

c. Drying is completed in 3-4 days as against 5-6 days required in the traditional practice.

d. Extraneous impurities like dust are removed from the berries.

8.1.4 Drying

Green pepper has moisture content of 60% to 70% at harvest, which should be brought to safer levels of less than 10% by adequate drying. The green colour of matured green pepper is due to the presence of chlorophyll pigment. During drying, enzymatic browning sets in and the phenolic compounds are oxidized by atmospheric oxygen under the catalytic influence of the enzyme phenolase and eventually turn black. Sun drying is the conventional method followed for drying of black pepper. The despiked berries are dried in the sun for 3-5 days to bring the moisture content below 10%. After drying, if the moisture content is too high, black pepper will be susceptible to fungal attack. Mycotoxins secreted by the fungi render the pepper unfit for human consumption. In order to achieve quality dry product, green pepper berries are spread on clean dry concrete floor / bamboo mats pasted with fenugreek paste / PVC sheets and dried in the sun for a period of 4 to 6 days. The average dry recovery varies between 33%-37%.

8.1.5 Cleaning and grading

The dried black pepper has extraneous matter like spent spikes, pinheads, stones, soil particles, etc. mixed with it. Cleaning and grading are basic operations that enhance the value of the produce and help to get higher returns. On a small scale, winnowing and hand picking remove most of these impurities. Such unit consists of a fan/ blower and a feeding assembly. The fan is placed at the rear end of the hopper. Cleaning is achieved by feeding the material through the hopper into a stream of air blowing in perpendicular direction. The heavier fractions (dust, immature berries, pin heads and spent spikes) are blown away. The cleaning and grading units available are given in the Table 7.

Table 7. Cleaner cum graders for black pepper developed by various agencies

Agency	Capacity	Mechanism	Efficiency (%)
KAU	-	Rotary type with three concentric cylindrical sieves	99.6
TNAU	50 kg/h	Inclined belt separator	98.0
CFTRI	100 kg/h	Aspirator, destoner, rotary washer, continuous fluidized bed dryer	100.0
Export house	Continuous	Pneumatic separators with magnetic gadgets, multiple sieve cum air classifier	100.0

Source: Dhas & Korikanthaimath, 2003

Using sieves, cleaned black pepper, is sifted into different grades based on size. These standards are being implemented rigorously under the Compulsory Agmark Grading Scheme for exports from India (Table 8). The grades are as follows:

a. Malabar Garbled (MG Grades 1 and 2) Black Pepper
b. Malabar Ungarbled (MUG Grades 1 and 2) Black Pepper
c. Tellicherry Garbled Black Pepper Special Extra Bold (TGSEB)
d. Tellicherry Garbled Extra Bold (TGEB)
e. Tellicherry Garbled (TG)
f. Pin Heads (PH Grade special and Grade 1)
g. Garbled Light Pepper (GL Special, GL Grades 1 and 2)
h. Ungarbled Light Pepper (UGL Special, UGL Grades 1 and 2)
i. Black Pepper (Non-specified)

Table 8. Agmark grade specifications of black pepper

Grade	Trade name	Size, diameter of sieveholes on which retained (mm)	Extraneous matter not exceeding (% w/w)	Light berries not exceeding (%w/w)	Moisture content not exceeding (%w/w)
Tellicherry Garbled black pepper					
TGSEB	Special Extra Bold	4.75	0.5	3	11
TGEB	Extra Bold	4.25	0.5	3	11
TG	Tellicherry Garbled	4.25 (50% min) 4.00 (50% max)	0.5	3	11
Garbled Malabar Black Pepper					
MG Grade 1	-	-	0.5	2	11
MG Grade 2	-	-	0.5	5	11
Un-Garbled Malabar Black Pepper					
MUG Grade 1	-	-	2	7	12
MUG Grade 2	-	-	2	10	12
MUG Grade 3 L	-	-	3	15	12
MUG Grade 4 L	-		4	20	12

Source: Spices Board, 2001

8.1.6 Packaging

Mixing of organically produced black or white pepper with non organic products or storing or transporting together is to be avoided. The dried black or white pepper is to be packed in eco-friendly packing materials such as clean gunny bags. Use of polythene bags should be minimized to the extent possible. Recyclable /reusable packing materials shall be used wherever possible.

8.1.7 Storage

Black pepper is hygroscopic in nature and absorption of moisture from air, notably during rainy season with high humidity may result in mould and insect infestations. Before storage it is to be dried to less than 10% moisture. The graded produce is bulk packed separately in multi layer paper bags or woven polypropylene bags provided with food grade liners for export or in jute bags.

8.2 Hygiene of environment and personnel during processing

The premises used to dry and store black pepper should be clean and hygienic and free from dust, cobwebs, undesirable foreign matter like excreta of animals and birds, bits of stone, etc and should be cleaned daily. The premises should be fenced using nylon or other netting to keep off birds. The storage area should be properly plastered to avoid the entry of rodents and openings in the walls should be covered with metal grills. Persons engaged in threshing, sifting, etc. of black pepper should clean their hands and feet with soap and water before they start their work. Vessels and utensils used in processing and storage should always be kept clean. They should never be allowed to be soiled by dust or other impurities or contaminated by pests. Gutters and other water outlets connected to platforms used for drying black peppershould always be maintained in a clean condition. Only clean bamboo trays should be used for sifting black pepper. These trays should never be smeared with cow dung. Precautions should be taken to see that the clean and sifted black pepper is not re-contaminated by dust and other impurities.

8.3 Marketing and export

In India, cultivation of black pepper is mainly confined to Kerala, Karnataka and Tamil Nadu, and to a certain extent in Andaman and Nicobar Islands and north-eastern states. Kerala alone contributes about 95% of total production of black pepper. In terms of productivity the all India average is 315 kg/ha while that of Kerala is 323 kg/ha. The crop is major source of income and employment (> 40 million man days) for rural households.

Organic black pepper is one of the novel products prepared by Indonesia, India, Srilanka. The basic stand for production of organic black pepper is the guide lines provided by International federation of organic agriculture movement (IFOAM). The world demand for organically produced black pepper is growing rapidly in European Union, USA, Japan and Australia. Other developed and some developing countries are likely to source organic pepper in coming years.

9.0 Conclusion

The overuse of chemicals, fertilizers and pesticides has ruined the mother earth, polluting food, drinking water and air. Organic mode of production is the only way to save ecology, environment, and produce hazard- free foods for the people, conserving natural resources. Organic production is a commercially viable option for farmers, as organic foods/ products fetch a premium price, at less cost for production, as it avoids costly chemical fertilizers and pesticides. Let us return back our mother earth sufficient quantity of organic materials to the soil to increase, balance its humus content and nutrients. Organic farming is the much needed oxygen now and let us breath and continue to breath.

REFERENCES

Anandaraj, M. (1997) Ecology of *Phytophthora capsici*. Causal organism of foot rot disease of black pepper (*Piper nigrum* L.). *Ph.D. Thesis,* University of Calicut, Calicut, 154 pp.

Anandaraj, M. and Leela, N.K. (1996) Toxic effect of some plant extracts on *Phytophthora capsici,* the foot rot pathogen of black pepper. *Indian phtopath.,* 49, 181-184.

Anandaraj, M. and Sarma., Y.R. (1994) Effect of Vesicular Arbuscular Mycorrhiza on rooting of black pepper (*Piper nigrum* L). *J. Spices and Aromatic crops.* 3, 39-42.

Anon, (2004) *Guideline for production of organic spices in India.* Spices Board, Cochin, 281 pp.

Anon, (2005) *Pepper production guide.* International Pepper Community. Jakartha. Indonesia.

Anon, (2007) Pepper. *In: Arecanut and Spices data base.* Premaja P & Manojkumar (Eds). Directorate of Arecanut and Spices Development Calicut.

Bavappa, K.V.A. and Gurusinghe, P.De S. (1978) Rapid multiplication of black pepper for commercial planting. *J. Plantn. Crops* 6, 92-95.

Bopaiah, B.M. and Khader, K.B.A. (1989) Effect of biofertilizers on growth of black pepper (*Piper nigrum*) *Indian J. Agric. Sci.,* 59, 682-683.

Creech, J.L. (1955) Propagation of black pepper. *Econ. Bot.* 11, 233-242.

De Waard, P.W.F. (1969) *Foliar diagnosis, Nutrition and yield stability of black pepper (Piper nigrum L.) in Sarawak.* Communication No. 58. Department of Agricultural Research, Koninklijk Instituut voor de Tropen, Amsterdam.

Devasahayam, S. and Anandaraj, M. (1997) Black pepper (Piper nigrum L.) Research. In S.S. Narwal, P. Tauro and S.S. Bisla (eds.), *Neem in sustainable Agriculture.* Scientific publishers, Jodhpur, pp.117-122.

Devasahayam, S., Anandaraj, M., Thankamani, C.K., Saji, K.V. and Jayashree, E. (2006) Black pepper. *In: Major Spices- Production and Processing.* Parthasarathy, V.A. and Rajeev, P. (Eds). Indian Institute of Spices Research, Calicut.

Dhas, P.H.A. and Korikanthaimath, V.S. (2003) Processing and quality of black pepper-a review. *J. Spices Aromatic Crops* 12, 1-13.

Duarte M L R and Alburquerque T C (1988) *Fusarium* disease of black pepper in Brazil. *In:* Sarma Y R & Premkumar T (Eds). *IPC workshop on joint research for the control of foot rot Disease,* National Research Center for Spices, Calicut. pp. 39-54.

George, C.K. (1981) A review of systems of pepper cultivation in different countries. *In Work shop on processing of pepper and pepper products.* Central food Technology Research and Training Institute, Mysore, India, p.3-5.

Govindan, M. and Chandy, K.C. (1985) Utilization of the diazotroph, Azospirillum for inducing rooting in pepper cuttings (*Piper nigrum*). *Curr .sci.,* 54, 1186-1188.

Govindarajan, V.S. (1979) Pepper - Chemistry, Technology and Quality Evaluation. CRC Critical Rev. *Food Sci. Nutr.* 9, 115-225.

Indian Institute of Spices Research (IISR) (1998) *Annual Report* 1998, Indian Institute of Spices Research, Calicut.

Jayasree, S., Wahid P. A. and Kamalam, N. V. (1988) Absorption of soil applied radio-phosphorus by black pepper vine and support tree in relation to their root activities. *J. Plant. Crops,* 16: 73-87.

Kandiannan, K, Sivaraman, K, Anandaraj, M. and Krishnamurthy, K.S. (2000) Growth and nutrient content of black pepper (*Piper nigrum* L.) cuttings as influenced by inoculation with biofertilzers. *J. Spices and Aromatic Crops,* 9, 145-147.

Kueh, T.K. and Sim, S.L. (1992) Etiology and control of *Phytophthora* foot rot of black pepper in Sarawak, Malaysia. In. P. Wahid, D. Sitepu, S. Deciyanto and U Superman (eds). *The international workshop on Black pepper Diseases, Bander, lampung,* Indonesia, PP.144-154.

Kurien, S. and Nair, P.C.S. (1988) Effect of pruning on yield in pepper (*Piper nigrum* L.) *Agric. Res. J. Kerala* 26, 137-139.

Kurien, S. Babu, N.M., Cheeran, A. and Nair, B.P. (1985) A note on crop standard interaction in pepper. *Indian Cocoa Arecanut Spices J.* 9 (2), 35-36.

Litzenberge, S. C. and Lip, H.J. (1961) Utilising *Eupatorium odoratum* to improve crop yields in combodia. *Agronomy J.* 55, 321-324.

Loganathan, S. (1990) Effect of certain tillage practices and amendments on physiochemical properties of problem soils. *Madras Agric. J.*, 77, 204-208.

Manohara, D., Kasim, R. and Sitepu, D. (1992) *Current research status of foot rot disease in Indonesia.* In. P. Wahid, D. Siteup, S. Deciyanto and U. Superman (eds)., Proc. The international workshop on Black pepper Diseases, Bander, lampung, Indonesia, Institute of Spice and Medicinal Crops, Bogor, Indonesia, pp.144-154.

Mathew, P.G., Wahid, P.A. and Nair Sreekandan, G. (1995) Soil fertility and nutrient requirement in relation to productivity in black pepper. *J. Plantn. Crops* 23, 109-115.

Mathew, T., Kumar, M.B. and Babu, S.K.V. (1996) Evaluation of some live standards for black pepper. *J. Plantn. Crops* 24, 86-91.

Menon, K.S., Nair, M.K., and Sharma, O.P. (1992) Preliminary report on the performance of black Pepper on non living standard. *Indian Spices*, 19, 5-6.

Palaniappan, S.P. and Natarajan, K. (1993) Practical aspects of organic matter maintenance in soil. In: Thampan, P.K. (Ed).*Organics in soil Health and Crop production* (pp.23-41), Peekay Tree Crops Development Foundation, Cochin.

Pillai, V.S. (1992) Management of improved cultivars of black pepper. In: Y.R. Sarma, S. Devasahayam and M. Anandaraj (Eds.) *Black pepper and cardamom- problems and prospects*, Indian Soc. Spices Calicut, pp. 17-19.

Pillai, V.S., Sasikumaran, S. and Ibrahim, K.K (1988) Effect of rainfall pattern on the yield of black pepper. In: *Agrometeorology of Plantation Crops*. Kerala Agricultural University, Trichur. pp. 152-159.

Prakasam, V., Subbarajan, K.T. and Bhaktavatsalu, C.M. (1990) Mosaic disease- a new record in black pepper. *Indian J. Agric. Sci.* 47, 59-60.

Rajagopalan, A. and Mammootty, K.P. (1996) Performance of black pepper cv. Karimunda grown on different tree standards. *Spice India* 9 (8), 13-14.

Randombage, S. and Bandra, J.M.R.S. (1984) Little leaf disease of *Piper nigrum* in Srilanka. *Plant Pathol.* 33, 479-482.

Ravikumar, K. and Krishnamoorthy, K.K. (1983) Effect of different organics and crop residues on physical and chemical properties. *Madras Agric. J.* 70(3): 41-43.

Reddy, B. N., Sadanandan, A. K., Sivaraman, K. and Abraham, J. (1992) Effect of plant density on the yield and nutrient availability of black pepper (*Piper nigrum* L.) varieties. *J. Plantn. Crops* 20 *(Suppl.)* : 10-13.

Sadanandan, A.K. (1993) Water management of spice crops-Problems and Prospects. *In: Proceedings of Seminar on water management for plantation crops: Problems and prospects.*

Satheesan K.V. (Ed).Center for Water Resource Development and Management. Kozhikode. pp. 53-64.

Sadanandan, A.K. (2000) Agronomy and nutrition of black pepper. *In: Black Pepper (Piper nigrum)*. Ravindran, P.N. (Ed.). Harwood Academic Publishers, Amsterdam, pp. 163-223.

Sarma, Y.R., Solomon, J.J., Ramachandran, N. and Anandaraj, M. (1988) Phyllody disease of black pepper (*Piper nigrum* L). *J. Plantation Crops*, 16, 69-72.

Satheesan, K.N., Rajagopalan, A., Sukumarapillai, V., Jagadeesh Kumar, T.N., Mammooty, K.P. and Unnikrishnan Nair, P.K. (1997b) Influence of irrigation levels on Yield, quality, and incidence of pests and diseases in back pepper (*Piper nigrum* L). In: *National seminar on and water and nutrient management for sustainable production and quality of spices*. Sadanandan, A.K., Krishnamurthy, K.S., Kandiannan, K. and Korikanthimath, V.S. (Eds.) 6 October, 1997, Madikeri, Karnataka, India. pp. 107-110.

Satheesan, K.V., George Mammen, Krishna Kumar, K.N. and Sadanandan, A.K. (1997a) Influence of irrigation levels on yield, quality and incidence of pests and diseases in black pepper (*Piper nigrum* L.) In: Sadanandan, A.K., Krishnamurthy, K.S., Kandiannan, K. and Korikanthimath, V.S. (Eds.) *Proc. Nat. Sem. On Drought management in plantation crops*, 5-6 October, 1997, Madikeri, Karnataka, India, 107-110.

Senanayake, Y.D.A. and Kirthisinghae J.P. (1983) Effect of shade and irrigation on black pepper (*Piper nigrum* L.) cuttings. *J. Plantn. Crops* 11, 105-108.

Sivakumar, C. and Wahid, P.A. (1994) Effect of application of organic materials on growth and foliar nutrient contents of black pepper (*Piper nigrum* L.). *J. Spices and Aromatic Crops* 3, 135-141.

Sivaraman, K. Kandiannan, K. Peter, K.V. and Thankamani, C.K. (1999) Agronomy of black pepper (*Piper nigrum* l.) - a review. *J. Spices and Aromatic crops* 8(1), 1-18.

Srinivasan, V., Hamza, S., Sadanandan, A.K. (2005) Evaluation of composted coir pith with chemical and biofertilizers on nutrient availability, yield and quality of black pepper (*Piper nigrum* L.). *J. Spices and Aromatic Crops*, 14, 15-20.

Thankamani, C.K., Srinivasan, V., Hamza, S., .Kandiannan, K. and Mathew, P.A. (2006) Evaluating soil less mixture for the production of planting materials in Black pepper, *(Abstract). SYMSAC-III Crop and Product Diversification, Indian Institute of Spices* Research, Calicut.

Tiwari, R.S. and Agarwal, A. (2005) *Production technology of spices*. International Book Distributing Co, Lucknow, India, p. 487.

Wahid, P. and Sitepu, D. (1987) *Current status and future prospect of pepper development in Indonesia*. Food and Agricultural Organization, Regional Office for Asia and Pacific, Bangkok, Thailand.

Zadocks, J.C. and Van denbosch, F. (1994) On the spread of plant disease: A theory of foci. *Ann. Rev. Phytopathol.* 32, 503-521.

Zaubin, R. and Manohara, D. (2004) *A strategy* for fertilizer use on black pepper (*Piper nigrum* L.) in Lampung. *Focus on pepper* (*Piper nigrum* L.) 1, 17-24.

Zulkifly, E. (1996) *Prospects of implementing organic farming on pepper*. Int. Pepper News Bull. 20 (2), 24-30.

9

Cardamom

D. Prasath and M.N. Venugopal

1.0 INTRODUCTION

Cardamom *(Elettaria cardamomum* Maton.) acclaimed as the 'Queen of Spices' is the true cardamom belonging to the family *Zingiberaceae*. The cardamom of commerce is the dried ripe fruit and is one of the most important valued spices next only to saffron. It is native of evergreen forests of Western Ghats of South India. Cardamom is a shade loving plant (Pseophyte) thriving well in elevations up to 600-1200 m above MSL under an average annual rainfall of 1500-4000 mm and temperature range of 10-35^{0}C. India has been the world's largest producer of cardamom until 1979-80. Later, Guatemala emerged as world's premier producer and the exporter of cardamom accounting for about 90 per cent of the global trade. Cardamom is also grown lesser extent in Sri Lanka, Papua New Guinea and Tanzania.

1.1 Area and Production

In India, area under cardamom has been fluctuating over the years. It was 93,950 ha during 1982-83 and remained static for three years (1984-1985 to 1986-87), then it shot up to 1,05,000 ha in 1987-88. The area came down to 73,590 ha during 1996-97 and has remained more or less static till 2004-05. The production has gone up from 2,900 mt during 1987-88 to 11,420 mt in 2004 - 05. During the period, the productivity has increased from almost 41 kg/ha to 208 kg/ha. Cardamom cultivation is confined to three South Indian states namely, Kerala, Karnataka and Tamil Nadu. In 2004-05, Kerala has 56 per cent of area and contributes 75 per cent production; Karnataka with 37 per cent of area and 16 per cent production, while Tamil Nadu has 7 per cent of area and contributes 9 per cent production (Table 1).

Table 1. State-wise area and production of cardamom in India (area in ha, production in MT)

States	1970-71		1980 - 81		1990 - 91		2000 - 01		2004 - 05	
	Area	Prodn	Area	Prodn	Area	Prodn	Area	Prodn	Area	Prodn
Kerala	55190	2130	56380	3100	43826	3450	41288	7580	41378	8616
Karnataka	28220	1000	28220	1000	31605	800	25947	2100	27094	1879
Tamil Nadu	8070	235	9350	300	6123	500	5085	800	5253	920
Total	**91480**	**3170**	**93950**	**4400**	**81554**	**4750**	**72320**	**10480**	**73725**	**11415**

The productivity increase in recent years is due to the use of high yielding varieties and improved agro techniques developed from different research organizations working on cardamom in India. However the export has touched rock bottom during the same period. The maximum export of 3,272 mt was during the year 1985-86 while the least (173 mt) was during the year 1989-90 and the export during 2004 - 05 was 650 mt. The export earnings have come down from Rs. 64.00 crores in 1984-85 to Rs. 23.81 crores during 2004-05. India has been the world's largest cardamom producer until 1979-80, when Guatemala came to the picture as the major competitor and world leader in cardamom production. Presently, 90 per cent of world trade in cardamom is from Guatemala.

The other cardamom producing countries are Tanzania and Sri Lanka in small quantities and the production level in these countries is around 125 mt (Tanzania) and 75 mt (Sri Lanka) per year (Ravindran, 2002).

1.2 Composition and uses

1.2.1 Composition cardamom (seeds)

The composition of cardamom varies slightly with the variety, region and age of the product.

Components	Percentage	Components	Percentage
Moisture	7-10	Volatile oil	5.5-10.5
Total ash	3.8-6.9	Alkalinity of ash	0.4-2.4
Water soluble ash	1.3-5.05	Acid insoluble ash	1.3-5.0
Crude fibre	6.7-12.8	Crude protein	7.0-14.0
Starch	39.0-49.9	Calcium	0.3
Phosphorus	0.215	Sodium	0.015
Potassium	1.2	Iron	0.012

(Pruthi, 1987)

Vitamins (mg/100g)

Vit B1 (thiamine)	0.18	Vitamin C	12.0
Vit B2 (Riboflavin)	0.235	Vitamin A	175IU/100g
Niacin	2.3		

(Pruthi, 1987)

1.2.2 Composition of cardamom capsules

Components	Percentage	Components	Percentage
Moisture	20.0	Protein	10.2
Ether extract	2.2	Volatile oil	7.4
Mineral matter (total ash)	5.4	Crude fibre	20.1
Carbohydrates	42.1	Calcium	0.13
Phosphorus	0.16	Iron	5 mg/100g.

(Pruthi, 1987)

1.2.3 Main components of cardamom volatile oil

Main components of cardamom volatile oil are listed below. The volatile oil is extracted from the seeds and the husk hardly gives only 0.2 per cent oil.

Components	**Percentage**	
1,8 Cineole	38.8	49.0
α-terpinyl acetate	24.6	24.7
B- pinene	6.2	-
Linalool	5.7	5.8
Sabinene	4.2	1.4
Limonene	4.0	0.9
α -terpineol	3.4	2.7
α -pinene	2.3	0.4
Terpinen4 -ol	2.0	1.9
Linayl acetate	1.0	4.6
Geranyl acetate	0.8	-
Geraniol	0.2	2.5
Nerolidol	trace	1.2
Myrcene	-	1.2
Nerol	-	0.9
B caryophyllene	-	0.4

Gopalakrishnan, (1994); Variyar and Bandyopadhyay, (1995)

Besides the usual terpene hydrocarbon and alcohols as minor compounds and the dominance of 1,8 - cineole and α-terpinyl acetate, it is significant that methyl eugenol also has been identified (Lawrence, 1979). The basic cardamom aroma is produced by a combination of the major components, α-terpinyl acetate and 1,8-cineole.

1.2.4 Uses

Cardamom is an ancient spice and has the longest influence in India, its birthplace. Since 4th century BC, it has been used in medicine and cooking. Cardamom has long been used in Middle East. Cardamom was grown in the royal gardens of Babylon in 721 BC. It was mentioned in an Egyptian papyrus as early as 1550 BC, which described its numerous medical properties. The ancient Greeks and Romans also used it in food, medicines and perfumes. The Vikings, who discovered it on a trip to India, enjoyed cardamom in festival cakes. Cardamoms were first imported to Europe in 1214, where it was used in pomanders, and ascribed aphrodisiac properties.

Cardamom is widely used in Indian cooking. It is an essential ingredient in garam masala and also used as a breath freshener. It is common to pop the whole pod or a seed or two and chew on them after a spicy meal. They also believe this aid in digestion. In fact, recent studies suggest that cardamom may prevent teeth cavities. In the Middle East, cardamom is widely considered as an aphrodisiac. Scandinavians still use cardamom to spice their "Danish pastry" and other deserts, as well as meat dishes like Swedish meatballs. Cardamom also features prominently in German cookies. In south Asia and South East Asia, cardamom is an ingredient in betel nut chewing.

It is very common to use tinctures of cardamom in medicines for windiness or stomachic. Powdered cardamom seeds are invariably mixed with ground ginger, cloves and caraway and used mainly for combating digestive ailments. It is used as a powerful pleasant aromatic stimulant, carminative, stomachic and diuretic. Use of cardamom checks nausea and vomiting. In the present day stress prone population cardamom is used invariably as cardiac stimulant.

A Popular nasal application is prepared by using extracts of cardamom, neem and myrobalans along with animal fat and camphor. Cardamom seeds are chewed to prevent unpleasant smell in mouth, indigestion, nausea and vomiting due to sickness and to prevent pyrosis (excessive watering in mouth). Gargling with the infusion of cardamom and cinnamon is known to cure pharyngitis, sour throat, and hoarseness during the infective stage of flue. Powdered seeds of cardamom boiled with tea-water imparts very pleasant aroma to the tea and it can be used as medicine for scanty urination, diarrhea, decently, palpitation of the heart, exhaustion due to over work, depression etc. It is believed that eating cardamom capsule daily along with a tablespoon of honey improves eye site, strengthens the nervous system and thus improves health of the person. Krishnamurthy (1991) described Ela class (27 plant group including cardamom) are capable of destroying Kapa and poison, promote skin complexion, destroy itching, pustules and akotha (a kind of leprosy or ring worm with large round spots).

The terpenoid constituents are responsible for the antifungal and antibacterial effects. Mishra *et al.*, (1991) and Mishra and Dubey, (1990), studied the effect of cardamom on *Aspergillus flavus*, which produces aflatoxin B1. Mycostatic activity was observed at 400 ppm level. The flavour components also showed antibacterial effects against several food-born microorganisms (Kubo *et al.*, 1991). Banerjee *et al.*, (1994) has found that cardamom oil enhances glutathione-s-transferase enzyme and acid soluble sufhydril activities. These enzymes mediate the oxidation and detoxification of xenobiotic. Cardamom oil suppresses DNA adduct-formation by aflatoxin B1 in a dose dependant manner (Hashim *et al.*, 1994)

Cardamom in action, it is aromatic, cardiac, carminative, deodorant, digestive, diuretic, expectorant, purgative, stimulant, thirst reliever and tonic. Useful in asthma, burning sensation, cold and cough, diseases of bladder and kidney, flatulence, heart weakness, indigestion, scanty urine and piles (Dey, 1994). The seeds are aromatic, acrid, sweet, cooling, stimulant, carminative, stomachic, diuretic, cardiotonic, abortifacient and are useful in bronchitis, hemorrhoids, stangury, renal and vesical calculi, anorexia, dyspepsia, gastropathy and vitiated condition of vata (Anon, 1994). The seed is fragrant tonic to the heart, stomachic, laxative, lessens inflammation, useful in headache (Govil, 1998).

2.0 BOTANY AND VARIETIES

2.1 Botany

Elettaria is a small genus with 3-4 species in East and South-East Asia. The cultivars are grouped as 'Malabar' (prostrate panicles), 'Mysore' (erect panicles) and 'Vazhukka' (semi erect panicles), based on the adaptability, nature of the panicle, shape and size of fruits.

It is a herbaceous perennial (2-5 m in height) having underground rhizomes with aerial pseudostems made of leaf sheaths. The leaves are long, most pubescent/puberelent and lanceolate in shape. Inflorescence is a long panicle with racemose clusters arising from the underground stem, but comes up above the soil. The growth habit of the panicles and the shape as well as the size of the capsules vary in different cultivated varieties/types of cardamoms. The flowers are bisexual, bracts linear, oblong and persistent, raceme, sepals 3, petals 3, unequal, lip longer than corolla tube shortly exerted, white sheathed with violet carpels 3, style 1, ovary-trilocular, axile placentation, ovules-numerous in each carpel. Fruit is a trilocular, fragrant capsule round to oblong in shape containing 10-18 seeds.

Flowers are bisexual, zygomorphic and open in succession from base towards apex in racemes. Bracts linear, oblong and persistent, sepals 3, petals 3, unequal, lip longer than corolla tube shortly exerted, white sheathed with violet carpels 3, style 1, ovary-trilocular, axile placentation, ovules-numerous in each carpel. Fruit is a trilocular, fragrant capsule round to oblong in shape containing 10-18 seeds. Structure of cardamom flower is pre-disposed for insect pollination as indicated by the prominent labellum, stigma positioning above anthers, and the presence of nectar glands. Generally 2 to 4 panicles emerge from the swollen base of tillers. Flower initiation takes place in March-April and from initiation to full bloom it takes 25 to 35 days and from bloom to maturity, it takes 110 to 140 days. Flowering is spread over a period of 6 - 7 months from April to continue up to October in India. Capsules start maturing in August-September and continue further up to December-January. Maximum number of flowers open during early hours (3.30 a.m) and continues till 7.30 a.m. of the day followed by the anther dehiscence (7.30 to 8.30 a.m.). Flowers wither on the same day. The pollen grains are round and mostly found in single and measured 87.6μ on an average. Though apparently 85.2 per cent of the pollen grains appear fertile, germination tests showed that the maximum of 70.1 per cent germination in artificial media containing 20 per cent sucrose and 1 per cent agar solution. Studies on the viability of the pollen grains indicated that only 6.5 per cent of pollen grains remained viable after 2 hours of storage, but after 6-8 hours of storage, the percentage of viability was practically zero. With respect of pollen morphology of the three varieties of cardamom, it was noticed that all the three varieties of cardamom are round in shape and appeared as a creamy powder. The stigma remains receptive from 4 am on the day of the flower opening depending on the environment conditions. Receptivity of stigma was highest between 8 am and 12 noon giving about 72 per cent fruit set. Thereafter, the stigma receptivity decreases gradually resulting in the minimum fruit set of 24 per cent at 4 PM. Though cardamom has bisexual flowers and is self-compatible, cross-pollination is the rule, since self-pollination is hindered in the flowers due to the slight protrusion of the stigma above the stamens. Cardamom flowers remain in bloom for 15-18 hours and stigma receptivity and pollen viability were reported to be maximum during morning hours.

2.2 Varieties

Selection of cardamom is highly location specific in their agro-ecological requirements. The improved selections are much superior to the local clones with regard to yield, capsule characters (Table 2) and tolerance to biotic stresses (Table 3).

Table 2. Salient features of cardamom varieties

Variety	Source	Average yield (dry kg/ha)	Suitable areas
IISR Suvasini	Indian Institute of Spices Research (ICAR), Calicut, Kerala	745	Adoptive to all cardamom-growing areas of Karnataka, Wayanad in Kerala and Lower Pulneys in Tamil Nadu and highly suitable for High Production Technology by following intensive agro practices
PV-1	KAU, Cardamom Research Centre, Pampadumpara, Idukki, Kerala	260	All cardamom tracts of Kerala and Karnataka
ICRI 1	Indian Cardamom Research Institute (Spices Board), Myladumpara, Idukki, Kerala	325	South Idukki zone of Kerala
ICRI 2	Indian Cardamom Research Institute (Spices Board), Myladumpara, Idukki, Kerala	375	Vandanmettu and Nelliampathi zones of Kerala
ICRI 3	Indian Cardamom Research Institute (Spices Board), Myladumpara, Idukki, Kerala	439	Hill zones of Karnataka
ICRI 4	Indian Cardamom Research Institute (Spices Board), Myladumpara, Idukki, Kerala	455	Lower Pulneys in Tamil Nadu
ICRI 5	Indian Cardamom Research Institute (Spices Board), Myladumpara, Idukki, Kerala	1543	Kerala (900-1200 m MSL with annual rainfall of 2000 mm) and parts of Tamil Nadu
Mudigere 2	UAS, Regional Research Station, Mudigere, Karnataka	476	Traditional cardamom growing tracts of hills of Karnataka

Table 3. Characteristic features of biotic and abiotic tolerant varieties

Variety	Characteristic features	Average yield	Reference
IISR Avinash	It is highly suitable for planting in valleys and rhizome rot prone areas	847 kg/ha	Venugopal and Prasath, (2006)
IISR Vijetha	Recommended to moderate rainfall with moderate to high shaded mosaic (cardamom mosaic virus) infected areas	643 kg/ha	Venugopal and Prasath, (2006)
ICRI 6	Moderately tolerant to major pests (thrips and shoot borer)	1886 kg/ha (potential yield)	Madhusoodanan *et al*., (2006)
Mudigere - 1	Tolerant to hairy caterpillar and white grubs	275 kg/ha	Madhusoodanan *et al*., (2002)
PV - 2	Tolerant to thrips, capsule rot and clump rot	982 kg/ha	Madhussodanan *et al*., (2002)

Few progressive cardamom planters by their own interest and observation over a considerable period of time, selected following high yielding races and these races have attracted the cardamom planting community.

2.2.1 Farmer's varieties

Njallaniyil green gold, Vander cardamom, Panikulangara-1, Panukulangara-2, Vali green bold, Palakudi, Gayathri, PNRS, Krishna, HYT-1, HYT-2, HYT-3, Elarani-1, Elarani-2, Elarani-3, Kaniparampil-1, Poovathumkal, Ponnampara and Kalarickal white (ICRI, 2006)

Fig. 1: Some important popular cultivars of cardamom (a) IISR Avinash (b) IISR Suvasini (c) IISR Vijetha (d) Njallaniyil Green Gold (*Source:* Director, IISR, Calicut)

3.0 CLIMATE AND SOIL

3.1 Climate

Cardamom is highly sensitive to elevation and the wrong choice of cultivar or location can severely affect the growth and productivity. The optimum altitude for growing cardamom is 600 to 1500 MSL (Anon, 1970). The 'Malabar', types productive under lower elevations of 500-700 MSL. The other two cardamom types, 'Mysore' and 'Vazhukka' are suitable for cultivation in lower altitudes ie. 700 MSL and below (Abraham and Tulasidas, 1958). Vegetative growth is satisfactory but fruit production is poor at lower elevations. In Guatemala, cardamom is grown at varying altitudes, ranging from 90 to 1200 m above MSL.

Cardamom is very sensitive to temperature fluctuations (Jacob *et al.*, 1995). In India, optimum growth and development is observed in the warm and humid conditions at a temperature range of 10-35°C (Anon, 1976). The upper temperature limit is usually around 31-35°C. It is noticed that rate of spread of '*katte*' disease is more during summer than in rainy season and cold conditions result in low capsule setting. Thus, extremes of temperature and wide diurnal variations are not conducive for healthy growth of cardamom plants. In hills, especially removal of canopy cover may lead to adverse effect on soil temperature than in the plains. Under shade less condition, elevated soil temperature (36°C) and the strong wind movement increases evapo-transpiration and the soil dries up quickly. All these inter related changes in microclimate would affect the growth and the development of cardamom (Murugan *et al.*, 2000).

Moderate rainfall areas (2000 to 2500 mm/annuum) with filtered natural shade should be selected for growing cardamom. In South India, 75-80 per cent of the cardamom cultivation is under rainfed conditions (Charles, 1986). It is grown under rainfall conditions ranging from 1500 to 5750 mm. Nearly 75 to 78 per cent of total rainfall is received during South- West monsoon which lasts for only three months (June-August). North-east monsoon and pre monsoon showers (summer rains) are not well distributed and dependable. Hence, these crops suffer both from excessive moisture and moisture stress. Dry spells in these tracts usually occur during December - May. Even though all the characters of rainfall were positively correlated with the cardamom production, a significant correlation was observed only between total number of rainy days and yield. An annual rainfall of 2000 mm with uniform distribution was considered ideal for cardamom (Murugan *et al.*, 2000).

In Guatemala, rainfall conditions are much more favourable than in India. Here rainfall varies from 2000 to 5000 mm per annum and distributed evenly through out the year, though two peaks of high rainfall occur. Because of the absence of heat and drought stress, cardamom yields are much better in Guatemala and similar situation also occurs in Papua New Guinea (Krishna, 1997).

3.2 Soil

Cardamom is cultivated under shade trees and the soils in general have high fertility status due to addition of leaf litter (Zachariah, 1978). In India, soils of major cardamom growing areas come under the order alfisols, formed under alternate wet and dry conditions and the sub order ustalfs derived from schists, granite and gnesis and lateritic in nature (Sadanandan *et al.*, 1990). Cardamom generally grows well in forest loamy soils with good drainage. The clay fraction is predominantly kaolinitic and hence there is some fixation of potassium in these soils. The cardamom growing soils of Karnataka are mostly clay loam (Kulkarni *et al.*, 1971). These soils are generally acidic in nature with pH from 4.2 to 6.8, high in organic matter and nitrogen, low to medium in available phosphorus and medium to high in available potassium.

The soils were acidic (5.0 to 6.6 pH) in nature with high organic carbon content, which decreased with depth of the soil. The soils were medium (56.82%) to high (43.18%) in available nitrogen, low (79.55%) in available phosphorus and low (38.63%) to medium (52.27%) in available potassium (Korikanthimath *et al.* 2000b). The soil profile collected from cardamom fields indicated relatively higher level of available nitrogen and organic matter than from uncultivated area due to recycling of organic materials. Studies on uptake pattern revealed that N, P and K uptake increased with increase in dry matter production. The total K uptake was more followed by N and P in both stages of crop and it suggests that crop requires more of K than N and P.Among the different plant parts, highest contribution to the total N uptake was by green leaves (44.69%) and tillers (29.6%) in one-year-old and prepotents respectively. The highest P and K uptake was by roots and grown up tillers in one year–old plants respectively Korikanthimath *et al.* (2000a).

Sulphur content ranged from traces to 36 ppm in Kerala, traces to 27.5 ppm in Karnataka and 15-36 ppm in Tamil Nadu. 43.3 per cent of Kerala samples and 50 per cent of Karnataka samples showed deficiency of sulphur nutrient (Srinivasan *et al.* 2002). It was observed that in the South Indian soils, 68 per cent of sampled area was deficient in available zinc, 49 per cent in available boron, 28 per cent in available molybdenum and 9 per cent in available manganese (Srinivasan *et al.* 1993b).

4.0 NURSERY MANAGEMENT

Cardamom is propagated either through seeds or vegetative means and being a cross-pollinated crop the seedling population is not uniform. Only 35 per cent of the plants are good yielders in a plantation raised from seedling population (Krishnamurthy *et al.*, 1989). Hence, vegetative propagation is normally adopted for multiplication of elite clones. Vegetative propagation can be either through suckers or tissue culture (Kumar *et al.*, 1985).

Initially the seeds can be collected from any elite plantation even if they are not grown organically. However, the methods followed for raising seedlings should confirm to the organic standards. If rhizomes are to be used as planting material the plantation should have been following organic methods of production at least one year prior to collection (Anon, 1998).

Seeds are collected from fully ripe capsules preferably from second to third round of harvest in September and are then washed in water and sown immediately or mixed with wood ash and dried for 2 to 9 days in room temperature. Storage of seeds results in loss of viability and delay in germination. Immediate sowing resulted in maximum germination. Germination was highest (71.8 per cent) when sown in September (Pattanshetty and Prasad, 1973; Pattanashetty *et al.*, 1978). Gradual decline in germination (56.7, 51.0, 46.4, 34.1, 32.5 and 29.6 per cent) was observed when sown on 1st August, 15th August, 30th August, 14th September, 29th September and 14th October, respectively (CPCRI, 1982). Cardamom seeds sown immediately after harvest (September) germinated uniformly, and seedlings were ready for transplanting at the end of 10 months. The ideal sowing season has been reported to be November - January for Kerala and Tamil Nadu and September for Karnataka (Anon, 1970 and 1979).

The low temperature prevailing in the cardamom growing area reduces germination per cent and prolongs days taken for germination (Krishnamurthy *et al.*, 1989). Germination is significantly correlated with maximum and minimum temperature prevalent in the area (Gurumurthy and Hegde, 1987). The cardamom seeds fail to germinate at temperature less than 15°C and greater than 35°C. Germination was optimum at 30°C (70.9-73.0% RH). The highest per cent of germination was observed under ambient temperature during September-October (Siddagangaiah *et al.*, 1993).

Mulching of seedbeds influence the germination of seeds (Abraham, 1958). The beds covered with paddy straw recorded highest germination (40 per cent), when seeds were sown in September and was on par with dry leaves of rose wood (37 per cent), goose berry leaves (37 per cent) and wild fern (38 per cent) (CPCRI, 1980). Mulching with coir dust, paddy straw and goose berry leaves enhanced the germination (Sulikeri and Kologi, 1978; CPCRI, 1983).

Seedlings of four to five leaf stages from the primary nursery beds are generally transplanted in the secondary nursery at a distance of 20 to 25 cm or in polythene bags (20 x 20 cm). In Karnataka, where seeds are sown during August - September, transplanting in the secondary nursery is done during

November - January. In Kerala and Tamil Nadu, seedlings from primary beds are transplanted to secondary nursery beds at a spacing of 20 x 20 cm during June-July. Rate of mortality was high when transplanted in second leaf stage and can be minimized by transplanting at the four-five leaf stage (Ankegowda, 2003).

Acid treatment of seeds should be avoided, treatment of seeds with *Trichoderma* culture (50 ml spore suspension for 100 g of seed) is desirable as a prophylactic measure for managing nursery rot diseases (Anon, 1998). At the time of preparation of beds, incorporation of VAM multiplied in recommended organic medium is to be incorporated. For raising polybag seedlings (preferably bio-degradable polybags), potting mixture may be prepared by using 3:1:1 soil rich in organic matter, well rotten cow dung or vermicompost and sand. VAM and *Trichoderma* can also be added to the potting mixture (250 g mixed with 25 kg of well rotten cow dung). If growth of the seedlings is not adequate, spraying vermiwash once in a month is desirable (20 ml per plant). The diseases in the nurseries may be managed by regular surveillance and adopting phytosanitary measures. Restricted application of Bordeaux mixture 1% may be taken up to control rot disease at the initial stage itself. Changing the nursery site is benefited to ward off pests and diseases and for vigorous growth of seedlings (Anon, 1998).

Suckers from elite clones can be used for establishing plantations capable of high productivity. Plants raised from suckers come to bearing earlier than suckers. Suckers should not be used in areas where *'katte'* and other virus diseases (such as *Kokke kandu* and Niligiri necrosis) are common.

5.0 PLANTING AND AFTER 1CARE

5.1 Conversion period and Isolation

Conversion period of minimum three years and isolation of 25 m wide space should be provided between organic and inorganic plots. If organically produced planting materials are used and if at least two years have elapsed without use of any inorganic inputs in the field prior to planting, the yield from such a crop shall be considered as organic. In case of cultivation in virgin lands and farms wherein no chemical inputs have been applied in the past, the conversion period can be relaxed. In the case of cardamom plants grown wild in the forest, the entire produce can be considered as organic (Anon, 1998).

5.2 Site selection and Shade requirement

The shade canopy provides suitable environment by maintaining humidity and evaporation at suitable level (Abraham, 1965). Shade requirements vary from place to place and depend on the lay of land, soil type, rainfall pattern and crop combination etc (Korikanthimath, 1991). In Guatemala, which receives well distributed rainfall and has cool climate round the year, cardamom is grown practically in open areas with either no shade or having only very sparse shade (Anon, 1977). Cardamom yield level was high under 65-70 per cent light intensity (Sulikeri, 1986). Total biomass, chlorophyll and protein contents, photochemical activity were adversely affected in scorched seedlings grown under full light (100 per cent), when compared to healthy seedlings grown under medium light (45-55 per cent) (Ravindran and Kulandaivelu, 1998a). Forty percent reduction in growth of plants exposed to full light was observed compared to those exposed to medium light. Plants under low light had growth and yield values intermediate between those in medium and full light (Ravindran and Kulandaivelu, 1998b). Light requirement for cardamom nursery was found to be 55 per cent (Ranjithakumari *et al.*, 1993) and this intensity was found to be optimum for better growth and development of seedlings in the nursery. Laboratory studies on photosynthetic efficiency in cardamom indicated that efficiency was greater under low light intensities (30.6 to 106.63 micro Einstein per sec per sq meter) than at higher ones. Photochemical processes were favoured by low light compensation point (Vasanthkumar *et al.*, 1989).

5.3 Preparation of main field

Cardamom is generally cultivated in forest area. The initial work involves clearing the forest and thinning out overhead canopy in order to get one umbrella canopy to get filtered shade. If land is sloppy, it is advisable to start cleaning from the top and work down wards. In sloppy areas, adequate soil and water conservation measures are necessary while preparing the land for planting. Planting in trenches across the slopes in low rainfall areas, diagonal planting and mulching the soil will help in soil and water conservation.

5.4 Spacing

Spacing should be decided based on variety and duration of crop in the field. 'Mysore' and 'Vazhukka' types are vigorous and need wider spacing. 'Malabar' types need closer spacing as comparatively smaller plants.

Korikanthimath and Venugopal (1989) reported higher yield in cardamom seedlings planted at 2 x 1m in hill slopes and 2.1 x 2.1m in flat lands. The trench method of planting with a spacing of 2 x 1 m resulted in better growth and yield (Siddagangaiah et al., 1998a). In spacing trial conducted at Yercaud, Tamil Nadu, it was found that close spacing (1 x 1 and 1.5 x 1.5 m) recorded higher yield than wider spacing of 2.5 x 2.5 m and 2 x 2 m. In general, 'Mysore' and 'Vazhukka' types planted at a spacing of 3 x 3 m or 2.4 x 2.4 m in high rainfall or irrigated areas. A spacing of 1.8 x 1.8 m or 2 x 2 m is suitable for cardamom growing areas of Karnataka (Anon, 2003).

5.5 Method of planting

The trench method of planting (60 x 30 cm) with a spacing of 2 x 1 m resulted in the best growth and yield (Siddagangaiah *et al.*, 1998b). Greater moisture retention was reported under trench system than the pit system of planting (NRCS, 1989). In some places seedlings are planted in holes, just scooped out at the time of planting. In other areas, considerable care is taken in the preparation of planting pits and they are filled with surface soil mixed with leaf mould, compost or cattle manure (Subbaiah, 1940). Mayne (1951) reported that a pit size of 60 x 60 x 45 cm; 90 cm^3; or even 120 x 120 x 30 cm was suitable for Kerala. 'Mysore' and 'Vazhukka' types, usually planted in the pits of 90 x 90 x 45 cm or 90 cm^3, where as, 45 x 45 x 45 cm pits are generally opened during April-May after pre-monsoon showers are used for planting 'Malabar' types. Pits are filled with mixture of topsoil and compost or well rotten farmyard manure and 40 g of rock phosphate (Anon, 2003).

5.6 Planting season

Planting season is decided based on topography and rainfall pattern. Normally planting is done in June-July. In areas experiencing heavy south-west monsoon, planting is either completed before July or is taken up in August-September, so as to avoid heavy rains in July. Early planting gives better establishment and growth than late plantings (Mayne, 1951). Planting in low laying areas (such as valleys) should be done only after heavy rains by July (CPCRI, 1980). Better establishment and growth of seedlings were recorded in August planting in Mudigere of Karnataka (Pattanshetty and Prasad, 1972). Cardamom suckers are planted from June to August on the surface or 15-20 cm deep below the surface. Suckers planted in August survived best, with a mean mortality of 25 per cent and those planted on the surface had the lowest mortality rate of 17.5 per cent (Pattana shetty *et al.*, 1974). Nanjan *et al.* (1981) found that July planting gave the best establishment (87.92 per cent) followed by August, September, October and November planting (77.9, 75.4, 63.7 and 61.6 per cent respectively). During the month of planting, minimum of 322 mm rainfall along with temperature range of 15.5 - 17.5° C (minimum) and 19.5 - 25° C (maximum) is required for better establishment.

5.7 Planting

Ten or eighteen month old cardamom seedlings were planted in pits by taking small portion of filled soil and adding rock phosphate (40 g) in the centre of pit and seedling is placed in the centre of depression. Soil is then replaced taking care that the roots are distributed in their normal position and well pressed around the base of the clump. Planting need to be done just above the ground level to avoid rotting in heavy rains. Seedlings are normally planted in a pit at acute angle for promoting shoot production. In eighteen months old seedlings light root pruning need to be taken up. In case planting suckers, suckers may be placed in the slanting position and cover the base of rhizome with soil. After planting, cross-staking need to given immediately after planting to avoid damage due to wind. Mulch the plant base to reduce evapo transpiration and rain damages. After transplanting, care should be taken to offset the transplanting shock and to save the seedlings from heavy rains. The newly planted area should be inspected regularly and gaps may be filled immediately, if the climate is favourable.

5.8 Mulching

Fallen leaves of the shade trees are utilized for mulching. Sufficient mulch should be applied during November - December to reduce the ill effect of drought which prevails for nearly 4-5 months during summer. Zachariah (1976) reported that mulching is a practical solution for conserving soil moisture and has been acclaimed as the most important cultural operation for the overall improvement of soil and yield of cardamom. Among the mulches tried, leaf mulch and phoenix leaves were on par and significantly superior to other mulches in sucker production (Raghothama, 1979).

Demulching is also equally important during May after the pre monsoon showers for facilitating honey bee movement to obtain better pollination and capsule setting and also provide better aeration and minimizes incidence of clump rot or rhizome rot. The practice of uncovering the panicles shortly after the commencement of flowering improves fruit set in cardamom. The average number of capsules per panicle is 27.4 and 2.1 in case of exposed and covered panicles respectively in 'Malabar' (Pattanshetty and Prasad, 1974).

5.9 Weeding and trashing

Clean weeding is to be limited to the plant bases (50 cm) and the inter rows are to be maintained by slash weeding (Anon, 1998). The weeded materials should be used for mulching. Trashed materials and fallen leaves may also used for mulching. Trashing the dry leaves, leaf sheaths and removal of yielded old suckers along with rhizomes may be carried out once in a year about a month after completion of final harvest, which can be used, for composting. Trashing thrice a year and cleaning the plant bases during monsoon months helps in better pollination, elimination of breeding sites of sucking pests and reduce rhizome rot incidence.

5.10 Irrigation management

Cardamom is generally grown as a rainfed crop and cardamom tracts of India experience a dry spell of about 5-6 months. Increased denudation of forests, deterioration in forest ecology, coupled with erratic trends of rainfall, leads to aridity effects, adversely affecting cardamom production (Rethinam and Korikanthimath, 1985). Since the roots of cardamom are shallow, moisture at the root zone does not last long and hence it is essential to irrigate the crop during January to May (Sivanappan, 1985). Irrigation at 75 per cent available soil moisture (ASM) recorded favourable growth parameters compared to 25 per cent ASM (Raghothama, 1979; Raju, 1981). Further, it was reported that, higher level of irrigation produced maximum capsules and weight compared to irrigation at lower levels.

Cardamom irrigated at 25 per cent ASM and 75 per cent ASM produced 2.01 and 2.14 times more capsules than those without irrigation. The yield computed on hectare basis was 1832 kg and 1977 kg in plants irrigated at 25 and 75 per cent ASM respectively as against 856 kg in plants without any irrigation. Irrigation at 75 per cent ASM combined with high light resulted in the highest yield of 2618 kg/ha. The lowest yield of 645 kg per hectare was obtained under high light intensity and without irrigation (Sulikeri, 1986).

Water not contaminated with insecticides, fungicides, other chemicals and fertilizer leachates should only be used for irrigation under organic system of cultivation. This implies that the watersheds for irrigation source should also be maintained following organic methods of production (Anon, 1998).

Sprinkler irrigation equivalent to 4 cm rainfall at every 15 days would be quite sufficient for cardamom crop (John and Mathew, 1977; Saleem, 1978, Vasanth Kumar and Sheela, 1970). Sprinkler irrigation (25 mm rain) once in 12 days recorded higher yield compared to sprinkler irrigation once in 15 days and control (IISR, 2004). Drip irrigation daily @ 8 litres/plant from 15th January onwards recorded significantly higher yield and this was followed by sprinkler irrigation once in 12 days (IISR, 2004).

5.10.1 Water harvesting

Though cardamom tracts receive heavy rainfall (1500-4500 mm), the availability of soil moisture during summer months is limiting factor due to undulating topography of the plantations and 75 per cent of the total rainfall fall is within three months (June, July and August) (Rajendra Hegde and Korikanthimath, 1996). In low to medium rainfall areas (1200 to 2000mm), trench system of planting was found to conserve soil moisture (NRCS, 1987). In areas having more than 7 per cent slope, of all along the contour, bench terracing of about 5-6 feet width has to be created for taking up cardamom planting. The weeds removed should be kept between the rows across the slopes, which act as barrier to soil erosion and for effective soil and moisture conservation (NRCS, 1989). Agronomical practices like shade regulation, trench, terrace system of planting, provision of shelter belts, mulching, earthing up, vegetative barriers and intercropping help in better growth and development of cardamom besides effective soil and water conservation measures (Ankegowda *et al.,* 2002).

5.10.2 Drainage

Although cardamom requires high moisture level, it is very sensitive to high water table and consequent waterlogged situations (Sulikeri *et al.* 1978). Drains should be opened at regular intervals so as to keep water table 30 cm below the ground surface for better growth. In valleys and high rainfall areas with medium slopes, suitable drains (45 cm depth and 30 cm width) have to be provided while planting in between two rows of cardamom. These drains have to be cleaned regularly to avoid stagnation of water during rainy season. This will also avoid rhizome and root rot to considerable extent.

5.11 Shade regulation

Shade requirement studies indicated that cardamom does not tolerate direct sunlight at the same time too much shade affects metabolic activities of plants. Hence, removal of excess shade is also essential so as to allow sufficient penetration. Shade has to be regulated based on the topography of land, moisture content etc. so as to get about 50 per cent filtered sunlight for proper growth and flowering.

Shade regulation is one of the important practices in cardamom plantations. It should be done during summer (March-April) in the new planting areas and during May-June after the receipt of summer showers in the existing plantation. Normally all kinds of trees that exist in forest are maintained as shade trees. Trees belonging to 32 families of Angiosperms constitute the major tree flora in cardamom hills of Kerala, Karnataka and Tamil Nadu (Shankar, 1980). An ideal shade tree in cardamom plantation was described by George *et al.* (1984). It is desirable to maintain a mixed population of medium sized shade trees, that facilitate shade regulation and to maintain more or less optimum conditions throughout the year. Common shade trees are Balangi (*Artocarpus fraxinifolius* Wt), Nili (*Bisochofia javanica* Blum), Jack (*Artocarpus heterophyllus* Lamk), Red cedar (*Cedrella toona* Roxb), Karimaram (*Diospyros ebenum* Koenig), Karna (*Vernonia monocis* C.B. Darke), Nandi (*Lagerstroemia lanceolata*) and Spanish cherry (*Mimusops elangi*) are desirable as shade trees for cardamom (Rai, 1978; Abraham, 1957). *Maesopsis eminii*, an introduction from Africa, is also a good shade tree for cardamom cultivation (CPCRI, 1983). George *et al.*, (1984) reported that cardamom grown under *Diospyros ebenum*, yielded 40-50 per more than those plants under Jack, Nandi and Spanish cherry.

Under organic system of cultivation, to facilitate penetration of sufficient light, restricted lopping of shade tree branches may be made. However, even under such situations no treetop shall be cut. In areas, which are overexposed, planting of shade trees is an essential operation and while doing so, maximum bio-diversity suited to the local situation may be considered. Trees having desirable characters such as defoliation during rainy season, self pruning habit, flowering during summer and medicinal value may be considered. If such trees belong to leguminous species they are preferred. Restricted loppings and leaf litter may be used for green leaf manuring or composting (Anon, 1998).

5.12 Pollination management

Though cardamom has bisexual flowers and is self compatible, cross pollination is the rule. Self pollination is hindered in due to the slight protrusion of the stigma above the stamens. In cardamom, cross pollination is mediated by activity of the bees (*Apis cerana, Apis indica and Apis dorsata*) as pollinators. Cardamom flowers remain in bloom for 15-18 hours. Stigma receptivity and pollen viability are maximum during morning hours. The receptivity is maximum between 8 AM and 10 AM when 72% of the opened flowers set fruit. After 10 AM the stigma receptivity decreases gradually and only 24 per cent of the flowers opening at 4 PM set fruit. It is reported that receptivity of stigma is highest between 8 AM and 12 noon (Krishnamurthy *et al.*, 1989; Kuruvilla and Madhusoodanan, 1988). Parvathi *et al.* (1993) and Belavadi *et al.* (2000) noticed that the peak receptivity around 12 noon and coincides with peak pollinator activity. The active foraging of bees is seen in the morning hours and is instrumental in increasing the fruit set in cardamom.

In cardamom, pollination is effected by the activity of bees, two species namely *Apis cerena indica* F. (Indian hive bee) and *Apis dorsata* F. (Rock bee) constitute more than 95 percent of all the flower visitors and are the true pollinating agents. The active foraging of bees are seen in the morning hours of the day which help in increasing the fruit set of cardamom. Honeybees visit cardamom flowers during flowering season for collecting nectar and pollen and they do help in attaining over 90 per cent pollination.

The mean number of flowers open per day per plant is 34.5. The optimum bee visit for pollination is 3 to 5. In general, bee activity starts by around 7.30 AM and continues till 6.30 PM in the evening, the peak is between 11.00 AM to 1.00 PM. On a clear day a flower is visited as many as 120 times, on a cloudy day 57 times and a rainy day 20 times on an average (Belavadi *et al.*, 1993, 1997, 1998; Parvathi *et al.*, 1993).

The extent of fruit set recorded in different months indicates fruit set is high (50 to 59 per cent) during June, July, August and September because of high prevailing humidity. During the dry season from December to March, there was practically no fruit set. Bee pollinated fruits were found to contain average of 11 seeds (Chandran *et al.*, 1983) in South India while in Papua New Guinea the average number of seeds per capsule was 13.8 (Stone and Willmer., 1989)

- The number of honeybee colonies required for effective pollination of cardamom is minimum of four colonies (about 5000 foragers per colony) per hectare (3000 plants). In cardamom, by maintaining the beehives 9 to 13 percent increased yield were reported.
- Mulching during flowering period hinders bee pollination resulted in 33 to 44 per cent reduction in fruit set. Thus, demulching after the pre monsoon showers to facilitate better honey bee movement and capsule setting
- Bee pollination resulted in better quality capsules of uniform size and shape.
- Promoting use of beekeeping for pollination of cash crops will be of benefit to both the beekeeper who will receive money for the pollination services of his honeybees and harvest honey and to the farmer whose income will be increased through boosting crop productivity as a result of pollination services of bees.

5.13 Cropping system

In the inter-cropping system plants compete for light, water, soil and space. Competition will occur when the immediate supply of a single necessary factor falls below the combined demands of the plants. Better utilization of environmental resources is the prime objective of these planting systems. In mixed cropping, even if all other conditions are favorable for growth of mixed crops, light is the limiting factor since light penetration of plant stand is reduced through interception and absorption by the taller canopy (Balasimha and Yadukumar, 1993). In this situation, shade loving species which require low light intensities are likely to perform better under high density multispecies cropping system (Rethinam and Venugopal, 1994). Srinivasan *et al.* (1999) conducted survey and economic analysis of various cardamom plantations adopting mixed cropping systems in Kerala and Karnataka indicated that the inclusion of black pepper, coffee and arecanut was highly remunerative giving higher cost benefit ratios. Mixed cropping of cardamom in arecanut gardens given higher income than arecanut monocropping (Korikanthimath *et al.* 1997a,b).

Cardamom is a valuable source of income for numerous resource-poor farmers inhabiting the East Usambara Mountains in Tanzania. On-farm research demonstrated that organic cardamom could be profitability grown in home gardens and by implementing agro-forestry systems, the harmful impact of cardamom cultivation on the forests might be dramatically reduced (Reyes *et al.*, 2006).

6.0 NUTRIENT MANAGEMENT

6.1 Nutrient recycling

Biomass regeneration and utilization as a source of organic input has been a feasible approach towards upgrading productivity of land and crop. Cardamom being sciophyte, requires filtered light for its growth and yield. It is commonly cultivated beneath the shade tress, which play dominant role in recycling of nutrients through their leaves and other parts. Such tree crop based land use systems are more efficient in maintaining soil fertility than annual cropping system. Nair *et al.* (1980) observed that shade trees help to maintain a high fertility status and soil pH in surface soils. Even though on an average about 5.5 t/ha of organic materials, leaf litter, weeds and pruned plant parts are recycled in a cardamom plantation in a year, the nutrients are mainly in the organic form and are available to the crop only by the process of mineralization. As the rate of mineralization is always low, the nutrients that become available to the crop will be able to sustain only average growth and production (Srinivasan *et al.*, 1993a). Dinesh Kumar and Babitha (2006) studied the rate of leaf fall from some common shade trees present in cardamom production system and characterized the same based on major nutrients at Mudigere of Karnataka. Most shade trees shed leaves during September to December. Highest leaf shed was noticed in *Syzygium cumini* followed by *Terminalia arjuna*. As far as major nutrients are concerned, *Erythrina indica* recorded higher contents of nitrogen and phosphorous, while *Lagerstroemia lanceolata* contained more potassium.

6.2 Manuring

Cardamom is generally grown in the rich fertile soils of the forest eco-system. Being a perennial crop sucker production is throughout the year initiation of panicles and development of capsules spread over a period of eight to nine months in a year (Korikanthimath, 1984). A steady absorption and utilization of plant nutrients take place throughout the life cycle of cardamom and hence a regular application of nutrients should be followed for higher yields (Mohankumar and Hegde, 1983). Application of compost or farm yard manure @ 5 kg/plant/year in May-June is recommended (Anon, 1997). Application of 3 - 4 t of well rotten cattle or sheep manure apart from 30 kg each of N, P_20_5 and 40 kgs K_2O/acre greatly enhanced the yield in cardamom (Shanmugavelu and Madhava Rao, 1977). Thimmarayappa *et al.*, (2000) reported that integrated nutrient management with 25% organic manure + 75 % inorganic fertilizers and 50 % organic manure + 50 % inorganic fertilizers also recorded yields at par with 75 % organic manure + 25 % inorganic fertilizers. Field studies in cardamom has shown that application of recommended NPK nutrients as organic fertilizers (50 per cent each as FYM and neem cake + 50 per cent P each as bone meal and rock phosphate + 50 per cent K as wood ash) as effective in increasing the yield and quality of cardamom (Sadanandan and Hamza, 2006).

Application of organic manures such as neem cake @ 1 kg or poultry manure/farmyard manure/compost/vermicompost @ 2kg per plant is recommended once in a year during May-June. Mussorie rock phosphate or bone meal may be applied based on soil analysis (Anon, 1998). Nearly 70 per cent of cardamom roots were confined to shallow depth of 5 - 40 cm and 30 - 50 cm radius (Khader and Sayed, 1977; Nair, 1988). Therefore, for the maximum efficiency of applied manures it would be necessary to apply at a radius within 50 cm and being a surface feeder, deep placement of manures are not advisable.

6.3 Bio fertilizers

Three species of vesicular arbuscular mycorrhizal (VAM) fungi (*Glomus macrocarpum, G. fasciculatum* and *Gigaspora coralloidea)* were recorded in the roots of cardamom plants. Per cent colonization in roots varied from 40 – 100 per cent in pre monsoon samples and 63 - 94 per cent in post monsoon samples. Degree of infection varied from 4.5 to 76.0 in pre monsoon and 30.0 to 70.0 in post monsoon samples. The increasing doses of fertilizer tended to decrease the intensity and incidence of the VAM population in soil and roots (Rohini Iyer *et al.*, 1989).

Sreeramulu and Bagyaraj (1998) tested the growth response of cardamom seedlings ('Malabar') to 13 different vesicular arbuscular mycorrhizal fungi. In general seedlings inoculated with the fungi grew taller, had more leaves and tillers, and greater seedling biomass and uptake of nutrients than control seedlings. Among the various mycorrhizal fungi tested, seedlings inoculated with *Gigaspora margarita* and *Glomus monosporum* exhibited significantly greater growth with increased uptake of nutrients.

7.0 DISEASES AND PEST MANAGEMENT

7.1 Fungal diseases

Major fungal diseases causing considerable crop loss are 'azhukal' (capsule rot), rhizome rot (clump rot), seedling rot (damping off), leaf blight, leaf blotch and nursery leaf spots.

7.1.1 Capsule rot (Azhukal disease)

Azhukal is a local term meaning rotting in Malayalam. Besides capsules (fruits), disease can also be seen in other parts of plant such as leaf, panicle, inflorescence and rhizome. The disease is particularly important in Kerala (Idukki and Wynad districts) with losses upto 30% (Nambiar and Sarma, 1974). Large circular or irregular water soaked lesions on leaves or capsules are the first visible symptoms. These lesions enlarge and affected portions rot. In some instances long lesions extending over the entire length of the leaf were observed. The exposed portions of tender, unopened leaves remain rotted and decomposed without opening within 2 to 4 days. The leaves become shredded and remain attached to the pseudostem. Infected capsules show water soaked spots, which enlarge and turn grayish brown. Such affected capsules rot and drop down prematurely during rains emitting foul smell (Thomas *et al.*, 1991a; 1989; Thomas and Bhai, 2002). From the capsules, infection spreads to panicle and whole inflorescence is affected (Alagianagalingam and Kandaswamy, 1982). *Phytophthora meadi* and *P. nicotianae* var *nicotianae* are the primary fungi associated with the disease (Nair, 1979; Bhai, 1998). Associations of *P. palmivora, Pythium vexans* and a *Fusarium* sp. have also been reported (Radha and Joseph, 1974; Nambiar and Sarma, 1976).

The spraying of 1 % Bordeaux mixture is recommended to control the disease. Since the occurrence and spread of 'azhukal' disease is directly correlated to weather conditions, an integrated disease management strategy involving plant sanitation and biological methods are more suitable. Plant sanitation including removal and destruction of diseased plant parts and regulation of shade in plantations are to be carried out before the onset of monsoon. The first round of prophylactic fungicidal application should be done before the onset of monsoon, usually in May. Since water stagnation aggravates disease, better drainage in the plantation should be assured. Plant sanitation coupled with timely application of Bordeaux mixture (1%) two to three times per year effectively controls the disease (Thomas *et al.*, 1989). Application of biocontrol agents such as *Trichoderma viride* and *T. harzianum* twice during May and July with one kg of multiplied biocontrol agent in carrier media having a cfu of 23 X 10^8/ g was successful in the management of the disease under field conditions (Bhai *et al.*, 1993; Bhai, 1998).

7.1.2 Rhizome / root rot

This disease also referred to as clump rot is reported from both Karnataka and Kerala states. The disease is also seen in nurseries where it is referred to as damping off. The disease affects roots and rhizomes causing decline and death of plants. Incidence of disease up to 20% has been reported. Affected plants show pale and yellow foliage leading to premature death of older leaves. Root and rhizome of affected plants rot. Rotting symptom develop at the collar region as it turns soft and brownish. Aerial stem of the affected clump break off or comes off with a gentle pull. In nursery, soon after emergence, seedlings wither away causing damping off. Seedlings of 2-4 leaf stage show root rot and wilting and subsequent drying of leaves.

The disease is caused by *Pythium vexans* (Subba Rao, 1938), *Rhizoctonia solani* (Ramakrishnan, 1949) and *Fusarium oxysporum*. In a few cases nematode, *Meloidogyne incognita* also causes damage to root and in combination with *P. vexans* and *R. solani*, the damage is more. Similarly, root grub infestation also reported to pre-dispose the plant to infection by *P. vexans* and *R. solani*. All these pathogens are soil borne and disease appears with the onset of south-west monsoon during June.

Removal of infected plants and change of nursery site would reduce inoculum build up. Raising nursery in solarized is recommended to ensure pathogen free seedlings (Anand Rau and Venkataramani, 1949). Soil drenching with 1% Bordeaux mixture or neem oil cake @ 500g/plant was effective in managing the disease. Application of biocontrol agents such as *Trichoderma harzianum* and *T. viride* were reported to reduce disease incidence and formulation of *T. harzianum* was developed for field application (Thomas *et al.*, 1991b; 1997). In order to get effective control, integration of methods such as phytosanitation, providing good drainage and soil application of biocontrol agents is necessary.

7.1.3 Chenthal or Leaf blight

This disease also locally known as 'chenthal' was first reported from Idukki district of Kerala (George *et al.*, 1976). Disease appears post monsoon and becomes severe during summer. The symptoms of the disease appear as elongated water soaked lesions on the abaxial surface of young leaves. In advanced stages, the lesions become brown to dark brown with a pale yellow halo. The leaves wither resulting in wilting of pseudostems. The disease affected gardens present a burnt appearance. Flowers produced after disease incidence fail to form capsules (George and Jayasankar, 1979; Thomas and Bhai, 2002). Though *Corynebacterium* was reported earlier (George and Jayasankar, 1977), detailed investigations carried out in later years showed that *Colletotrichum gloeosporioides* is the pathogen responsible for the disease (Govindaraju *et al.*, 1996). Providing adequate shade in the plantation and mulching the plants to conserve soil moisture etc. ensure disease control to a certain extent. Spraying of Bordeaux mixture (1%) was found to limit the further spread of infection (Govindaraju *et al.*, 1996).

7.1.4 Leaf blight

In recent years, leaf blight caused by foliar infections by *Phytophthora meadii* and *Colletotrichum gloeosporides* is gaining importance. The disease appears during mid-monsoon, becomes severe during late monsoon periods (October-November) and declines by March. The symptoms develop as brownish spots and patches on the leaf lamina, often extend to large continuous areas and finally these portions or the entire affected leaves shred and dry. The disease is found only in certain localities. It has been observed that intermittent rain and continued mist formation in plantations favours the incidence and spread of the disease. The disease can be controlled by one or two rounds of Bordeaux mixture spray (1%).

7.1.5 Nursery leaf spot

This is a most important nursery disease caused by *Phyllosticta elettariae* (Chowdhary, 1958). The disease which occurs mainly on tender leaves of seedling starts as water soaked lesions on the lower surface of the leaves that turn necrotic surrounded with water soaked areas. Mature lesions appear circular to oval and grayish white. In severe cases, rotting of leaves is also observed followed by reduced tillering. In wet weather lesions coalesce forming large necrotic patches leading to leaf shredding. In dry weather, spots remain isolated and bound by a faint halo. Younger seedlings are more susceptible, susceptibility gradually decline with their age (Agnihothrudu, 1987). The disease is more severe and widespread in nurseries in Karnataka. The fungus survives in leaves for one year and can serve as primary source of inoculum (Naidu, 1981). Disease incidence was less in nurseries raised under watersheds compared to the exposed nursery. Spraying with Bordeaux mixture (1%) is effective in managing *Phyllosticta* leaf spots (Rao and Naidu, 1974).

7.2 Viral diseases

Cardamom is a perennial herbaceous rhizomatous crop and it is susceptible to attack of viral diseases at any stage of the crop. Because of perennial nature of the spice infection by viral diseases reduce crop to uneconomical level within 1-3 years of infection. Already five viral diseases are reported on cardamom.

Viral diseases of cardamom

Disease	Virus	Spread	Prevalence
CdMV / katte	Maclura virus (+ssRNA)	Vegetative, aphid	India, Guatemala
CdVCV/ kokke kandu	Poty virus? (+ssRNA)	Vegetative, aphid	India
CdNV/ Nilgiri necrosis	Carlavirus (+ssRNA)	Vegetative	India
Infectious variegation virus	?	?	India
Banana Bract mosaic virus B Br MV	Poty virus (+ssRNA)	Vegetative, aphid	India

7.2.1 Mosaic (Katte or marble) disease

'Katte' or mosaic or marble virus disease caused by *Cardamom mosaic virus* (CdMV) has been responsible for low yields and rapid decline of cardamom in Guatemala, India and Sri Lanka (Dimitman, 1981; Dimitman *et al.*, 1984; Gonsalves *et al.*, 1986). In India, *katte* is widely distributed in all cardamom growing tracts with incidence ranging from 0.01-99 per cent (Mayne 1951; Venugopal, 2002). Loss in yield due to the disease depends on growth stage at the time of infection. In early infected plants loss will be almost total. Late infection results in gradual decline in productivity. Crop losses of 10-60%, 26-91% and 82-92% were reported under cardamom-areca mixed cropping in first, second and third years of production respectively (Varma, 1962). In monocrop situations, a yield reduction of 38, 62 and 68.7% were reported for the first, second and third year of infection (Venugopal and Naidu, 1987;

Venugopal, 1995). In general total decline of plants occur within 3-5 years of infection. The first experimental transmission of *katte* disease in India was obtained with banana aphid (*Pentalonia nigronervosa* Coq) (Uppal *et al.*, 1945). So far 13 aphid species were reported to transmit the disease (Rao and Naidu, 1974).

Fig. 2: Katte disease affected cardamom leaves showing mosaic symptoms
(*Source:* Director, IISR, Calicut)

7.2.2 Cardamom vein clearing virus or Kokke Kandu disease

This disease reported so far only from India is of relatively recent origin has become a threat to cardamom cultivation in few endemic pockets of Karnataka (Venugopal and Govindaraju, 1993; Govindaraju *et al.*, 1994). Because of its characteristic symptom, it is locally referred to as *kokke kandu*, meaning hook-like tiller. The disease is reported from Kodagu, Hassan, Chickmagalur, Shimoga and North Canara districts of Karnataka. This disease may occur either singly or mixed infections with *katte*. Unlike *katte* disease, the plants affected with this disease decline rapidly with yield reduction upto 62-84% in the first year of peak crop (IISR, 1995; 1996; 1997). The virus is not transmitted through seed, soil or sap. The disease is transmitted through cardamom aphid, *P. nigronervosa* f. *caladii* in a semi persistent manner (Anand *et al.*, 1998).

7.2.3 Viral disease management

Management of viral diseases is much more difficult than that of diseases caused by other pathogens as viral diseases have a complex disease cycle, efficient vector transmission and lack of effective chemical (viricide). Most of the procedures that can be used effectively involve measures designed to reduce sources of infections inside and outside the crop, to limit spread by vectors and to minimize the effect of infection on yield (Hull, 2002; Varma *et al.*, 2002).

Elimination of sources of inoculum

In management of plant viruses, phytosanitation involves the detection and elimination of virus sources present within and outside the crop. Efficiency of phytosanitaion in management of viral diseases of cardamom is centered on prompt inspection, detection and elimination of virus sources. Rouging is effective in minimising the spread and enhancing economic life of the plantation. It is more appropriate to adopt rouging as an effective means of containing disease spread if the disease is less than 10 per cent. Removal of infected volunteers in replanted area and total avoidance of volunteers for nursery activity in hot spots are most important requisites for producing virus free planting material.

Since *'katte'*, *'kokke kandu'* and Nilgiri necrosis are not seed transmitted and lack of long distance movement by the vector beyond 2000 m, raising nursery in isolated places is necessary to produce healthy seedlings. For *'katte'*, isolation of more than 200 m from virus sources is adequate and for *'kokke kandu'*, isolation of more than 2000 m is necessary (Venugopal, 1999; 2002).

Removal of breeding sites

The photophobic vector breeds in the senile, concealed parts of the host. Periodical removal of old parts of the rhizomatous crop is effective in reducing the aphid population and the spread of mosaic. Other plants like *Colocasia* sp. and *Caladium* are the other natural hosts of cardamom aphid which are naturally found in the swampy areas of cardamom plantations. Removal of the breeding sites and other hosts are necessary to check reinfestation of aphid apart from undertaking vector control measurers.

Use of biopesticides

Extracts of many botanicals were found to be effective in reducing the breeding of aphid vector. Neem products significantly reduced the settlement of the aphids on the cardamom leaves even at 0.1 per cent concentration and lethal to aphids at higher concentrations. Aqueous extracts of *Acorus calamus*

L. (dried rhizome), *Annona squamosa* L. (seeds) and *Lawsonia inermis* L. (leaves) reduced the settling percentage of aphids. Vapours of *A. calamus* are highly toxic to aphids and cause total mortality. Essential oil of turmeric (*Curcuma longa* L.) was also found to possess repellency against the aphid (Naidu *et al.*, 1985; Agnihothrudu, 1987; Venugopal, 2002).

Use of resistant varieties

Survey of hot spot areas of *katte* disease led to the collection of 134 disease escapes. The escapes were planted in sick plots and screening continued for six years. 23 of them remained totally free from the symptoms (Venugopal, 1999). These lines when planted in hot-spot areas, 19 lines remained free from disease of which 17 had satisfactory agronomic characters. These 17 lines were further tested after interplanting with known susceptible checks, and they remained free of disease. The mechanism of resistance in the above lines is not understood (Venugopal, 2002).

7.3 Pest management

Pest problem is a major production constraint in cardamom. Hence management of pest infestation at various stages of crop is an important factor, which contributes to the high cost of cultivation and low yield in cardamom. As many 60 insect and non-insect pests attack cardamom, but only 5 or 6 are the major problems. The most important pests are thrips, shoot and capsule borers, root grubs, hairy caterpillar and shoot fly (Kumareshan and Varadarasan, 1987; Kumaresan *et al.*, 1988; Premakumar *et al.*, 1994).

7.3.1 Cardamom thrips (Sciothrips cardamomi Ramk*)*

The cardamom thrips is the most destructive pest of cardamom and was first described from Anamalai hills and later from other cardamom growing areas (Ayyar, 1935). The pest causes 10-90% damage to the crop in various areas. The adults and larvae lacerate and feed on leaves, shoots, inflorescences and capsules. When panicles are infested, it results in shedding of flowers and immature capsules. The feeding activity on tender capsules results in the formation of corky, scab-like encrustations on them. These later become malformed and shriveled with slits on the outer skin; the seeds within them are poorly developed and lack the usual aroma (Nair, 1975; Nambiar *et al.*, 1975; Varadarasan, 1992).

The thrips breed within unopened leaf spindles, leaf sheaths, flower bracts and flower tubes. The adults are grayish brown and measure 1.25-1.50 mm in length. Reproduction is by sexual and parthenogenetic means. The egg period lasts for 9-12 days. There are two larval, a prepupal and pupal stage.

The life cycle from egg to adult is completed in 3-4 weeks (Nair, 1975; Kumaresan *et al.*, 1988). The population of thrips is generally high during summer months and declines with the onset of rains. The Mysore and Vazhukka types of cardamom are highly susceptible to thrips infestation. The pest infestation was also higher with closer spacing (Chandrasekhar and Balu, 1993).

Varieties and thrips attack

Persistence of thrips infestation is the major handicap for organic cultivation. Variations in the adherence and persistence of bracts around the flowers and in fruits and variations in the length of pedicle play a determining role to thrips injury (Mayne, 1951). Mayne (1951) reported the comparative freedom of thrips in prostrate panicle types.

Fig. 3: Cardamom capsules damaged by thrips (*Source:* Director, IISR, Calicut)

7.3.2 Shoot and capsule borer (Conogethes punctiferalis Guen.)

The shoot and capsule borer is a serious pest in nurseries and main fields in Kerala, Karnataka and Tamil Nadu. The larvae bore into pseudo stems and feed on the internal contents leading to formation of 'dead heart' symptoms. The larvae also bore into the panicles resulting in drying up of the same and into the capsules feeding on the seeds thus making capsules empty. Yield losses of 70-80% have been recorded in severely infested plantations (Vardarasan *et al.*, 1989).

The adult is a small moth with a wing span of about 24 mm; the wings are orange yellow with small black spots. The earlier instars of larvae bore into panicles, flower buds and capsules and the late instar larvae bore into pseudo stems (Vardarasan *et al.*, 1989). Studies on the life cycle of the pest conducted at Idukki (Kerala) indicated that the total period from egg to adult took 41-57 days during summer (March-May) and 84-123 days during winter (November-February). The pest infestation was generally observed throughout the year but was higher during January, March, June, August and October (Vardarasan *et al.*, 1989; 1991).

Fig 4: Cardamom plant infested with shoot borer (*Source:* Director, IISR, Calicut)

7.3.3 Root grub (Basilepta fulvicorne Jacoby)

Root grubs are major pests of cardamom in nurseries and main fields and widely distributed in Kerala, Karnataka and Tamil Nadu. In Karnataka, the pest is generally serious in primary and secondary nurseries causing 16 to 38% seedling mortality (Thyagaraj *et al.*, 1991). In plantations, infestation is observed twice a year (March-April and August-October). The grubs feed and damage roots and portions of rhizomes; some times the entire root system is eaten away. The infested plants turn yellow and are stunted; severely infested plants die. Crop losses of 29-66% have been reported (Varadarasan, 1995).

The adults are shiny metallic blue, green or brown beetles measuring 4-6 x 2-4 mm. Males are smaller than females. The adults are polyphagous feeding on a number of plants that are present around cardamom plantations. The peak periods of adult emergence at Idukki (Kerala) was during April and September. The females lay eggs in soil around the base of clumps and fully grown grubs are pale white, stout and 'C' shaped, and measure about 1 cm in length. At Idukki, the egg, larval and pupal stages lasted for 13-19, 45-60 and 10-17 days, respectively. At Mudigere (Karnataka) the egg, larval and pupal stages lasted for 10-15, 78-80 and 20-25 days, respectively (Thyagaraj *et al.*, 1991; Gopakumar *et al.*, 1987; 1991; Gopakumar and Chandrasekar, 2002).

7.3.4 Pest Management

Nursery stage

In nursery stage seedlings are affected by cut worm, shoot borer, root grub, leaf thrips and root knot nematode. Infestation by root grubs and leaf thrips poses more problems in two season nurseries. To manage whole pest complex, the following measures are to be followed at various stages of nursery.

Cultural practices

- Raise nurseries away from main plantations to reduce possibilities of infestation and reinfestation.
- Provide sufficient organic fertilizers to encourage better vegetative growth.
- Shift nurseries sites regularly to overcome soil borne pest infestation like root knot nematode and root grubs.
- Catch and kill the beetles of root grubs through insect nets in 2 season nurseries.
- Collect the cut worms hiding in the mulch in the affected spot and destroy.

Prebearing stage

Shoot fly, root grubs and root knot nematode are the main pests in the prebearing stage which lasts for 18-30 months. Response of plants, tillering, vegetative vigour and uptake of nutritional inputs depend on the health of the root system. The maiden peak harvest totally depends on the proper management of these three main insect pests. To manage these pests, following measures can be adopted.

- Catch root grub beetles through insect nets and destroy.
- Trace the maggots and larvae of shoot borer from the small tillers showing dead heart symptoms and destroy.

Bearing stage

Thrips, shoot and capsule borers, root grubs, root knot nematodes and occasionally defoliators are the main pest during the bearing period. To manage the complex pest problems, integration of cultural methods and biological control methods are essential. Heavy (more than 60 mm) over lapping rain is also detrimental to most of the pests.

- Undertake earthing up, up to the collar region immediately after last rain to protect active exposed feeder roots.
- Undertake trashing thrice a year during early monsoon, mid-monsoon and late monsoon periods to remove pest breeding sites.
- Provide adequate mulch to protect feeder zone.
- Remove the weeds and other herbaceous cover which act as hosts for many potential pests.
- Collect the adult beetles of root grubs during the peak periods of emergence i.e. April-May and September-October using hand nets and kill them. The beetles, which are brightly coloured in metallic shades, could be easily spotted during day time. Being weak fliers, they could be easily caught using sweep nets. In the same way the adult moths of shoot and capsule borer, which are found resting on the lower surface of cardamom leaves can be collected by net and killed.
- Cardamom tillers harbouring the larvae and pupae of shoot and capsule borers as indicated by the presence of fresh excreta should be cut and destroyed. This will greatly reduce the emergence of the subsequent generation
- The hairy caterpillars are gregarious in habit and they congregate on the trunk of shade trees during daytime. These congregations should be traced, collected and killed mechanically.
- The adults of hairy caterpillars can be lured and captivated by operating light traps at night. These can be collected later and killed.

As soon as bore holes of stem borer (*Conogethes punctiferalis*) are noticed, injection of *Bacillus thurigensis* preparation into the bore hole (0.5 ml in 10 ml water) to kill the larva, so that subsequent resurgence is reduced. Wherever organic methods of cultivation are adopted out break of white flies (*Dialeurodes cardamomi*) is seldom observed. However in the event of such outbreak, collection of adults using yellow sticky trap and control of nymphs by spraying neem oil with soft soap made out of minimum caustic soda (500 ml neem and 500 g soft soap in 100 l water) is to be followed, In areas prone for nematodes (*Meloidogine* sp.) application of crushed neem seed can take care of the problems. Application of fish oil rosin soap may be made for managing thrips (*Sciothrips cardamomi*). 'Malabar' varieties are found to be tolerant to thrips to a certain extent. Regular surveillance is absolutely essential for timely detection and adoption of remedial measures against the pests affecting cardamom (Anon, 1998).

7.3.5 Nematode pests

Plant parasitic nematodes belonging to 28 species and 19 genera are reported. Among them, root knot nematodes (*Meloidogyne incognita* and *Meloidogyne javanica*) pose serious problems in nurseries and plantations in India (Ramana and Eapen, 1992 and 1995).

Nematode problems are more severe in cardamom nurseries and hence utmost importance may be given for production and distribution of nematode free seedlings. Soil fumigation or soil solarization of nursery sites can minimize nematode problems to a greater extent (Ali and Koshy 1982). Cardamom soils are generally rich in organic content and the status of these soils has to be preserved or enhanced. Application of neem oil cake reduced nematode population and increased the yield of cardamom (Ali, 1987). A wide range of soil amendments have been used for control of nematode population. In cardamom neem oil application twice at 500 g/plant during pre monsoon and late monsoon period reduced nematode population and increased the yield (Ali, 1987).

Paecilomyces lilacinus which is a common opportunistic fungus prevalent in many soils has been tried in cardamom nurseries. The fungus multiplied on coffee husk or neem oil cake, reduced root knot nematode multiplication by 48.5 to 57% (Eapen, 1995). In a field trial *P. lilacinus* was applied in sick cardamom nurseries at the time of sowing, rhizome formation and tillering stage. To study its affects on root knot nematode and rhizome rot in beds treated with P. *lilacinus*, the nematode population was reduced by 74.2% and rhizome rot by 19.7% (Eapen and Venugopal 1995). *Trichoderma* sp. was more effective in controlling both root knot nematodes and diseases like damping off / rhizome rot especially in solarized soil. *Trichoderma* sp. controlled root knot nematode, multiplication rate and enhanced the growth of cardamom seedlings under field conditions. Number of quality standard seedlings was more and incidence of rhizome rot was reduced to 26% (Eapen *et al.,* 2000). *T. harzianum* reduced nematode population by 60 – 87% in sterile soil and 50-82% in non sterile soil.

The natural occurrence of *Pasteuria penetrans* on *M. incognita* infesting cardamom was also reported from Idukki, Kerala (Sosamma and Koshi, 1997). Two VAM fungi viz., G. *fasciculatum* and *Gigaspora margarita* were also found to minimize root knot nematode problems in cardamom seedlings (Thomas *et al.* 1989).

8.0 HARVEST AND POST HARVEST HANDLING

Cardamom suckers or seedlings start bearing two years after planting. Panicles appear from the base of plant from January onwards and the flowering is from April to December, which may extend further. Peak flowering is observed during June to August and fruits mature in about 120 days after flowering. Fruits have trilocular capsules containing 15-20 seeds and on maturity seed turn black in colour. The Indian cardamom is known for its rich colour, flavour and aroma.

8.1 Time and stage of harvesting

Due to staggered and prolonged flowering, cardamom capsules mature and ready for harvesting successively at 10 - 15 days intervals over an extended period of 8 months (August to March). Generally harvesting is carried out at an interval of 15-30 days and completed in 8-9 rounds. Fruits (capsules) that are fully matured and retaining dark green colour of rind (peel) with black coloured seeds are ready for harvesting. Harvesting at over ripe stage leads to loss of green colour and capsule splitting during curing. Per cent recovery of dry capsules varies from 20-24 % in ripe capsules, 18-20 % in physiologically mature stage and 14-16 % in immature stage. Hence it is ideal to harvest cardamom at physiologically mature stage (green colour rind and black colour seeds). After harvesting, the freshly harvested capsules need to be cleaned from dirt. Cardamom capsules are cured by reducing the moisture content from 80 % to 8-12 % in an optimum temperature by retaining green colour to the maximum extent. Curing is generally undertaken in two methods.

8.2. Curing

8.2.1 Sun drying

Cardamom capsules are directly dried under the sunlight and generally require 5-6 days. This practice is followed only in some parts of Karnataka. The disadvantages of this method are, not dependable during rainy season and also not possible to obtain good green colour cured capsules.

8.2.2 Conventional curing

The most commonly adopted method for curing cardamom and requires a structure fitted with furnace, flue pipes, chimney, ventilators etc. It is a masonry structure consisting of two apartments, a curing room and a furnace room. Curing room is a tall one provided with veiling at the roof and fitted with wire gauge on the beams at the middle of the room parallel to the ground floor, making the room in to two compartments. Flue pipes having a radius of about 25 cm made by galvanized iron sheets are provided in the ground floor from one end to the other from the furnace to chimney pipe to expel the smoke through the roof. Racks holding rectangular trays are also fitted to the side walls for accommodating larger quantities of cardamom.

Capsules are spread in a single layer on the racks and trays. After spreading, the curing room is closed and heating is done by burning fire wood in the furnace and the heat produced is conducted. Only fallen trees and lopped branches should be used as fuel. The host smoke passes through the pipes bringing the room temperature to 45 to 50°C. This temperature is maintained for 3 to 4 hours. At this stage capsules sweat and give off moisture. Ventilators are then opened for sudden cooling and sweeping out vapour from the drying capsules. Ventilators are closed after vapour is escaped completely and temperature is maintained at 40°C for about 24 to 30 hours. Temperature is raised again to 45°C for one hour. The whole process of curing takes about 28 to 36 hours. In general, quality of capsules cured by this method is very good. Community curing is cheap and less polluting.

8.2.3 Moisture content

Moisture content in cardamom market samples ranges from 7 to 20 % depending on region and method of curing. It is found that 10 % moisture is ideal for the retention of green colour and well-dried capsules produce a typical sound on shaking.

8.3 Specifications for Indian cardamom (based on physical characters)

<table>
<tr><th>Grade</th><th>Description</th><th>Size (mm)</th><th>Weight (g/l)min</th><th>Colour</th><th>General characteristics</th></tr>
<tr><td colspan="6">AG Alleppey Green</td></tr>
<tr><td>AGB</td><td>Extra Bold</td><td>7</td><td>435</td><td rowspan="2">Green</td><td rowspan="4">Kiln dried, 3 cornered and with ribbed appearance</td></tr>
<tr><td>AGS</td><td>Superior</td><td>5</td><td>385</td></tr>
<tr><td>AGS 1</td><td>Shipment</td><td>4</td><td>320-350</td><td rowspan="2">Light green</td></tr>
<tr><td>AGL</td><td>Light</td><td>3.5</td><td>260</td></tr>
<tr><td colspan="6">CG Coorg Green</td></tr>
<tr><td>CGEB</td><td>Extra Bold</td><td>8</td><td>450</td><td rowspan="2">Golden to light green</td><td rowspan="6">Round ribbed or smooth skin</td></tr>
<tr><td>CGB</td><td>Bold</td><td>7.5</td><td>435</td></tr>
<tr><td>CG-1</td><td>Superior</td><td>6.5</td><td>415</td><td>Light green</td></tr>
<tr><td>CG-2</td><td>Mota, Green</td><td>6</td><td>385</td><td>Green</td></tr>
<tr><td>CG-3</td><td>Shipment</td><td>5.5</td><td>350</td><td>Cream</td></tr>
<tr><td>CG-4</td><td>Light</td><td>3.5</td><td>280</td><td>Brown</td></tr>
</table>

Bleached					
BL-1		8.5	340	Pale	Fully developed round/3 cornered ribbed or smooth skin
BL-2		7	340	Creamy	
BL-3		5	300	Dull white	

Source: Indian Standards Institution, New Delhi-1

8.4 Packing

Cardamom is a high value spice and all care should be given for efficient packing. The cured capsules are graded using different sieves of 8, 7.5, 7 and 6 mm. Cardamom need to be stored over a period of time and is normally packed in double lined polythene bags. Storage rooms should be free from insect damage. Cardamom samples dried and stored at or below 10 % moisture content retains original parrot green colour with less mould growth. It is advisable to make use of dried cardamom capsules preferably within 12 months of curing.

9.0 HOLISTIC PACKAGE FOR ORGANIC CARDAMOM PRODUCTION

Till 1940's around 3500 mt of cardamom produced in India was free from chemical fertilizers and pesticides. Well-developed mulch and well-marked layer of humus accumulation is the common feature in all the areas where cardamom development and production was good. In the past growers, importance was given to maintenance of mulch, prevention of soil erosion, run off and conservation of soil moisture for successful cultivation. This practice is being followed in all traditional high rainfall cardamom-growing areas. However, there is no specific package to give incite into organic cardamom production. The following package gives general accepted practices for organic cultivation of cardamom.

Organic packages	Advantages
Nursery management	
Changing of nursery site	Avoidance of soil borne nematodes and fungal pathogens
Selection of nursery site: Minimum 200 m away from infected cardamom plantation	Avoidance of viral infection
Solarization of nursery beds/ nursery mixtures	Soil borne nematode and fungal pathogens Minimum weed growth

Select uniformly high yielding plants with bold capsules for seed collection	High percentage of high yielders in seedling progenies and freedom from virus infection
Early sowing (September)	High germination percentage, Well-grown uniform seedlings (tolerant to nursery leaf spot).
Thin sowing in raised nursery beds with fertile soil	Sturdy seedlings and higher percentage of standard seedlings, avoidance of damping off
Provide 50 per cent shade in nursery	Less incidence of leaf spot and damping off. Sturdy seedlings and higher percentage of standard seedlings
Catch and kill the beetles of root grubs through insect nets in 2 season nurseries.	Collect the cut worms hiding in the mulch in the affected spot and destroy.
Collect larvae from shoot borer affected shoots and destroy	Effective control over major nursery pests
Application of neem oil cake at rapid tillering stage (1 kg/ m^2)	Suppression of root knot nematode population and root grubs
Application of *Trichoderma* through organic carrier (pre sowing, transplanting and tillering)	Management of soil borne fungal pathogens and suppression of root knot nematodes
Spraying with 10% cow urine, 10% cow dung slurry/ vermiwash (farmers practice)	Uniform vigorous and dark green seedlings
Main plantation	
Selection of suitable growing areas	Moderate to high rainfall well drained areas are suitable for organic cardamom cultivation as pest incidence (thrips and borers) are reduced due to rainfall.
Choice of planting material : Malabar types Genotypes with longer panicles	 Less incidence of thrips Less incidence of thrips
Shade regulation: Provide 50-60% filtered light	 Better establishment, tillering and less incidence of rhizome rot
Provide drains/ staggered drains	Less incidence of panicle and rhizome rot

Organic nutrition	
Neem cake application (1/2 kg/plant twice during pre monsoon and mid monsoon)	Suppression of root knot nematode population and root grubs
Application of *Trichoderma* through organic carrier during early monsoon period	Management of soil borne fungal pathogens and suppression of root knot nematodes
FYM/Compost (10 kg/plant) at the time of planting and pre monsoon and post monsoon application of 10 kg each	Enhanced tillering, growth and sustained yield
Application of agriculture lime/dolomite (based on soil test)	Maintenance of favourable conditions for organic recycling, maintenance of optimum soil pH, and micronutrient status
Spraying with 10% cow urine, 10% cow dung slurry/vermiwash at three quarterly interval (farmers practice)	Enhanced tillering, growth and sustained yield
Mulching during winter and summer	Conservation of soil moisture and protection of feeder roots
Demulching, thrashing (thrice - June, August and September) during monsoon	Reduced incidence of rhizome rot, uninterrupted pollinators movement, better green colour development of capsules Population reduction of aphid vector and thrips
Maintenance of 4 honey bee colonies /ha	Higher fruit set and sustainable income
Tracing of virus infected plants at regular intervals	Management of viral disease
Catching of root grub beetles through insect nets	Management of root grubs
Collection and destruction of shoot borer larvae	Suppression of overlapping population of shoot borer
Harvesting of only matured and ripened capsules at 15 - 20 days interval	Minimum damage by mammalian pests Higher recovery and market price
Drying in kilns within 24 h of harvest	Less splits, uniform attractive colour and premium market price

REFERENCES

Abraham, P. (1957) 'Karuna' is excellent as shade tree for cardamom. *Indian Farming*, 7(9): 14-16

Abraham, P. (1958) New technology for cardamom growers. *Indian Farming*, 8(2): 34-38

Abraham, P. (1965) The cardamom in India. *ICAR Publication*, New Delhi, pp. 1-46.

Abraham, P. and Tulasidas, G. (1958) South Indian cardamom and their agricultural value. *ICAR Tech Bull.*79, ICAR, New Delhi. pp. 1-27.

Agnihothrudu, V. (1987) Diseases of small and large cardamom. *Review of Tropical Plant Pathology*, 4: 127-147.

Alagianagalinagam, M.N. and Kandaswamy, T.K. (1982) Control of capsule rot and rhizome rot of cardamom (Elettaria cardamomum Maton.). *Madras Agriculture J.* 68: 564-567.

Ali, S.S. (1987). "A preliminary observations on the effect of some synthetic nematicides and neem oil cakes in cardamom field infested with root knot nematode". In: *Proceedings of PLACROSYM-VI* 1984, Indian Society for Plantation Crops, Kasargod, India, pp. 215-223.

Ali, S.S. and Koshy, P. K. (1982) A note on chemical control of root knot nematodes in cardamom nurseries. *Indian J. Nematology*, 12: 147-150.

Anand, T., Govindaraju, C., Sudharshan, M.R. and Srinivasulu, P. (1998) Epidemiology of vein-clearing virus of small cardamom. *Indian J. Virology*, 14: 105-109.

Ananda Rau, M. and Venkataramani, K.S. (1949) *Mycological report of the Botanist* (1948-49). Planters Association of South India, pp 15-21.

Ankegowda, S.J. (2003). "Optimum leaf stage for transplanting small cardamom seedlings from primary nursery to secondary nursery". In: *Abstracts of National Seminar on Physiological Interventions for Improved Crop Productivity and Quality: Opportunities and Constraints*, December 12-14, 2003, S.V. College, Tirupati, pp.184.

Ankegowda, S.J. Chethan, C.V. and Korikanthimath, V.S. (2002) "Soil and water conservation measures in cardamom cultivation". In: *National Seminar on Soil and Water Conservation measures and sustainable land use systems with special reference to the Western Ghats region*, 16-17 November, 2002 ICAR Research Complex for Goa, Old Goa, India. pp.47-49.

Anonymous (1970) Cardamom culture - Part I. *Cardamom News*, 4(6): 2-3.

Anonymous (1976) *Cardamom in Karnataka, Tech. Series No. 14*, University of Agricultural Sciences, Bangalore. P. 1-10.

Anonymous (1977) Report of the Indian cardamom delegation to Guatemala. *Cardamom*, 6: 3-9

Anonymous (1979) Propagation of cardamom (Nursery practices of cardamom). University of Agricultural Sciences, Regional Research Station, Mudigere, p. 5.

Anonymous, 1998 "Concepts, Principles, Basic standards, Production guidelines, Documentation, Inspection and Certification". In: *Production of Organic spices*. Spices Board, Cochin.

Anonymous (1994) *Topics in Indian medicinal plants a compendium of 500 species*, Kottakal Aryavaidya sala, Orient longman, New Delhi, Vo2.

Anonymous (1996) *Tissue Cultured Cardamom*: Product Plan, Final Report, Spices Board, India.

Anonymous (2003) *Package of practices cardamom*, Spices Board, Cochin

Anonymous (2007). State/ spice wise area and production of spices in India. Spices Board. Available from http:// www.indianspices.com/ pdf/ state_prd.pdf (Accessed 24 th April 2007).

Balasimha, D. and Yadukumar, N. (1993) Effect of plant density on photosynthesis in cashew. *J. Plant Physiology*, 36: 5-7

Banerjee, S., Sharma, R., Kale, R.K and Rao, A.R. (1994). Influence of certain essential oils on carcinogen-metabolising enzymes. *Nutr. Cancer*, 21, 263-269.

Belavadi, V.V., Chandrappa, H.M., Shadakshari, Y.G and Parvathi, C. (1993) "Isolation distance for seed gardens of cardamom (*Elettaria cardamomum* Maton.)". In: Veeresh, G.K., Umashanker, R. and Ganesan, K.N. (eds.) *Pollination in tropics*, IUSSI, Bangalore, pp. 241-243.

Belavadi, V.V., Parvathi, C. and Raju, B. (1998) "Optimal foraging by honey bees on cardamom". In: *Abstracts of XIII Plantation Crops Research Symposium*, Dec. 16-18, Coimbatore.

Belavadi, V.V., Parvathi, C. and Raju, B. (2000) "Optimal foraging by honeybees on cardamom". In: *Recent Advances in Plantations Crop Research*, pp. 318-322.

Belavadi, V.V., Venkateshalu and Vivek, H.R (1997) Significance of style in cardamom corolla tubes for honeybee pollination. *Current Science*, 73: 287-290.

Bhai, R.S, Thomas, J. and Naidu, R. (1993) Biological control of 'Azhukal' disease of small cardamom caused by *P. meadii* Mc. Rae. *J. Plantn. Crops*, 21 (Suppl.): 134-139.

Bhai, R.S. (1998) Studies on capsule rot disease of cardamom. *Ph.D Thesis*, University of Calicut, Calicut, Kerala.

Chandran, K., Raja, P., Joseph, D and Suryanarayana, M.C (1983) "Studies on the role of honeybees in the pollination of cardamom". In: *Proceedings of second International Conference on Tropical Climate*, pp. 497-504.

Chandrasekhar, S.S. and Balu, A. (1993) Vertical distribution of thrips *Sciothrips cardamomi* (Ramk.) in cardamom. *J. Plantn. Crops*, 21 (Suppl.): 227-230.

Charles, J.K (1986) Productivity in cardamom-an insight. *Cardamom Journal.*, 19 (12): 13-20

Chowdhary, S. (1958) Notes on fungi from Assam. *Lloydia*, 21: 152-156.

CRCRI (1980) *Annual Report 1979-80*, Central Plantation Crops Research Institute, Kasargod, Kerala

CPCRI (1982) *Annual Report 1981-82*, Central Plantation Crops Research Institute, Kasargod, Kerala

CPCRI (1983) *Annual Report 1982-83*, Central Plantation Crops Research Institute, Kasargod, Kerala

Dey, A.C. (1994) Topics in Indian medicinal plants used ayurvedic preparations, Bishen Sing Mahendra Pal, Dehradun.

Dimitiman, J. E. (1981) An aphid transmitted virus of cardamom in Guatemala (Abstr.). *Phytopathol.* 71: 104-105.

Dimitman, J.E., Floras, A. and Nikoloff, J. A. (1984) Cardamom mosaic – a member of the potyvirus group in Guatemala (Abstr.). *Phytopathol.* 74: 844.

Dinesh Kumar, M. and Babitha, J. (2006) Rates of leaf fall and nutrient recycling of shade trees in coffee (Coffea arabica L.), cardamom (*Elettaria cardamomum* Maton.) and black pepper (*Piper nigrum* L.) production systems of mudigere. Karnataka. *J. Spices and Aromatic Crops*, 15(2): 108-114.

Eapen, S. J. (1994) Pathogenicity of root knot nematodes on small cardamom (*Elettaria cardamomum* Maton.). *Indian J. Nematology*, 24: 31-37.

Eapen, S.J. and Venugopal, M.N. (1995) Field evaluation of *Trichoderma* species and *Paecilomyces lilacinus* for control of root knot nematode and fungal diseases in cardamom nurseries. *Indian J. Nematology*, 25: 115-16.

Eapen, S.J., Venugopal, M.N., Ramana, K.V and Sarma, Y.R. (2000) "*Trichoderma* Spp. For the management of root knot nematode and rhizome rot disease in cardamom nurseries". In: Muraleedharan, N and Raj Kumar, R. (eds.) *Recent Advances in Plantation Crops Research*, (PLACROSYM XIII). pp. 382-386.

George M. and Jayasankar, N.P. (1977) Control of chenthal (Bacterial blight) disease of cardamom with penicillin. *Current Science*, 46: 237.

George M. and Jayasankar, N.P. (1979) "Distribution and factors influencing chenthal disease of cardamom". In: *Proceeding of PLACROSYM-II*, CPCRI, Kasaragod, pp. 343-347.

George, M., Joseph T., Potty, and Jayasankar, N.P. (1976) A bacterial blight disease of cardamom. *J. Plantn. Crops*, 4: 23-24.

George, M.V., Mohammed Sayed, A.A. and Korikanthimath, V.S. (1984) *Diospyros ebenum* Koenig, an ideal shade tree for cardamom. *J. Plantn. Crops*, 12(2): 160-163.

Gonsalves, D., Trujillo, E. and Hoch, H.C. (1986) Purification and some properties of a virus associated with cardamom mosaic, a new member of the potyvirus group. *Plant Disease*, 70: 65-69.

Gopakumar, B. and Chandrasekar, S.S. (2002). "Insect pests of cardamom". In: Ravindran, P.N. and Madhusoodanan, K.J., (eds.). Cardamom -The genus *Elettaria*. Taylor & Frances, London, pp. 180-206.

Gopakumar, B. and Kumaresan, D.and Varadarasan, S. (1987) Management strategy of root grub in cardamom. *Cardamom*, 20: 15-17.

Gopakumar, B., Kumaresan, D., Varadarasan, S. and Naidu, R. (1991) "Bioecology, damage potential and management of root grub *Basilepta fulvicorne*". In: Veeresh, G.K., Rajagopal, D. and Virktamath, C.A. (eds.) *Advances in Management and conservation of soil fauna*, Oxford & IBH Publishing Co. Ltd., New Delhi, India.

Gopalakrishnan, N. (1994) Studies on the storage quality of carbon dioxide extracted cardamom and clove oils. *J. Agric. Food Chem.* 42: 796-798.

Govil, J.N. (1998) *Glimpses in plant research, Medicinal plants part I*, Today and Tomorrow's printers, Vol XII.

Gurumurthy, B.R. and Hegde, M.V. (1987) Effect of temperature on germination of cardamom (*Elettaria cardamomum* Maton.). J. Plantn. Crops, 15: 5-8

Hashim, B., Aboobaker, V.S., Madhubala, R.K and Rao, A.R (1994). Modulatory effects of essential oils from spices on the formation of DNA adducts by aflatoxin B1 in vitro. *Nutrition and Cancer*, 21, 169-175.

Hull, R. (2002) *Matthews' Plant Virology*, Fourth edition, Academic Press, New York, USA.

ICRI (2006). *Annual Research Report 2006*, Indian Cardamom Research Institute, Myladumpara, Idukki, Kerala, India.

IISR (1997) *Annual Report 1996-97*, Indian Institute of Spices Research, Calicut, Kerala India.

IISR (1995) *Annual Report 1994-1995*, Indian Institute of Spices Research, Calicut, Kerala, India.

IISR (1996) *Annual Report 1995-1996*, Indian Institute of Spices Research, Calicut, Kerala, India.

IISR (1997) *Annual Report 1996-1997*, Indian Institute of Spices Research, Calicut, Kerala, India.

IISR (2004) *Annual Report 2004*, Indian Institute of Spices Research, Calicut, Kerala, India.

Jacob, J.A., Ancy Joseph and Murugan, M (1995). *Spice India*, 8(6):12-14

John, J.M. and Mathew (1977) Sprinkler irrigation in cardamom – a sure step to increase production. *Cardamom*, 9(6): 17-21

Khader, K.B.A. and Sayed, A.A.M. (1977) Fertilizing cardamom – its importance. *Cardamom*, 9(12): 13-14.

Korikanthimath, V.S. (1984) Nutrition of cardamom. *Cardamom* J. 17(12): 3-7.

Korikanthimath, V.S (1991) Shade management for high productivity in cardamom (*Elettaria cardamomum* Maton.). *Spice India*, 4(2): 15-21.

Korikanthimath, V.S., Ankegowda, S.J., Yadukumar, N., Rajendra Hegde, Hosmani, M.M. (2000a) Microclimatic and photosynthetic characteristics in arecanut and cardamom mixed cropping systems. *J. Spices and Aromatic Crops*, 9(1): 61-63.

Korikanthimath, V.S., Gaddi, A.V. and Ankegowda, S.J. (2000b) Status of major nutrients in soils of cardamom (*Elettaria cardamomum* Maton) plantations in Kodagu District, Karnataka, India. *J. Spices and Aromatic Crops*, 9 (2): 117-122.

Korikanthimath, V.S., Hiremath, G.M., Rajendra Hegde, Ravindra Mulge, Hosmani, M.M., (1997a) Comparative input utilization pattern in mono v/s mixed cropping of arecanut and cardamom. *J. Maharashtra Agric. Univ.* 22(2): 210-212.

Korikanthimath, V.S., Kiresur, V., Hiremath, G.M., Rajendra Hegde, Ravindra Mulge, Hosmani, M.M. (1997c) Economics of mixed cropping of cardamom (*Elettaria cardamomum* Maton) in arecanut (*Areca catechu* L.) gardens. *J. Spices and Aromatic Crops*, 6(2): 107-113

Korikathimath, V.S and Venugopal, M.N. (1989) *High production technology in cardamom*. Technical Bulletin, National Research Centre for Spices, 1988-89, Calicut, Kerala.

Krishna, K.V.S. (1997) Cardamom plantations in Papua New Guinea. *Spice India*, 10(7): 23-24.

Krishnamurthy, K., Khan, M.M., Avadhani, K.K., Venkatesh, J., Siddaramaiah, A.L., Chakravarthy, A.K. and Gurumurthy, S.R. (1989) *Three decades of Cardamom Research at Regional Research Station, Mudigere* (1958-1988), Technical Bulletin No. 2, Regional Research Station, Mudigere Karnataka, India, p. 94.

Krishnamurthy, K.H. (1991) Wealth of Susruta, International Institute of Ayurveda, Coimbatore, p.338.

Kubo, I., Himejima, M and Murari, H. (1991) Antimicrobial activity of flavour components of cardamom. *J. Agric Food Chemistry*, 39, 1984-1986.

Kulkarni, D.S., Kulkarni, S,V., Suryanarayana Reddy, B.G. and Pattanashetty, H.V. (1971) "Nutrient uptake by cardamom (*Elettaria cardamomum* Maton.)". In: *Proceedings of International Symposium on Soil Fertility Evaluation*, New Delhi, Vol.1. pp. 293-296.

Kumar, K.B., Prakash Kumar, P., Balachandran, S.M. and Iyer, R.D. (1985) Development of clonal plantlets from immature panicles of cardamom. *J. Plantn. Crops*, 13(1): 31-34

Kumaresan, D. and Regupathy, A. and Bhaskaran, P. (1988) *Pests of spices*. Rajalakshmi Publications, Nagarcoil, Tamil Nadu, India, p. 241.

Kumaresan, D. and Varadarasan, S. (1987) "Review and current status of research on insect pest control in cardamom cropping system". In: *Workshop on Insect Pest Management of Coffee, Tea, Cardamom cropping systems*, Central Coffee Research Station, Chikmagalur, India.

Kuruvilla, K.M and Madhusoodanan, K.J (1988) Effective pollination for better fruit set in cardamom. *Spice India*, 1(6): 19-21.

Lawrence, B.M. (1979). "Major tropical spices - cardamom (*Elettaria cardamomum* Maton.)". In: Essential Oils, Allured Publishers, Wheaton, III, 104.

Madhusoodanan, K.J., Pradip Kumar, K and Ravindran.P.N. (2002) "Botany, crop improvement and biotechnology of cardamom". In: Ravindran, P.N and Madhusoodanan, K.J. (eds.) *Cardamom, the genus Elettraia*. Taylor and Francis, London, pp. 11-68.

Madhusoodanan, K.J., Radhakrishnan, V.V., Kuruvilla, K.M., Vadivel, V and Thomas, J (2006) "MCC 73 – a high yielding and high quality selection of cardamom (*Elettaria cardamomum* Maton.)". In: Ramakrishnan Nair, R., Kandiannan, K. and Jhon Zachariah T. (eds.) *Abstracts of papers, SYMSAC III*: Crop and product diversification, Indian Society of Spices, Calicut, p. 1.

Mayne, W.W., (1951) Report on cardamom cultivation in South India. *ICAR Bull* 50: 87.

Mishra, A.K and Dubey, N.K. (1990) Fungitoxicity against *Aspergillus flavus. Economic Botany*, 44: 350-533.

Mishra, A.K., Dwivedi, S.K., Kishore, N and Dubey, N.K. (1991) Fungistatic properties of essential oils of cardamom. *International J. Pharacognosy*, 29, 259-262.

Mohankumar, G.N. and Hegde, G.S. (1983) Nutrition of cardamom for better yields. *Cardamom* J. 15(10): 3-9.

Murugan, M., Miniraj, N. and Carmel Rani Joseph (2000) Changes in climatic elements and their impact on production of cardamom (*Elettaria cardamomum* Maton.) in the Cardamom Hills of Kerala, India. *J. Spices and Aromatic Crops*, 9(2): 157-160.

Naidu, R. (1981) Morphology, parasitism and epidemiology of *Phyllosticta elettariae* Chaudhury. causing nursery leaf spot diseases. *J. Plantn. Crops*, 9: 23-27.

Naidu, R., Venugopal, M.N. and Rajan, P., (1985) "Investigations on strainal variation, epidemiology and characterization of 'katte' virus agent of small cardamom". In: *Final Report of Research Project*, Central Plantation Crops Research Institute, Kasargod, Kerala.

Nair, C.K. (1988) "Phosphatic fertilizers of small cardamom". In: *Proceedings of the Seminar on the use of Rock Phosphate in West Coast Soils.* UAS Bangalore and PPCL, p. 79.

Nair, C.K., Srinivasan, K. and Sivadasan, C.R. (1988) Soil acidity in cardamom plantations. *Spice India,* 1(6): 40-41.

Nair, M.R.G.K. (1975) *Insects and Mites of Crops in India.* ICAR, New Delhi, p. 408.

Nair, R.R. (1979) Investigations on fungal diseases of cardamom. *Ph.D. Thesis,* Kerala Agricultural University, Vellanikara, Trissur, pp. 161.

Nambiar K.K.N. and Sarma, Y.R. (1974). Chemical control of capsule rot of cardamom. J. *Plantn. Crops,* 2: 30-31.

Nambiar, K.K.N. and Sarma, Y.R. (1976) Capsule rot of cardamom Pythium vexans de Bary as casual agent. *J. Plantn. Crops,* 4: 21-22.

Nambiar, M.C., Pillai, G.B. and Nambiar, K.K.N (1975) Diseases and pests of cardamom - a resume of research in India. *Pesticides Annual,* pp. 122-127.

Nanjan, K., Muthuswami, S., Thangarajan, T. and Sundarajan (1981) Time of planting of cardamom at Shevaroy hills. *Cardamom,* 13(1): 13-15.

NRCS (1989) *Annual Report, 1989,* NRC for Spices, Calicut, Kerala, India, pp: 24-26.

Parvathi, C., Shadakshari, Y.G., Belavadi, V.V and Chandrappa, H.M (1993) "Foraging behaviour of honey bee on cardamom (*Elettaria cardamomum* Maton.)". In: Veeresh, G.K., Umashanker R. and Ganesan K.N. (eds.) *Pollination in tropics, IUSSI,* Bangalore, pp. 99-103.

Pattanshetty, H.V. and Prasad, A.B. N. (1974) Exposing the cardamom panicle from a layer of leaf mulch to open pollination by bees and thereby improving the fruit set. *Current Research,* 3(8): 90

Pattanshetty, H.V. and Prasad, A.B.N. (1972) Early August is the most suitable time for planting cardamom in Mudigere area. *Current Research,* 1(9): 60-61.

Pattanshetty, H.V. and Prasad, A.B.N. (1973) September is the most suitable month for sowing cardamom seeds. *Current Research,* 2(5): 26.

Pattanshetty, H.V., Nusrath, R., Sulikeri, G.S. and Prasad, A.B.N. (1974) Suitable season and depth of planting cardamom suckers. *Current Research,* 3(8): 84.

Pattanshetty, H.V., Rafeeq, M. and Prasad, A.B.N. (1978) "Influence of length and type of storage on the germination of cardamom seeds". In: Nelliat, E.V. (ed.) *Proceedings of First Annual symposium on plantation crops,* Indian Society for Plantations Crops, Central Plantation Crop Research Institute, Kasaragod, pp. 267-274.

Premkumar, T., Devasahayam, S. and Koya, K.M. A. (1994) "*Pest of spice crops*". In: Chadha, K.L. and Rethinam, P. (eds.). Adv. Hort. Vol. 10: *Plantation and Spice Crops,* Malhothra Publishing House, New Delhi, India.

Pruthi, J.S. (1987) *Spices and Continents.* National Book Trust, New Delhi, India.

Radha, K. and Joseph, T. (1974) *"Investigations on the bud rot disease of coconut"*. In: Final Report of PL. 480, CPCRI, Kayamkulam, pp 30

Raghothama, K.G. (1979) Effect of mulches and irrigation on the production of suckers in cardamom (*Elettaria cardamomum* Maton.). *M.Sc. Agri. Thesis,* University of Agricultural Sciences, Bangalore.

Rai, S.N. (1978) Nursery and planting of some tropical evergreen and semi evergreen species. *Tech. Bull.,* Karnataka Forest Department, p. 49.

Rajendra Hegde and V.S. Korikanthimath, 1996 "Agronomic Approaches for the drought management in cardamom". In: Satheesan, K.V. (eds.) *Proceedings of the National Seminar on Drought Management in Plantation Crops,* 7-8 February, 1996, Kottayam, Kerala pp. 75-79.

Raju, B. (1981) Effect of mulches and irrigation on growth and yield of cardamom (*Elettaria cardamomum* Maton.) *M.Sc. Agri. Thesis,* University of Agricultural Sciences, Bangalore.

Ramakrishnan, T.S. (1949) The occurrence of *Pythium vexans* de Bary in South India. *Indian Phytopathol.* 2: 27-30.

Ramana, K.V. and Eapen, S.J. (1992) "Plant parasitic nematodes of black pepper and cardamom". In: *Proceedings of the National Seminar on Black pepper and Cardamom,* 17-18th May, 1992, Calicut, Kerala, pp. 43-47.

Ramana, K.V. and Eapen, S.J. (1995) Parasitic nematodes and their management in major spices. *J. Spices Aromatic Crops,* 4(1): 1-16, 1995.

Ranjithakumari, B.D., Kuriachen, P.M., Madhusoodanan, K.J. and Naidu, R. (1993) Studies on light requirement of cardamom seedlings in nursery. *J. Plantn. Crops, 21* (Suppl): 360-362.

Rao D.G. and Naidu, R. (1974) Chemical control of nursery leaf spot disease of cardamom caused by *Phyllosticta elletariae. J. Plantn. Crops,* 2: 14-16.

Ravindran, K.C. and Kulandaivelu, G. (1998a) Effect of light intensity on growth and yield of cardamom. *J. Plantn. Crops,* 26(1): 87-88

Ravindran, K.C. and Kulandaivelu, G. (1998b) Effect of light intensity on leaf scorching in nursery seedlings of cardamom. *J. Spices and Aromatic Crops,* 7(2): 129-131.

Ravindran, P.N. (2002) "Introduction". In: Ravindran, P.N. and Madhusoodanan, K.J., (eds.) *Cardamom-The genus Elettaria,* Taylor & Francis, London, pp. 1-10.

Rethinam, B.P. and Korikanthimath, V.S. (1985) Frequency and probability of dry spells at Mercara. Geobios, 12: 224-227

Rethinam, P. and Venugopal, K. (1994) *"Cropping systems in plantation crops"*. In: Chadha, K.L. and Rethinam, P. (eds.). Adv. Hort. Vol.9: *Plantation and Spice Crops,* Part 1., Malhotra Publishing House, New Delhi. pp. 603-620.

Reyes, T., Luukkanen, O. and Quiroz, R. (2006) Small cardamom - Precious for people, Harmful for mountain forests. Possibilities for sustainable cultivation in the East Usambaras, Tanzania. *Mountain Research and Development*, 26(2): 131-137.

Rohini Iyer, Sastry, R.K. and Korikanthimath, V.S. (1989) Effect of fertilizer levels on vesicular arbuscular mycorrhizae in cardamom. *J. Plantn. Crops*, 16 (Suppl): 429-432.

Sadanandan, A.K. and Hamza, S. (2006) "Effect of organic farming on soil quality, nutrient uptake, yield and quality of Indian spice" In: *Abstracts of 18 th world congress of soil science*, July 9-15, 2006, Philadelphia, USA.

Sadanandan, A.K., Korikanthimath, V.S. and Hamza, S. (1990) "Potassium in soils of cardamom plantations". In: *Seminar on Potassium in Plantations Crops*, University of Agricultural Sciences, Bangalore.

Saleem, C.P. (1978) Sprinkler irrigation doubles cardamom yield. *Cardamom*, 10 (6): 17-20.

Shankar, B.D. (1980) Shade trees and their role in cardamom cultivation. *Cardamom*, 12(3): 3-11.

Shanmugavelu, K.G. and Madhava Rao, V.N. (1977) *Spices and Plantation crops*, Popular book Depot, Madras.

Siddagangaiah, Krishnakumar, V. and Sudarshan, M.R. (1998a) Effect of different methods of planting and spacing on growth and yield of cardamom (*Elettaria cardamomum* Maton.). *J. Spices and Aromatic Crops*, 7(2):81-84.

Siddagangaiah, Krishnakumar, V., Vadiraj, B.A. and Sudharshan, M.R. (1998b) "Influence of cultural practices on growth and yield of cardamom". In: *Proceedings of the PLACROSYM XII*, Kottayam, India, November 27-29, 1996. Allied Publishers, New Delhi, pp.195-198

Siddagangaiah, Srinivasan, K., Sudharshan, M.R. and Naidu, R. (1993) Effect of seed size and incubation temperature on seed germination of cardamom (*Elettaria cardamomum* Maton.). *J. Spices and Aromatic Crops*, 2(1/2): 47-52.

Sivanappan, R.K. (1985) Soil conservation measures and water management for cardamom. *Cardamom*, 18(6): 3-8.

Sosamma, V.K. and Koshy, P.K. (1997) Biological control of Meloidogyne incognita on black pepper by *Pasteuria penetrans* and *Paecilomyces lilacinus* (J.). *J. Plantation Crops*, 25(1): 72-76.

Sreeramulu, K. R. and Bagyaraj, D.J. (1998) Response of cardamom seedlings to vesicular arbuscular mycorrhizal fungi. *J. Spices and Aromatic Crops*, 7(2): 89-94.

Srinivasan, K, Siddagangaiah, Biswas, A.K. and Krishnakumar, V. (1999) Mixed cropping ssyetems in cardamom - An analysis. *J. Spices and Aromatic Crops*, 8(1): 63-66.

Srinivasan, K., Rama Rao, K.V.V. and Naidu, R. (1993a) Organic matter addition and its nutrient contribution in cardamom plantations. *J. Tropical Agric.* 31:119-121

Srinivasan, K., Vadiraj, B.A., Krishnakumar, V. and Naidu, R. (1993b) Micronutrient status of cardamom growing soils of South India. *J. Plantn. Crops,* 21 (Suppl.): 67-74.

Srinivasan, K., Vadiraj, B.A., Krishnakumar, V., Bindumol, G.B. and Potty S.N. (2002) "Available sulphur status of cardamom growing soils of South India". In: *Proceedings of PLACROSYM XIV*, pp. 342-344.

Stone, G.N. and Willmer, P.G (1989) Pollination of cardamom in Papua New Guinea. *J. Agriculture Research*, 28(4), 228.

Subba Rao, M.K. (1938). *Report of the Mycologist (1937-38)*. United Planters Association of South India, pp. 28-37.

Subbaiah, M.S. (1940) Cardamom cultivation in the Bodi hills. *Madras Agricultural J.* 28(10): 279-288.

Sulikeri, G.S. (1986) Effect of light intensity and soil moisture on growth and yield of cardamom (*Elettaria cardamomum* Maton.). *Ph.D. (Hort.) Thesis*, University of Agricultural Sciences, Bangalaore.

Sulikeri, G.S. and Kololgi, S.D. (1977) Seed viability in cardamom. *Current Research*, 6(10): 163-164.

Sulikeri, G.S. and Kololgi, S.D. (1978) *Phyllanthes emblica* L. leaves – a suitable mulch for cardamom nursery beds. *Current Research*, 7(1): 3-4.

Sulikeri, G.S., Pattanshetty, H.V. and Kololgi, S.D. (1978) "Influence of the level of water table on the growth and development of cardamom (*Elettaria cardamomum* Maton.)". In: Nelliat, E.V. (eds). *Proceedings of the First annual symposium on plantation crops, Kottayam*, Kerala, pp.216-219, CPCRI, Kasargod.

Thimmarayappa, M., Shivashankar, K.T. and Shanthaveerabhadraiah, S.M. (2000) Effect of organic manure and inorganic fertilizers on growth, yield attributes and yield of cardamom (*Elettaria cardamomum* Maton.). *J. Spices and Aromatic Crops*, 9: 57-59.

Thomas, J. and Bhai, R.S. (2002) "Diseases of cardamom". In: Raivindran, P.N. and Madhusoodhanan, K.J., (eds.). *Cardamom – The genus Elettaria.* Taylor and Francis, London, pp 160-179.

Thomas, J. and Suseela Bhai, R., Dhanapal, K. and Vijayan, A.K. (1997) "Integrated management of rot diseases of cardamom". In: *International Conference on Integrated Plant Disease Management for Sustainable Agriculture*, November, 10-15, New Delhi.

Thomas, J. Suseela Bhai, R., Vijayan, A.K. and Naidu, R. (1991b) "Management of rot diseases of cardamom through bio-agents". In: *Abstracts of the National Seminar on Biological Control in Plantation Crops*, RRI, Kottayam, Kerala, pp 21.

Thomas, J., Bhai, R.S, and Naidu, R. (1989) Comparative efficacy of fungicides against Phytophthora rot of small cardamom. *Pesticides*, 40-42.

Thomas, J., Suseela Bhai, R. and Naidu, R. (1991a) Capsule rot disease of cardamom (*Elettaria cardamomum* Maton.) and its control. *J. Plantn. Crops*, 18 (suppl.): 264-268.

Thyagaraj, N.E., Chakravarthy, A.K., Rajagopal, D., Sudharsan, M.R, (1991) Bioecology of cardamom root grub (*Basilepta fulvicorne* Jacoby). *J. Plantn. Crops*, 18: 316-319.

Uppal, B.N, Varma, P.M.and Kapoor, S.P. (1945) A mosaic disease of cardamom. *Current Science*, 14: 208-209.

Varadarasan, S. (1992) "Pest management in cardamom". In: Sarma, Y.R., Devasahayam, S. and Anandaraj, M. (eds.). *Black pepper and cardamom - Problems and prospects*, Indian Society for Spices, Calicut, pp 51-54.

Varadarasan, S. Kumaresan, D. and Gopakumar, B. (1989) Bioecology and management of cardamom shoot/panicle/capsule borer (*Conogethes punctiferalis* Guen.). *Spice India*, 2: 15-17.

Varadarasan, S., Gopakumar, B. and Kumaresan, D (1991) Dynamics of the life cycle of cardamom shoot borer (*Conogethes punctiferalis* Guen.). *J. Plantn. Crops*, 18 (suppl): 302-304.

Varadasan, S. (1995) "Biological control of insect pests of cardamom". In: Ananthakrishnana, A.N. (ed.). *Biological Control of Social Forest and Plantation Crops Insects*, Oxford & IBH Publ.Co. Ltd., New Delhi, India, pp. 109-119.

Variyar, P.S and Badyopadhyay, C. (1995) On the volatiles of clove, cardamom, Nutmeg and mace. *PAFAI J.* 17(1): 19-25.

Varma , P.M., (1962) Control of katte or mosaic disease of cardamom in North Kanara. *Arecanut* J. 13: 79-78.

Varma, A., Jain, R.K. and Bhat, A.I., (2002) Virus resistant transgenic plants for environmentally safe management of viral disease. *Indian J. Biotechnology*, 1: 73-86.

Vasanthkumar, K. and Sheela, V.C. (1970) Sprinkler irrigation in cardamom plantations. *Cardamom*, 23(3) ; 5-9.

Vasanthkumar, K., Mohankumaran, N. and Wahid, P.A. (1989) Rate of photosynthesis and translocation of photosynthates in cardamom. *J. Plantn. Crops*, 17(2): 96-100

Venugopal, M.N and Prasath D. (2006) New Cardamom varieties. *Indian Horticulture*, 51(4): 15.

Venugopal, M.N. (1999) Natural disease escapes as sources of resistance against cardamom mosaic virus causing katte disease of cardamom. *J. Spices and Aromatic Crops*, 8(2): 145-151.

Venugopal, M.N. (2002) "Viral disease of cardamom". In: Ravindran, P.N. and Madhusoodanana, K.J., (eds.). *Cardamom - The genus Elettaria,* Taylor & Francis, London, pp 143-159.

Venugopal, M.N. and Govindaraju, C., (1993) Cardamom vein clearing virus - a new threat to cardamom. *Spice India,* 5: 6-7.

Venugopal, M.N., (1995) Viral diseases of cardamom (*Elettaria cardamomum* Maton.) and their management. *J. Spices and Aromatic Crops,* 4: 32-39.

Venugopal, M.N.and Naidu, R. (1987) "Effect of natural infection of katte disease on yield of cardamom". In: *Proceedings of the PLACROSYM-VI,* Oxford and IBH, New Delhi, pp 115-119.

Venugopal, M.N, Prasath, D and Ravindra Mulge. (2006) Rhizome rot resistant and high yielding cardamom variety (*Elettaria cardamomum* Maton) - IISR Avinash. *J. Spices and Aromatic crops,* 15(1): 14 – 18.

Zachariah, P.K. (1976) Mulching in cardamom plantations. *Cardamom News,* 8(1): 3-9.

Zachariah, P.K. (1978) "Fertilizer management for cardamom". In: *Proceedings of the First Annual Symposium on Plantations Crops,* Kottayam, pp.141-156.

10

Ginger

V. Srinivasan, K. N. Shiva and A. Kumar

1. INTRODUCTION

Ginger (*Zingiber officinale* Rosc.) is a perennial herb and the economic part is underground stem (rhizome), which is used as a spice. It is native to tropical South-East Asia, introduced into the West Indies, African countries and other tropical belt of the world. The ginger rhizome is harvested between 6 and 12 months after planting and used as spice in the fresh condition, peeled and split dried or powedered forms. Ginger is utilized widely as a spice, for pickles, candies and as a medicinal herb. The anti-inflammatory and anti-nausea qualities of ginger have applications in the pharmaceutical industry. Ginger and supplements derived from ginger have received attention for the treatment of chronic inflammation. Administration of ginger has resulted in decreased symptoms of rheumatoid arthritis (Srivastava and Mustafa, 1992) and gingerol (a component of ginger) has been reported to have anti-inflammatory actions (Kiuchi *et al.*, 1992; Tjendraputra *et al.*, 2001).

1.1 Area and Production

Ginger production is mainly distributed over India, China, Nigeria, Indonesia, Bangladesh, Thailand, Philippines, Nepal etc. It is also grown in countries such as Cameroon, Korea, Sri Lanka, Malaysia, Hawaii, and West Indies. Nigeria ranks first with respect to area under ginger covering about 52 percent of world total area followed by India (24%), China (7%), Indonesia (5%), Philippines (4%), Nepal (3%) and Bangladesh (2%). India ranks first with respect to ginger production contributing about 29 percent of total world's ginger production followed by China (26%), Indonesia (14%) and Nigeria (10%). The Indian production of ginger during the year 2004 was 3.03 lakh tonnes from an area of 85.90 thousand hactares. The productivity in India was 3969 kg ha^{-1} during 2004-05 and the world average ginger productivity is 3025 kg ha^{-1} (2004). In India, states of Kerala, Karnataka, Sikkim, Himachal Pradesh, Orissa, West Bengal, Nagaland, Assam and other Northeastern states cultivate this crop very extensively (Table 1) (Premaja and Manojkumar, 2007). However, states namely Kerala, Meghalaya, Arunachal Pradesh, Mizoram, Nagaland and Orissa together contribute 70 per cent to the country's total production. Geographical Information System (GIS) aided studies showed that ginger has wider adaptability with many areas in Maharashtra such as Rajnandgaon, Thane, Nasik, Satara and Pune; Surat in Gujarat, Belgaum in Karnataka, Bastar and Betul in MP, Koraput, Ganjam are best. Majority of the areas in Assam and other Northeastern States except parts of Sikkim and Arunachal Pradesh are ideal for ginger cultivation (Utpala *et al.* 2006). The crop exhibited an annual growth rate of 4.6 per cent in area, 7.4 per cent in production and 2.7 per cent in productivity (Manjunath, 2004).

Table 1. State-wise area, production and productivity of ginger in India (2004-05)

State	Area ('000 ha)	Production ('000 t)	Productivity (kg/ha)
Kerala	10.0	45.3	4535
Orissa	15.7	30.4	1936
Meghalaya	9.2	47.1	5120
West Bengal	9.1	18.8	2058
Arunachal Pradesh	4.5	32.9	7311
Mizoram	4.5	29.6	6578
Karnataka	9.1	11.9	1308
Madhya Pradesh	5.4	6.2	1148
Sikkim	6.5	34.7	5338
Andhra Pradesh	1.9	10.9	5823
Himachal Pradesh	2.0	14.7	7350
Maharashtra	1.0	1.3	1300
Tripura	1.4	2.8	2000
Gujarat	1.9	4.0	2105

Contd....

Uttar Pradesh	1.0	2.8	2800
Nagaland	10.2	63.5	6225
Bihar	0.4	0.5	1250
Tamil Nadu	0.6	4.6	7666
Manipur	1.0	2.4	2400
Andaman & Nicobar Islands	0.5	1.8	3600
Rajasthan	0.1	0.2	2000
Uttaranchal	0.8	6.1	7625
Chhattisgarh	1.3	1.4	1077

Source: Premaja and Manojkumar (2007)

1.2 Scope for organic ginger

The total Indian spice export in 2005-06 was 3.20 lakh tonnes valued at Rs. 2295.2 crores. Quantity wise, this is only about 10% of India's production envisaged. Among spices exported from India during 2005-06, ginger constituted only 2% in quantity and value. The export of ginger from India was about 4 per cent of the domestic production during 2004-05 worth of Rs. 595.0 millions. During 2005-06, India exported 7250 t of ginger worth of 407.5 millions. Export of ginger showed a highly fluctuating trend during the last decade. India is exporting ginger mainly to Pakistan, Bangladesh, Saudi Arabia, UAE, Netherlands, Spain, Japan, Morocco, UK and USA. Indian ginger has high esteem in the global market because of its characteristic lemon like flavour. Indian dry ginger is known in the world market as 'Cochin Ginger' (NUGC) and 'Calicut Ginger' (NUGK). In the recent years, Indian ginger is facing stiff competition from other producing countries like China, Nigeria and Thailand. Any new entrant will face severe competition, by way of established brand loyalty, wide variety and range of herbs and spices, catering to consumer tastes, and highly organized marketing capability. However, some effort by the spice-producing countries in developing premium brands is indicated in view of high added value for the pre-packed spices and the rapid growth expected in this market with rising affluence in many countries.

In recent years, organic agriculture has been gaining considerable importance and many farmers are switching to this traditional method of cultivation as a means to produce safe foodstuffs and preserve the environment. Organic farming favors lower input costs, conserves non-renewable resources, high-value markets and boosts farm income. Estimate by SOEL-Survey shows that India has 41000 ha (0.03% of total agricultural area) under organic cropping, producing agricultural crops like plantation, spices, pulses, fruits, vegetables and oil seeds etc. Since spices like ginger form part of many of ethnic medicines, the demand for organically produced ginger is also increasing considerably in the importing countries. India as a largest producer of ginger has capabilities to cater to the requirements of the consuming countries in terms of quantity and quality matching to international standards (Selvan and Manoj, 2002; Peter and Nybe, 2002).

From India, during 2004-05 around 6800 tonns of organic products valued at Rs. 7120 lakhs were exported making around 85% of total organic crop production. India's organic spices export has enhanced form 100 t in 2001-02 to 300 t valued at Rs. 845 lakhs in 2005-06. Major export market for Indian organic producers are Australia, Belgium, Canada, France, Germany, Italy, Japan, Netherlands, Sweden, Singapore, South Africa, Saudi Arabia, UAE, UK, and USA. Since organic foods are free from chemical contaminants, the demand for these products should steadily increase in the coming years. Even though the current market size is small, demand is expected to grow at an annual rate of 15-20%.

India has great potential to capture the expanding global organic spice market as per capita consumption of fertilizers and pesticides in our country is far below that of developed countries and it is very easy for Indian farmers to embrace organic spice farming in its true sense. The NE states *viz.*, Meghalaya, Arunachal Pradesh, Mizoram and Nagaland have large area under ginger and contribute around 50% of nation's production. As most of these NE states are having virgin land without any commercial cultivation and cultivation is practiced under shade trees, which are available naturally as evergreen forest and left out Jhum land, they are very much suitable for organic cultivation.

1.3 Composition and Uses

The average chemical composition of some of the traditional ginger cultivars and varieties are listed in Table 2. Ginger is used in dry as well as in fresh (raw) form. The dry ginger is the basic material for ginger powder, which is major ingredient in all the curry powders. Ginger powder is used in confectionary, pharmaceutical preparations, neutroceuticals, canning industry etc. Oil extracted from ginger is used for flavoring liquors, soft drinks, culinary products etc. Oleoresin extracted from the powder is used as a substitute in catering industry as it imparts total flavor of the spice.

Crude fibre: In fully matured ginger it varies from 3-8%. Fibre separates at the last phase of maturity. Rainfall, soil texture and other agro-climatic factors also influence crude fibre content. It is estimated by acid and alkali digestion of the ginger powder and whatever remains is considered as fibre.

Dry Ginger: It is the form in which ginger can be stored after drying. Dry recovery varies from 18-25% depending on location and climatic factors. Dry ginger is used for preparing ginger powder, extracting ginger oil, ginger oleoresin etc.

Ginger oil: Dry ginger on distillation yields about 1.5 to 2.5% volatile oil (15 to 25 ml per kg dry ginger) that contributes aroma to oil. The main constituent in the oil is zingiberene. Technology is now available to extract oil from fresh ginger. Ginger from North Eastern States may contain more oil due to the low temperature in the final phase of maturity.

Ginger oleoresin: Dry ginger powder on treating with organic solvents like acetone, alcohol, ethyl acetate etc. yield a viscous mass that attribute the total taste and smell of the spice. The major non-volatile principle in oleoresin is gingerol. Technology is now available to extract oleoresin from fresh ginger. Oleoresin content varies from 4 to 10% (40 to 100 gm per kg.). Generally North East region climate helps in recovery of high oleoresin as the prevailing low temperature reduce synthesis of starch, which lead to more recovery of oleoresin per kg.

Table 2. Composition of some ginger cultivars in India

Variety	Oil	Fibre	Oleoresin
Bajpai	1.7	5.5	5.8
Jorhat	2.0	5.7	5.7
Karakal	2.0	6.4	6.4
Nadia	1.0	5.3	4.4
Narasapattom	1.6	6.4	5.2
Riode Janeiro	1.7	6.7	9.8
Thingpui	1.4	8.1	6.2
Thinladium	1.7	6.7	4.5
Wynad local	1.9	7.2	7.3
IISR Varada	1.7	3.3	6.7
IISR Rejatha	2.4	4.0	4.5
IISR Mahima	1.7	3.3	6.2
Suprabha	1.9	4.4	8.9
Suruchi	2.0	3.8	10.0
Suravi	2.1	4.0	10.2

Source : Kumar *et al.*, 2006

2.0 BOTANY AND VARIETIES

2.1 Botany

The genus *Zingiber* of the family Zingiberaceae consists of about 150 species and the commercially cultivated species is *Zingiber officinale* Rosc. The ginger plant a herbaceous perennial grown as an annual, has many fibrous roots, aerial shoots (pseudostem) with leaves and underground stem, rhizome. The first developing branch grows from the apical bud and subsequent growth of the rhizome is due to the development of axillary buds. The number of primary branches may vary from two to four and a mature rhizome may have 6-26 axillary branches. Ginger flowers are fragile and short lived, produced in peduncle spikes arising directly from the rhizomes. In general ginger dose not flower under sub-tropical or sub-temperate climatic conditions and is shown to be a quantitative short day plant (Ravindran *et al.*, 2005). As ginger rarely sets seed, the general mode of propagation is asexual. Crop improvement is mainly through introduction and selection.

2.2 Varieties

Several commercial cultivars of ginger are cultivated throughout the world especially in India, and many land races and improved cultivars which excel in yield and one or more quality traits are available (Ravindran *et al.* 1994). Yield, percentage recovery of dry ginger from fresh ginger and fibre content are the criteria used to differentiate these types (Aiyadurai 1966). The performance of various varieties under different locations in India evaluated by Thomas and Kannan (1969), Nybe and Nair (1981), Thangaraj *et al.* (1983), Rattan *et al.* (1988), Mohanty *et al.* (1990a) and NRCS (1992). Potential yield and quality of genotypes may vary with agroclimatic conditions, soil fertility and agronomic practices.

Ginger has rich cultivar diversity and specific to the major cultivating tracts. Several cultivars of ginger are grown in different ginger growing areas in India with majority in Kerala and Northeastern region. Some of the prominent indigenous cultivars are Maran, Kuruppampadi, Ernad, Wayanad, Himachal and Nadia. Exotic cultivars such as Rio-de-Janeiro have also become very popular among cultivators (Table 3). Besides, improved varieties like Varada, Mahima, Rejatha and Suprabha, Suruchi, Suravi are also released from Indian Institute of Spices Research and Orissa University of Agriculture and Technology, India.

Table 3. Popular cultivars in major ginger growing states in India

State	Cultivar
Assam	Nadia, Chekerella
Himachal Pradesh	Himachal, Rio de Janeiro
Karnataka	Himachal, Jorhat, Wyanad local
Kerala	Rio de Janeiro, Burdwan, Jamaica, Nadia, Maran, Bajpai, Narasapattam, Kuruppampadi, Thingpui, Assam, China, Ernadan, Mananthavady
Maharashtra	Mahim, Poona, Wayanad
Meghalaya	Tura, Thingpui, Nadia
Mizoram	Thingpui, Thinglaidum, Thingria
Nagaland	Thinglaidum, Nadia, Khasi local
Orissa	Suprava, Suruchi, Suravi
Sikkim	Bhaisey, Gorubathaney
Tamil Nadu	Maran, Thodupuzha, Himachal, Rio de Janeiro
Tripura	Nadia
West Bengal	Gorubathan, Nadia

Under regulations of organic farming, all seeds and plant material used should be certified organic. Moreover, emphasis should be given to varieties adapted to the soil and climatic conditions of the location, be resistant to pests and diseases and in the choice of varieties genetic diversity should be taken into consideration. Some of the ginger cultivars/ varieties that are having high yield potential, resistance/ tolerance to pest and diseases and suitable for cultivation under organic farming system are given below :

Cultivar/ Variety	Character/ Triat
Rio de Janeiro	Tolerant to Shoot borer (Nybe *et al.*, 1980)
IISR Varada, Anamika	Resistance to storage pest (Sasikumar *et al.* 1994)
Valluvanad, Tura	Least infestation with root knot nematode (Charles and Kuriyan, 1982)
Jorhat, Maran, Narasapattam, Anamika, Poona, Himachal	Least susceptible to rhizhome rot (Nybe and Nair, 1979; Mohanty, 1984; Sasikumar *et al.* 1994)

Contd...

Maran, Nadia, Bajpai, Kunduli local	Tolerant/ less susceptible to leaf spot (Nybe and Nair, 1979; Sasikumar *et al.* 1994)
Rio de Janeiro, Nadia, Maran, Bajpai, Narasapattam, Thingpui, China, Wayanad local, Himachal, Suprabha, Suruchi, Suravi, Varada, Mahima	High fresh and dry yielders under rainfed conditions (Sasikumar *et al.* 1994; Mohanty and Panda, 1994)

China also has the richest cultivar diversity in ginger. Some of the Chinese cultivars that are having average yield of 30-45 tonnes ha^{-1} under moderate management are Laiwu slice ginger and Laiwu big ginger. Dense-ringed delicate fleshy ginger and Yellow claw ginger are cultivars with better disease and drought tolerance with moderate rhizome yield and suitable for intercropping (Ravindran *et al.*, 2005). "Buderim Gold" is the first commercial tetraploid developed in Australia.

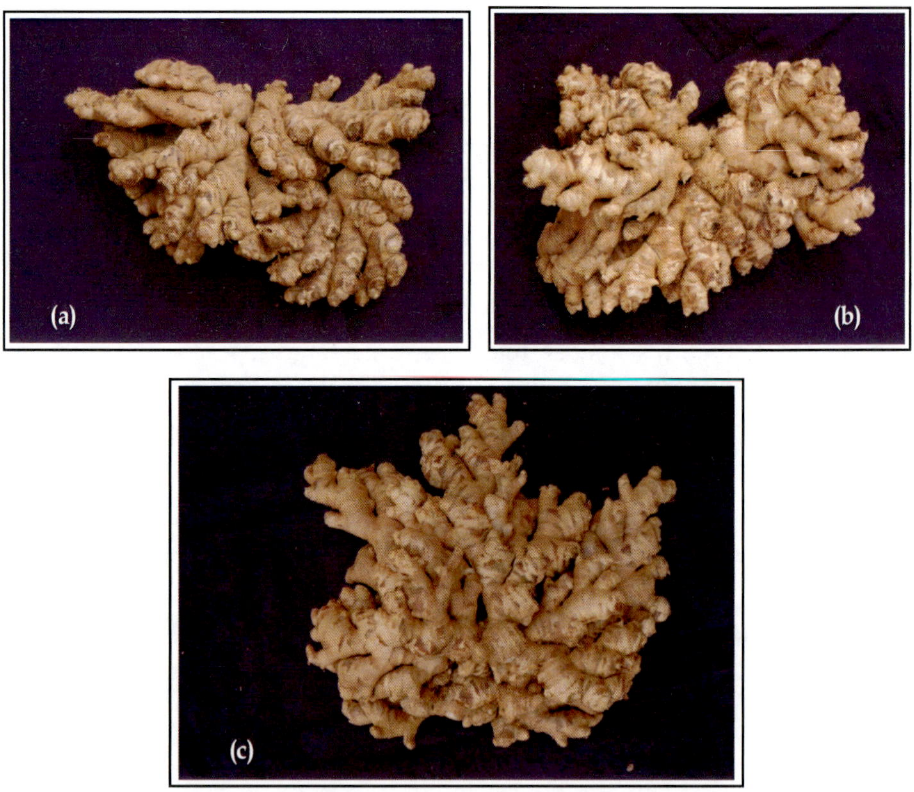

Fig. 1: Some high yielding varieties of ginger. (a) IISR-Mahima (b) IISR Rejetha (c) IISR Varada (*Source:* Director, IISR, Calicut)

3.0 CLIMATE AND SOIL

3.1 Climate

Ginger requires a tropical or subtropical climate, being a shade loving plant. It prefers a warm, humid climate and cannot withstand very low temperature. It comes up well up to an altitude of 1500 m above MSL, the optimum being 300-900 m. The base temperature requirement is 13°C with an optimum range of 19-28°C. The optimum soil temperature of 25-26°C is preferable for germination of rhizomes and 27.5°C for growth and development (Evenson *et al.*, 1978). Xizhen *et al.* (1998) categorized the suitable temperatures as 22 to 28°C at seedling and early growth stages and 25°C at rhizome enlargement stage. And below 15°C, ginger stops growing. In Kerala, the crop is being cultivated in the temperature range of 28-35°C. Warm sunny days are preferred but extremely high temperature causes sunburn and low temperature induces dormancy. In Australia, full emergence is delayed up to 6 weeks because often, the soil temperature at top 10 cm layer takes several months to reach optimum after the normal planting season (Weiss, 1997). The crop requires short or long day length for its growth and development. With the increase in day length from 10 to 16 h, vegetative growth enhances and while the day length decreases from 16 to 10 h rhizome swelling is promoted. Good sunshine, heavy rainfall and high relative humidity are necessary for getting good yield. A well-distributed rainfall of 1500-3000 mm over a span of 8-10 months produces good crop. The crop is sensitive to water logging, frost and salinity. Ginger requires partial shade for better rhizome yield. Frost is injurious to foliage and rhizomes of ginger. Hence, in areas where frost is common as that of hill regions of North/North eastern India, harvesting is done before the frost.

3.2 Soil

Ginger can be grown on a wide variety of soils such as sandy loams, clay loams, alluvial and lateritic soils. However, it is mainly grown in red and laterite soils of Kerala, Karnataka, Orissa, West Bengal, Maharashtra and Northeastern states. Well-drained soil with at least 30 cm depth, loose and friable is preferable for cultivation since the rhizomes and roots proliferate in the top 25 cm. By adopting cultural practices such as bedding and surface mulching, shallow soil can be utilized. Alluvial soils and drained paddy fields or well-drained marshy areas can also be utilized for ginger cultivation as it is commonly practiced in China, Taiwan, and Japan (Weiss, 1997). Deep soils with rich organic matter content and nutrient availability are more suitable for cultivation (Cho *et al.* 1987). However, it performs well on medium loams having enough quantity of humus. Even tough virgin forest soils after deforestation are found ideal (Paulose, 1973), Ridely (1912) quoted about possibility of growing ginger for longer years in the same patch and it is therefore quite unnecessary to destroy forests of great value. Compact clay soils characterized by water logging or coarse sands with poor water holding capacity, gravelly soils or soils with hard pan are not suitable (Lawrence, 1984). Deep slopes in hilly areas are not advisable for ginger cultivation as it leads to soil erosion during heavy rainfall. It is grown in volcanic soils in Mauritius and other Indian Ocean Islands and on silty clay loams in Madagascar. In West Africa, forest soils are preferred. In Jamaica, ginger is grown mainly on hilly terrains having clay loam over limestone or conglomerates. In Australia, moderate to heavy soils are utilized.

Growth, development and maturity of ginger are also affected by soil type. For example, oil content was highest in cv. Rio-de Janeiro at 180-210 days, when it was grown in calcareous soils in north-east India, compared to those in clayey soils of Southern India. Soil hardiness less than 15.7 mm is optimum. The yield decreases with increase in soil clay content and decrease in pH. The optimum soil pH preferred for ginger is 5-7 and if the pH is more than 8, growth is retarded. Maximum rhizome yield can be achieved in sandy loam soil having minimum bulk density (1.20 g/cc), moderately acidity (pH 5.7), high organic matter and available potassium and yield decreased with increasing clay content (Sahu and Mitra, 1992). The soil should be relatively free of root knot nematodes and soil-borne diseases causing rhizome rot and bacterial wilt. Soil less aeroponic cultivation of ginger to get high-quality rhizomes that are free from pesticides and nematodes in mild-winter greenhouses also was tried by Hayden *et al.* (2004) in Arizona. The unique aeroponic growing units incorporated a "rhizome compartment" separated and elevated above an aeroponic spray chamber. Accelerated growth and faster maturity was observed in plants receiving bottom heat on perlite medium.

4.0 PLANTING MATERIAL

4.1 Seed storage

Ginger is propagated vegetatively by portions of rhizomes known as seed rhizomes and among inputs, seed material alone accounts for about 40% of total cost of production. In order to obtain good germination, proper storage of early season seed rhizomes is essential. As per organic production standards, organic seed or planting materials of appropriate varieties and quality shall be used. When organically grown seed rhizomes are not available, conventional materials may be used provided that they have not been treated with pesticides; not otherwise permitted by the standards. If untreated conventional seed rhizomes are not available, chemically treated plant material may be used, where the certification body shall establish time limits and conditions for permitting their use.

The seed rhizomes should be stored appropriately so that rotting, shriveling, dehydration and sprouting can be avoided until the next season. Maintaining a storage temperature of 22-25°C make the growing buds fat and strong and temperature higher than 28°C in the long run make the buds thin and weak. If the storage humidity is too low, ginger rhizome epidermis may also loose water and wrinkle and the sprouting speed and bud quality may be affected (Xizhen *et al.* 2005). Ginger can be stored in pits (1 x 1 x 1 m size), inner walls lined with stones/ bricks. The bottom is filled with 10 cm thick dry sand. Disease free bold rhizomes are selected after harvest, cleaned and treatment with 1:1:100 Bordeaux mixture for 20 min is recommended for seed rhizomes (Xizhen *et al.* 2005). Prestorage steeping of rhizomes in *Trichoderma hamatum* or *T. viride* also showed inhibition against *Fusarium equiseti* infection (Bhardwaj *et al.* 1988). Treated rhizomes are placed in pits leaving 10-15 cm space on the top, covered with wooden plank to have space for aeration and plastered with cow dung. Covering the seed material with a layer of *Glycosmix pentaphylla* leaves is also beneficial (Nybe and Miniraj, 2005). In China, cellars, holes dug out of the earth in shady or covered places, are used commonly for storage of seed rhizomes. Zero energy cool chamber (ZECC), is found ideal for storing fresh ginger. The loss in weight of rhizomes was only 23% after storing for 4 months in this chamber, while the ginger stored in open conditions was shrunken in 4 months (IISR, 2002). Studies on storage of "seed pieces" of ginger showed that, in general, the number of days to germination decreased with length of storage period while percentage germination and yield increased from 0 to 42 days storage. However, germination and yield were consistently lower after 35 days storage. This anomalous behaviour may be due to secondary dormancy during which the seed pieces lost their dormancy up to 21 days of storage but regained or entered into secondary dormancy at 35 days and again lost dormancy after 42 days (Timpo and Oduro, 1977).

The stored rhizomes must be examined at monthly intervals and remove the rotten ones to keep the pathogen inoculum level low. In Nigeria, delayed harvesting with grass mulching up to 15 cm thick in the field itself or storing the harvested rhizomes in shade on layers of sand or grass cover are some alternate ways of seed ginger storage (Anonymous, 1970).

4.2 Seed size and rate

Seed rhizome weight and the length used for sowing vary from place to place and variety to variety. Size of seed rhizome has a direct influence on the final yield (Whiley, 1974). The research results also showed that the stored nutrients in the ginger seed were partly transferred to new plants in the whole process of growth. A certain proportion remained in the ginger seed itself. To obtain high yield and better returns, rhizome pieces weighing 0.5-2.0 ounces with two three sprouts are recommended. In general, seed rhizomes are cut into small pieces of 2.5 – 5.0 cm length, weighing 20-25 g with one or two viable buds. Different size of rhizome pieces namely, 15-20g (Mohanthy *et al.*, 1990b), 20-25 g (AICRPS, 1992), 20-30 g (CSIR, 1976), 40-45 g (Lee *et al.*, 1981), 50-80 g (Whiley, 1974) were reported/ adopted under field situations. Higher yield and profit can be obtained by planting larger seed pieces of 35-80g (Aiyadurai, 1966; Whiley, 1981; Sengupta *et al.*, 1986; Okwuowulu, 1988a; Roy and Wamanan, 1989). Whiley (1990) observed that larger 'seed piece' size (85.5 ± 2 g) increased both 'knob' size (12%) and rhizome yield (26%). Planting of full mother rhizomes (50-60 g) has been found to result in the highest yield and lowest disease severity. In China, even though, large seed ginger size (100+or-5g) produced vigorous growth at the early stage; highest yield was obtained with 75+or-5 g. However, neither the largest nor the smallest seed sizes are recommended (Wang *et al.*, 2003). Use of 2.5 to 5 cm long rhizome bits is recommended for field planting by Furutani *et al.* (1985) under African conditions. The cost benefit analysis indicated that the use of small pieces weighing 10g was more profitable (Nizam and Jayachandran, 2001).

Seed rate varies with variety and soil fertility. In general, seed rate varies from 1500 to 2500 kg per hectare depends on seed size and spacing in India. An optimum seed rate of 1250 kg/ha with each seed rhizome weighing 30g was suggested by Randhawa and Nandpuri (1970). The optimal size of seed size and rate recommended for different places are given in Table 4. For plains and lower altitudes, 1500-1800 kg and for higher altitudes (> 1000m), 2000-2500 kg is recommended (Aiyadurai, 1966; NRCS, 1989). In Queensland estates, about 8-10 t ha^{-1} of setts are required on fully mechanized conditions. Whiley (1974) predicted 10 t ha^{-1} as economical but farmers use 4-6t/ha. Approximately 6 t ha^{-1} corresponding to 1,40,000 plant population was adopted by Lee *et al.* (1981). In China, seed pieces of 50-75g are used at the seed rate of 5.25 to 7.5 t ha^{-1}. However, under less intensive conditions (India/ Sri Lanka), 1.5 - 4 t ha^{-1} is sufficient in the raised bed system (Weiss, 1997). In Kerala, seed rate of 800-1100kg/ha is adopted (Mirchandani, 1971), whereas in Himachal Pradesh, 2.3-3.5 t ha^{-1} is used. Rhizomes which remain un-decomposed at crop maturity can be detached during crop growth without affecting yield (Okwuowulu, 1988b). In northeastern states, farmers plant whole rhizomes and unearth the mother rhizome when the crop reaches 30-

35 cm height and 94.6 per cent recovery was obtained at 3 months after planting (Singh, 1982). By this method farmers get back 60 to 70 per cent of the seed cost. Thus, detaching and recycling the setts in the same season or later, provides a means of achieving rapid seed ginger multiplication and higher aggregate yield.

4.3 Seed treatment

The aim of seed treatment is to prevent seed-borne pathogens and pests and to induce and ensure uniform germination. Farmers in Kerala dip seed rhizome, in cowdung solution. The cut end of the seed may provide entry for pathogens and to prevent this, the cut rhizome pieces are dipped in fresh and clean plant ash to seal off the wounds (Xizhen *et al.*, 2005). Treating the seed rhizomes by steeping in spore suspension of *Trichoderma viride* or *T. hamatum* found effective against *Pythium aphanidermatum* (Bhardwaj *et al.* 1988) and with *T. harzianum, T. aureoviride, Gliocladium virens* and *T. viride* reduced rhizome rot caused by *Fusarium solani* or *P. myriotylum* and significantly increased the growth, development and yield of ginger (Ram *et al.* 2000).

5.0 PLANTING AND AFTER CARE

5.1 Preparation of land and planting

5.1.1 Land preparation

Land areas that are otherwise naturally away from contamination sources and having good drainage are suitable for organic cultivation. A buffer zone with suitable physical/ tree barriers or buffer crops all around the organic plantation to avoid contamination by drift from neighboring conventional plantations is necessary. A buffer crop of ginger itself can also be raised in the isolation zone, but that cannot be sold as organic. The area of buffer zone may vary with the landscape and will be specified and certified by the certification agencies based on the local agro situations.

Land preparation may vary with soil type, slope and irrigation. The soil should be thoroughly broken up and pulverized. The land is ploughed several times or dug thoroughly with the receipt of early summer showers to get a fine tilth. In Jamaica, clay lands are forked and left to dry for three months before planting, forked and drained by means of drenches (Graham, 1936) and ridges and furrows method is used. In Queensland, after initial ploughing during November-December, a cover crop of maize is planted during summer to build up organic matter in the soil or land is fallowed. Poultry manure or press mud (by product from sugarcane industry) is incorporated before cover crops are sown to provide additional organic matter (Whiley, 1974).

In Kerala and Tamil Nadu, the land is ploughed six times and the soil is brought to a fine tilth. However, no added advantage in ploughing the land over the minimum requirement of 3 to 5 times was reported (Aiyadurai, 1966). Solarization of beds for 40 days by using transparent polythene sheets with *Trichoderma* application are recommended for the areas prone to rhizome rot and nematode diseases. In China, tillage is started after the harvest of the pervious crop, during autumn or winter. Ploughing 25-30 cm deep promotes root penetration and enlarges absorption area. After harrowing once or twice, 75-120 t ha^{-1} good quality manure is applied and flattened (Xizhen *et al.*, 2005). Ploughing followed by harrowing several times at 5-7 days intervals to kill weed seed germination and rotovating is common in African region (Okwuowulu, 2005).

In India, beds of about 1 m width, 15 cm height and of convenient length varying from 3 to 6 m are prepared with a width of 30 cm for the channels between the beds. In Bengal, the field is ploughed 12 to 13 times during March-April after the first rain, leveled and water channels are made to irrigate the ground. In hill slopes, the beds are formed along the contours to reduce soil erosion. For irrigated crop, ridges and furrows method is followed with spacing at 40 cm apart.

Two distinct methods of cultivation are adopted in India. In the Malabar system, beds of 3 x 1 m are laid out at 30-45 cm apart from each other with small shallow pits on beds for planting at required spacing and handful of cattle manure is applied to each of these pits. In sloppy areas, smaller beds are formed. In the South Kanara system, no beds are laid out. A mixture of manure and burnt earth is applied in the form of a 5 cm thick ridge in between the rows 100-200 cm apart from each other and the seed rhizomes are placed in the rows and earthed up to make the ridges 15-20 cm high (CSIR, 1976). Soon after sowing, the field should be irrigated. Sufficient moisture at the time of planting is ensured through irrigation after basal application of manures and about one hour before planting, in China. Planting of ginger is recommended on raised beds in order to facilitate drainage. Planting irrigated ginger in raised beds gave higher yields as compared to planting in ridges, furrows and flat ground in research trials. Raising ginger in flat beds in sandy loam soil and on raised beds in clay loam soil, followed by earthing up with the application of fertilizers are most suited for successful cultivation (Singh *et al.*, 2003).

5.1.2 Planting time

The time of planting is important as it affects the yield and quality of ginger. The main factors to be considered while planting are time of planting, spacing and depth of seed placement. The planting time depends on the onset of the monsoon. In India, generally planting is done during March-June (Table 4) (Randhawa *et al.*, 1972; Jha *et al.*, 1986; Phogat and Pandey, 1988; Mohanty *et al.*, 1998). Studies have shown that planting during April gives better growth and development of rhizomes and less incidences of diseases. It is planted during September at Australia and Fiji (Whiley, 1974, Haynes *et al.* 1973), March- April at, West Indies (Ridely, 1912), mid-April at South Eastern Nigeria (Okwuowulu *et al.*, 1989), May - June at Jamiaca (Prentice, 1959), February-April at Taiwan (Lawrence, 1984), April-May at Sierra Leone and Hawaii (Furutani *et al.*, 1985), June at Ghana (Lawrence, 1984) and January-April at China (Xizhen *et al.*, 2005). However, under irrigated condition ginger can be planted at any time of the year, the best being middle of February. In India, ginger is planted with commencement of South West monsoon. A yield increase up to 200 percent was recorded by planting during the first week of April (Aiyadurai, 1966), with the receipt of summer showers, than the normal planting in May-June. Early planting of ginger is beneficial as the crop can grow sufficiently to withstand heavy rains (Nair and Varma, 1970). Though transplanting in ginger is not conventional, it is found profitable in Himachal Pradesh (Solan), India.

5.1.3 Spacing

Plant population is an important factor decided by spacing and it varies with soil type, variety, climate and management practices. Closer spacing gives higher yield with maximum plant density (Aiyadurai, 1966; Nair, 1982). Different spacings (15-45 x 15-45 cm) are recommended for various states in India (Table 4). Various spacings of 40 x 15 cm (Lee *et al.*, 1981) and 60 x 11.8 cm (Whiley, 1981) at Australia, 38.1 x 38.1 cm at Mauritius and 45 x 40 cm at Trinidad, West Indies (Wilson and Ovid, 1993) have been adopted. In China, 60-65 x 19-20 cm spacing with a planting density of 82,500 to 90,000 plants ha^{-1} for Lai Wu big ginger and 50-55 x 18-20 cm spacing with a density of 105,000 to 112,500 plants ha^{-1} for Lai Wu slice ginger are found to give better yield (Xizhen *et al.*, 2005). In Africa, spacing of 20 x 20 cm for ware-ginger production and 10-15 x 10-15 cm for seed ginger production through minisett technique is followed (Okwuowulu, 2005). By adopting a correct row and column spacing of 20 x 20 cm and by timely planting weed density could be reduced to a bare minimum. Whiley (1990) observed that greater planting densities with intra-row spacings 11.2 and 17.0 cm increased rhizome yield by up to 43% and reduced the time to final first harvest maturity (35% fibre free rhizome) by about 10 days.

Seed rhizomes are planted at optimum depth for its proper and early germination/emergence. Planting depth may vary depending upon seed size, soil type and soil moisture content. In general, bolder seed rhizomes are planted deeper and smaller ones at shallow depths. Based on trials, planting on raised beds at a spacing of 20-25 x 20-25 cm and a depth of 4-10 cm with the viable bud facing upward is recommended (Kannan and Nair, 1965; Paulose, 1973; Lee *et al.*, 1981; Wilson and Ovid, 1993). Okwuowulu (1992) reported that planting at 5 cm depth predisposes the rhizome to significant desiccation loss, if the harvest is delayed and recommended seed placement at 10 cm depth in the areas where delayed harvesting is done. Light irrigation after planting is beneficial for better germination. Ginger takes 10-15 days for sprouting after planting under ideal conditions, which may be prolonged up to 2 months.

Table 4. Planting time, seed rate and spacing recommended at different places

Country / State	Planting time	Seed rate (kg ha^{-1})	Seed size (g)	Spacing (cm)
Kerala	April – June	1500 to 1800	15	20-25 x 20 -25 at 5 cm depth
Orissa	April	1800	15 - 20	25 x 20 at 5 cm depth
Himachal Pradesh	March – April	2500 to 3500	50 - 100	30 x 20
Sikkim	Feb – April	3000 to 6000	75 - 150	45-60 x 30
Bihar	April - June	1800	18-20	25 x 20
Meghalaya	March - April	2500 to 3500	25 - 50	30 x 30
Andhra Pradesh	April – June	1700	20 - 25	30 x 20
Australia	September	8000 to 10000	50 - 80	40-60 x 15
China	January – April	5250 to 7250	75 - 100	60-65 x 20; 50-55 x 20
Jamaica	May - June	4000 to 6000	45 - 60 (2.5 -5 cm long)	15-20 x 15-20

5.2 Mulching

Mulching enhances germination, conserves moisture, prevents run-off, increases infiltration, regulates temperature, suppresses weed growth and improves soil fertility by adding organic matter. Mulching can alter the physical and chemical properties of the soil and increase the availability of phosphorus and potassium and demonstrated to reduce nematode infestation in Fiji (Haynes *et al.* 1973). The quantity o1f mulch to be applied varies with availability of material. In general, 10 to 30 t ha^{-1} is applied twice or thrice, one at planting, second at 45th day and third at 90th day after planting. In plains mulching with 30 t ha-1 of green leaves is recommended. Mulching the beds with green leaves at 15 t ha^{-1} after planting followed by two mulching (at 7.5 t ha^{-1}) at 45 and 90 days after planting is an essential practice (Nybe and Miniraj, 2005).

Commonly used mulch materials are green and dry forest leaves, residues like sugar cane trash, paddy, wheat, finger millet, barely straws and coconut leaves, banana leaves, dry sal leaves and vegetation of the locality. Mishra and Mishra (1982) reported that mulching with dry leaves suppressed weed growth and increased emergence, growth and yield. Das (1999) showed that mulching with maha neem leaves (*Melia azadirachta*) reduced the rhizome rot (*Pythium*) infection. Farm yard manure (FYM) and compost are also used as mulches. Live mulches like sun hemp, green gram, horse gram, black gram, niger, sesbania, cluster bean, French bean, soybean, cowpea, daincha and red gram can also be grown as intercrop and mulched *in situ* between 45-60 days after planting (Kandiannan *et al.*, 1996). Growing green manures crops like *Sesbania rostrata, S. aculeate, S. speciosa* or fodder cowpea amongst ginger and using as second or third mulch provided biomass of 4 kg green leaves per 1.5 m^2 (Valsala *et al.* 1990), reduced weed growth and increased the ginger yield (Kurian *et al.* 1997). In Sikkim, ginger beds are covered by leaves and twigs of various forest trees after sowing. Under Meghalaya conditions, application of locally available organic mulches like paddy straw and dry leaf mulch of *Schima wallichii* at 16 t ha^{-1} increased the yield of ginger. One-fourth quantity of the recommended dose (30 t ha^{-1}) of green mulch could be saved if ginger is intercropped under coconut plantations with 25% shade (Babu and Jayachandran, 1997). In Tamil Nadu, daincha and sunhemp are raised as intercrops in alley spaces and incorporated as mulch.

In African countries, green leaves of guinea grass (*Panicum maximum*), Elizabeth weed (*Chromolaena odoratum*) and rubber at about 10 t ha^{-1}, decomposed sawdust (with low C:N ratio) at 30 t ha^{-1} or well rotted dip-litter poultry manure or cowdung at 20 t ha^{-1} are commonly used for mulching (Okwuowulu, 2005). In China, mulching is commonly done with a layer of 3-5 cm thick wheat straw, about 300-4500 kg ha^{-1} to decrease the soil temperature, hold the soil moisture and to improve the field microclimate. Plastic film mulching, stretching a double layer of black plastic film tightly over the ridge, is also popular among ginger growers as it is better material than straw with 8-30 percent increased yield and economic benefit (Xianchang *et al.* 1996; Zhifeng *et al.* 2000). Polythene as mulch material gave 19.9 t of fresh rhizome per ha compared to 12.0 t in unmulched plots (Mohanty *et al.*, 1990b). Only products based on polyethylene and polypropylene or other polycarbonates are allowed under organic farming standards and the use of polychloride based products is prohibited. These shall be removed from the soil after use and shall not be burnt on the farmland.

Fig. 2: Mulching of beds after planting (*Source:* Director, IISR, Calicut)

5.3 Weed Management

Weeds are serious problems in ginger cultivation and reduce the yield considerably. The traditional method of hand weeding is the common practice and is done just before manure application and mulching. Two to three weedings are required depending on the intensity of weed growth. The first weeding is done on 45 DAP and the second weeding during 120-135 DAP. Experimental studies revealed that highest fresh ginger yield can be obtained with 4 hand weedings. Soil solarization is also effective in suppressing the weed population.

Earthing up may be combined with hand hoeing (weeding) and mulching. Soil stirring and earthing up are essential to break the hardpan formed by rain or irrigation and it also helps in checking weeds, conserving moisture, mixing manures with soil, enlargement of daughter rhizomes and provides adequate aeration for roots and protects the rhizome from scale insects (Panigrahi and Patro, 1985).

5.4 Irrigation

Normally, ginger is cultivated both under rainfed and irrigated conditions. In areas receiving less rainfall, the crop needs regular irrigation. According to Queensland Irrigation and Water Supply Commission, ginger requires 1320-1520 mm of water during its complete crop cycle. Irrigation is given at fortnightly intervals, usually during the middle of September to middle of

November (Aiyadurai, 1966). Increased moisture supply increases the yield of rhizomes and essential oil content. Irrigation can be given from 10 am to 3 pm during the time of sunburn weather to establish a microclimate favorable for the crop (Whiley, 1974). Ginger crop raised in the first week of May in Orissa may need two - four irrigations at an interval of 7 days depending on soil type and climatic conditions and after the receipt of monsoon, the crop thrives well till the end of September (Panigrahi and Patro, 1985). Critical stages of water requirement are germination stage, rhizome initiation stage (90 DAP) and rhizome development stage (135 DAP). First irrigation is given to ginger immediately after planting and subsequent irrigations at an interval of 10 days in order to meet a total water requirement of 90-100 cm in 16-18 irrigations (Vaidya *et al.*, 1972). During dry spell, a fortnightly irrigation helps significantly to improve yield and quality of the produce. Maximum rhizome yield could be obtained by scheduling of irrigation at 60 mm cumulative pan evaporation (CPE) and IW/CPE ratio of 1.0 (KAU, 1994). In China, irrigations are given once in four to six days during summer or drought months to keep the soil moisture at about 70-80% (Xizhen *et al.*, 2005).

5.5 Shade management

Even though shade is not absolutely necessary, ginger prefers light shade. Shading is helpful in reducing water loss and provides a micro climate suitable for the plant. Under full sunlight, the crop became shorter with fewer leaves per tiller and shade increased height, tiller number (Wilson and Ovid, 1993), net assimilation rate and chlorophyll content. Dry matter production, nutrient uptake yield and quality are also higher under low to medium shade (25 per cent) (Beena *et al.*, 1994; Ancy and Jayachandran, 1993; Sreekala and Jayachandarn, 2002). A heavier shade (beyond 50 per cent) decreases number of tillers and yield (Aclan and Quisumbing, 1976). Ginger can also be gown under full sun (open condition). Ginger can be grown in homestead garden and also suitable for intercropping under coconut and other perennial crops, where low to medium shade intensity exists. The incidence of *Phyllosticta* leaf spot is also much less under shade.

Varieties like Himachal is found to perform well both in the open and low-light intensities (75% shade) with respect to dry matter accumulation and uptake of N, P and K (George *et al.*, 1998a, b). Ginger grown under low to medium shade (25 percent) produced good quality parameters (volatile oil and starch contents) with less crude fibre (Ajithkumar and Jayachandran, 2003), compared to those grown under open and heavy shade under Kerala conditions. Under low and medium shade, nutrient up take and rhizome yield were more compared to open and high shade (Ancy and Jayachandran,

1996 and 1998). The cultivars Jamaica, Valluvanad, Jorhat and Rio de Janeiro are evaluated and found suitable for intercropping in plantations with 50% or more shade levels.

5.6 Cropping system

Ginger can be grown as a sole crop (under open or shade) or as a component in inter or mixed cropping systems. Cropping system helps in building up of depleted nutrients removed by ginger, maintaining soil fertility and reduction in the pathogenic inoculums of rhizome rot causing organism. It also provides additional income to the farmers. Moreover, there is a better utilization of environmental factors, greater yield stability, a useful variability in the food supply and ensurance against any failure of one component crop. Ginger can be intercropped with vegetables (cabbage, tomato, chillies, french bean and lady's finger), pulses (pigeon pea, black gram and horse gram), cereals (maize, finger millet), oilseeds (caster, soybean, sunflower and niger) and other crops (sesbania, tobacco and pineapple) (Kandiannan *et al.*, 1996; Okwuowulu, 2005). The component crops are also to be raised by organic means only.

Ginger being a partial shade loving crop, it can be grown as an intercrop in coconut, arecanut, rubber, orange, guava, mango, papaya, coffee and popular plantations with effective recycling of farm wastes. Leaves and bunch wastes of coconut, arecanut, oil palm, coffee husk have been found to give very high bioconversion efficiency. The multiple cropping system results in the continuous uptake, addition of biomass and higher level of nutrient recycling that would lead to a positive influence on the physico-chemical and biological properties of soil (Table 5).

Table 5. Biomass available in plantations for recycling in India

Crop	Biomass
Coconut	11.2 mt
Arecanut	0.43 mt
Cocoa - Leaf shed	360 t
• Prunings	12056.3 t
• Pod husk	32900 t
Tea/ Coffee wastes	0.40 mt
Shade trees	8 - 10 mt

Source: Parthasarathy *et al.* 2006

In a cropping system, it is desirable that organic method of production is followed in the entire farm, but in the case of large extent of area, the transition can be done in a phased manner for which a conversion plan is to be prepared. In coffee based spices multi-storied cropping system, local cultivars of ginger grown as an intercrop gave a gross income of Rs.79,800 during the fourth year (Korikanthimath *et al.*, 1995). It is profitable to have ginger in coconut based cropping system with a gross return of Rs. 1,00,500 ha^{-1}. As intercrop in arecanut plantation, ginger gave higher yield with bigger rhizome and taller plants but the duration of crop was more (198.5 days) than in the open conditions (Hegde *et al.*, 2000).

Green manure crops like daincha (*Sesbania aculeata*) can be raised successfully in inter-spaces of beds in a row, which adds to 50 per cent of the green leaves required for mulching, suppresses weed growth and reduces cost of production. Ginger can also be grown as mixed crop with castor, redgram, fingermillet and maize. Sharma and Bajaj (1998) recommended intercropping with bell pepper (*Capsicum annuum*) for reducing nematode population and infection thereby with equal number of ginger and pepper plants and also 76% control of ginger yellows disease (Dohroo and Sharma, 1997). In Bangladesh, ginger produced a high yield (4.33 t/ha) and benefit cost ratio (1:6.2) when grown under juvenile mango trees (8-10 years old) (Haque *et al.*, 2004). Ginger is also the most favoured crop component under agroforestry. Growing fodder tree, *Quercus leucotrichophora* with ginger is found to be the most ideal and remunerative silvi-horticultural combination in hills of U.P (Bisht *et al.*, 2000). Studies of Thomas *et al.* (1998) suggested that competition between the shade trees and inter crop ginger for nutrients applied to the tree component is unlikely in the Ailanthus-ginger mixed species system studied. Therefore, for a better crop management, it is better to fertilize ginger under the mixed species system adequately, as it will also benefit the tree component. Nutrient use efficiency may be higher under such situations. Ginger in the interspaces of five year-old ailanthus trees with 52% mean daily PAR flux or 73% midday PAR flux exhibited better growth and highest rhizome yield compared to sole crop and also Ailanthus + ginger combinations improved the site nutrient capital (Kumar *et al.* 2001). The rhizome length, rhizome breadth and yield per ha in ginger exceeded under poplars (*Populus deltoides* 'G-3' Marsh.) spacing 5×4 m but showed a drastic reduction under the closest poplar spacing (Jaswal *et al.* 1993). Ginger gave high yields when intercropped in 7-year old plantations of *Paulownia elongata* in Eastern China and was found as an ideal shade crop for these systems (Newman *et al.* 1997).

Crop rotation with green manures and legumes is essential, as ginger depletes more nutrients in the soil coupled with rhizome rot problem under continuous monoculture. Ginger is rotated with tapioca, seasame, little millet and dry paddy in rainfed conditions and vegetables under irrigated conditions. Pegg *et al.* (1974) also recommended beans, cucurbits and strawberries as suitable crops for rotation to minimize nematode problems. However, crop rotation using tomato, potato, chillies, brinjal and peanut should be avoided, as these plants are hosts for the wilt causing organism, *Ralstonia solanacearum*.

Fig. 3: Ginger mixed cropped with (a) black pepper and coffee (b) maize (c) coconut (*Source:* Director, IISR, Calicut)

6.0 NUTRIENT MANAGEMENT

6.1 Nutrient removal

Ginger is a nutrient exhaustive crop and responds well to the application of organic manures and fertilizers. Ginger rhizomes are mainly N and K exhausting, intermediary in P and Mg removal and the least in Ca removal (Nagarajan and Pillai, 1979). Xu *et al.* (2004) observed that ginger shoots and leaves were growth centers at the seedling stage, and 80.7 % of the carbon assimilation was transferred to these parts. Later, the distribution rate for shoots and leaves decreased gradually with the growth, whereas the distribution rate for the rhizome increased. Up to the vigorous growth stage of rhizome, carbon assimilation is mainly transported from leaves into the rhizome, thus the rhizome is the growth center. The absorption and utilization of nitrogen were the same as carbon assimilates. 48.41 % of the nitrogen absorbed from fertilizer applied at seedling stage was distributed to the shoots and leaves. While 65.43 % of the nitrogen derived from fertilizer applied at vigorous growth of rhizome was distributed into rhizomes, only 32.04 % distributed into shoots and leaves.

The results indicate that the rate of fertilizer-N utilization increase with the delay of application. The exchange of carbon and nitrogen nutrition between the above ground parts and underground seed of ginger would be the characteristics of their transportation and distribution during ginger growth ensuring that the seed-ginger will not shrivel.

Nutrient uptake of a crop varies considerably with variety, soil type, climatic conditions etc. The development of ginger can be classified into three distinct growth phases namely, active growth (90-120 days after planting (DAP)), slow vegetative growth (120-180 DAP) and senescence (180 DAP) in which the rhizome development continues till harvest. The uptake of N, P and K in leaf and pseudostem increase up to 180th day and then decrease, whereas that of rhizome uptake steadily increases till the harvest. An average dry yield of 4.0 t ha^{-1} ginger rhizomes removes 70 kg nitrogen, 17 kg P_2O_5, 117 kg K_2O, 8.6 kg Ca, 9.1 kg Mg, 1.8 kg Fe, 500 g Mn, 130 g Zn and 40 g Cu per ha. For healthy growth of ginger, very low external calcium is required. The leaves of healthy plants contained 1.1 to 1.3% of Ca and concentrations as low as 2 ppm is sufficient to achieve 90 per cent of maximum yield. In order to diagnose the foliar nutrient deficiencies of N, P and K, Johnson (1978) recommended sampling of 5th pair of leaf as index leaf during 90th to 120th DAP.

6.2 Manuring

Under organic farming, the certification programme shall set limitations to the total amount of biodegradable material of microbial, plant or animal origin brought onto the farm unit, taking into account local conditions and the specific nature of the crops. The requirement of organic matter may be met from various sources such as green/organic manures and mulches. The importance of organic manures is evidenced by good performance of ginger crop in high fertility conditions of Wayanad, Kerala supplied with 10-30 t of organic manure and 15 t of green leaf mulch per hectare, without any fertilizer application. Ginger performs well with good supply of humus and organic matter has positive correlation with yield (Cho *et al.*, 1987).

Farm Yard Manure, vermi compost, coir pith compost, poultry manure, green leaf, municipal composts, press mud, oil cakes, biofertilizers, cow urine are various sources of organics. The quantity of organics applied may vary between 5-30 t ha^{-1} depending on its availability and its nutrient content. Organics are mostly applied as basal doses and in certain places it is also applied after the emergence of crop as mulch. Maharashtra farmers apply heavy doses of FYM, about 40 to 50 t ha^{-1}. In Kerala the recommended dose of organic manure is 25-30 t ha^{-1} of FYM and 30 t ha^{-1} of green leaf mulch is applied in three splits. Application of FYM up to 48 t ha^{-1} resulted in highest yield and benefit: cost ratio under Kerala conditions (Chengat, 1997).

Application of organic cakes (neem cake @ 2 t ha^{-1} or groundnut cake @ 1 t ha^{-1}) and coir pith compost @ 2.5 t ha^{-1} also increases nutrient availability, improves physical condition of the soil, increases the yield (Table 6) and reduces the rhizome rot incidence (Sadanandan and Hamza, 1998). In addition to neem cake, phosphobacteria and rock phosphate application increases the availability of P, Ca, Mg, Zn and Mn in soil (IISR, 2002). Studies at Kerala Agricultural Univarsity (KAU), Ambalavoyal showed that organic amendments such as coconut oil cake, ground nut cake, seasamum cake and cashew shell when applied to the soil to give nitrogen requirement of 50 kg/ ha were found to reduce the soft rot incidence of ginger caused by *Pythium aphanidermatum*. Lime application is practices in acidic soils to correct the imbalance of other nutrient availability. Under Kerala conditions where the soils are lateritic with high P fixing capacity, available P increases with the application of rock phosphates incubated with FYM and among the sources, Rajphos was superior. The apparent phosphate recovery, agronomic efficiency of the applied P and percentage yield response was higher for Gafsaphos, followed by Rajphos incubated with FYM in ginger (Srinivasan *et al.*, 2000). Sharma et al (1997) found that inoculation with VAM (*Glomus mosseae)* gave better growth parameters and higher yield under Himachal Pradesh conditions. Soil application of Gigaspora at the time of planting (2.5g per rhizome) also was found to increase the yield similar to pine needle organic amendment and seed treatment with *Trichoderma harzianum*. As ginger is a heavy feeder of potassium, supply through permitted source like powdered sulphate of potash can be ensured.

Table 6. Nutrient availability, yield and quality of ginger as influenced by organic manures/ cakes

Treatment	Org. matter %	P (mg kg^{-1})	K (mg kg^{-1})	Dry yield (kg/ha)	Oleoresin (kg/ha)
Check	3.07	7	58	2787	209
FYM	3.22	8	68	3599	281
Neem cake	3.40	10	71	3946	320
Cotton cake	3.30	10	70	3887	278
Brassica cake	3.47	9	70	3413	242
Groundnut cake	3.65	15	85	4077	302
Gingily cake	3.44	13	79	3851	266
CD at 5%	0.13	3.3	13.9	301	22.8

Source: Sadanandan and Hamza (1998)

Studies are conducted at the Indian Institute of Spices Research (IISR), Calicut on organic farming of ginger with application of different composts and manures for nutrient management (20 t FYM, 2 t Neem cake, 1 t Ash, 4 t Vermicompost per ha) along with biofertilizers *Azospirillum,* Phosphate solubilizing bacteria. Seed treatment/ soil application with bio control agents *Pseudomonas* sp. and *Trichoderma* sp. and need based spray of Bordeaux mixture for foliar diseases control and cultural method of cutting the infected shoots followed by neem oil sprays for pest control and other necessary sanitation and regulatory procedures were followed for the organic production. The observations made on soil nutrient buildup indicated that organic carbon content was high (2.33%) in organic system on par with integrated system and among the different systems organic management systems recorded the highest concentration of soil phosphorus, calcium, magnesium, zinc and copper availability. The effect of different cropping systems on microbial population in soil under ginger also showed that the population of *Pseudomonas fluorescens, Azospirillum and* Phosphobacteria was highest in the organic system. The activities of enzymes like dehydrogenase, acid phosphatase, alkaline phosphatase, cellulose, ureasae that are indirect measure of active microbial community was significantly higher under organic management as compared to integrated and inorganic systems (IISR, 2006).

Among different combinations of composts/ manures tried to supplement the nutrient requirement of ginger, significantly highest fresh rhizome yield of ginger (14 t ha-1) was observed in FYM (10 t), neem cake (2 t) and vermicompost (4 t) combination per ha followed by 10 t Coir compost + 8 t vermicompost per ha, over control. These combinations also resulted in higher oleoresin and lower fiber contents in ginger rhizomes. Trials conducted through AICRP on spices at various locations revealed that, combined application of different organic sources like FYM + pongamia oil cake + Neem oil cake + stera meal + rock phosphate + wood ash has yielded on par to the conventional practice in addition with high quality. The experiment on the effects of organic manure on the yield components of ginger conducted at Nigeria showed that organic manures gave significantly higher rhizome yield of between 114.3 and 250.6% relative to the control. Highest yield of 11.42 t ha^{-1} was obtained from plots applied with 30t ha^{-1} cow dung and 10t ha^{-1} poultry litter. The cow dung and poultry litter application increased the soil pH, organic matter, N, available P, exchangeable K, Ca and Mg relative to control (Ayuba *et al.* 2005).

6.3 Micronutrients

The order for micronutrients removal by ginger is Fe> Mn> Zn> B> Cu. Application of micronutrient increases the rhizome yield. Soil and plant analysis of samples across the country revealed 49% mean deficiency of zinc (Zn) in Indian soils with acid soils of Meghalaya having the highest deficiency rate of 57%. Since very large quantities of FYM and green leaf mulch are applied to ginger the micro nutrient requirement will be met from them and deficiencies of micronutrients if appear can be corrected with need based application of approved micronutrients under organic farming. The optimum fertilizer rate for obtaining maximum rhizome yield was determined as 6 kg Zn ha^{-1} for Inceptisols of Kerala with the maximum limit of soil DTPA-extractable zinc as 3.4 mg kg^{-1} (Srinivasan *et al.*, 2004). Foliar application of Zn @ 0.25% twice at May-June and August-September increases the rhizome yield as compared to soil application. Combined spraying of Zn (0.3%) + Fe (0.2%) + B (0.2%), twice at 45 and 75 DAP boost the rhizome yield in deficient areas (Roy *et al.* 1992).

Studies at Solan on the effect of micronutrients showed significant increase in the yield when NPK along with two sprays of Jagromin - chelated form of micronutrients (0.7%), once at rhizome initiation and again one month after the first spray was applied (Ramana *et al.*, 2003).

7.0 DIESEASES AND PEST MANAGEMENT

In organic system, pests and diseases should be controlled by a number of preventive cultural techniques which limit their development, e.g. suitable rotations, green manures, early and predrilling seedbed preparations, mulching, mechanical control and the disturbance of pest development cycles. The natural enemies of pests and diseases should be protected and encouraged through proper habitat management of alternate hosts and weeds etc.

7.1 Diseases

Ginger is affected by many pest and diseases. Of these, soft rot, bacterial wilt, yellows, *Phyllosticta* leaf spot, and storage rots are major diseases that cause economic losses (Table 7). In India, bacterial wilt and soft rot are

Table 7. Diseases of ginger and the causal organisms

Disease	Pathogen	Distribution
Soft rot	*Pythium* species *P. aphanidermatum (Edson)* Fitz *(syn: P. butleri* Subram *l, P. gracile (deBary)* Schrent *P. deliense* *P. myriotylum, P. ultimum* *P. pleroticum, P. vexans*	Andhra Pradesh, Assam, Bihar, Gujarat, Kerala, Karnataka, Maharashtra, Madhya Pradesh, West Bengal Madhya Pradesh Himachal Pradesh, Kerala, Maharashtra, Rajasthan Himachal Pradesh
Bacterial wilt	*Ralstonia solanacearum*	All over the country
Yellows	*Fusarium* spp *F. oxysporum f.* sp. *zingiberi* *F. solani,F. moniliforme* *F. graminearum,F. equiseti*	Himachal Pradesh, Rajasthan
Leaf spots	*Phyllosticta zingiberis* *Helminthosporium maydis* *Colletotrichum zingiberis* *Pyricularia zingiberis* *Leptosphaeria zingiberis* *Coniothyrium zingiberis* *Curvularia lunata* *Vermicularia zingiberis* *Septoria zingiberis*	Kerala, Karnataka, Himachal Pradesh Bihar Andhra Pradesh Assam Meghalaya Assam Bihar Andhra Pradesh
		Andhra Pradesh
Leaf blight/Dry rot	*Rhizoctonia solani* *R. bataticola*	Himachal Pradesh Haryana, Kerala
Thread blight	*Pellicularia filamentosa*	Kerala
Basal rot	*Sclerotium rolfsii* *(Corticium rolfsii)*	Maharashtra
Sheath rot	*Fusarium* sp.	Maharashtra
Virus diseases	*Cucumber mosaic virus* (CMV) *Chlorotic Fleck virus* (GCFV) Chirke virus	Kerala Assam

prevalent in all major ginger-growing areas. *Pythium* spp. *Fusarium oxysporum, Ralstonia solanacearum* and *Pratylenchus coffeae* are potent pathogens causing soft rot, yellows, bacterial wilt, and dry rot, respectively, in the field or in storage (Kumar *et al.* 2006).

7.1.1 Soft Rot

Soft rot is also called *rhizome* rot. Butler (1907) recorded the incidence of this disease for the first time from Surat (Gujarat) in India. The disease is prevalent in all the ginger growing countries across the world such as India, Japan, China, Nigeria, Fiji, Taiwan, Australia, Hawaii, Sri Lanka, and Korea. *P. aphanidermatum, P. myriotylum, P. vexans, P. ultimum* and *P. deliense* have been recorded as causal agent of soft rot of ginger.

Symptoms

The disease is predisposed by water logging conditions and is caused by several species of *Pythium.* The plant is susceptible to *Pythium* infection at all stages of growth. All the underground parts like root, stem and emerging sprouts are susceptible to this disease. The buds, roots, developing under ground stem the rhizome, and collar regions are the portals of infection. When the seed rhizomes are infected, they fail to sprout due to the rotting of young buds. After sprouting, the infection takes place through root or through the collar region, finally reaching the rhizome. Symptoms appear initially as water-soaked patches at the collar region. These patches enlarge and the collar region becomes soft and watery, and then rots.

In grownup plants, collar infection leads to yellowing of leaves. This yellowing starts from the leaf tip and spreads downward, mainly along the margins resulting in death of leaves. The dead leaves droop and hang down the pseudostem until the entire shoot becomes dry. The basal portion of the plant exhibits a pale translucent coloration. This area later becomes water soaked and soft to such an extent that the whole shoot either topples or can easily be pulled out. Rhizomes first turn brown and gradually decompose, forming a watery mass of putrefying tissue enclosed by the tough skin of the rhizome. The fibro vascular strands are not affected and remain isolated within the decaying mass. Roots arising form the affected regions of the rhizome become soft and rot. The rotten parts emit a foul smell. Rotting attracts opportunistic fungi, bacteria, and insects particularly the scavengers.

Management

1. *Seed rhizome selection and treatment:* Infected rhizomes are the primary source of infection and spread of soft rot in the field. The best method to

manage the disease is by the use of disease-free rhizomes for planting. Use apparently good looking bold and healthy rhizomes for planting.

2. *Cultural methods:* One of the predisposing factors for soft rot of ginger is ill drained field in continuous wet weather. Proper drainage in sandy loam soil for cultivation ensures healthy crop of ginger.

3. *Soil Solarization:* Soil solarization is a soil disinfection practice achieved by covering moist soil with transparent polythene film during the period of high temperature and intense solar radiation. Wherever possible this practice can be adopted.

Fig. 4: Symptoms of soft rot on ginger (*Source:* Director, IISR, Calicut)

Fig. 5: Soil solarization (*Source:* Director, IISR, Calicut)

4. *Biological control:* Antagonistic fungi namely, *Trichoderma harzianum, T. hamatum, T. virens* and bacterial isolates *Bacillus and Pseudomonas fluorescens* have been reported to be suppressive to soil borne pathogens of ginger. Vilasini (1996) reported that soil solarization for 30 days and incorporation of Trichoderma (125 g m^{-2}) amended with neem cake (500 g m^{-2}) controlled the soft rot disease and increased the rhizome yield considerably.

7.1.2 Bacterial wilt

Bacterial wilt disease of ginger is one of the most important production constraints in tropical, sub tropical and warm temperature region of the world (Hayward, 1991). Bacterial wilt of ginger inflicts serious economic losses to small and marginal farmers who depend on this crop for their livelihood. Once introduced in an area the soil become unsuitable for further cultivation of ginger. The severity of the disease is evident from its rapid and quick spread in the field when the environmental conditions (high rainfall and warm weather) are favorable for the disease development. Each of the infected plant is capable of releasing hundreds of thousands of bacterial cells in the form of bacterial ooze.

Bacterial wilt of ginger is caused by a bacterium, *Ralstonia solanacearum* biovar III (Smith) Yabuuchi, is one of the important rhizome borne diseases affecting ginger crop in the field. In India biovar 3 causes rapid wilt in ginger in 5-7 days under artificial stem inoculation and in 7-10 days under soil inoculation of the pathogen (Kumar and Sarma 2004). Traditionally ginger is cultivated in previously fallowed soil or virgin soil. Incidence of bacterial wilt noticed in such field is one of the indirect evidences of rhizome borne nature of *R. solanacearum* in ginger. Being a vascular pathogen it is presumed that the pathogen *Ralstonia solanacearum* is surviving in ginger at very low inoculum level without affecting the normal state of the ginger.

Management

Various control measures have been evaluated to combat the disease with limited success. Bacterial wilt is a major problem in the production of ginger and other vegetable crops owing to its wider host range & genetic variability it exhibits. Besides the pathogen is endowed with multiple modes of survival and fast lateral transmission within and between fields. The strategies for management of bacterial wilt are

1. Selection of healthy rhizome material from disease free area
2. Selection of field with no history of bacterial wilt in the past

Fig. 6: A bacterial wilt affected plant showing yellowing on leaves (*Source:* Director, IISR, Calicut)

3. Pre-plant rhizome treatment by heat or rhizome solarization
4. Strict phytosanitation in the field including restrictions on movement of farm workers and irrigation water across the field
5. Clean cultivation and Minimum tillage
6. Crop rotation with non-host plants like cereals such as paddy, maize etc
7. Insect pest and nematode control in the field
8. Soil amendments including biological control agents

Studies at KAU reported that bacterial wilt of ginger can be effectively controlled by oil drenching and foliar spraying with 1% Bordeaux mixture at monthly intervals two months after planting.

7.1.3 Fusarium yellows or dry rot

This disease is caused by *Fusarium oxysporum* f. sp. *zingiberis*, predisposed by nematode infestation by *Pratylenelus coffeae*. Ginger yellow is originally reported from Queensland and subsequently from Hawaii (USA) and India (Haware and Joshi, 1973).

Symptoms

On leaves, symptoms appear as yellowing of the margins of the lower leaves, which gradually spread over the entire leaf. Older leaves dry up first, followed by the younger ones. Plants may also show premature drooping, wilting, and drying in patches in the field or in the whole bed. Plants, generally, do not lodge on the ground as noticed in soft rot or bacterial wilt. In rhizomes, a cream to brown discoloration accompanied by shriveling is commonly seen. Vascular rot is also prominent. In final stages, only the fibrous tissue remains within the rhizomes. A white cottony fungal growth may develop on the surface of stored rhizomes. This disease along with nematode infestation severely reduces the marketability of the rhizomes. When used as seed rhizome the disease may affect the germination of the rhizomes.

Management

1. *Healthy rhizome selection:* As the disease spreads through contaminated rhizomes, selection of healthy rhizomes has been found to be an effective preventive measure for the disease.
2. Rhizome treatment with hot water at 51°C for 10 min is recommended in places where the disease occurs in endemic proposition
3. *Biological control:* Biocontrol agents like *T. harzianum, T. hamatum,* and *G. virens* as seed treatment and soil application are found to control the disease

7.1.4 *Phyllosticta* Leaf Spot

This disease is widespread in most ginger-growing countries including India. Leaf spot disease of ginger is reported for the first time in Godavari and Malabar regions of India by Ramakrishnan (1942).

Symptoms

A phyllosphere fungus *Phyllosticta zingiberi* causes this disease. Small, spindle to oval or elongated spots appear on younger leaves. The spots have white papery centers and dark brown margins surrounded by yellowish halos. The spots later increase in size and coalesce to form large spots, which eventually decrease the effective photosynthetic area on the leaf surface. As the plants put forth fresh leaves, they subsequently become infected. Such infected areas often dry up at the center, forming holes. In the case of a severe attack, the entire leaf dries up. The crop looks a grayish disheveled look as a result of infection.

Management

1. Crop rotation for 2 - 3 years or more.
2. *Site selection:* Planting in high elevated field that is well drained.

3. Seed rhizome selection: The seed rhizomes should be selected from disease free area as the disease can spread through rhizomes that are looking normal. The cultivars like Narasapattom, Tura, Nadia and Thingpui are moderately resistant with a disease index less than 5 percent.
4. Rhizome treatment: Seed rhizome can be treated with biocontrol agents like *T. harzianum*
5. Growing ginger under shade trees such as coconut as shade.
6. Spraying one or two rounds of Bordeaux mixture (1%) before the start of heavy rains.

Fig. 7: Leaf spot caused by *Phyllosticata* sp. (*Source:* Director, IISR, Calicut)

7.1.5 Thread Blight

Sundram (1954) reported this disease for the first time in Malabar region of Kerala state. The disease is not of much significance and occurs very rarely during heavy rainfall. This disease is caused by *Pellicularia filamentosa.*

Symptoms

Small water-soaked lesions appear on the leaf margins or other parts of the leaf during the initial stage of this disease. Later on, the infected leaves lose their turgidity, wilt, and may get detached from the sheath. Fine hyphal threads spread over the infected parts, and small brown sclerotia are present on the lower surface. The infected portion turns white and papery upon drying.

Management

Protective spraying with Bordeaux mixture (1%) before the start of heavy rains.

7.1.6 *Nematode Diseases*

Several plant parasitic nematodes on ginger and among them *Meloidogyne* spp., *Radopholus similis* and *Pratylenchus* spp. have been reported are the major ones of economic importance in that they cause significant damage to ginger plants.

Management

Pre-planting application of neem cake (2 t/ha) is recommended for the control of *M. incognita*. Mulching with green leaves of mahaneem (*Melia azadirachta*), karanji (*Pongamia glabra*) and mango (*Mangifera indica*) at 2.5 kg m^{-2} indicated possible nematicidal properties keeping nematode populations at low levels (Das, 1999). Hot water treatment at 45°C is also suggested for disinfecting the seed ginger rhizomes. Studies by Stirling and Nikulin (1998) showed that root-knot nematode on ginger can be controlled by alternating ginger with a green manure crop and applying at least 150 m^3 ha^{-1} $annum^{-1}$ of poultry manure, with or without additional sawdust. But, when ginger is grown in successive years, nematode populations increased to high densities despite the use of organic amendments.

7.1.7 *Viral Diseases*

Mosaic Disease

The symptoms appear as a yellow and dark green mosaic pattern on leaves. The affected plants show stunting. The virus causing mosaic disease in ginger has spherical particles with a diameter of 23 to 38 nm. It shows a positive serological reaction with antiserum to cucumber mosaic virus (CMV). The virus is known to be transmitted by sap to different plants known to be hosts of CMV.

Chlorotic fleck Disease

This is viral disease first described by Thomas (1986). The geographical distribution of the virus is uncertain, but is thought to occur in India, Malaysia, and Mauritius. The ginger chlorotic fleck virus (GCFV) has isometric particles approximately 30 nm in diameter with single-stranded RNA of MW 1.5×10^6 Daltons, and a major coat protein, mw 29×10^3daltons. The virus is mechanically transmitted by *Myzus persicae, Pentalonia nigronervosa, Rhopalosiphum maidis,* or *R. padi* (Thomas, 1986).

Big Bud

Pegg *et al.,* (1974) reported the disease in for the first time in Queensland and suspected it to be a *Phytoplasma* disease. The tomato big bud organism causes this disease in ginger. The affected plants cease to grow and leaves become bunched at the top of the stem. As the disease advances, plants turn yellow and die. The pathogen has a wide host range, and the disease is transmitted by leafhoppers. In seed production areas, affected plants are removed and destroyed carefully.

Chirke Virus

Raychaudhuri and Ganguli (1965) reported Chirke virus on ginger, which is a known disease in large cardamom in India.

Control of viral diseases

Avoid using rhizome from diseased plants for seed purpose

7.1.8 Post-harvest diseases

As ginger undergoes three months of dormancy in storage during Feb-April, it is important to protect it from various storage losses due to microorganisms and insect pests apart from abiotic stress like heat buildup. Under storage, different fungi and bacteria have been found associated with ginger rhizomes that result in rotting and decaying of the rhizomes. In storage, the fungi such as *Fusarium oxysporum, Pythium deliense, Pythium myriotylum, Geotrichum candidum, Aspergillus flavus, Cladosporium lennissimum, Gliocladium roseum, Graphium album, Mucor racemosus, Stachybotrys sansevieriae, Thanatephorus cucumeris,* and *Verticillium chlamydosporium* are known to affect the ginger rhizomes. Bacteria such as *Erwinia caratovora* and *Enterobacter cloacae* are known to affect ginger during storage. The fungus *A. flavus* in association with ginger rhizomes was implicated in the production of carcinogenic aflatoxin.

Control Measures (Can be used for seed rhizome selection and storage)

1. Select healthy plots (which did not show any disease during the growing season) when foliage is completely dried out: this is usually in January.
2. Do not leave the seed in the soil beyond January, because of the winter dew.
3. Carefully harvest, removing all those that have rots, insect damage or are blemished in other ways.
4. For seed purposes, chose only rhizomes with roots.

5. Take out rotten rhizomes, burn or bury in a pit far away from ginger fields and streams.

6. As temperature rise in the rhizomes during storage is very important for predisposing the rhizomes for post harvest losses, give proper ventilation and keep rhizomes at temperature of 24-28°C until they are planted. Choose dry shed or room, in a field clamp, covered with grass, and/or other dry vegetation, or above ground protected from rain. Use sawdust or leaf from *Glycosmix* for proper aeration for the rhizomes.

7. Pile carefully and inspect regularly (every fortnight) for any sign of rot.

8. Rhizomes can be pre treated with biological control agents such as *Trichoderma* or *Pseudomonas fluorescens* for avoiding the losses due to post harvest diseases.

7.1.9 Minor diseases

Dry Root by *Macrophomina phaseolina* (Tassi) Goid, basal sheath rot by *Aphelenchus* (nematode) and a *Fusarium* sp, Basal rot by *Sclerotium rolfsii* (*Corticium rolfsii*), violet rot pathogen, *Helicobasidium mompa* Tanaka, Black Root by *Rosellinia zingiberis* Stevens, leaf spot by *Leptosphaeria zingiberis, Coniothyrium zingiberis* and *Cercoseptoria* sp, *Curvularia lunata (Cochliobolus lunata), Vemicularia zingibereae, Pyricularia zingiberis, Colletotrichum zingiberis, Septoria zingiberis, Helminthosporium* sp was found associated with ginger leaves or rhizomes. *Rhizoctonia (Corticium) solani* causing pseudostems rot and *Rhizoctonia bataticola (Macrophomina)* causing leaf blight are other diseases affects ginger at limited scale. There is no need to have separate control strategies for these diseases as they appear very scantly.

7.2 Pests

7.2.1 Shoot borer

The shoot borer *(Conogethes punctiferalis)* is the most serious insect pest of ginger. The larvae bore into shoots and feed on the internal tissues resulting in yellowing and drying of infested shoots. The presence of bore-holes on the shoots through which frass is extruded and the withered central shoot is a characteristic symptom of the pest infestation.

The adult is a medium-sized moth with a wingspan of about 20 mm; the wings and body are orange-yellow with minute black spots. The eggs are laid on the topmost unopened leaf and the newly hatched larva feeds on the unopened leaf and later they bore into the shoots. Fully-grown larvae are light brown with sparse hairs and measure 15-25 mm in length. The pest is observed in the field throughout the crop season and its population is higher during September-October. The shoot borer is highly polyphagous and has been recorded on more than 30 host plants including several economically important plants in India.

Fig. 8: Shoot borer infestation of ginger (*Source:* Director, IISR, Calicut)

Management

An integrated schedule including pruning of freshly infested shoots (as indicated by the extrusion of frass) at fortnightly intervals during July- August and spraying neem gold 0.6% during September-October (at monthly intervals) or Dipel (formulation of *Bacillus thuringiensis*) 0.3% during July to October is effective against the shoot borer of ginger. The spraying should be initiated as soon as the first symptom of pest infestation is noticed in the field to achieve higher levels of control. Among the various predators and parasites documented on the pest in the field, the entomophagous nematode *Hexamermis* sp. and the hymenopteran parasitoid *Apanteles taragammae* are major natural enemies.

7.2.2 Rhizome scale

The rhizome scale (*Aspidiella hartii*) infests rhizomes of ginger in the field (especially at later stages) and storage. The pest feeds on plant sap and when the rhizomes are severely infested, they become shriveled and desiccated and fail to germinate. Adult (female) scales are minute (about 1 mm in diameter), circular and light brown to grey and appear as encrustations on the rhizomes.

Management

Severely infested rhizomes are to be discarded before storage of rhizomes. Storage of rhizomes in dried leaves of *Strychnos nux-vomica* + sawdust in 1 : 1 proportion helps in keeping the seed rhizomes free of scale infestation. Among the parasites *Physcus comperei* brought down the population of the rhizome scale by 80% and a predatory beetle *Cocobius* sp. was also observed.

Fig. 9: Ginger rhizome infested with rhizome scale (*Source:* Director, IISR, Calicut)

7.2.3 Root grubs

Root grubs (*Holotrichia* spp.) sometimes cause serious damage to ginger plants especially in Sikkim and parts of Kerala. The grubs feed on roots and newly formed rhizomes. The pest infestation leads to yellowing of leaves and in severe infestations the pseudostems may be cut at the basal region. The entire crop may be lost in severely infested plantations. The adults of *H. seticollis* commonly occurring in Sikkim are dark brown beetles measuring about 2.3 x 1.3 cm in size. The adults emerge in large numbers with the receipt of summer showers during April / May.

Management

Mechanical collection and destruction of adults during their peak periods of emergence and application of the entomophagous fungus *Metarrhizium anisopliae* mixed with fine cow dung is effective for the management of root grubs.

8.0 HARVEST AND POST HARVEST HANDLING

8.1 Harvesting

In India normally harvesting is done from January to April, varying with the locations. Fresh and dry yields of rhizomes increase steadily up to 210 days. The quality of ginger is affected by the stage of the harvest. The fibre content will be low till 210 days and increases with rhizome maturity. For vegetable purpose, the crop has to be harvested from sixth month onwards. For dry ginger, matured rhizomes are to be harvested after eight months. Studies at Solan (H.P.) indicated that maximum fresh yield and dry recovery with highest oleoresin are obtained at 225-240 days after sowing. However, harvesting at 210 DAP gives minimum fibre with maximum essential oil. Irrigation is stopped one month before harvest, allowing the tops to dry, leaves and stem are cut to the ground and the rhizomes are dug out by hoeing or plowing.

In the plains where the average temperature is between 30-35°C the crop is ready for harvest in about 8 months after planting when the turn yellow, and start drying up gradually. The clumps are lifted carefully with a spade or digging fork and the rhizomes are separated form the dried up leaves, roots and adhering soil. In Sikkim and Darjeeling extraction of mother rhizome after 2-3 months of sowing is a local practice. By this practice, farmers get back their investment on seed rhizome even if there is a severe outbreak of rhizome rot disease. The mother rhizome has equal market value as that of freshly harvested ginger because of the large size of rhizomes (100-500 g) planted. Late harvest is also practiced, as the crop dose not deteriorates by leaving it for some more months underground.

8.1.1 Yield

An initial decline in yields after discarding synthetic inputs and converting in to organic production is usually experienced. The degree of yield loss varies, and depends on inherent biological attributes of the farm, the extent to which synthetic inputs were used under previous management, and the state of natural resources. Organic farming experiments on ginger at IISR, Calicut also revealed that organic management produced yield (14.3 t ha^{-1}) which was 22.8% and 25% lower compared to chemical and integrated farming in the first year (IISR, 2005). Similar trials in AICRP on spices at various locations showed that during initial years, 15-20% reduction in yield under organic system as compared to integrated management.

As per regulatory standards, a conversion period of at least two years under certified organic is necessary for marketing the produce as organically grown. The accredited inspection and certification agency may decide in certain cases (such as idle use for two years or more/ fresh plantation) to extend or reduce the conversion period in the light of previous status of the land but the period will be equal or exceed twelve months. But the certification programme may allow plant products to be sold as "produce of organic agriculture in process of conversion" or a similar description during the conversion period of the farm.

8.2 Processing

Ginger is harvested at full maturity when the pseudo stem turns yellow. The maturity varies between 210-240 days. Traditionally, ginger has been prepared in three ways:

1. *Coated ginger* : dried with the skin whole or partly scraped
2. *Uncoated ginger* : skin completely peeled and dried to provide a light-colored smooth-looking product.
3. *Bleached ginger* : coated with lime, with or without skin and dried.

Except for ginger from Jamaica (and formerly from Cochin), the rest of the ginger produced is largely dried with the skin. Soon after harvest, the rhizomes are trimmed to remove roots and are washed to remove the adhering earth; they are then spread on clean cemented yards to dry. The layers are turned occasionally and heaped into lots every evening. The drying is continued for 7 to 10 days, depending on weather conditions. This adversely affects the appearance of the product, which is generally used for extraction. The care taken during cleaning and prolonged sun drying will be reflected in the final product. The variety and maturity at harvest determine the bold or shriveled appearance. Some fleshy varieties, such as China and Rio de Janeiro, are not suitable raw materials as they take longer (12 to 14 days) to dry and yield a shriveled product (Kumar *et al.* 2006). Jamaica traditionally produces peeled, dried whole ginger, which is valued in the retail market for its bold, clean, light buff appearance with a mild aroma.

8.2.1 Peeling

Hand-peeling is a skilled and time-consuming operation, especially peeling the skin between the fingers (branches) of the rhizome. As the essential oil and resin-bearing cells are reported to be chiefly located under the epidermal tissues, excessive scraping substantially depreciates the aroma quality. The skilled labor required for the delicate operation is reported to be becoming scarce and costly, and production is, therefore, decreasing to the point of becoming uneconomical.

To aid peeling, early attempts were made to use a short lye-dip (Strong alkali), but this has generally been unsatisfactory, as the treatment interfered with the quality of flavor. Recently, abrasive peeling has been attempted in pilot experiments with equipment normally used for potato peeling. Abrasive peeling for 60 sec gave a good grade of dried ginger with satisfactory drying characteristics and essential oil content compared to the laboriously hand-peeled sample. However, this product was not as clean as the hand-peeled product because the area between the fingers cannot be reached in mechanical peeling. Further improvements on such mechanical devices are worth development as they have edge in producing a high-quality clean product.

8.2.2 Drying

After harvest the rhizomes are cleaned to remove the soil, soaked in water for 3-4 hours, outer skin is peeled and put for sun drying on a clean surface. It takes about 14 days for complete drying of rhizome to a moisture level of 10%. Black polyethylene sheet or cement floor can be used for drying. Dry recovery of ginger is $1/5^{th}$ of fresh rhizomes. Dry ginger is used for making ginger powder, which serves as base for various end products. Crude fibre content of dry ginger varies between 3-8%. After harvest matured ginger can be held without deterioration for some weeks at ambient temperature. Early-harvest rhizomes are fleshy and are used for dehydration as whole dry ginger. With later harvests, the skin hardens, becoming impervious to moisture transfer from inside to outside for drying. In Australia and the Solomon Islands, where sun drying is not possible due to light showers at the harvest period and high humidity, mechanical drying has been studied to find optimal design and conditions. The cost compared to sun drying would be high, but a consistently high-quality product can be expected which is valuable to flavor houses.

The lipid content of ginger varies greatly among varieties of ginger, from 5.8 to 11.0%, but is generally around 7%. The lipids of ginger are high in unsaturated acids and could create problems of rancidity and off-aroma in powdered ginger and extracts on storage.

Only a few early proximate analyses of world varieties of ginger are available. The fibre content is an index of maturity, and variety varies from 2.3 to 4.7%, being low in the scraped and limed varieties and higher in the whole African ginger. Low fibre content is important for making syrup or crystallized ginger; for this purpose, ginger is harvested at early maturity. Crude protein content is around 8% and starch, the dominant constituent and an index of maturity, is around 55%. Ash content, an indication of good processing and handling while preparing for market, is around 3.5%. The results on organic farming experiments conducted at IISR, Calicut, India indicated that certain quality parameters registered an increase in organically managed plots compared to integrated and chemically managed ginger. In ginger (Table 8), parameters like oil content, and oleoresin did not vary significantly among the treatments. However, higher starch content and reduced fibre content were observed in organic system compared to inorganic and integrated system.

Table 8. Effect of different management systems on quality of ginger

Management systems	Oil	Oleoresin	Starch	Fibre
			%	
Organic	1.20	3.96	70.07	1.69
Inorganic	1.20	3.15	62.21	1.90
Integrated	1.25	3.36	55.82	1.90
CD at 5%	NS	0.23	9.89	0.077

Gas chromatography in conjunction with mass spectrometry analysis of organically grown samples of fresh Chinese white and Japanese yellow varieties of ginger resulted in the detection of 20 hitherto unknown natural products and 31 compounds previously reported as ginger constituents (Jolad *et al.* 2004). These include paradols, dihydroparadols, gingerols, acetyl derivatives of gingerols, shogaols, 3-dihydroshogaols, gingerdiols, mono- and diacetyl derivatives of gingerdiols, 1-dehydrogingerdiones, diarylheptanoids, and methyl ether derivatives of some of these compounds.

8.2.3 Storage and Packing

Dry ginger, usually packed in gunny bags, is highly susceptible to insect infestation during warehouse storage. It is preferable to use polythene laminated gunny bags for packing dried ginger. Dried ginger should be stored ensuring protecting it from dampness. Dunnage of wooden crates should be used to stack the packaged bags to prevent moisture ingress from the floor. Care should be taken to stack the bags 50 to 60 cm away from the walls. Insects, rodents and other animals should be effectively prevented from getting access to the premises where ginger is stored. The most important storage pests feeding on dry ginger are *Stegobium paniceum* L., commonly known as the drugstore beetle, *Lasioderma serricorne* F., the cigarette beetle, and *Araceus fasciculates* Deg. To prevent pest attack fumigation with neem leaves can be used. Disinfestations of spice and spice powders could be achieved by heat, but the risk of altering or loss of flavor is high. Heat treatment for bulk quantities is also time consuming and costly. Prolonged storage of ginger would result in deterioration of its aroma, flavour and pungency. Fully dried rhizomes can be packed in airtight containers such as high-density polyethylene or similar packing materials. Organic ginger has to be packaged in reusable, recycled, recyclable, and biodegradable packaging whenever possible and the packaging material should not also contaminate organic food. Packaging materials, and storage containers, or bins that contain a synthetic fungicide, preservative, or fumigant are prohibited. Organic produce should not be packaged in reused bags or containers that have been in contact with any substance likely to compromise the organic integrity of product or ingredient placed in those containers.

8.3 Marketing

Indian dry ginger is known in the world market as 'Cochin Ginger' (NUGC) & 'Calicut Ginger' (NUGK). Indian ginger especially Cochin ginger is considered one of the best in the world. India is exporting ginger mainly to Pakistan, Bangladesh, Saudi Arabia, UAE, Morocco, and USA. In the recent years Indian ginger is facing stiff competition from other producing countries like China, Nigeria and Thailand.

The primary products of ginger traded in the world market are fresh ginger, preserved ginger in syrup or brine and dried ginger. Fresh ginger is the form in which it is consumed domestically in the producing countries itself while the preserved and dried forms find a place in the international trade.

Commercial ginger in India is graded according to the region of production, number of fingers contained in the rhizome, size, colour and fibre content. The first grades are usually with bold and round bits of rhizomes having maximum dry matter content and low fibre content. Some of the known grades in the local market are 'Gola' (superior grade in Himachal Pradesh) and 'Gatti' (inferior grade). In the international market, known grades like Jamican ginger, African ginger, Calicut ginger, Cochin ginger and Japanese ginger are reported. Dried ginger is traditionally traded internationally in the whole or split forms and grinding is done only in the consuming countries.

REFERENCES

Aclan, F. and Quisumbing, F.C. (1976) Fertilizer requirement, mulch and light attenuation on the yield and quality of ginger. *Philippine Agriculturist*, 60: 183-191.

Aiyadurai, S.G. (1966) *A review of research on spice and cashewnut in India*. Regional Office (Spices and Cashewnut). Indian Council of Agricultural Research. Ermalum, pp. 228.

Ajithkumar, K. and Jayachandran, B. K. (2003) Influence of shade regimes on yield and quality of ginger (*Zingiber officinale* Rosc.) *J. Spices and Aromatic Crops*, 12: 29-33.

All India Co-ordinated Research Project on Spices (AICRPS) (1992) *Research Highlights*: 1982-91. National Research Centre for Spices, Calicut.

Ancy, J. and Jayachandran, B. K. (1993) Effect of shade and fertilizers on the quality of ginger (*Zingiber officinale* R.). *South Indian Hort.*, 41(4): 219-222.

Ancy, J. and Jayachandran, B. K. (1996) Nutrient requirement of ginger under shade. *Indian Cocoa Arecanut and Spices J.*, 20 (4): 115-116.

Ancy, J. and Jayachandran, B. K. (1998) Enhanced nutrient uptake and rhizome yield in ginger (*Zingiber officinale* Rosc.) under shade. In: Mathew, N. M. and C. Kuruvilla Jacob (eds). *Development in Plantation Crops Research*, (PLACROSIM XII) Allied Publishers Ltd., New Delhi, pp. 236-238.

Anonymous. 1970. *Guide to production of ginger*. Extension guide No. 7, AERALS, ABU, Nigeria.

Ayuba S A, John C and Obasi M O. 2005. Effects of organic manure on soil chemical properties and yield of ginger - research note. *Nigerian J. Soil Sci.* 15: 136-138.

Babu, P. and Jayachandran, B. K. (1997) Mulch requirement of ginger (*Zingiber officinale* Rosc.) under shade. *J. Spices and Aromatic Crops*, 6(2): 141-143.

Beena, E. G., Sreedevi, P. and Nair, R. V. (1994) Performance of ginger cultivars as influenced by shade levels (Abstract). In: XIII National Symposium on Integrated Input Management for Efficient Crop Production, Indian Society of Agronomy, New Delhi and Tamil Nadu Agricultural University, Coimbatore.

Bhardwaj, S. S., Gupta, P. K., Dohroo, N. P. and Shyam, K. R. (1988) Biological control of rhizome rot of ginger in storage. *Indian J. Plant Pathol.*, 6(1): 56-58.

Bisht, J. K., Chandra, S., Chauhan, V. S., Singh, R. D. (2000) Performance of ginger (*Zingiber officinale*) and turmeric (*Curcuma longa*) with fodder tree based silvi-horti system in hills. *Indian J. Agric. Sci.*, 70 (7): 431-433.

Butler, E.J. (1907) An account of genus *Pythium* and some Chytridiaceae. Mem. Dep. Agric. India, *Bot. Ser.*, 1 (5): 70.

Charles, J. S. and Kuriyan, K. J. (1982) Relative susceptibiltiy of ginger cultivars to the root knot nematode, *Meloidogyne incognita*. In: Nair, M. K., Premkumar, T., Ravindran, P. N. and Sarma, Y. R. (eds) *Proceedings, National Seminar on Ginger and Turmeric*, Central Plantation Crops Research Institute, Kasaragod, India, pp. 133-134.

Chengat, T. (1997) Influence of organic manures and Azospirillum on growth, yield and quality of ginger (*Zingiber officinale*). *M.Sc. (Hort.) thesis*, Kerala Agricultural University, Kerala, India.

Cho, G. H., Yoo, C. H., Choi, J. W., Park, K. H., Hari, S. S. and Kim, S. J. (1987) Research Report. Rural Development Administration, *Plant Environment Mycology and Farm Products Utilisation*, Korea Repulic, 29 (2): 30-42.

Council of Scientific and Industrial Research (CSIR) (1976) *The Wealth of India*, Raw materials Vol. II. Council of Scientific and Industrial Research, New Delhi.

Das N. (1999). Effect of organic mulching on root knot nematode population, rhizome rot incidence and yield of ginger. *Ann. Plant Protect. Sci.*, 7(1): 112-114.

Dohroo, N. P. and Sharma, S. (1997) Effect of organic amendments and intercropping on spore density of VAM fungi and yellows of ginger. *Plant Dis. Res.*, 12(1): 46-48.

Evenson, J. P., Bryant, P. J. and Asher, O. J. (1978) Germination and early growth of ginger (*Zingiber officinale* Rosc) effect of constant and fluctuation soil temperature. *Trop. Agric. (Trinidad)*, 55: 1-7.

Furutani, S. C., Villanueva, J. and Tanaka, M. J. (1985) Effect of ethephon and heat on the growth and yield of edible ginger. *Hort. Sci.*, 20: 392-393.

George, B. E., Sreedevi, P. and Vikraman Nair, R. (1998a) Effect of shade on plant characters and net assimilation rate of ginger cultivars. *J. Trop. Agric.*, 36 (1&2): 65-68.

George, B. E., Sreedevi, P. and Vikraman Nair, R. (1998b) Effect of shade on nutrient uptake and dry matter accumulation on ginger *J. Trop. Agric.*, 36 (1&2): 69-72.

Graham, J. A. (1936) Methods of ginger cultivation in Jamaica. *J. Jamaica Agric. Soc.*, 40: 231-232.

Haque, M.E., Roy, A. K. and Sikdar, B. (2004) Performance of ginger, turmeric and mukhi kachu under shade of mango orchard. *Hort. J.* 17: 101-107.

Haware, M. P. and Joshi, L. K. (1973) Basal rot of ginger (*Zingiber officinale*) caused by Sclerotium rolfsii from Madhya Pradesh. *Indian Phyto Pathol.* 26(3): 575-576.

Hayden, A. L., Brigham, L. A. and Giacomelli, G. A. (2004) Aeroponic cultivation of ginger (*Zingiber officinale*) rhizomes. *Acta Horticulturae*, 659: 397-402.

Haynes P H, Partridge I J and Sivan P. (1973) Ginger production in Fiji. *Fiji Agric. Res. J.* 35: 51-56.

Hayward, A.C. (1991) Biology and epidemiology of bacterial wilt caused by *Pseudomonas solanacearum*. *Annual Rev. Phytopathol.*, 29: 67-87.

Hegde, N. K., Sulikeri, G. S. and Rethinum, B. P. (2000) Distribution of photosynthetically active radiation (PAR) and performance of ginger under arecanut shade. In: (eds.) Ramana, K. V., Santhosh J. Eapen, Nirmal Babu, K., Krishnamoorthy, K. S. and Kumar, *A. Centannial conference on Spices and Aromatic Plants*, Indian Society of Spices, Calicut, pp. 107-112.

Indian Institute of Spices Research (IISR). (2006). *Annual Report, 2005-06*, Indian Institute of Spices Research, Calicut. pp 45-66.

Indian Institute of Spices Research (IISR) (2002). *Research Highlights 2001-02*. Indian Institute of Spices Research, Calicut, Kerala, 17 p.

Indian Institute of Spices Research (IISR). (2005) Annual Report, 2004-05, Indian Institute of Spices Research, Calicut. pp. 38-39.

Jaswal, S. C., Mishra, V. K. and Verma, K. S. (1993) Intercropping ginger and turmeric with poplar (*Populus deltoides* 'G-3' Marsh.). *Agroforestry Syst.* 22(2): 111-117.

Jha, R.C., Maurya, K.R. and Pandey, R. P. (1986) Influence of mulches on the yield of ginger in Bihar. *Indian Coco, Arecanut and Spices J.* 9(4): 87-90.

Johnson, P.T. (1978) Foliar Diagnosis, yield and quality of ginger in relation to N, P and K. M.Sc. (Agri.) thesis, Kerala Agricultural University, Kerala, India.

Jolad, S. D., Lantz, R. C., Solyom, A. M., Chen, G. J., Bates, R. B. and Timmermann, B. N. (2004) Fresh organically grown ginger (*Zingiber officinale*): composition and effects on LPS-induced PGE2 production, *Phytochem.*, 65: 1937–1954.

Kandiannan, K., Sivaraman, K. and Thankamani, C.K. (1996) Agronomy of ginger (*Zingiber officinale* Rosc.) – a review. *J. Spices and Aromatic Crops,* 5: 1-27.

Kannan, K. and Nair, K. P. V. (1965) Ginger (*Zingiber officinale* Rosc.) in Kerala. *Madras Agric. J.*, 52: 168-76.

Kerala Agricultural University (KAU) (1994). Scheduling irrigation to ginger under varying nitrogen levels. Sixteenth Zonal Research and Extension Advisory Council Meeting, Regional Agricultural Research Station, Kerala Agricultural University, Pattambi, India.

Kiuchi, F., Iwakami, S., Shibuya, M., Hanaoka, F. and Sankawa, U. (1992) Inhibition of prostaglandin and leukotriene biosynthesis by gingerols and diarylheptanoids. *Chem. Pharm. Bull.* 40: 387–391.

Korikanthimath, V. S., Rajendra Hegde, Sivaraman, K. (1995) Integrated input management in coffee based spices multistoried cropping system. *Indian Coffee,* 59(2): 3-6.

Kumar, A. and Sarma, Y. R. (2004) Characterization of Ralstonia solanacearum causing bacterial wilt of ginger, *Indian Phytopathol.* 57: 12-17.

Kumar, A., Sasikumar, B., Srinivasan, V., John Zachariah, T. and Rajeev, P. (2006) Ginger. In: (eds) Parthasarathy, V. A. and Rajeev, P. *Major spices – Production and Processing,* Indian Institute of Spices Research, Calicut, Kerala, pp. 115-151.

Kumar, B. M. Thomas, J. and Fisher, R. F. (2001) *Ailanthus triphysa* at different density and fertiliser levels in Kerala, India: tree growth, light transmittance and understorey ginger yield. *Agroforestry Syst.* 52 (2): 133-144.

Kurian, A., Valsala, P. A. and Nair, G. S. (1997) *In situ* green manure production as mulch material in ginger. *J. Trop. Agric.* 35: 50-58.

Lawrence, B. M. (1984) Major tropical spices - ginger (*Zingiber officinale* Rosc.). *Perfumer and Flavorist,* 9 (5): 1-40.

Lee, M. T., Edward, D. G and Asher, C. J. (1981) Nitrogen nutrition of ginger (*Zingiber officinale* Rosc.) II: Establishment of leaf analysis test. *Field Crops. Res.* 4(1): 69-81.

Manjunath, K (2004) Production and export of ginger in India (online). http://www.toenre.ac.In/seminar/sem1-manjuk.htm (accessed on 17th January 2007).

Mirchandani, (1971) Investigations into Methods and Practices of Farming in Various States. Indian Council of Agricultural Research, New Delhi.

Mishra, S, and Mishra, S. S. (1982) Effect of mulching and weedicides on growth and fresh rhizome yield of ginger. In: Abstracts of Papers, *Annual Conference of Indian Society of Weed Science*, Dholi, Bihar, India.

Mohanty, D. C. (1984) Germplasm evaluation and genetic improvement in ginger. *Ph. D Thesis*, Orissa Univ. Agricultural Technology, Bhubaneswar, India.

Mohanty, D. C. and Panda, B. S. (1994) Genetic resources of ginger. In: Chaddha, K. L and Rethinam, P. (eds.). *Advances in Horticulture*. Vol. 9: Plantation crops and Spices - Part 2. Malhotra Publishing House, New Delhi. pp. 151-168.

Mohanty, D. C., Naik, B. S., Dash, D. K., Sarma, Y. N. and Panda, B. S. (1990a) Genetic resource and their evaluation in ginger. *Indian J. Plant Genet. Res.* 3 (8): 16-18.

Mohanty, D. C., Naik, B. S. and Panda, B. S. (1990b) Ginger research in Orissa with reference to its varietal and cultural improvement. *Indian Cocoa, Arecanut and Spices J.* 14(2): 61-65.

Mohanty, D. C., Sarma, Y. N., Naik, B. S. and Panda, B. S. (1998) Influence of fertilizer - cum - seed rates on yield and performance of selected ginger clones in the Eastern Ghats Highland Zone of Orissa. *Indian Cocoa Arecanut and Spices J.* 12: 1-3.

Nagarajan, M. and Pillai, N.G. (1979) A note on nutrient removal by ginger and turmeric rhizomes. *Madras Agric. J.* 66: 56-59.

Nair, P. C. S. (1982) Agronomy of ginger and turmeric. In: Nair, M. K., Premkumar, T., Ravindran, P. N. and Sarma, Y. R. (eds) Proceedings, *National Seminar on Ginger and Turmeric*, Central Plantation Crops Research Institute, Kasaragod, India, pp. 63-68.

Nair, P. C. S. and Varma, A. S. (1970) Ginger in Kerala: Steps towards increased production. *Indian Farming*, 20 (3): 37-39.

National Research Centre for Spices (NRCS) (1989) *Ginger: Package of Practices*. National Research Centre for Spices, Calicut, Kerala.

National Research Centre for Spices (NRCS) (1992) *Annual Report: 1991-92*. National Research Centre for Spices, Calicut.

Newman, S. M, Bennett, K. and Wu, Y. (1997) Performance of maize, beans and ginger as intercrops in Paulownia plantations in China. *Agroforestry Syst.* 39(1): 23-30.

Nizam, S. A. and Jayachandran, B. K. (2001) Performance of mini seed rhizome as planting material in ginger (*Zingiber officinale* Rosc.) Proc. 13th Kerala Science Congress, STEC, Trivandrum, Kerala, India. pp. 457-458.

Nybe, E. V. and Mini Raj, N. (2005) Ginger production in India and other South Asian countries. *Ginger: the genus Zingiber* (eds.) Ravindran, P. N. and Nirmal Babu, K. Medicinal and aromatic plants - industrial profiles, CRC Press, Washington, 41. pp. 211-240.

Nybe, E. V. and Nair, P. C. S. (1979) Studies on the morphology of ginger types. *Indian Cocoa Arecanut and Spices J.* 3: 7-13.

Nybe, E. V. and Nair, P. C. S. (1981) Varietal screening in ginger. *Agric. Res. J. Kerala,* 19 : 5-9.

Nybe, E. V., Nair, P. C. S. and Mohankumar, N. (1980) Assessment of yield and quality components in ginger. In: Nair, M. K., Premkumar, T., Ravindran, P. N. and Sarma, Y. R. (eds.) *Proc. National Seminar on Ginger and Turmeric,* Central Plantation Crops Research Institute, Kasaragod, India, pp. 24-29.

Okwuowulu P A. (1992). Effects of depth of seed placement and age at harvest on the tuberous stem yield and primary losses in edible ginger (*Zingiber officinale* Rosc.) in south eastern Nigeria. *Trop. Landwirtsch Vet. Med.* 2: 163-168.

Okwuowulu, P. A. (1988a). Effect of seed ginger weight on flowering and rhizome yield of field grown edible ginger (*Zingiber officinale* Rosc.) in Nigeria. *Trop. Sci.* 28: 171-176.

Okwuowulu, P. A. (1988b) Parent sett re-cycling in edible ginger (*Zingiber officinale* Rosc) production in southeastern Nigeria. *Trop. Sci.* 28: 177-184.

Okwuowulu, P. A. (2005) Ginger in Afrtica and the Pacific Ocean islands. *Ginger: the genus Zingiber* (eds.) Ravindran, P. N. and Nirmal Babu, K. Medicinal and aromatic plants - industrial profiles, CRC Press, Washington, 41. pp. 279-303.

Okwuowulu, P. A., Ene, L. S. O., Odurukwe, S. O. and Ojinaka, T. I. (1989) Effect of time of planting and age at harvest on yield of stem tuber and shoots in ginger (*Zingiber officinale* Rosc.) in the rain forest zone of southeastern Nigeria. *Exptl. Agric.* 26: 209-212.

Panigrahi, U. C. and Patro, G. K. (1985) Ginger cultivation in Orissa. *Indian Farming,* 35 (5): 3-4.

Parthasarathy, V. A., Srinivasan, V. and Dinesh, R. (2006) Organic Spices- present status and scope. *National Seminar on Appropriate Technologies for Sustainable Horticulture Souvenir,* Annamalai University, Tamil Nadu. pp. 34-42.

Paulose, T. T. (1973) Ginger cultivation in India. *In: Proceedings, Conference on Spices.* Tropical Products Institute, London, pp. 117-121.

Pegg, K. G., Melda, L., Meffett, M. L. and Colbran, R. C. (1974) Diseases of ginger in Queensland. *Queensland Agric. J.* 5: 611-614.

Peter, K. V and Nybe, E. V. (2002) Dominating Global Markets. *The Hindu Survey of Indian agriculture*: 87-97.

Phogat, K. P. S. and Pandey, D. (1988) Influence of the time of planting on the growth and yield of ginger var. Rio-de-Janeriro. *Progressive Hort.* 20: 46-48.

Premaja, P. and Manojkumar, K. (2007) *Arecanut and Spices database.* Directorate of Arecanut and Spices Development, Calicut, Kerala, India. 110p.

Prentice, A. 1959. Ginger in Jamaica. *World Crops,* 11: 25-26.

Ram, D., Mathur, K., Lodha, B. C. and Webster, J. (2000) Evaluation of resident biocontrol agents as seed treatments against ginger rhizome rot. *Indian Phytopathol.* 53(4): 450-454.

Ramakrishnan, T. S. (1942) A leaf spot disease of Zingiber officinale caused by *Phyllosticta zingiberi n.* sp. *Proc. Indian Acad. Sci. Sect. B.,* 20 (4), 167-171.

Ramana, K.V., Shiva, K.N. and Johny A. K. (2003) Production of Quality Planting Materials of Ginger. *In: National Consultative Meeting for Improvement in Productivity and Utilization of Ginger,* Aizawl, Mizoram, pp. 37-45.

Randhawa, K. S. and Nandpuri, K. S. (1970) Ginger (*Zingiber officinale* Rosc.) in India - a review. *Punjab Hort. J.* 10: 111-112.

Randhawa, K. S., Nandapuri, K. S. and Bajwa, M. S. (1972) The growth and yield of ginger (*Zingiber officinale* Rosc.) as influenced by different dates of sowing. *J. Res. Punjab Agric. Univ.* 9: 32-34.

Rattan, R. S., Korla, B. N. and Dohroo, N. P. (1988) Performance of ginger varieties under Solan condition of Himachal Pradesh. In : Satyanarayanan G, Reddy M S, Rao M R, Azam K M and Naidu R (eds.) *Proceedings, National Seminar on Chillies, Ginger and Turmeric,* Hyderabad. Spices Board, Cochin and Andra Pradesh Agricultural University, Hyderabad, pp. 71-73.

Ravindarn, P. N. Nirmalbabu, K. and Shiva, K. N. (2005) Botany and crop improvement of ginger. *In: Ginger: the genus Zingiber* (eds.) Ravindran, P. N. and Nirmal Babu, K. Medicinal and aromatic plants – industrial profiles, CRC Press, Washington, 41. pp. 15-85.

Ravindran, P. N., Sasikumar, B., Johnson K. George, Ratnambal, M. J., Nirmal Babu, K., Zachariah, J. T. and Nair, R. R. (1994) Genetic Resources of ginger (*Zingiber officinale* Rosc.) and its conservation. *Plant Genetic Res. Newsletter,* 98: 1-4.

Raychaudhuri, S. P. and Ganguly, B. (1965) Further studies on Chirke disease of large cardamon (*Amomum subulatum* Roxb.) *Indian Phytopathol.* 18 (4): 67-69.

Ridley, H. N. (1912) Ginger. *In : Spices.* Macmillan Co., London, pp. 389-421.

Roy, A. R. and Wamanan. P. P. (1989) Effect of seed-rhizome size on growth and yield of ginger. *Adv. Plant Sci.,* 2: 62-66.

Roy, A., Chatterjee, R., Hassan, A. and Mitra, S. K. (1992) Effect of Zn, Fe, and B on growth, yield and nutrient content in leaf of ginger. *Indian Cocoa Arecanut and Spices J.* 15: 99-101.

Sadanandan, A. K. and Hamza, S. (1998) Effect of organic farming on nutrient uptake, yield and quality of ginger (*Zingiber officinale* R.). In: (eds.) Sadanandan, A. K., Krishnamurthy, K. S., Kandiannan, K. and Korikanthimath, V. S,), *Water and Nutrient Management for sustainable Production and Quality of Spices,* Indian Society of Spices Research, Calicut, India, pp. 89-94.

Sahu, S. K. and Mitra, G. N. (1992) Influence of physicochemical properties of soil on yield of ginger and turmeric. *Fert. News,* 37 (10): 59-63.

Sasikumar, B., Ravindran, P. N. and George, K. J. (1994) Breeding ginger and turmeric. *Indian Cocoa, Arecanut and Spices J.* 18 (1): 10-12.

Selvan, M. T. and Manojkumar, K. (2002) Production and export of ginger. *Indian J. Arecanut, Spices and Medicinal Plants,* 4 (3): 109-116.

Senguptha, D. K., Maity, T. K., Som, M. G. and Bose, T. K. (1986) Effect of different rhizome size on the growth and yield of ginger (*Zingiber officinale* Rosc.). *Indian Agric.* 30: 201-204.

Sharma, G. C. and Bajaj, B. K. (1998) Effect of intercropping bell pepper with ginger on plant parasitic nematode populations and crop yields. *Ann. Appl. Bio.* 133(2): 199-205.

Sharma, S., Dohroo, N. P. and Korla, B. N. (1997) Effect of VAM inoculation and other field practices on growth parameters of ginger. *J. Hill Res.* 10 (2): 74-76.

Singh, A., Singh, B., Singh, A. and Singh, P. (2003) Response of ginger (*Zingiber officinale*) to methods of planting and levels of phosphorus in a rehabilitated forest developed in sodic land. *J. Spices and Aromatic Crops,* 12: 63-66.

Singh, B. S. (1982) Cultivation of ginger and turmeric in Meghalaya. In: Nair, M. K., Premkumar, T., Ravindran, P. N. and Sarma, Y. R. (eds) *Proceedings, National Seminar on Ginger and Turmeric,* Central Plantation Crops Research Institute, Kasaragod, India, pp. 224-226.

Sreekala, G. S. and Jayachandran, B. K. (2002) Influence of shade regimes on the physiological parameters of ginger. *J. Spices Aromatic Crops,* 11: 30-34.

Srinivasan, V., Hamza, S., Krishnamurthy, K. S and Thankamani, C. K. (2004) Threshold level of soil zinc for optimum production of ginger (*Zingiber officinale* Rosc.). *J. Spices and Aromatic crops,* 13 (1): 55-57.

Srinivasan, V., Sadanandan, A. K. and Hamza, S. (2000) Efficiency of rock phosphate sources on ginger and turmeric in an *Ustic Humitropept. J. Indian Soc. Soil Sci.* 48 (3): 532-536.

Srivastava, K.C., Mustafa, T. (1992) Ginger (*Zingiber officinale*) in rheumatism and musculoskeletal disorders. *Med. Hypotheses*, 39: 342–348.

Stirling, G. R. and Nikulin, A. (1998) Crop rotation, organic amendments and nematicides for control of root-knot nematodes (*Meloidogyne incognita*) on ginger. *Australasian Plant Pathol.* 27(4): 234 - 243.

Sundram, N. V. (1954) Thread blight of ginger. *Indian Phytopathol.* 6(2): 80-85.

Thangaraj, T., Muthuswamy, S., Muthukrishnan, C. R., Khader, J. B. M. M. A. (1983) Performance of ginger (*Zingiber officinale* Rosc) varieties at Coimbatore. *South Indian Hort.* 31 : 45-46.

Thomas, J. E. (1986) Purification and properties of ginger chlorotic fleck virus. *Ann. Appl. Biol.* 108(1): 43-50.

Thomas, J., Kumar, B. M., Wahid, P. A., Kamalam, N. V. and Fisher, R. F. (1998) Root competition for phosphorus between ginger and *Ailanthus triphysa* in Kerala, India. *Agroforestry syst.* 41 (3): 293-305.

Thomas, K. M. and Kannan, (1969) Comparative yield performance of difference types of ginger. *Madras Agric. J.* 52 : 512-515.

Timpo, G. M. and Oduro, T. A. (1977) The effect of storage on growth and yield of ginger (*Zingiber officinale* Rosc). *Acta Horticulturae*, 53: 337-340.

Tjendraputra, E., Tran, V.H., Liu-Brennan, D., Roufogalis, B.D. and Duke, C.C. (2001) Effect of ginger constituents and synthetic analogues on cyclooxygenase-2 enzyme in intact cells. *Bioorganic Chem.*, 29: 156–163.

Utpala, P., Johny, A. K., Parthasarathy, V. A., Jayarajan, K. and Madan, M. S. 2006. The diversity of ginger cultivation in India – a GIS study. *J. Spices Aromatic Crops*, 15: 94-98.

Vaidya, V. G., Sahasrabudhhe, K. R. and Khuspe, V. S. (1972) *Crop Production and Field Experimentation*. Continental Prakashan, Poona, India.

Valsala, P. A., Amma, S. P. and Sudhadevi, P. K. (1990) Feasibility of growing daincha in the interspaces of ginger beds. *Indian Cocoa, Arecanut and Spices J.* 14(2): 65-66.

Vilasini, T. N. (1996) Effectiveness of soil solarization for the control of soft rot disease in ginger. *M.Sc. (Ag) thesis*, Kerala Agricultural University, Kerala, India.

Wang, G., Xu-Kun and Zhang, Y. (2003). Effect of seed ginger size on growth and yield of ginger (*Zingiber officinale*). *China Veg.* 1: 13-15.

Weiss, E. A. (1997) *Essential oil crops*. CAB international, Wallingford, U.K., pp. 539-567.

Whiley, A. W. (1974) Ginger growing in Queensland. *Queensland Agric. J.* 91: 100.

Whiley, A. W. (1981) Effect of plant density on time of first harvest, maturity, knob size and yield in two cultivars of ginger (*Zingiber officinale* Rosc) grown in southeast Queensland. *Trop. Agric. (Trinidad).* 8: 245-251.

Whiley, A.W. (1990) Effect of 'seed piece' size and planting density on harvested 'knob' size and yield in two cultivars of ginger (*Zingiber officinale* Rosc.) grown in Southeast Queensland, *Acta Horticulturae*, 275: 167-172.

Wilson, H. and Ovid, A. (1993) Growth and yield responses of ginger (*Zingiber officinale* Rosc.) as affected by shade and fertilizer applications. *J. Plant Nutr.* 16: 1539-1546.

Xianchang, Y., Zhifend, C. and Xiheng, A. (1996) Effect of plastic mulch on growth and yield of ginger production, *China Veg.* 6: 15-16.

Xizhen, A., Jinfeng, S. and Xia, X. (2005) Ginger production in Southeast Asia. *Ginger: the genus Zingiber* (eds.) Ravindran, P. N. and Nirmal Babu, K. Medicinal and aromatic plants - industrial profiles, CRC Press, Washington, 41. pp. 241-278.

Xizhen, A., Zhenxian, Z. and Shaohui, W. (1998) Effect of temperature on photosynthetic characteristics of ginger leaves. *China Veg.* 3: 1-3.

Xu, K. Guo, Y. Y. Wang, X. F. (2004) Transportation and distribution of carbon and nitrogen nutrition in ginger, *Acta Horticulturae*, 629: 347-353.

Zhifeng, C. Xizhen, A. and Lianying, Z. (2000) Yield effect on ginger in plastic house. *China Veg.* 3: 14-16.

11

Turmeric

K. Kandiannan, B. Chempakam, K.S. Krishnamurthy

1.0 INTRODUCTION

Turmeric (*Curcuma domestica* Val. (syn. *C. longa* Koenig *non* L.) is an important spice, a native of India and South East Asia. Chikarmane *et al.* (2003) by analyzing historical patterns of movement of South Asian and Southeast Asian people known to have used turmeric suggested that turmeric was introduced into Southeast Asia via Indian colonization in the early Christian centuries. An increase in colonization events from South India during this time was because of discoveries in navigational science and desires to expand civilizations, rather than military conquests. The cultural expansion from South India supports the introduction of turmeric cultivation to Southeast Asia via human means. These human migratory patterns are compared to a phylogenetic analysis of DNA sequence data within *Curcuma,* and suggested that key locations such as the Isthmus of Kra were important not only to human movement but to species evolution in *Curcuma* (Chikarmane *et al.,* 2003).

In many languages, the name of turmeric just mean "yellow root"; in Tamil, Manjal is synonymous to the Tamil word for 'yellow', so are Haldi in Hindi, Dutch geelwortel, German Gelbwurz. English turmeric is derived from the (now obsolete) French terre-mérite (Latin terra merita, "meritorious earth"), probably because ground turmeric resembles mineral pigments (ocher). In most contemporary European languages, the name of turmeric is derived more or less directly from Latin curcuma (Venkateswaran, 2003). In many languages turmeric is simply named as "yellow root". The other names of turmeric in common parlance are Indian Saffron, Poor Man's Saffron, Yellow Ginger. Names of turmeric in other languages are curcuma, saffron des Indes (*French*), wong geung fun(*Chinese*), haldee, haldi, huldee, huldie (*Indian*), fa nwin *(Burmese)*, kunjit, kunyit *(Indonesian)*, Gelbwurz *(German)*, curcuma *(Spanish, Italian)*, kharkoum (*Arabic*), kunjit (*Malay*), kaha *(Sinhalese)*, kamin *(Thai)*. Indian vernacular names for turmeric are Halud, Pitras, Haldi (Assamese), Hazlud, Haldi (Bengali), Haldi, Haladi (Hindi & Urdu), Haldhar, Haldi (Gujarat, Punjabi), Arishina (Kannada), Lidar (Kashmiri), Manjal (Malayalam, Tamil), Halad (Marathi), Haldi (Oriya), Haldi, Haldhar, Haridra, Harita Rajani, Gauri, Nisha, Varavat Ela (Sanskrit), Pasupu (Telugu). WHO monographs on medicinal plants, lists 70 selected vernacular names in different languages for turmeric (WHO, 1999), showing the popularity of this plant among people of different origin.

1.1 Area and production

Turmeric is grown in countries like India, Pakistan, Myanmar, Chile, Peru, Haiti, El Salvador, Japan, China, Sri Lanka, Bangladesh, Indonesia, Malaysia, Thailand, Vietnam, Taiwan, Bhutan, Nepal, Jamaica, West Indies, Java, Nigeria, Fiji, Papua New Guinea, the Solomon Islands, and Australia. India is a major producer, consumer and exporter. However, there is no data available on global area and production of this crop.

In India (a major player in turmeric trade), turmeric occupies fifth place next to chillies, cumin, coriander, and black pepper , in terms of area among spices and in production it ranks second, next to chillies. It occupies 6.3% of the spice area and shares 18.8% of spice production (Table 1). In India, it is cultivated in more than 20 states with an area of 1, 61, 200 ha with a production of 7, 16, 800 tonnes during 2004-05(DASD, 2007). Andhra Pradesh, Tamil Nadu, Orissa, Karnataka, West Bengal, Gujarat, Meghalaya, Maharashtra, Assam are some of the important states cultivating turmeric. However, Andhra Pradesh alone occupies 38.0% of area and 58.5% of production. The national productivity of the crop is 4446 kg ha^{-1} and wide variations in productivity is seen among the states. The area, production and yield increased over the years. The rate of increase is low in area compared to yield and production increase.

1.2 Prospect of organic

In terms of quantity and value of export, among spices it ranks second, next to chillies during 2004-05, it even pushed the pepper to third place, it imply that there is a lot of scope to export this spice. In terms of quantity and

Table 1. Percentage of turmeric in relation to total spices in India

Item	Spices	Turmeric	% Share
Area ('000 ha)	2571.77	161.2	6.3
Production ('000 tonnes)	3817.9	716.8	18.8
Export ('000 tonnes)	335.49	46.5	13.9
Value (Rs. in lakhs)	220000	15300	7.0

value of export of spices, it shares 13.9% and 7.0%, respectively. India exports only 6.48% of it's production of turmeric to more than 50 countries. The turmeric export and foreign exchange of India enhanced considerably over time. The importing capacities of countries vary with their need. Some countries are re-exporting after value addition or processing. Important importing countries are Bangladesh, U.A.E., U.S.A., Japan, U.K., Malaysia, The Netherlands, South Africa, Sri Lanka, Germany, Saudi Arabia, Egypt and Morocco. During 2004-05, India exported 43, 000 tonnes valued Rs. 15, 650.00 Lakhs. It is exported mainly in two forms, dry produce (63%) and powder form (37%). However, International Trade Center (ITC) during 2004 estimated that ground turmeric constitutes around 80 percent of the total world trade in turmeric (Vinning *et al.*, 2004). Trade in the ground form is expected to increase for three reasons:

- Economies of transportation favour movement in the ground form.
- Grinding is cheaper in the producing countries.
- Grinding is essentially dusty and unpleasant work and thus packers in importing countries are prepared to pay a small margin to have the product ground off-shore.

Intra-Asian trade, especially between near neighbors, tends to be in the fresh form. Surprisingly, India imports turmeric. It imported 83, 120, 1313, 3005, 1620 and 4025 tonnes during 2000-01, 2001-02, 2002-03, 2003-04, 2004-05, 2005-06, respectively. The value for the same is Rs. 18.10, Rs.11.74, Rs. 250.37, Rs. 841.29, Rs. 704.53 and Rs. 1677.00 Lakhs (DASD, 2007). It might be for value addition and re-export.

In turmeric trade Alleppey, Madras, Rajpuri/Rajpore, West Indian and Chinese are the popular names. In India, Alleppey from Kerala, Madras Finger and Erode from Tamil Nadu, Rajapore and Sangli from Maharashtra and Nizamabad Bulb from Andhra Pradesh are well-accepted trade forms. The share of organic turmeric is only 11 per cent compared to conventional turmeric. There is a great demand for organic turmeric in USA, Germany, France and Japan. At present there is no separate data for area of crop under organic / production and organic export. There is a growing demand for organic spices in the market (ITC, 2004). Growing demand for natural colours in industry, fast food chains, pharmaceuticals offer a potential scope for organic production of turmeric.

Turmeric requires heavy manure. Like any other crop, before 19th century turmeric was organically produced, still in many places the crop is produced organically in small and marginal holdings under high rainfall areas as a rainfed crop, whereas, in irrigated production, invariably fertilized with external chemical fertilizers and pesticides. The easy availability and handling of manufactured fertilizers could be the reason for this. Such intensive crop management practice in turmeric producing centers with a limited land and water resources would pollute the soil and water. In order to switch over to organic production from chemical treated production system, a minimum prescribed conversion period is two years (Spices Board, 2004). During this conversion time organic system of production has to be adapted and the produce not treated as organic. However, the seed material taken from this harvest can be used for starting of organic production. A buffer zone of 25 to 50 feet is essential, if the neighboring farms are non-organic. The produce from this buffer zone is not considered as organic.

1.3 Composition and uses

Turmeric contains protein (6.3%), fat (5.1%), minerals (3.5%), carbohydrates (69.4%) of moisture (13.1%). The essential oil (5.8%) obtained by steam distillation of rhizomes has α-phellandrene (1%), sabinene (0.6%), cineol (1%), borneol (0.5%), zingiberene (25%) and sesquiterpines (53%)(Kapoor, 1990). Curcumin (diferuloylmethane) (3–4%) is responsible for the yellow colour, and is comprised of curcumin I (94%), curcumin II (6%) and curcumin III (0.3%) (Ruby *et al.*, 1995).

The use of turmeric dates back nearly 4000 years, to the Vedic culture in India where it was used as a culinary spice and had some religious significance (Venkateswaran, 2003; Vinning *et al.*, 2004; Hossain *et al.*, 2005a; Rao *et al.*, 2006; Stix, 2007; Ravindran *et al.*, 2007). A Barcelona photographer Carlos (2006), admits that *Curcuma* to be a wonderful subject for a colorful picture.

According to Ridley (1912), turmeric is not seem to have been known as early as ginger, and never ranked in importance as high as ginger, and chiefly valued for color rather than spice. Whereas, Purseglove *et al.* (1981), stated that turmeric is an important spice among the rice-eating peoples of India, South East Asia and Indonesia and is indispensable in the preparation of curry powder. Curry powder usually contains about 24% of turmeric powder. In addition to its use as a spice, it has other uses that are prominent in the life of the people of southern Asia, and connected with birth, marriage, death and in agriculture. Turmeric is an important yellow dye in southern Asia; it also used in Europe before the discovery of aniline dyes. Cultivars with harder, darker-colored, less aromatic rhizomes are preferred for dyes. The rhizomes are boiled with little water and made into a paste. With lime or other alkalis, it produces a red or reddish-brown color. It is used for dying cotton, silk or wool without a mordant, but the color is fugitive (Purseglove *et al.*, 1981). Turmeric powder in water is used as cosmetics in India and elsewhere in South East Asia. Rice, colored yellow with turmeric, is used on ceremonial occasions. Turmeric is used as a protective charm and in primitive medicine, both externally and internally, in southern Asia. It is used as coloring matter in pharmacy, confectionary, and food industries. Yellow turmeric paper is used as a test for alkalinity, which turns it brown.

The major quantity of turmeric is utilized as a condiment and a small quantity is used in medicines and cosmetics in India. In many Asian countries, turmeric is used as a food adjunct in vegetables, meat and fish preparations. It is used to flavour and colour butter, cheese, margarine, pickles, mustard, liquor, fruit drinks, cakes, table jellies, and other food stuff. It also acts as an appetizer and aids in digestion. A pinch of turmeric powder is often added to most of our savouries to impart simultaneously an agreeable flavour, colour and to improve the keeping quality. Turmeric oil and oleoresin are also used to impart the flavour in food and perfume industries.

According to Ishita *et al.* (2004), it has been used in traditional medicine as a household remedy for various diseases, including biliary disorders, anorexia, cough, diabetic wounds, hepatic disorders, rheumatism and sinusitis. For the last few decades, extensive work has been done to establish the biological activities and pharmacological actions of turmeric and its extracts. Curcumin (diferuloylmethane), the main yellow bioactive component of turmeric has been shown to have a wide spectrum of biological actions. These include its anti-inflammatory, antioxidant, anti-carcinogenic, anti-mutagenic, anticoagulant, anti-fertility, anti-diabetic, antibacterial, antifungal, anti-protozoal, antiviral, anti-fibrotic, anti-venom, anti-ulcer, hypotensive and hypocholesteremic activities (Table 2). Its anticancer effect is mainly mediated through induction of apoptosis. Its anti-inflammatory, anticancer and antioxidant roles may be clinically exploited to control rheumatism, carcinogenesis and oxidative stress-related pathogenesis. Clinically, curcumin has already been used to reduce post-operative inflammation. Safety evaluation studies indicate that both turmeric and curcumin are well tolerated at a very high dose without any toxic effects. Thus, both turmeric and curcumin have the potential for the development of modern medicine for the treatment of various diseases.The Curcumin is active ingredient of turmeric. It is a polyphenol with a molecular formula $C_{21}H_{20}O_6$. Curcumin can exist in at least two tautomeric forms, keto and enol (Kolev *et al.*, 2005). The keto form is preferred in solid phase and the enol form in solution. Systematic name of curcumin: (1*E*, 6*E*)-1,7-bis (4-hydroxy- 3-methoxyphenyl)-1,6- heptadiene-3,5-dione.

Modern science has revealed that curcumin mediates its effects by modulation of several important molecular targets, including transcription factors (e.g., NF-B, AP-1, Egr-1, β-catenin, and PPAR-), enzymes (e.g., COX2, 5-LOX, iNOS, and hemeoxygenase-1), cell cycle proteins (e.g., cyclin D1 and p21), cytokines (e.g., TNF, IL-1, IL-6, and chemokines), receptors (e.g., EGFR and HER2), and cell surface adhesion molecules. Because it can modulate the expression of these targets, curcumin is now being used to treat cancer, arthritis, diabetes, Crohn's disease, cardiovascular diseases, osteoporosis, Alzheimer's disease, psoriasis, and other pathologies (Shishodia *et al.*, 2005).

Table 2. Biological activity of turmeric and its compounds

Compound/extract	Biological activity
Turmeric powder	Wound-healing
Ethanol extract	Antiinflammatory, Hypolipemic, Antitumour, Antiprotozoan
Petroleum ether extract	Antiinflammatory, Antifertility
Alcoholic extract	Antibacterial
Crude ether extract	Antifungal
Chloroform extract	Antifungal
Aqueous extract	Antifertility
Volatile oil	Antiinflammatory, Antibacterial, Antifungal
Curcumin	Antibacterial, Antiprotozoan, Antiviral, Hypolipemic, Hypoglycemic Anticoagulant, Antioxidant, Antitumour, Anticarcinogenic
Ar-turmerone	Antivenom
Methylcurcumin	Antiprotozoan
Demethoxycurcumin	Antioxidant
Bisdemethoxycurcumin	Antioxidant
Sodium curcuminate	Antiinflammatory, Antibacterial

Source: Ishita *et al.* (2004)

2.0 BOTANY AND VARIETIES

For *C. longa* of Koenig, the true turmeric, Valeton proposed new name in 1918 i.e., *Curcuma domestica* Valeton (Purseglove *et al.*, 1981). It is synonymous with *Curcuma longa* L., *C. rotunda* L., *C. xanthorrhiza* Naves, *Amomum curcuma* Jacq.) belonging to the family *Zingiberaceae*. The family *Zingiberaceae* has 40 genera and 400 tropical species in the Old World (Rao and Rao, 1994). Genus *Curcuma* has about 117 species whereas from India around 40 species are reported (Velayudhan *et al.*, 1999). Members of *Curcuma* are strong erect herbs with rhizome bearing habit. Among *Curcuma* species, only *Curcuma domestica*(*C. longa*) is commercially cultivated for the production of turmeric in large scale whereas *C. aromatica* Salisb (Kasturi types), *C. caesia* (Black turmeric), *C. amada* (Mango ginger), *C. zedoaria, C. purpuresecens, C. mangga, C. heyneana, C. xanthorrhiza, C. aeruginosa, C. phaeocaulis* and *C. petiolata* are cultivated in limited scale for cosmetic and pickling purpose at different regions of India (Velayudhan *et al.*, 1999).

Pseudo stem is tall, robust with oblong / elliptic leaves narrowed at the base. Plant height ranges from 67-69 cm in short and 67-83 cm in long duration types, 8-10 leaves in long and 7-8 in short duration types. The leaf length may range from 30-45 cm and breadth from 14-16 cm with petiole equaling the blade. Chromosome number is 2n = 62. Rhizomes are bigger with a stout mother rhizome with branching primary and secondary fingers exhibiting yellow to bright orange yellow core. Most of the high yielding varieties are long duration types which take 8-9 months for maturity (Rao and Rao, 1994).

A number of cultivars are available in the country and are known mostly by the name of locality where they are cultivated. Some of the popular cultivars are Duggirala, Tekkurpet, Sugandham, Amalapuram, Erode local, Alleppey, Moovattupuzha, and Lakdong. These traditional varieties adapted to the local soil and climatic conditions need to be promoted under organic farming (Parthasarathy *et al.*, 2006). Poduval *et al.* (2001) tested the location suitability of 13 cultivars of turmeric under West Bengal conditions and found that only Mohitnagar and G. L. Puram performed with highest percentage of curcumin (9.361%) and Rajendra Sonia produced the highest yield of 23.63 tonnes/ha, with moderate percentage of curcumin (6.110). PCT 13, PCT 11, GL Puram and PCT 15 with higher yields and Lakadong with highest curcumin (7.33%) content were found to be more suitable for mid hill conditions of Meghalaya. (Chandra *et al.*, 1999).

Earlier, turmeric was considered as sterile triploid and crop improvement was based on selection or mutation breeding. Later it was proved that *curcuma* species set seed (Nazeem *et al.*, 1993; Sasikumar *et al.*, 1994). Good number of improved varieties of turmeric are available in India for different states (Table 3). However, there is no studies are on performance of these varieties under organic production system in an organized way.

Table 3. Characteristics of improved turmeric varieties

S. No.	Variety	Mean yield (fresh) (t/ha)	Crop duration (days)	Dry recovery (%)	Curcumin (%)	Oleoresin (%)	Essential oil (%)
1.	Suvarna	17.4	200	20.0	4.3	13.5	7.0
2.	Suguna	29.3	190	12.0	7.3	13.5	6.0
3.	Sudarsana	28.8	190	12.0	5.3	15.0	7.0
4.	IISR Prabha	37.5	195	19.5	6.5	15.0	6.5
5.	IISR Prathibha	39.1	188	18.5	6.2	16.2	6.2
6.	Co-1	30.0	285	19.5	3.2	6.7	3.2
7.	BSR-1	30.7	285	20.5	4.2	4.0	3.7
8.	Krishna	9.2	240	16.4	2.8	3.8	2.0
9.	Sugandham	15.0	210	23.3	3.1	11.0	2.7
10.	Roma	20.7	250	31.0	9.3	13.2	4.2
11.	Suroma	20.0	255	26.0	9.3	13.1	4.4
12.	Ranga	29.0	250	24.8	6.3	13.5	4.4
13.	Rasmi	31.3	240	23.0	6.4	13.4	4.4
14.	Rajendra Sonia	26.7	225	18.0	8.4	-	5.0
15.	IISR Alleppey Supreme	35.4	210	19.3	6.0	16.0	-
16.	IISR Kedaram	34.5	210	18.9	5.5	13.6	-

Source : Chempakam *et al.* (2006)

3.0 SOIL AND CLIMATE

3.1 Soil

Well drained, deep loamy to clay loam soils with good organic status are well suited for this crop. Red loamy or alluvial soils are ideal. Very coarse and heavy soils are unsuitable for rhizome development. Turmeric is grown in wide range of soil types of varying fertility grouped under Inceptisols, Entisoils, Vertisols, Alfisols and Ultisols. Soils that are having high organic carbon, base saturation, major and secondary nutrients are suitable for turmeric cultivation.

(a)

(b)

(c)

Fig. 1: Some high yielding varieties of turmeric (a) Alleppey Supreme (b) IISR Kedaram (c) IISR Prathibha (*Source:* Director, IISR, Calicut)

The crop can thrive well in the soil pH range of 4.3 to 7.5. Sahu and Mitra (1992) found that PTS-9 recorded higher yield (27 t /ha) in sandy loam whereas PTS-21 recorded 31.7 t/ha in clay loam soil. Both varieties recorded low yield in clay soil.

3.2 Climate

The suitability of crop for a particular region is decided by the climate and soil of the region. Turmeric is cultivated in the tropical and subtropical humid climate. It is successfully cultivated from subtropical dry to wet through dry to wet tropical zone at an altitude of 1200 m above MSL, with an optimum range of 450-900 m. In India it is widely grown in warm to hot, per humid to humid eco sub regions. It tolerates an annual rainfall of 640 to 4290 mm. Moderate rainfall of 1500 mm at sowing, fairly heavy and well distributed rain during growing period and dry weather about one month before harvest are much suitable. The temperature range of 18.2 - 27.4°C is optimum. The crop is raised rainfed where rainfall is bimodal and with irrigation in plains where rainfall is uni-modal and low.

Panigrahi *et al.*, (1987) observed the air temperatures of 30-35°C, 25-30°C, 20-25°C and 18-20°C during germination, tillering, rhizome initiation and bulking, respectively as optimum. Kandiannan *et al.* (2001) has developed a regression model using most influencing climatic parameters to predict the yield of turmeric and total production in Coimbatore district of Tamil Nadu. In general, mean minimum air temperature, total rainfall, number of rainy days, mean minimum relative humidity showed positive and mean evaporation, mean sunshine hours, mean solar radiation, mean maximum air temperature showed negative relationship with yield (Kandiannan *et al.*, 2002). In the organic production system also similar influence is expected. Organic addition to the soil would act as a buffer for many adverse conditions. Particularly application of mulch is essential in the rainfed condition. Length of growing period determines the moisture availability. The right time of planting would harness the full potential of favourable climate for early growth and enhanced yield. Climate driven irrigation ensures optimum use of water under organic system. Right time of cultural operations is essential to reap the maximum benefit from organic production.

4.0 PLANTING MATERIALS

4.1 Seeds and Storage

Turmeric is traditionally propagated vegetatively through rhizome (underground modified stem). Although seed set was reported (Nazeem *et al.*, 1993; Sasikumar *et al.*, 1994), it is useful in breeding programmes. For commercial production rhizomes are used as seed. Turmeric rhizomes are mainly two types *viz.*, oval shape mother rhizome (first formed rhizome from main pseudo-stem) and finger rhizomes developed from mother rhizome. The fingers may be first order (primary rhizome), secondary or tertiary order. In the literature, the term 'bulb' or 'corms' is also used to denote rhizomes. All the three will give new plants after planting.

According to Mudaliar (1960), mother rhizome is used for planting in Krishna and Guntur districts, and fingers alone are planted in Cuddapah district of Andhra Pradesh. Both mother and fingers are used in Tamil Nadu, but are used separately in the field and not mixed together. Aiyadurai (1966) states that mother rhizomes were found to be better than finger rhizomes. The mothers split longitudinally into two halves and fingers are broken into pieces of 5 to 10 cm length weighing approximately 50 to 100 gram with one or two buds are used for planting. In general, in most of the turmeric areas primary fingers used for seed (Rao *et al.* 2006). Transplanting of sprouted buds also attempted in turmeric but Aiyadurai (1966) and Randhawa *et al.* (1984) reported direct planting was much superior to transplanting. Direct - planted crop yielded 49.7 percent more than the transplanted crop. Planting mother rhizome was found to produce more yield compared to primary or secondary rhizome (Rashid *et al.* 1996) (Table 4).

Table 4. Effect of planting materials on the growth and yield of turmeric

Planting material	Height (cm)	No. of leaves per plant	Tillers per clump	Yield per plant (g)	Fresh yield (t/ha)
Mother	110.91	27.18	5.15	628.68	28.67
Primary	100.69	23.69	4.72	548.73	22.32
Secondary	94.89	23.08	4.60	499.28	19.43
LSD (0.05)	-	3.56	0.42	-	-
LSD (0.01)	6.97	-	-	78.70	4.65

Source: Rashid *et al.* (1996)

The micropropagated plants raised using young vegetative buds from sprouting rhizomes can also be used as planting material. They showed a significant increase in shoot length, number of tillers, number and length of leaves, number of fingers and total fresh rhizome weight per plant compared with conventionally propagated plants (Salvi *et al.*, 2002). However, this may not be permitted under organic production.

Initially, to start organic production, getting seed from certified organic source may be difficult; hence, it is advised to use seed from available sources.

The seed should be healthy and disease free. Such rhizomes selected for seed should be stored on moist sand under shade or in ground pits.

Rhizomes for seed purpose are generally stored by heaping in well-ventilated rooms and covered with turmeric leaves. The seed rhizomes can also be stored in pits with saw dust, sand, leaves of *Glycosmis pentaphylla*, and *Srtychnos nux-vomica*. The pits are to be covered with wooden planks with one or two openings for aeration. Adequate aeration is required during storage. The highest recovery (94.88%) of healthy seed rhizomes at the end of storage was recorded when stored in zero energy cool chamber, followed by 250 gauge polyethylene bags with 0.5% area of vents with more sprouting. Storing in heaps under a tree canopy covered with leaves had lowest sprouting (64.41%), highest rotting (21.94%) caused by *Pythium* spp. (Wet rot) and *Aspergillus* spp. (dry rot) and insect damage caused by *Lasioderma serricorne* beetles (Kirankumar *et al.*, 2002). Hot water treatment at 50°C for 30 min. without affecting germination eradicates all fungi associated with turmeric seed rhizomes.

4.2 Seed rate

Turmeric seed rate generally vary between 1000 and 1200 kg ha^{-1} (Agnihotri, 1949). When mother rhizome was used, rate was 1800 kg ha^{-1} while it was 1200 kg ha^{-1} for fingers (Rao, 1957). Indian Institute of Spices Research, Calicut recommends 2500 kg ha^{-1} (IISR. 2005a). Govinden and Cheong (1995) reported an optimum planting rate of about 3000 kg ha^{-1}. The highest yields are obtained from seed rhizome pieces with 2 to 3 eyes. The turmeric seed rhizome of 30-40 g with a larger diameter or seed mother rhizome weighing 25-34 g (Philip, 1983), free from daughter rhizomes can be used. Shanmugam *et al.*, (2000) observed that 14-day old sprouted rhizomes as planting materials produced plants with significantly higher number, length, fresh weight and dry weight of mother, primary and secondary rhizomes with 27% more yield of fresh rhizomes (40 t ha^{-1}). The curcumin, oleoresin, and oil yield per hectare were also highest in the rhizomes developed from the 14-day-old sprouted rhizomes.

Increasing tuber size from 20-29 to 80-89 g increased yields (Aoi, 1992). Hossain *et al.*, (2005a) observed that plants from 30 g, 40 g and 50 g of daughter rhizomes had a significantly larger shoot biomass and higher yield than those from smaller daughter rhizomes in both the greenhouse and field experiments. The shoot biomass and yield were highest in the plants grown directly from mother rhizomes when compared to the plants from daughter rhizomes attached to that of mother rhizomes. Although the mother rhizomes are better yielders, Tayde and Deshmukh (1986) suggested that during shortages of mother rhizomes, secondary rhizomes could be used successfully with additional manuring for obtaining higher yields.

Planting of full mother rhizome of 80-100 g resulted in minimum leaf blotch incidence caused by *Colletotrichum capsici* and maximum rhizome yield (220 q/ha) followed by half mother rhizome of 50-80 g (Archanakumari *et al.*, 2002). Whole mother rhizomes (70-80g) planted at 50x20 cm gave highest yield in Haryana.

4.3 Seed treatment

Hot water treatment at 50°C for 30 min without affecting germination eradicates all fungi associated with turmeric seed rhizomes.

5.0 PLANTING AND AFTER CARE

5.1 Field preparation

The land is prepared with the receipt of early monsoon showers. The soil is brought to a fine tilth by giving appropriate ploughing depending upon soil types. Mudaliar (1960) mentioned that number of ploughing varies between two in the Mydukur region to 16 in the Rajampet area, both in the Cuddapah district; the average is 6 to 8. Earlier bullock / cattle drawn country plough of varied size was used for ploughing operation. Farmers are now switching over to tractor ploughing as maintaining bullocks are costly affair. In organic production both are acceptable as it may not affect the produce quality. However, traffic of tractor would influence the soil compactness. There is no evidence to show the influence of soil compactness on rhizome yield. It is usual to apply bulky organics like FYM or compost during this preparatory tillage so as to ensure uniform mixing in the soil. In sloppy lands ploughing may not possible. Only digging with spade or other suitable tools are used then raised beds are formed. Similarly, by making raised beds, shallow soil also is effectively utilized for turmeric production. Under high rainfall areas low pH is a problem. Application of amendment like hydrated lime enhance the pH level to near neutral.

5.2 Planting

Turmeric is an annual crop with 8 to 9 months duration. The places in India like Kerala, North Eastern States, part of West Bengal, Bihar, Orissa and Andaman and Nicobar Islands receive rain from both the monsoon (north east and south west monsoon) where crop is mainly rain fed, it is supplemented with one or two irrigation. Under low rainfall, crop needs irrigation regularly; hence, it is grown where unlimited and assured water supply is available. Areas fed with tank, canal or dam are suitable. In India, generally planting depends upon receipt of rain / irrigation. In canal command areas planting will start early in the head end and it will be delayed in the tail end as it will take sometime for water to reach the lower region.

5.2.1 Planting time

Planting starts with the commencement of southwest monsoon rain in most parts of India. Planting is also advanced for better utilization of summer showers as in Kerala where the crop is rain fed. In general, planting is done between April and August in India. Normally, turmeric is planted from middle of March to early April in Punjab, between 15th April and 15th May in Kerala (Nair, 1982), first week of May to June in Orissa (Panigrahi *et al.*,1987). Balashanmugam *et al.* (1989) reported that planting season is spread over from first week of May to the first week of August in Tamil Nadu. Kandiannan and Chandaragiri (2006) reported that May 15th was better than June 15th and July 15th planting in terms of yield and quality at Bhavanisagar, Tamil Nadu. The rhizome yield decreased with delayed planting from March to May (Aoi, 1992). Even though planting turmeric later than the normal period decreases leaf blotch (caused by *Taphrina maculans*) intensity, the yield reduction will also be high. The yield of fresh and dry rhizomes increased with delay in planting from 20th to 30th April but decreased with further delay up to first week of June (Randhawa *et al.*, 1984). This was also confirmed by Anjaneyulu and Krishnamurthy (1979b). Planting at the end of April gave the highest yields. It is suggested that under delayed planting, turmeric should be planted in a nursery and then transplanted rather than direct planting in the field. However, Aiyadurai (1966) reported that normal method of dibbling cut-mother rhizomes was much better than transplanting detached sprouts from the germinated mother rhizomes. In Orissa, early planting at third week of May was observed to increase the plant height, tillers and weight of fresh rhizome plant^{-1} in eight varieties of turmeric than delayed planting (Mishra *et al.*, 1997). Barholia *et al.* (1992) obtained the higher fresh rhizome yield with May 16th planting by using finger and mother rhizomes as seed material, compared to April 16th and June 16th plantings under Madhya Pradesh conditions.

In Brazil planting is done during October - December and higher production was obtained in 20th November planting with spacing of 0.30 m (Cecilio *et al.*, 2004). Planting of turmeric in September at a spacing of 0.45 x 0.10 m is recommended for higher yields in South Africa followed by November planting as November planting, reduced risk of sun scorch (Nel, 1988). Planting time varies with locations and planting at the commencement of the season (which may vary with location) recorded the best yield than advance or delayed plantings (Table 5).

Table 5. Time of planting of turmeric at different production centres.

State/ Country	Time Suggested	Reference
Andhra Pradesh, India	First week of July	Sarma and Murthy (1965)
Punjab, India	April end	Randhawa *et al.* (1984)
Bihar, India	May 23rd	Jha *et al.* (1983)
Madhya Pradesh, India	May 16th	Barholia *et al.* (1992)
Orissa, India	May third week	Misra *et al.* (1997)
Tamil Nadu, India	May 15th	Kandiannan and Chandaragiri (2006)
Kerala, India	April- May	IISR (2005a)
Japan	March	Aoi (1992)
Taiwan	March 15 to May 15	Hu *et al.* (1996)
Brazil	October - December	Cecilio *et al.* (2004)
South Africa	September	Nel (1988)

5.2.2 Planting method

The methods of planting vary considerably at different production centers. It is mainly depends on soil type (depth), rainfall, irrigation facility, cropping system to be followed etc. In Malabar, Kerala, rainfall is high and crop grown is under rain fed where raised bed system is adopted with beds of 1.0-1.5 m width, 15 cm height and of convenient length are prepared with spacing of 50 cm between beds. Small pits at a distance of 20 cm to 25 cm apart on either side are dug on the beds and filled with well decomposed cattle manure or compost and seed placed over it, then covered with soil. The raised beds allow free drainage. In irrigated condition, this system is not at all followed. In Meghalaya, India, turmeric is grown by bun method of cultivation, where a series of beds (called buns) are formed along hill slopes (Ghadge *et al.*, 2001). Planting is done on ridges and furrows and broad ridges under irrigated condition. The ridges are one and half to two feet apart in loamy soils and planting is done on the crests of the ridges whereas, it is two to two and a half feet in clayey soils and planting is done on both the sides of ridges (Mudaliar, 1960). He also mentioned that the rhizomes sometimes scattered evenly over the field and pressed into the soil with the hand. In another method rhizomes are dropped in the plough furrows and covered by soil when adjacent furrows opened. However, it may not be possible when tractors are used. Gill *et al.* (2004) found that flat beds are preferred in red soils and raised beds in black dry loams at Punjab. Turmeric fresh rhizome yield was significantly higher in ridge method of planting than flat method of planting. It produced significantly taller plants and more number and weight of primary and secondary rhizomes at Ludhiana, India (Gill *et al.*, 2004). Planting primary fingers 20 cm apart in broad ridges (80 cm long, 10 cm wide, 20 cm high with 30 cm wide irrigation channels) gave 52% higher yield than the ridges and furrow method (Anjaneyulu and Krishnamurthy, 1979a). In Japan, turmeric is planted in a triangular pattern to obtain higher dry yield of shoot and rhizomes (Hossain *et al.*, 2005b). Singh *et al.* (2002) found flat bed followed by earthing as the best planting method for higher rhizome yield. The conventional methods of planting also are adopted under organic system. It is essential to provide good drainage, otherwise under ground rhizome will be infected with rot diseases.

5.2.3 Planting depth and spacing

Plant population per unit area influences the final yield. Spacing to be adopted depends on soil type, fertility level, season, cultivar and irrigation facility or rainfall distribution. Planting is done in small pits with spacing of 25 cm x 30 cm and covered with soil or dry powdered cattle manure. The optimum spacing of 25 x 25 cm for 'flat bed' method and 45 to 60 cm between rows and 25 cm between plants in the case of 'ridge and furrow' method was suggested by Aiyadurai (1966). A planting distance of 6 inches at a depth of 2 inches tends to be the optimum. When turmeric is planted at a closer spacing, rhizome cannot not expand properly, which ultimately results in the smaller rhizome compared with that planted with a larger spacing.

Between-plant spacing of 20 cm was found to be optimum for higher rhizome yield by Carvalho *et al.* (2001). The closest spacing of 30 x 45 cm was found best for higher yields by Rajput *et al.* (1982). Whereas, Ponnuswamy and Muthuswami (1981) suggested recommended 45x20 cm spacing. Shankaraiah and Reddy (1988) found 30x15 cm spacing as optimum for cultivars Armoor local and Duggirala. Ramachandran and Muthuswami (1984) found that closer spacing (30 x 15 cm) registered higher fresh rhizome yield over wider spacing (50 x 22.5 cm). Sundararaj and Thulasidas (1976) suggested 15 to 20 cm spacing as optimum, while Rao (1979) reported the spacing of 30 x 20 cm. In Kerala, 30 x 15 cm was recommended under 'flat bed' system (Nair, 1982). Anjaneyulu and Krishnamurthy (1979c) recommended planting with whole mother rhizomes with 30 x 22.5 cm for getting higher yields. Even though 30x30 and 30x45 cm spacing gave the maximum plant height, highest number of tillers and leaves per clump, the highest fresh rhizome yield and cured yield was obtained with 30x15 cm spacing by Manjunathgoud *et al.* (2002).

Planting turmeric at 45x45 cm spacing (49383 plants per hectare) significantly reduced leaf blotch intensity without any significant reduction in yield (Prasadji *et al.,* 2002). Shashidhar and Sulikeri (1996) reported that curing percentage was higher at medium spacing (45 x 22.5 cm) but, Valsala *et al.* (1998) observed that planting of turmeric at different spacings did not influence the curing percentage and curcumin content.

Kandiannan and Chandaragiri (2006) also recorded that closer spacing of 30 x 15 cm gave maximum yield compared to wider spacings (45cm x 15 cm and 60 cm x 15 cm). They also reported that spacing did not influence the quality (curcumin and dry recovery). Reduction in rhizome yield was noticed when within-row spacing increased from 0.05 m (25 t/ha) to 0.40 m (18 t/ ha) but this increased productivity obtained from smaller within-row spacing could not overcome the higher expenses due to respective planting (Silva *et al.,* 2004). On the non availability of mother rhizomes, using primary fingers spaced at 22.5 cm with 45 cm between the rows is recommended.

Yields generally decreased as the planting depth increased from 15 to 20 and 30 cm. Ishimine *et al.* (2003) suggested for planting of turmeric rhizomes at a depth of 8 to 12 cm in dark red soil for a higher yield and lower weed competition, in Japan as 30% of turmeric in the field planted at a 4-cm depth was uprooted by a typhoon, but not at the depths of 8, 12 and 16 cm. The rhizome development was also earlier with significantly higher shoot biomass and yield when planted at 8, 12 and 16cm depths than at 4-cm. Hossain (2005) reported that harvest from deep planted crop was difficult. Organic system adopts spacing and depth of planting as that of traditional methods of a particular location.

5.3 Mulching

Mulching is an integral part of turmeric production under high rainfall areas like Malabar region of Kerala or North Eastern region. There are several advantages of mulching reported. First of all it protect the rhizome from exposing to scorching sun by splashing rain drops. It is a food for several hetero-trope microbes which bring benefits to crop. It enriches the soil. Mulching with locally available leaves or straw is beneficial. The crop is to be mulched immediately after planting with green leaves @ 12-15 tonnes/ha and repeated after 45 days at the time of top dressing. In Orissa 12.5 t/ha green mulch materials are being applied thrice in a cropping season (at planting, 45 and 90 DAP) for maximizing yield Mishra *et al* (2000). Mulching with Daincha and Sunnhemp leaves was also found to be useful (Aiyadurai, 1966). Mulching improved the organic status of the soil, also hastened germination by a week and decreased the weed growth. Sugarcane trash mulch was found to increase the germination rate and rhizome yield in Punjab (Hussain *et al.*, 1969).

The mulch is effective in conserving soil moisture and increasing the growth and yield of turmeric under rainfed conditions. The application of mulches at 10 t ha^{-1} conserved more moisture and increased the yield of turmeric by 12% and paddy straw mulch resulted in 18% increase in yield over *Gliricidia* mulch (Kumar *et al.*, 2003a). But, no significant difference in rhizome yield and growth due to application of fresh sal leaf or paddy straw as mulch was observed by Mishra *et al.* (2000).

The application of straw mulch was more beneficial than intercropping turmeric with pigeon pea, maize or green gram in getting net returns. No improvement in the oil and curcumin contents was noticed by mulch application (Singh and Randhawa, 1988). Early planting (on 23 May) and mulching with *Dalbergia sissoo* leaves gave the highest yield of fresh rhizomes and the highest returns on non-irrigated calcareous soil compared to mulching with mango leaves or rice straw (Jha *et al.*, 1983). The highest yield and return per rupee investment were obtained with dry forest leaf litter mulch + intercropping with French bean (Rs 2.16), followed by dry forest leaf litter mulch alone (Mohanty *et al.*, 1991). The organic matters from plant or animal origin are allowed for organic system. The industrial organics should meet the certification standards of the organic system of particular agency or country.

5.4 Weed management

Turmeric as an annual crop, it's growth during initial phase is slow. Hence, weed management is must during this time. Traditionally three to four manual weeding is done. Due to non-availability labour, farmers are switching over to herbicides but herbicides are not in organic packages. Weed intensity vary with location. According to Hossain (2005), in Japan planting 30-40g seed of finger or mother rhizome in 30 cm triangular pattern at the depth of 8-12 cm on two - row ridges spaced 75-100 cm apart during March- April reduced the weed infestation. Mulching also would reduce the weed emergence. Intercropping with quick growing crops also smother the weeds. Besides any cultural practices of local importance are accepted under organic production.

5.5 Irrigation

Major turmeric areas are under flow irrigation from river and tank or dam systems. Water from well also supplemented. The number of irrigation varied with soil type. 15 to 20 irrigations are given for clayey soils and about 40 for sandy loams (Mudakliar, 1960). Rao *et al.* (1992) quantified the optimum water requirement for maximized net returns on a sandy-loam soil at Jagtial, Andhra Pradesh as 1.99 ha-m. The water use efficiency and rhizome yields of the crop were found to increase in drip irrigation, 40 or 80% of surface irrigation treatments by Selvaraj *et al.* (1997). Water use efficiency of turmeric under drip-irrigation system was highest under 60% evaporation replenishment (of USWB class A pan evaporation) with 60 cm lateral spacing and 100% evaporation replenishment with 60 cm lateral spacing produced highest yield (Singte *et al.*, 1997)(Table 6). In organic system quality of irrigation water is important, sewage water, waste water from industry are not allowed. Unless it meets the certification stadards, it is not allowed in organic production.

Table 6. Irrigation on green turmeric yield (q/ha)

Evaporation replenishment	1991-92 Lateral spacing		1992-93 Lateral spacing	
	60 cm	120 cm	60 cm	120 cm
100%	371.26	453.29	424.57	525.78
80%	396.63	416.63	400.63	479.12
60%	366.63	396.63	367.46	410.89
CD 5%	19.69		20.42	

Source : Singte *et al.* (1997)

5.6 Earthing up

Earthing up is yet another important cultural operation. It is normally done twice or thrice always accompany weeding and side dressing under irrigated condition whereas, in rainfed weeding, earthing and mulching are done simultaneously. Usually practiced during 45-60 days after planting (DAP) and 90-105 DAP, additional if required done on 120 to 135 DAP. This intercultural operation helps to form and enlarge finger rhizomes and also ensures adequate aeration to roots. It also protect rhizome from scale insects and check the weeds (Panigrahi *et al.*, 1987).

5.7 Cropping system

Turmeric as a best component crop in agri-horti and silvi-horti systems, recycling of farm waste can be effectively done when grown with coconut, arecanut, mango, *Leucaena*, rubber etc. As a mixed crop it can also be grown or rotated with green manure/ legume crops or trap crops enabling effective nutrient built up and pest or disease control.

Intensive cultivation in the form of inter-cropping to exploit slow growing turmeric during initial stages is a common practice. Turmeric can be grown as an intercrop in coconut and arecanut plantations and mango orchards. It can also be raised as a mixed crop with chillies, colocasia, onion, brinjal and cereals like maize, ragi, etc. Nutrient competition in intercropping system can be minimized by selecting crops of different rooting patterns and nutrient requirements by proper spacing. It is reported that second and third mulch can be avoided by inter-cropping a row of french bean in the inter row space of turmeric with a net profit of Rs. 3.3 per rupee invested (Mohanty *et al.*, 1991). Highest mean gross income was reported when two rows of turmeric were alternated with one row of maize (Shankaraiah *et al.*, 1987). Significant decrease in turmeric rhizome yield due to maize intercropping was also observed (Avilkumar and Reddy, 2000; Sivaraman and Palaniappan, 1994).

Intercropping turmeric with pigeon pea, maize or green gram reduced the availability of light and the rhizome yield. However, net returns were higher under intercropping, compared with monoculture (Singh and Randhawa, 1988). Intercropping with turmeric can be a substantial source of extra income in forest plantations. Turmeric can be grown as an effective intercrop (benefit:cost ratio of 2.51) in a silvi-horti system with fodder trees like bhimal (*Grewia optiva*), khark (*Celtis australis*), banj (*Quercus leucotrichophora*) and kachnar (*Bauhinia variegata*). The relationship between shade by the trees and yield of turmeric have a negative (r=-0.73) correlation. The highest rhizome yield of turmeric was recorded (18.70 t/ha) with *Q. leucotrichophora* (Bisht *et al.*, 2004) in Uttaranchal, India with a benefit:cost ratio of 3.21. Teak based Agroforestry and mango based agri horti systems can accommodate turmeric as a successful inter crop. A significantly high rice equivalent yield was obtained from turmeric-mango intercropping system by Singh and Rai (2003). A study conducted by Kerutagi *et al.* (2000) revealed that turmeric cultivation yielded a gross returns of Rs 208428 ha^{-1} in Belgaum district of Karnataka.

Turmeric is a component in arecanut based high-density multispecies cropping system (HDMSCS) (Ray and Reddy, 2001). *Leucaena leucocephala* and *Eucalyptus camaldulensis* are most compatible agroforestry systems for turmeric intercropping (Mishra and Pandey, 1998). Pradhan *et al.* (2003) recorded maximum rhizome yield of turmeric when grown within the rows of *Leucaena* planted 4.0 x 1.0 m apart and pruned 5 times/annum. The yield of intercrop per unit area slightly increased with increasing tree spacing and frequency of pruning whereas the yield of *Leucaena* decreased with both the increasing spacing and frequency of pruning. Pal *et al.* (2000) observed increased rhizome yield in sole cropping system compared to alley cropping of turmeric within two meter distance from five years old established tree plantation of *Leucaena leucocephala*.

The total yield and the yield attributing characters of turmeric are adversely affected under shade in comparison to open cultivation. Yield characters, such as number of tillers per clump and fresh rhizome yield were lowest under heavy shade as compared to open cultivation. Severity of *Colletotrichum* leaf spot was significantly lower (1.8%) in heavy and partial shade (4.5%) in comparison to open cultivation (23.7%) (Singh and Edison, 2003). Latha *et al.* (1995) observed higher fresh yield of turmeric rhizomes under partial shade and highest cured yield when grown in open condition due to high rate of dry matter production. Higher solar energy input under open condition helps in higher crop growth rate during bulking of rhizomes.

Multiple cropping systems in Assam involving coconut + betel vine + banana + Assam lemon + turmeric + colocasia improved the nutrient availability, soil fertility, population of *Azotobacter* and the net income per hectare (Sarma and Chowdhury, 2002). Pigeon pea-turmeric inter-cropping during the first 10 years of a poplar (*Populus deltoides*) plantation is also an economically viable system with a BC ratio of 6.68 in Bihar (Chaturvedi and Pandey, 2001). Turmeric can be grown efficiently as intercrop during the first three years after planting rubber (Sreenivasan *et al.*, 1987). Some results of intercropping studies are presented below (Table. 7). The choice of crops are driven by farmers need and market access.

6.0 NUTRIENT MANAGEMENT

6.1 Nutrient removal

The uptake of nutrients increase as the dry matter production increases. The maximum uptake was observed in active vegetative growth phase. Higher uptake of K upto third, N upto fourth and P upto fifth months of development was observed with subsequent decrease in their uptake. Use of mother rhizomes enhances the uptake of N, P and K as compared to primary or secondary rhizomes for planting. The uptake of nutrients by turmeric is in the order of potassium > nitrogen > magnesium > calcium (Table. 8). An average dry yield of 5.5 t ha^{-1} turmeric rhizomes removes 91 kg N, 16.9 kg P_2O_5, 245 kg K_2O (Sadanandan *et al.*,1998).

Table 7. Technologies for turmeric intercropping system

State	Station / Location	Cropping system	Results	Reference
Andhra Pradesh	Northern Telegana Zone	Turmeric + Maize Profitable Intercropping	Turmeric variety - Armoor, Duggirala, Mudukur Spacing (30X15cm) Maize variety – Poineer, Ganga -4, Deccan Spacing (60X30 cm)	Anon., 1987
	Regional Agricultural Research Station, Rudroor	Turmeric inter cropped with Maize, Okra, Caster, Chilli	Best are Turmeric+ Maize (2:1) Turmeric+Chilli (2:1)	Shankaraiah *et al.*, 1987
	Regional Agricultural Research Station, Jagatial	Turmeric + Maize intercropping	100% population of both crops at 2:1 ratio with fertilizer for both crops at recommended dose	Rao and Reddy, 1990
Karnataka	Gandhi Krishi Vigyana Kendra (UAS), Bangalore	12 lines tested under coconut garden	Acc.No. 10/8, Rajapuri, PCT 8 are the best	Hegde *et al.*,1998
Kerala	Agricultural Research Station, Mannuthy	14 lines tested under coconut garden	Amruthapani Kothapetta, Chayapa - subha, Amalapuram, Kasthurithanka, Kodur type, Kuchupudi, Mannuthy Local are the best	Pushkaran *et al.*, 1985
	Regional Agricultural Research Station, Pilicode	15 lines tested under coconut garden	VK-11, VK -77, VK -114, VK -116, PTS -10, Vontimitta are the best	Latha *et al.*, 1995
	College of Agriculture, Vellayani	Turmeric under coconut	Application of 120:120 kg N:K_2O along with boron (2 kg) and zinc (10 kg) ha^{-1} for turmeric recorded maximum economic yield. P application as per soil status.	Meerabai *et al.*, 2000

Maharashtra	Agricultural College, Pune	Turmeric + Maize	Adoption of broad ridge system of planting was the best
Orissa	High Altitute Research Station (OUA & T), Pottangi	Mulch and Intercropping (French bean, Red gram)	Sowing French bean at 45 days after turmeric planting serve as a cover crop and need not go for 2nd and 3rd mulching
Tamil Nadu	Tamil Nadu Agricultural University, Coimbatore	Double intercropping with turmeric, Intercrops tested - Bhendi, Coriander, Onion, Amaranthus, Radish, Fenugreek, Bush bean, Beet root	Turmeric + Fenugreek + Onion was the best treatment
	Agricultural Research Station, Bhavanisagar	Rely intercropping with turmeric Crop tested Coriander, Onion, Fenugreek, Fennel, Radish	Turmeric – Coriander Turmeric – Onion
	Agricultural Research Station, Bhavanisagar and Coimbatore	Turmeric-Maize and onion Intercropping	Raising turmeric with 50 % population of maize at 100 cm between maize rows and raising onion planted at 23 % population and application of 250 kg N ha^{-1} was the best
West Bengal	Bidhan Chandra Krishi Vishwa Vidyalaya, Kalyani	Turmeric intercropping with Pigeon pea, soybean, green gram, maize, rice	Short duration cereals like maize, rice or green gram provided the food to the subsistence farmers

Source: Kandiannan *et al.* 2005

Fig. 2: Turmeric intercropped with coconut (*Source:* Director, IISR, Calicut)

According to Kumar *et al.* (2003b) approximately 9% of the turmeric growing area in Tamil Nadu is severely limited by mineral nutrition and about 20% of samples were identified as having deficiencies. They worked out optimum levels for 12 nutrients (N, P, K, Ca, Mg, Na, S, B, Zn, Cu, Fe and Mn) in the leaves of turmeric using Diagnosis and Recommendation Integrated System (DRIS)/Modified Diagnosis and Recommendation Integrated System (MDRIS) and Compositional Nutrient Diagnosis (CND) approaches.

Table 8. Removal of nutrients by turmeric at harvest (kg ha^{-1})

Location/soil type	Nutrients					Reference
	N	P_2O_5	K_2O	Ca	Mg	
Kasaragod-Laterite	124	30	236	73	84	Nagarajan and Pillai (1979)
Vellanikkara-Lateritic	72-115	14-17	141-233	--	--	Rethinam *et al.* (1994)
Bhavanisagar-Sandy loam	166	37	285	--	--	Rethinam *et al.* (1994)
Coimbatore-Calayey loam	187	39	327	--	--	Rethinam *et al.* (1994)
Calicut – Lateritic	86	31	194			Sadanandan *et al.* (1998)

The nutrient imbalance was in the order S > B > Mg > Cu > P > Na > Ca > K > Zn > N > Fe > Mn based on DRIS, S > B > Cu > Ca > Na > Zn > Mg > P > Fe > Mn > K > N based on MDRIS, and S > Ca > Na > Mg > Cu > Zn > K > B > Mn > P > N > Fe based on CND. Among the various approaches, CND projected the nutrient imbalances better than DRIS. Boron (B) deficiency in turmeric results in reduced accumulation of sugars, amino acids, and organic acids at all leaf positions. Translocation of the metabolites towards rhizome and roots and photo assimilate partitioning to essential oil in leaf and to curcumin in rhizome decreases. Due to B deficiency, the overall rhizome yield and curcumin yield also decreases (Dixit *et al.*, 2002).

Dixit and Srivastava (2000) also observed decrease in plant growth, fresh weight, rhizome size, photosynthetic rate and chlorophyll content and significantly increased curcumin content in all the genotypes under Fe deficiency. N, Mn and Zn contents of rhizomes had a positive significant correlation with curcumin content, whereas P, Ca, Mg, S, Cu and Fe contents had no correlations (Kumar *et al.*, 2000).

Third leaf from the top has been identified as the index leaf nutrient status of the crop (Saifudeen, 1981) and the age of the plant between 90-120 days after planting (DAP) was suggested for sampling. Total Zn concentration in the third leaf was considered for delineation of deficiency that varied from 10 to 51.3 mg/kg. Zn concentration was significantly and positively correlated with leaf weight and rhizome yield and negatively with Fe: Zn and P: Zn concentration ratios. P: Zn ratio in the absence of added Zn proved as a good indicator of efficiency of response to added Zn (Singh *et al.*, 1986). Leaf P: Zn ratio had a positive correlation and soil P: Zn ratio had negative correlation with rhizome yield (IISR, 2005b).

6.2 Organic manuring

According to Ridley (1912), turmeric is manured in poor soil, he has mentioned that farmyard manure was the best procurable manure and used in Malay Peninsula. In Coimbatore, municipal sweepings and ashes were the favorite manure. It was also reported by him that burnt earth and wood ash as effective manures. He quoted from experiment conducted in Poona (Pune, India) that farmyard manure gave the best results in enhancing turmeric rhizome yield compared to effluent from septic tanks, (11 watering), poudrette 10 tons per acre and fresh night soil four dressings. Under organic system use of effluent is not allowed.

Manurial practices are more variable at different production centers. Mudaliar (1960) mentioned that manuring was done more heavily than for most other crops as a routine practice, particularly with locally available organic manures. Light soils are manured more heavily than heavy soils. He has documented following manurial practices in turmeric growing region of south India.

- Clay and silt are applied at 100 to 300 cart-loads to an acre in the garden lands of Guntur and Krishna districts. Tank and channel silt are preferred, if available nearby, to clay scrapped from the surface of wet lands.

- Cattle manure is applied at 20 to 50 cart-loads per acre, depending upon its availability. Turmeric is planted in small pits in Malabar and dry powdered cattle manure is applied at one double handful to each pit, before planting.
- Cattle are penned in the field during the summer months at 800 to 1,000 cattle for a night per acre, in Krishna and Guntur districts. The practice of keeping cattle in the open fields during the summer nights is a common feature in these districts. The cattle are more comfortable in the fields than in stalls, when the summer is very hot.
- Sheep are penned at 2000 to 4000 sheep for a night per acre extensively in turmeric regions, to supplement the cattle manure applied.
- Green manure crops are raised in areas where manure is in short supply and where the rainfall and the cropping permit green manure crops being raised during the preceding fallow period. Sunhemp is usually sown for the purpose in Coimbatore district and greengram in Repalli taluk of the Guntur district. Turmeric farmers give the green manures a value equal to that of 15 cart-loads of cattle manure per acre.
- Green leaves are regularly used for mulching turmeric fields in Malabar, once at the time of planting the crop and twice later at intervals of 10 weeks, and partially covered by earthing up. The twigs that do not decompose are removed at the time of the next mulching. In Cuddapah, green leaves are spread over the land at 20 cart-loads to an acre, after planting the crop.
- Organic refuse like groundnut husk, useless palmyra leaves from old dismantled roofs, *cumbu* (pearl millet) straw not suitable as fodder, composts etc., are suitable in heavy soils of Tamil Nadu.

The bulky organic manures are supplemented in certain cases with 4 to 8 bags of groundnut cake, 5 to 10 bags of castor cake per acre. These are broadcasted over the field at the time of the first weeding and covered by ploughing in between the turmeric rows, or by hand hoeing. These are also applied along the rows or near the base of individual plants. Pursegolve *et al.* (1981) also notes similar practices, they also stated by quoting De Geus (1973) that farmers apply 25 tonnes of FYM or compost at Maharashtra and Tamil Nadu.

Many experimental evidences show the beneficial effects of organic matter. Farm yard manure (FYM) or compost @ 40 tonnes/ha is applied by broadcasting and ploughing at the time of preparation of land or as basal dressing by spreading over the beds to cover the seed after planting. The increase in fresh yield was 37 per cent over control. Kerala Agricultural University (KAU) recommends 40t/ha compost or cattle manure as basal dressing, for Kerala conditions; where as in Orissa 15 kg is applied in 5m^2 bed area in three splits along. Organic manures like groundnut cake, vermicompost and neem cake can also be applied. In such cases the dosage of FYM can be reduced. Application of FYM increases the organic carbon, available P and effective cation exchange capacity content in the soil there by improving the fertility which intern significantly improves plant height, number and weight of mother, primary and secondary rhizomes and fresh rhizome yield. Nitrogen, phosphorus and potassium contents in leaves and rhizome improved with farmyard manure application (Gill *et al.*, 2004). Application of FYM at 50 t / ha increased the fresh rhizome yield up to 11.72 t/ha under Mizoram conditions (Saha, 1988).

Gill *et al.* (1999) observed significant increase in rhizome yield and curcumin content with increased FYM (up to 60 t ha^{-1}) and wheat straw mulch levels. The application of groundnut cake (1.1 t/ha) significantly increased the dry yield and resulted in the highest curcumin production (Table 9), on par to neem cake application (2.5 t/ha) (Sadanandan and Hamza, 1998). No significant differences in dry yield and curcumin content of turmeric rhizomes was observed with the combination of chicken manure and inorganic fertilizer by Hu *et al.* (2003) under China conditions. Fields that received neem cake as a dressing were free from turmeric pest Mimegralla fly (Reddy and Reddy, 2000).

Liming @ 400 kg ha^{-1} for acidic soils, application of 40 t ha^{-1} FYM as basal, 2 t of neem cake and 5-8 t of vermicompost and mulching with green leaves @ 12-15 t ha^{-1} at 45 days intervals are recommended. In turmeric, combined application of different organic sources like FYM + pongamia oil cake + Neem oil cake + Stera meal + Rock phosphate + wood ash yielded on par to the conventional practice in addition to high quality. But research results during initial years showed 15% and 23% reduction in yield in turmeric under organic system as compared to conventional and integrated managements, respectively. The quality parameters of turmeric (curcumin and starch) are found to be high under organic farming (IISR, 2005b). Among organics, highest fresh rhizome yield of 34.4 t ha^{-1} was recorded in combined application of vermicompost and coir pith compost (@ 5 t ha^{-1} each) followed by FYM, vermicompost and coir pith compost application. Application of vermi compost at 10 t ha^{-1} increased the rhizome yield from 6.7% - 25.5% (Vadiraj *et al.*, 1998).

Table 9. Yield and quality of turmeric as effected by organic and inorganic fertilizers

Treatment	Turmeric	
	Dry yield (Kg/ha)	**Curcumin** (Kg/ha)
Check	2374	169
FYM	2596	250
Neem cake	2602	287
Cotton cake	2640	284
Brassica cake	2784	243
Groundnut cake	2669	277
Gingily cake	2768	249
NPK fertilizer*	2480	268
NPK fertilizer + micro nutrients**	--	--
CD at 5%	--	32.2

* NPK @ 60, 50, 120, ** Zn, B, M0 @ 5, 2, 1 kg/ha.
Source : Sadanandan and Hamza, (1998).

6.3 Biofertilizers

Turmeric responds well to the application of organics and biofertilizers. Integrated application of coir compost (@ 2.5 t/ha) combined with FYM, biofertilizer (*Azospirillum*) and half recommended NPK significantly increased yield and quality of turmeric (Srinivasan *et al.*, 2000). In turmeric, 10% yield increase was reported with the inoculation of *Azosprillum* (Santhanakrishnan and Balashanmugam, 1993). Application of coir pith compost, *Azospirillum*, Phosphobacteria and VAM induced maximum IAA oxidase and peroxidase activity in turmeric (Velmurugan *et al.*, 2002). Several workers reported combined application of organic and inorganic along with *Azospirillum* for higher yield (Jena *et al.*, 1999; Selvarajan and Chezhiyan, 2001; Subramanian *et al.*, 2003), but inorganics are not allowed in organic farming, more over biofertilizer application is restricted to meet the standards of organic system.

The profound effect of zinc solubilizing bacteria (ZSB) on the increase of dry matter yield of turmeric was found to be 14.0, 14.3 and 18.1% for ZnSO4, Zn enriched FYM and Zn enriched coirpith along with ZSB, respectively than in treatments without the organism (Kumar *et al.*, 2004). All the genotypes of turmeric are found to have effective mycorrhizal association by extra-, intra- and intercellular hyphae of VAM fungi in their roots. Among cultivars, Suguna followed by Prabha and Sugandham, were observed to be the most heavily colonized with mycorrhizal fungi with mycelium, arbuscules, vesicles and spores of *Glomus*, *Gigaspora* and *Sclerocystis* with domination of *Glomus* population (Reddy *et al.*, 2003). The mycorrhizal inoculation is advantageous in improving plant growth and plants inoculated with different species of VAM fungi recorded a significant increase in growth compared to uninoculated plants (Kumar, 2004). The highest growth rate was found in plants inoculated with *Glomus fasciculatum*. Mohan *et al.* (2004) recorded a linear response of growth, yield and quality of turmeric with inoculation of *Azospirillum* in combination with N levels compared to that of *Azotobacter* under Karnataka conditions.

Turmeric requires heavy manuring. The nitrogen recommendation varied between 60 and 200 kg ha^{-1}; phosphorus dose between 50 and 125 kg ha^{-1} and potash between 50 and 250 kg ha^{-1}(Table 10). The inorganic fertilizer schedule for different centres is given below and should be supplemented through organic manure so as to ensure good yield.

Table 10. Optimum levels of N, P and K (kg ha^{-1}) recommended for different locations

Optimum N P K	Recommended Location	Reference
150:125:250 kg ha^{-1}	Karnataka	Venkatesha *et al.* (1998)
120: 60:120 kg ha^{-1}	Hillzone of Karnataka	Sheshagiri and Uthaiah (1994)
60: 50: 120 kg ha^{-1}	Calicut, Kerala	Sadanandan and Hamza (1996)
130: 90: 70 kg ha^{-1}	Arunachal Pradesh	Dubey and Yadav (2001)
200: 60: 200 kg ha^{-1}	Alluvial plains of West Bengal	Medda and Hore (2003)
75: 60: 150 kg ha^{-1}	Allahabad agro climate	Thomas *et al.* (2002)
100: 50: 50 kg ha^{-1}	Simla Hills	Randhawa *et al.* (1973)
200: 100: 100 kg ha^{-1}	Maharashtra	Yamgar *et al.* (2001)
100: 100: 100 kg ha^{-1}	Uttar Pradesh, India	Upadhyay and Misra (1999)

7.0 DISEASES AND PEST MANAGEMENT

7.1 Disease

The crop is susceptible to a number of fungal diseases. Bacterial and viral diseases are not so common and are of minor importance. The important fungal diseases affecting the crop are Rhizome rot, leaf blotch and leaf spots (Table 11).

Table 11. Diseases and pests of turmeric and management

Disease	Symptom	Managment
Rhizome rot	*Pythium* species are causal organisms. The disease is characterized by the appearance of water soaked lesions at the base of the pseudo stem and yellowing of the lower leaves. The root infection is manifested as browning and rotting of the roots which later advances into the rhizomes which become soft and rotten. The colour of the rhizome changes from bright orange to different shades of brown. The infection gradually spreads to all the fingers and mother rhizomes and eventually the plants die. When the affected rhizomes are split open, brown to dark brown fibro vascular tissues are seen. The infected plants show gradual drying up of leaves along margins and later the entire leaf dries.	Fungicides are not allowed in organic production. Fields that have received neem cake as a dressing were free from Mimegrella fly (Reddy and Reddy, 2000). Crop rotation is recommended to reduce the incidence of rhizome rot. PCT-13 and PCT-14 from Andhra Pradesh and Ca 69 and Shillong from Assam were found to be tolerant to rhizome rot.
Taphrina leaf spot / Leaf blotch	The disease is caused by *Taphrina maculans.* The disease is characterized by the appearance of small scattered oily looking translucent spots on the lower leaves when the plants are in 3-4 leaf stage. The leaf spots later turn dirty yellow and deepen to colour of gold and some times to bay shade. The adjacent individual leaf spots of 1-2mm in diameter coalesce forming reddish brown blotches leading to varying degrees of leaf blight. In severe case of attack, hundreds of spots appear on both the sides of leaves. The lower leaves are more prone to infection than upper leaves.	Since the infected plant material serve as the perpetuating media for the fungus, crop rotation is important to reduce the inoculum build up in the soil. Phytosanitation is the first and foremost practice in managing this disease. Cutting and burning of diseased leaves would greatly help in preventing the spread of the disease. All short duration cultivars are found resistant to *Taphrina maculans.* The varieties such as CLL 324, Amalapuram, Mydukur, Karhadi local, CLL 326, Ochira 24 and Alleppey among *C. longa* group and ca 68, ca 67, Dahgi and Kasthuri among *C. aromatica* types were reported to be resistant to infection.

Fig. 3: Symptoms of soft rot on turmeric (*Source:* Director, IISR, Calicut)

Fig. 4: Leaf spot caused by *Colletotrichum* sp. (*Source:* Director, IISR, Calicut)

7.2 Pests

The important pest and nematode and their management is given below (Table 12).

8.0 HARVEST AND POST HARVEST HANDLING

In India, at various places, different methods and equipments are used for processing of turmeric, but the traditional method consists of

- Harvesting
- Washing

Table 12. Pests and their management in turmeric

Shoot borer	The shoot borer (*Conogethes punctiferalis* Guen.) (Lepidoptera, Pyralidae) is the most serious insect pest of turmeric, being widely distributed in India. The larvae bore into pseudostems and feed on the central growing shoot resulting in yellowing and drying of infested shoots. The presence of a bore-hole in the pseudostem through which frass is extruded and the withered central shoot is a characteristic symptom of pest infestation. In ginger significant reduction in yield occurs when more than 50% of the pseudostems in a clump are infested by the pest (Koya*et al.*, 1996, Devasahayam and Koya, 2005; 2007).	The natural enemies, *Hexamermis* sp. (Mermithidae) and *Apanteles taragammae* (Braconidae) were identified as a potential biocontrol agents of the pest (Devasahayam, 1996). Commercial formulations of *Bacillus thuringiensis* such as Dipel 0.3% were effective in controlling the pest infestation on ginger and turmeric in the field when sprayed at 21 day intervals during July – October (Devasahayam, 2002).
Rhizome scale	The rhizome scale (*Aspidiella hartii* Ckll.) (Homoptera, Diaspididae) infests rhizomes of turmeric both in the field (especially during the later stages of the crop) and in storage. The pest infestation is generally seen in the form of encrustation on the rhizomes during the later stages of the crop and severely infested plants wither and dry. In storage, the pest infestation results in shriveling of buds and rhizomes affecting their sprouting (Devasahayam and Koya, 2005; 2007).	Hymenopterous parasitoids *Physcus comperei* Hayat (Aphelenidae) and *Adelencyrtus moderatus* Howard (Encyrtidae) were potential biocontrol agents of the pest (Devasahayam, 1996).
Nematode	*Meloidogyne incognita*– rooted knot *Practylenchus sp.* - dark brown necrotic lesions *Radopholus similis* - water-soaked, brownish areas and rotting of roots	Using hot water to treat the tubers at 55°C for 10min. or 45°C for 50min. killed the nematodes inside the tubers. *C. zedoaria* was found more resistant to nematode infestations.

Fig. 5: Shoot borer infestation of turmeric (*Source:* Director, IISR, Calicut)

- Boiling/Blanching/Cooking
- Drying
- Polishing
- Colouring
- Grinding/Powdering
- Packaging
- Marketing

8.1 Harvesting

Maturity of turmeric is indicated by the drying up of the plant including the base of the stem. Depending on the variety this takes 7-9 months after planting. Early variety matures in 7-8 months, medium varieties in 8-9 months and late ones after 9 months. In India normally harvesting is done from January to April, varying with the locations. Early harvesting in December is practiced in response to market requirements. Late harvest is also practiced, as the crop dose not deteriorate by leaving it for some more months underground. In irrigated condition, irrigation is stopped one month before harvest, allowing the tops to dry, leaves and stem are cut to the ground. The land is irrigated, if necessary ,to facilitate digging out the rhizomes, and ploughed in between the rows if the crop is planted on ridges. Otherwise a crowbar is used. The rhizome bunches are carefully lifted and adhering soil is removed by soaking in water and further cleaned of roots and scales before they are collected in the curing yard. The curing quality and the proportion of the cured and dried produce to the green produce depend mainly on the variety. Mother-rhizomes give a higher curing percentage than the fingers. The mother-rhizomes and fingers are separated. If need be, the former is kept for seed and the latter is cured for selling.

According to Govind (1987) and Umarani *et al.* (1982), for maximum rhizome yield crop should be harvested between 8-9 months (Fig. 1). When the crop is intended for curcumin extraction, it should be harvested between 7-8 months after planting.

Power tiller based turmeric harvester is available (Kathirvel & Manian, 2000). The advantages are

- Labour saving - 65% saving in cost.
- Timeliness - 90% saving time.
- Less damage to rhizome, 0.5% compared to manual (3%).
- Rhizome escape is minimum, 0.8% compared to manual (4.9%).
- Harvesting efficiency is 99%.

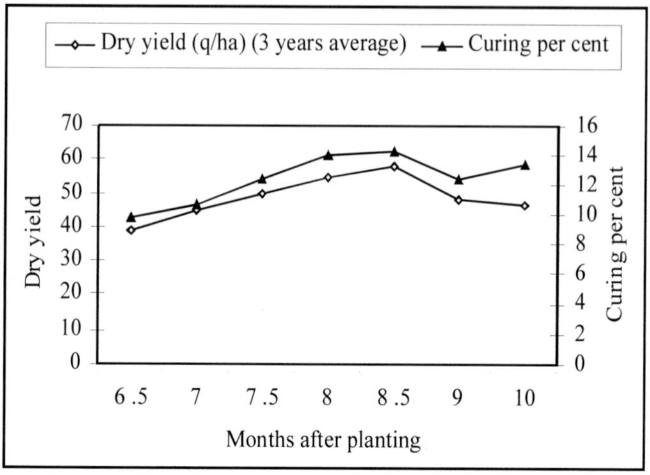

Fig 6. Time of harvest on turmeric yield

8.2 Post Harvest Processing

8.2.1 Washing

Rhizomes are separated after digging out from the soil. The globular mother rhizome and the longitudinal fingers are separated into lots. Root hairs and poorly developed and shriveled portions are then removed. These are washed, so that particles of soil, spray residues and non-useful particles attached to the rhizomes are removed. For this rhizomes are kept soaked in water throughout the night. Later on rhizomes are taken out and water is sprayed. This process can be achieved by soaking and spraying equipment. Spraying is done at low pressure and with wide angle jet or with high pressure jet. Mother rhizome is kept for seed while finger rhizomes are further processed and sold.

8.2.2 Boiling/Blanching/Cooking

The next step of processing is boiling or blanching. Traditionally boiling is done in metal or mud pots along with ¾ water. Top of the pots are covered with a lid or dry leaves. Boiling process is continued till foams and white foams start coming out. These come out with a special quality of flavor. Rhizomes are tested by pressing with fingers. If rhizomes are soft and inner color has become yellow instead of red then this process is said to be complete. The time required for the desired level of cooking varies from 1 hr to 4-6 hrs, depending on the quantity.

The bulbs as wholes or cut longitudinally into halves and the fingers are generally cured separately, as cooking time varies with difference in thickness.

Cooking helps in producing a product of fairly uniform color, due to the diffusion of the yellow pigment from the individual oil cells into the surrounding tissues. Boiling for periods varying from 30 minutes to 3 hrs, steaming upto 20 min., use of 0.1% sodium carbonate in the curing step influences the rate of drying, moisture, volatile oil and the curcumin content after drying. Boiling considerably reduced the drying time both in the sun and the mechanical drier, while the total color and the volatile oil remained practically the same. Steaming and sun drying tended to give slightly less curcumin. Slicing and drying gave a rate of drying equal to 1 hr boiled rhizome and additionally gave a product with the lowest moisture content.

By using developed method of boiling or blanching, both colour and quality are improved. Time taken is less. In the developed method, bulbs are treated with 0.1% Soda (Sodium Carbonate, Sodium bicarbonate or Ammonium Carbonate) and water solution. The time required for this process is 30 minutes to 6 hours. Boiler is used in the developed method, which is metal kettle open from the top. Soda solution is filled in it and it is heated from the bottom by electric heater or fuel oil. Perforated frame loaded with turmeric tubers or bulbs are sunk here. Water from all sides of perforations enters inside making turmeric tubers soft. After 30 or 40 minutes bulbs/tubers are taken out and tested along with the frame.

8.2.3 Drying

Sun drying in specially prepared toughened earth or cemented yards is the usual practice. The cooked material is spread two or three layers thick. Spreading in single layers is avoided to minimize the bleaching effect from the sun. The material is occasionally turned to ensure uniformity in drying. Drying is slow, taking 10-15 days for completion, when properly dried, the rhizomes become hard, almost horny, brittle and of uniform yellow color. The moisture content of the dried rhizomes is one of the lowest for spices; generally being less than 5%. The dry yield of the low moisture product is 20% on an average. Mother rhizome takes comparatively more time while finger rhizomes dry up quickly. Therefore these are dried separately also.

8.2.4 Polishing

Dried turmeric has a poor appearance and a rough and dull outer surface with scales and root bits. The appearance of dried rhizomes is improved by rubbing them against ground or below the foot to take out the hard layer over them and small roots are removed. By this process colour of turmeric becomes bright or shining. Later on removed roots, light garbage and thin layering are discarded. Another manual method is tumbling the dried rhizomes placed in bamboo reed baskets along with granite stones when the pieces are both smoothened and polished. The product is known in the trade as 'polished turmeric'. Manual methods give low output, around 20 kg for 8 hrs for two persons.

Mechanical polishing drums have been developed for handling large quantities. For this a drum having 0.9m diameter and 0.6m length is used. It is kept horizontally on a shaft and operated by a handle. Average capacity of this machine comes to 32 kg per batch. Dry turmeric 32 kg approx. is filled and polishing is done at least for 7 minutes. During this period water is also sprinkled which causes improvement in the color of turmeric. Mechanical power-oil or steam engines or electric motor is now used for large sized drums (which may be circular, hexagonal or octagonal in shape) when handling larger batches of dried rhizomes. During polishing, scales, rootlets and some of the epidermal layer are removed as dust through the sieve mesh and surrounding the polishing drums and the sieved dust is generally used as manure. The material loss during polishing varies from 2 to 8 % depending on the degree of polishing. The turmeric polisher available at IISR, Calicut, has a capacity of 100 kg per batch and takes 1 ½ h for fine polishing of the rhizomes.

8.2.5 Colouring

Exporting turmeric is given special colour by mixing yellow so that powder and processed materials can give better look and quality by a dry or wet process. In the dry process, turmeric powder is added to the polishing drum in the last 10 min. In the wet colouring process, turmeric powder is suspended in water and mixed inside by sprinkling inside the polishing basket. After colouring is complete, these are dried for one week. Later on these rhizomes are kept in sacks and closed for exporting.

8.2.6 Grinding/Powdering

Traditionally dried and polished turmeric are cut into pieces and beaten in mortar and pestle. After words, this is milled or ground with hand operated chakki. Hammer mill is also used for grinding. Powder should be so fine that it passes through 300 micron sieve and nothing is left over the sieve.

8.2.7 Packaging

Cured dried turmeric with a moisture content of 15-30% is transported in gunny bags to assembling centers where it is further dried, polished and colored, if necessary. Dried turmeric is graded according to Agmark specifications before bagging in strong gunny bags generally in 65-75 kg lots. The bags are then transported to warehouses and port centers. Fumigation and prophylactic treatments are routinely given during warehousing and before export. The color of turmeric has been found to be stable as long as it is not exposed to sunlight.

8.2.8 Standards and grade specifications

The Indian standard specification for turmeric whole and powder incorporating the country's Agmark grade specifications are summarized below in Table 13 &14.

Table 13. Indian Specification for turmeric-Grades

Grade designation	Pieces max;wt%	Foreign matter Max; wt%	Defectives Max: wt%	Bulbs Max ; wt%	Characteristics
Fingers(General)					Finger-like shape; breaks with a metallic twang; well set and close grained be perfectly dry; free from damage from weevils, over boiling etc.
Special	2.0	1.0	0.5	2.0	
Good	3.0	1.5	1.0	3.0	
Fair	5.0	2.0	1.5	5.0	
Fingers, Alleppy					As above
Good	5.0	1.0	3.0	4.0	
Fair	7.0	1.5	5.0	5.0	
Fingers, Rajapore	7.0				
Special		1.0	3.0	2.0	As above ;Admixture of other *Curcuma* varieties allowed at a maximum 2,5 and 10% in the three grades respectively.
Good		1.5	5.0	3.0	
Fair		2.0	7.0	5.0	
Bulbs(round)					
Special	-	1.0	1.0	-]	Be well developed, smooth, sound, free from rootlets The Rajapore variety has higher allowance of 3,5 and 7% defectives in the three grades respectively.
Good	-	1.5	3.0	-	
Fair	-	2.0	5.0	-	

Source: Marketing of turmeric in India, Agric. Market. Ser. No. 148, Directorate of Marketing and Inspection, Govt. of India, Nagpur. 1965

8.2.9 Processed products

India is the global leader in value-added products development and exports. Value added products from turmeric include curcuminoids, dehydrated turmeric powder, oils, and oleoresin. Turmeric like other spices is available as wholes, grinds and oleoresin. The institutional sector in West buys ground turmeric and oleoresins, while in the industrial sector, the spice

Table 14. Analytical specification for turmeric

Sample	Moisture (max) %	Total Ash (max) %	Acid insol. Ash (max) %	Starch (max) %	Crude fibre (max) %	Volatile oil (max) %	Color as curcumin %	Lead (max ppm)	Chromate test
Whole BP	8-10	6-9	-	-	4-6	2-5	-	-	-
US	9	7	0.5	-	6	4	5	-	-
DDR	-	7	-	-	-	2.6	3-4	-	-
Powder Indian	10	7	1.5	60.0	-	-	-	1.5	Negative
WHO	10	7	1.5	-	-	-	-	3	Negative

Source: Specification for turmeric powder, IS..2446 Indian Standards Institution, Delhi, India. 1961

grinder and flavour houses buy whole dry turmeric from importers. The nature of the material, the curcuminoids and the aromatic components are stable to moderate heat.

8.2.9.1 Turlmeric powder

Dried turmeric is powdered to fine mesh (60) to use in various end products.

8.2.9.2 Turmeric oil

The peculiar turmeric aroma is imparted by ar-turmerone, the major aroma principle in the oil and leaves are also used for extracting the volatile oil.

8.2.9.3 Turmeric Oleoresin

Turmeric oleoresin is the organic extract of turmeric, a ground powder from the root of the *Curcuma* plant, and is added to food items as a spice and coloring agent. Supercritical CO_2 extraction and molecular distillation to extract and separate turmeric is the latest high technology process followed to get quality product. Turmeric oleoresin is essentially used in institutional cooking in meat and fish products and certain products such as mustard, pickles and relish formulas, butter and cheese.

8.2.9.4 Curcumin

Curcumin or curcuminoids concentrate for use as a food color is not a regular article of commerce, because for most current uses the cheaper turmeric oleoresin has been found suitable. However, when demand for natural color increases in foods as in ice creams, gelatins and lemonades, the use of curcuminoids concentrate could become established.

8.3 Economics and marketing

The countries like Japan, USA, Iran, UK, Germany, Netherlands and Sri Lanka are the major importer of turmeric in the world. China, Pakistan, Bangladesh, Haiti, Jamaica, Peru and Taiwan are producing and exporting turmeric besides India. The increasing use of curry powder and spice mixes in the world and the trend in the developing countries towards shifting to natural colors in food, from artificial coloring matter, the prospects of increased trade in turmeric appears to be quite good. Turmeric is exported in the form of dry, powder and oleoresin.

In India, the arrivals of turmeric start from February and more than 70 % of the harvested crop is assembled in the markets during the next 3-4 months. It is utilized mostly after raw turmeric is processed and polished. The producers for seed purposes retain a small percentage of the raw turmeric produced. Only less than 15 % of the production is exported annually and the remaining quantity is consumed internally.

8.3.1 Marketing Practices

The system of agricultural marketing in India is saddled with a long chain of middlemen. Since the producers are mostly small and marginal farmers, they generally do not hold up the produce for long period since it is difficult for them to arrange for its storage. The general pattern of marketing is that the producers bring the dried produce to the primary markets and sell it to the village/itinerant merchants. The marketable produce in turmeric is either polished or unpolished in the form of fingers or mother rhizomes (bulbs). The village merchants fix the price based on quality factors such as hardiness, intensity of the core, polished or unpolished etc. Generally, village merchants have an upper hand in settling the price. When substantial stocks are accumulated, the commodity is carried to main assembling center. It has been estimated that the village merchants assemble more than 60 % of the production and the producers sell rest direct to the wholesalers at the assembling centers through commission agents who play an important role in the trade. Exporters generally have direct contact with foreign buyers and also with upcountry markets, purchase their requirements mostly through commission agents.

Due to price fluctuation and lack of price stability, producers are affected. The reasons for low price are attributed to (i) lack of holding power of the producers due to their poor financial position; (ii) loss in quality due to pest and diseases; (iii) competition among the growers to find buyers immediately after harvest and declining prices; (iv) poor bargaining power of the farmers; and (v) lack of scientific storage facility at village level.

8.3.2 Economics

A survey conducted at Nizamabad in Andhra Pradesh and Erode in Tamil Nadu found that farmer gets only a nominal profit and if the price falls further, he may not get any profit. Cost of cultivation for conventional method of turmeric is Rs 93, 000 / per ha whereas organic cultivation recorded less expense Rs. 85, 125 / per ha with more dry yield (7 to 9 MT /ha) compared to 6 to 7.5 MT / ha in conventional method. The cost of harvesting per hour was Rs. 5,500 with the harvester while it was Rs. 7,400 with manual harvesting. The saving in cost was 26 per cent by harvesting with mechanical harvester over manual harvesting. Economics of turmeric production vary at production centers and mainly depends on demand for produce in the market. Further cost of inputs like fertilizer, labor, and seed mainly influence the total cost of production.

REFERENCES

Agnihotri, B. N. (1949) Turmeric. *Indian J. of Hort.* 6(3 &4): 28-31.

Aiyadurai, S. G. (1966) A review of research on spices and cashewnut in India. Regional office (Spices and Cashew), ICAR, Ernakulam, India. p. 209.

Anjaneyulu, V. S. R. and Krishnamurthy, D. (1979a) The efficacy of broad ridge method of planting turmeric. *Indian Cocoa, Arecanut and Spices J.* 2(4): 116-118.

Anjaneyulu, V. S. R. and Krishnamurthy, D. (1979b). Efficacy of type of seed materials for different times of planting turmeric under Duggirala conditions. *Indian Cocoa, Arecanut and Spices J.* 2(4): 115-116.

Anjaneyulu, V. S. R., Krishnamurthy, D. (1979c). Spacing cum type of seed material trials on turmeric. *Indian Cocoa, Arecanut and Spices J.* 2(4): 119-120.

Anonymous. (1987) Maize and turmeric are profitable intercrops.*Indian Cocoa, Arecanut and Spices J.* 11(2): 63.

Aoi, K. (1992) The characteristics and cultivation methods of the medicinal plant *Curcuma domestica. Agriculture and Horticulture* 67(5): 507-511.

Avilkumar, K. and Reddy, M. D. (2000) Integrated weed management in maize+turmeric intercropping system. *Indian J. Weed Sci.* 32(1/2): 59-62.

Balashanmugam, P. V., Ahamed Shah and N. Chezhiyan. (1988) Studies on relay inter cropping in turmeric. *South Indian Hort.* 36(4): 207-208.

Balashanmugam, P. V., Chezhiyan, N. and Chamy, A. (1989) Effect of time of planting on growth and yield of turmeric. *Indian Cocoa, Arecanut and Spices J.* 12(4): 124-125.

Barholia, A. K., Bisen, A. L. and Mishra, K. K. (1992) Growth and yield of turmeric as influenced by time of planting and planting material. *Gujarat Agricultural University Res. J.* 17(2): 172-174.

Bisht, J. K., Chandra, S., Singh, R. D. (2004) Performance of taro (*Colocasia esculenta*) and turmeric (*Curcuma longa*) under fodder trees in a agri-silvi-horti system of western Himalaya. *Indian J. Agric. Sci.* 74(6): 291-294.

Carlos, L. (2006) Turmeric: A Spice Rooted in Catalan Medieval History. Available online at (*http://barcelonaphotoblog.blogspot.com/2006_11_01_archive.html*) (accessed online on 11-3-07)

Carvalho, C. M., Souza, R. J. de. and Cecilio Filho, A. B. (2001) Yield of turmeric (*Curcuma longa* L.) grown at different planting densities. *Cienciae Agrotecnologia* 25(2): 330-335.

Cecilio, F. A. B., Souza, R. J. de., Faquin, V. and Carvalho, C. M. de. (2004). Time and density of plantation on the turmeric production. *Ciencia Rural* 34(4): 1021-1026.

Chandra, R., Sheo Govind. and Desai, A. R. (1999) Growth, yield and quality performance of turmeric (*Curcuma longa* L.) genotypes in mid altitudes of Meghalaya. *J. Applied Hort.* 1(2): 142-144.

Chaturvedi, O. P. and Pandey, I. B. (2001) Yield and economics of *Populus deltoides* G3 Marsh based inter-cropping system in Eastern India. *Forests Trees and Livelihoods* 11(3): 207-216.

Chaugule, B. A. and Mohite, B. V. (1962) Yield of turmeric in the broad ridge and furrow lay-outs with and without maize as an intercrop. *Poona Agriculture College Magazine* 52(3&4): 14.

Chempakam,B., Suseela Bhai, R., Kandiannan, K., Krishnamurthy, K.S. and Parthasarathy, V.A. (2006) Turmeric. In: (Eds. V.A. Parthasarathy and P. Rajeev) *Major Spices – Production and Processing*. Indian Institute of Spices Research, Calicut, pp. 152-189.

Chikarmane, S., Rehse T., and Pryer, K.M. (2003) Tracing the cultural and botanical origins of turmeric (*Curcuma longa* L.)) (*Poster.http://www.2003.botanyconference.org/engine/search/ detail.php?aid=558.*) (available online, accessed on 11-3-07) .

DASD. (2007) *Arecanut and Spices Database*. Directorate of Arecanut and Spices Development, Calicut. p.110.

De Geus J.G. (1973) *Fertilizer Guide for the Tropics and Subtropics*, 2nd edn. Centre d' Étude de l'Azote, Zurich.

Devasahayam, S. (1996) Biological control of insect pests of spices. In: Anandaraj, M. and Peter, K. V. (eds.) *Biological control in spices*. Calicut; Indian Institute of Spices Research. pp. 33-45.

Devasahayam, S. (2002) Evaluation of biopesticides for the management of shoot borer (*Conogethes punctiferalis* Guen.) on turmeric (*Curcuma longa* L.). In: Rethinam, P., Khan, H. H., Reddy, V. M., Mandal, P. K. and Suresh, K. (eds.). *Plantation crops research and development in the new millennium*. Kochi; Coconut Development Board. pp. 489–490.

Devasahayam, S. and Koya, K. M. A. (2005) Insect pests of ginger. In : Ravindran, P. N. and Babu, K. N. (eds.) *Ginger. The genus Zingiber*. Washington; CRC Press. pp. 367–389.

Devasahayam, S. and Koya, K. M. A. (2007) Insect pests of turmeric. In : Ravindran, P. N., Babu, K. N. and Sivaraman, K. (eds.) *Turmeric. The genus Curcuma* . Washington; CRC Press.

Dixit, D. and Srivastava, N. K. (2000). Effect of iron deficiency stress on physiological and biochemical changes in turmeric (*Curcuma longa*) genotypes. *J. Medicinal and Aromatic Plant Sci.* 22 (1B): 652-658.

Dixit, D., Srivastava, N. K. and Sharma, S. (2002) Boron deficiency induced changes in translocation of $^{14}CO_2$-photosynthate into primary metabolites in relation to essential oil and curcumin accumulation in turmeric (*Curcuma longa* L.). *Photosynthetica* 40(1): 109-113.

Dubey, A. K. and Yadav, D. S. (2001) Response of turmeric (*Curcuma longa* L.) to NPK under foothill conditions of Arunachal Pradesh. *Indian J. Hill Farming* 14 (2): 144-146.

Ghadge, S. V., Agarwal, K. N., Singh, R. K. P. and Satapathy, K. K. (2001) Turmeric mechanization in Meghalaya - a case study of Shangpung village Jaintia Hills. *Indian J. Hill Farming* 14(2): 99-104.

Gill, B. S., Randhawa, R. S., Randhawa, G. S and Singh, J. (1999) Response of turmeric (*Curcuma longa*) to nitrogen in relation to application of farm yard manure and straw mulch. *J. Spices and Aromatic Crops* 8(2): 211-214.

Gill, B. S., Simerjit Kaur., Saini, S. S. (2004) Influence of planting methods, spacing and farmyard manure on growth, yield and nutrient content of turmeric (*Curcuma longa* L.). *J. Spices and Aromatic Crops* 13(2): 117-120.

Govind, S. (1987) Studies on optimum harvesting time of turmeric. *Haryana J. Hort. Sci.* 16 (3-4): 257-263.

Govinden, N. and Wong Yen Cheong, K. (1995) Planting material and optimum planting rate for turmeric. *Revue Agricole-et-Sucriere-de-l'Ile Maurice* 74(3): 1-8.

Hegde, S., Venkatesh, J. and Chandrappa, H. (1998) Performance of promising turmeric cultivars (*Curcuma domestica* Val.) under coconut cropping systems. In: (N.M.Mathew & C. Kuruvilla Jacob (Eds.) *Developments in Plantation Crops Research*. Allied Publishers, New Delhi, pp. 220-222.

Hossain, M A., Ishimine, Y., Akamine, H. and Motomura, K. (2005a) Effects of seed rhizome size on growth and yield of turmeric (*Curcuma longa* L.). Plant Production Science, 8(1): 86-94. Available online at *http://www.jstage.jst.go.jp/article/pps/8/1/86/_pdf* (accessed on 11-3-07).

Hossain, M. A., Ishimine,Y., Motomura, K. and Akamine, H. (2005b) Effects of planting pattern and planting distance on growth and yield of turmeric (*Curcuma longa* L.). *Plant Production Science* 8(1): 95-105.

Hossain, M.A. (2005) Agronomic practices for weed control in turmeric (*Curcuma longa* L.). *Weed Biology and Management* 5 (4) : 166-175.

Hu Min Fu, Jene, T.S., Fa, C.I., Yie, L.S. (2003) Effects of combined chicken compost and chemical fertilizer application on the yield and quality of turmeric (*Curcuma longa* L.). *J. Agric. Res. China* 52(4): 334-340.

Hu, M.F., Chiu, S.M., H.I. Liu, H.I., Lai, M.H. and Liu, S.Y. (1996) Effects of planting date and density on the rhizome yield and curcumin content of turmeric plant (*Curcuma aromatica* Salisb). *J. Agric. Res. China* 45(2): 164-173.

Hussain, A., Washeeda, A. and Zafar, M. A. (1969) Effect of mulches on germination, growth and yield of turmeric. *W. Pakistan J. Agric. Res.* 7: 153-157.

IISR. 2005a. *Turmeric* (Extension Pamphlet). Indian Institute of Spices Research, Calicut.

IISR. 2005b. *Annual Report*, 2005. Indian Institute of Spices Research (ICAR), Calicut, India. pp. 45-46.

Ishimine, Y., Md Amzad Hossain and Murayama, S. (2003) Optimal planting depth for turmeric (*Curcuma longa* L.) cultivation in dark red soil in Okinawa Island, Southern Japan. *Plant Production Science* 6(1): 83-89.

Ishita, C., Kaushik Biswas, Uday Bandyopadhyay and Ranajit K. Banerjee. (2004) Turmeric and curcumin: Biological actions and medicinal applications. *Current Sci.* 87 (1): 44 - 53.

ITC, (2004) Marketing Manual and Web Directory for Organic Spices, Herbs and Essential Oils. International Trade Centre (UNCTAD/WTO), Geneva.

Jena, M. K., Das, P. K. and Pattanaik, A. K. (1999) Integrated effect of microbial inoculants and fertilizer nitrogen on N-use efficiency and rhizome yield of turmeric (*Curcuma longa* L.). *Orissa J. Hort.* 27(2): 10-16.

Jha, R. C., Sharma, N. N. and Maurya, K. R. (1983) Effect of sowing dates and mulching on the yield and profitability of turmeric (*Curcuma longa*). *Bangladesh Hort.* 11 (1): 1-4.

Kandiannan, K. and Chandaragiri, K.K. (2006) Influence of varieties, dates of planting, spacing and nitrogen levels on growth, yield and quality of turmeric (*Curcuma longa*). *Indian J. Agric. Sci.* 76 (7) : 432 -434.

Kandiannan, K., Chandaragiri, K. K., Sankaran, N., Balasubramanian, T. N. and Kailasam, C. (2001) Empirical model for turmeric (*Curcuma domestica* Val.) yield prediction. J. *Spices and Aromatic Crops* 10(1): 59-61.

Kandiannan, K., Chandaragiri, K. K., Sankaran, N., Balasubramanian, T .N. and Kailasam, C. (2002) Weather during turmeric production and its relation to turmeric yield under tropical irrigated condition. *J. Plantn. Crops* 30(2): 45-49.

Kandiannan, K., Thankamani, C.K. and Shiva, K. N. (2005) Intercropping with spices – A special reference to turmeric. *Spice India* XVIII (9): 15-20.

Kapoor, L. D. (1990) Handbook of Ayurvedic Medicinal Plants, CRC Press, Boca Raton, Florida, p. 185.

Kathirvel, K. and Manian, R. (2000) Turmeric Harvesting. *Kisan World* 27 (11) : 48.

Kerutagi, M. G., Kotikal, Y. K., Hulamani, N. C and Hiremath, G .K. (2000) Costs and returns of turmeric production in Belgaum district of Karnataka. *Karnataka J. Agric. Sci.* 13(1): 209-211.

Kirankumar Gorabal., Rokhade, A.K. and Hanamashetti, S. I. (2002) Effect of methods of storage on post-harvest losses and viability of seed rhizomes in turmeric (*Curcuma longa* L.). *J. Plantn. Crops* 30(2): 68-70.

Kolev, T. M., Velcheva, E. A., Stamboliyska, B. A. and Spiteller, M. (2005) Properties, dynamics, and electronic structure of atoms and molecules DFT and experimental studies of the structure and vibrational spectra of curcumin. *International J. Quantum Chem.* 102 (6) : 1069 - 1079.

Koya, K. M. A., Devasahayam, S., Selvakumaran, S. and Kallil, M. (1996) Distribution and damage caused by scale insects and mealybugs associated with black pepper (*Piper nigrum* L.) in India. *J. Entomol. Res.* 20: 129–136. .

Kumar, D., Ranjeet Singh., Gadekar, H. and Patnaik, U. S. (2003a). Effect of different mulches on moisture conservation and productivity of rainfed turmeric. *Indian J. Soil Conservation* 31(1): 41-44.

Kumar, G. S. (2004) Association of VAM fungi in some spices and condiments. *J. Ecobiology* 16(2): 113-118.

Kumar, G. V. V., Khan, M. A. A. and Hameedunnisa Begum. (2000) Influence of mineral nutrient composition of turmeric rhizome on curcumin content of different cultivars. *International J. Trop. Agric.* 18(3): 265-269.

Kumar, P. S. S., Geetha, S. A., Savithri, P., Jagadeeswaran, R. and Mahendran, P. P. (2003b) Diagnosis of nutrient imbalances and derivation of new RPZI (Reference Population Zero Index) values using DRIS/MDRIS and CND approaches in the leaves of turmeric (*Curcuma longa*). *J. Applied Hort.* 5(1): 7-10.

Kumar, P. S. S., Geetha, S. A., Rajarajan, A., Savithri, P. and Raj, S. A. (2004) Evaluation of Zn use efficiency of Zn enriched organic manures and zinc solubilizing bacteria in turmeric crop and distribution of applied Zn among various Zn fractions of the soil using ^{65}Zn radiotracer technique. *Research on Crops.* 5(1): 118-125.

Kundu, A. L. and B. N. Chatterjee. (1982) Growth analysis of turmeric as a sole crop and in mixture with other crops. Indian J. Agric. Sci. 52(9): 584-589.

Latha, P., Giridharan, M. P and Jayaprakash Naik, B. (1995) Performance of turmeric (*Curcuma longa*) cultivars in open and partially shaded conditions under coconut. *J. Spices and Aromatic Crops* 4 (2): 139-144.

Manjunathgoud, B., Venkatesha, J. and Bhagavantagoudra, K. H. (2002). Studies on plant density and levels of NPK on growth, yield and quality of turmeric cv. Bangalore Local. *Mysore J. Agric. Sci.* 36(1): 31-35.

Medda, P. S. and Hore, J. K. (2003) Effects of N and K on the growth and yield of turmeric in alluvial plains of West Bengal. *Indian J. Hort.* 60(1): 84-88.

Meerabai, M., Jayachandran, B. K., Asha, K. R. and Geetha, V. (2000) Boosting spice production under coconut gardens of Kerala: maximizing yield of turmeric with balanced fertilization. *Better Crops International* 14(2): 10-12.

Mishra, M., Mishra, B. B. and Mishra, S. N. (1997) Effect of planting dates and varieties on yield of turmeric (*Curcuma longa*). *Indian J. Agron.* 42: 713-716.

Mishra, M., Mishra, S. N. and Patra, G. J. (2000). Effect of depth of planting and mulching on rainfed turmeric (*Curcuma longa*). *Indian J. Agron.* 45: 210-213.

Mishra, R. K. and Pandey, V. K. (1998) Intercropping of turmeric under different tree species and their planting pattern in agroforestry system. *Range Management and Agroforestry* 19(2): 199-202.

Mohan, E., Melanta, K. R., Guruprasad, T. R., Herle, P. S., Gowda, N. A. J. and Naik, C. M. (2004) Effects of graded levels of nitrogen and biofertilizers on growth, yield and quality in turmeric (*Curcuma domestica* Val.) cv. D K Local. *Environment and Ecology* 22(3): 715-719.

Mohanty, D. C., Sarma, Y. N and Panda B. S. (1991) Effect of mulch materials and intercrops on the yield of turmeric cv. Suroma under rainfed conditions. *Indian Cocoa Arecanut and Spices J.* 15(1): 8-11.

Mudaliar, V.T. S., (1960) Turmeric. *In : South Indian Field Crops.* Viswanathan S. and Anandam, V. (Eds.) Central Art Press, Madras (presently Chennai), pp. 221-228.

Nagarajan, M. and Pillai, N. G. (1979). A note on the nutrient removal by ginger and turmeric rhizomes. *Madras Agric. J.* 66: 56-59.

Nair, P. C. S. (1982) Agronomy of ginger and turmeric. In :(Eds. Nair, M K, Premkumar, T., Ravindran, P. N. and Sarma, Y. R.) *Ginger and Turmeric*, Central Plantation Crops Research Institute, Kasargod, India. pp. 63-68.

Nazeem, P.A., Menon, R. and Valsala. P.A. (1993) Blossom, biological and hybridization studies in turmeric (*Curcuma* spp.). *Indian Cocoa Arecanut and Spices J.* 16 : 106-109.

Nel, A. (1988) Turmeric trials. Important findings. Information Bulletin, Citrus and Subtropical Fruit Research Institute, South Africa, 188: 11-12.

Pal, S., Maiti, S. and Mandal, N. N. (2000) Effect of N and K fertilizer on turmeric under sole and alley cropping system. *J. Interacademicia* 4(2): 274-283.

Panigrahi, U. C., Patro, G. K and Mohanty, G. C. (1987) Package of practices for turmeric cultivation in Orissa. *Indian Farming* 37(4): 4-6.

Parthasarathy, V. A., Srinivasan, V. and Dinesh, R. (2006) Organic Spices- present status and scope. *In : National Seminar on Appropriate Technologies for Sustainable Horticulture.* Souvenir, Annamalai University, Tamil Nadu. pp. 34-42.

Philip, J. (1983) Effect of different planting materials on growth, yield and quality of turmeric. *Indian Cocoa Arecanut and Spices J.* 7(1): 8-11.

Poduval, M., Mathew, B., Hasan, M. A. and Chattopadhyay, P. K. (2001) Yield and curcumin content of different turmeric varieties and species. *Environment and Ecology* 19(4): 744-746.

Ponnuswamy, V. and Muthuswami, S. (1981) Influence of spacings on yield and yield components of turmeric (*Curcuma longa* L.). *South Indian Hort.* 29(4): 229-230.

Pradhan, U B, Maiti, S and Pal, S. (2003) Effect of frequency of pruning and tree spacing of Leucaena on the growth and productivity of turmeric when grown under alley cropping system with *Leucaena leucocephala. J. Interacademicia* 7(1): 11-20.

Prasadji, J. K., Rao, D. M., Murthy, K. V. M. K., Pandu, S. R. and Muralidharan, K. (2002) Effect of cultural practices on leaf blotch severity and yield in turmeric. *Indian J. Plant Protection* 30(2): 115-119.

Purseglove, J.W., Brown, E.G., Green, C.L. and Robbins, S.R.J. (1981) Turmeric. In : Spices. Volume 2. Longman Group Limited, London, pp. 532-580.

Pushkaran, K., Babylatha, A. K. and George, K. M. (1985) Comparative performance of turmeric varieties in coconut gardens. *South Indian Hort.* 33(4): 269-270.

Rajput, S. G., Patil, V. K., Warke, D. C., Ballal, A. L and Gunjkar, S. N. (1982) Effect of nitrogen and spacing on the yield of turmeric rhizomes. In :(Eds. Nair, M K, Premkumar, T., Ravindran, P. N. and Sarma, Y. R.) *Ginger and Turmeric,* Central Plantation Crops Research Institute, Kasargod, India. pp. 83-85.

Ramachandran, M. and Muthuswami, S. (1984) Studies on the influence of method of planting and spacing on yield and quality of turmeric. *South Indian Hort.* 32: 143-145.

Randhawa, K. S., Nandpuri, K. S and Bajwa, M. S. (1973) Response of turmeric (*Curcuma longa*) to NPK fertilization. *J. Res. Punjab Agricultural University.* 10(1): 45-48.

Randhawa,G. S., Mahey, R. K., Gill, S. R. S. and Sidhu, B. S. (1984) Performance of turmeric (*Curcuma longa* L.) under different dates and methods of sowing. *J. Res. Punjab Agric. Univ.* 21(4): 489-495.

Rao, A.M., Jagdeeshwar, R. and Sivaraman, K. (2006) Turmeric. In : (Eds. P.N. Ravindaran, K. Nirmal Babu, K.N. Shiva, Johny A. Kallupurackal). *Advances in Spices Research History* and *Achievements of Spices Research in India Since Independence.* Agrobios (India), Jodhpur, pp. 433-491.

Rao, C.H. (1957) Profitable intercrops in coconut plantations of East Godavari district. *Andhra Agricultural J.* 4(3) : 73-75.

Rao, M.R. and Rao, D.V.R. (1994) Genetic resources of turmeric. In : (Eds. K.L. Chadha and P. Rethinam) *Advances in Horticulture Vol 9. – Plantation and Spice Crops, Part 1.* Malhotra Publishing House, New Delhi.

Rao, T. S. (1979) Turmeric cultivation in Andhra Pradesh. *Indian Cocoa Arecanut and Spices J.* 2(2): 31-32.

Rao, A. M. and Reddy. L . (1990) Population and fertilizer requirement of maize in turmeric maize inter cropping system. *J. Plantn. Crops* 18(1): 44-49.

Rao, V. P., Reddy, M. S., Gopalakrishna, V. and Rao, K. S. (1992) Water-production function and optimal irrigation programme for turmeric (*Curcuma longa*). *Indian J. Agric. Sci.* 62(6): 405-407.

Ravindran, P.N., Nirmal Babu, K. and Sivaraman, K. (2007) Turmeric: The genus *Curcuma.* Taylor & Francis CRC Press, London

Ray, A. K. and Reddy, D. V. S. (2001) Performance of areca-based high-density multispecies cropping system under different levels of fertilizer. *Trop. Agric.* 78(3): 152-155.

Reddy, M. N., Devi, M. C and Sridevi, N. V. (2000) Evaluation of turmeric cultivars for VAM colonization. *Indian Phytopathol*. 56(4): 465-466.

Reddy, M. R. S. and Reddy, P. V. R. M. (2000) Preliminary observations on neem-cake, against rhizome fly of turmeric (*Curcuma longa*). *Insect Environment* 6(2): 62.

Rethinam, P., Sivaraman, K. and Sushama, P. K. (1994) Nutrition of turmeric. In : (Eds. K.L. Chadha and P. Rethinam) *Advances in Horticulture Vol 9. – Plantation and Spice Crops, Part 1.* Malhotra Publishing House, New Delhi, pp. 477-489.

Ridley, H.N. (1912) Turmeric. pp. 422-444. In : Spices. Macmillan and Co. Ltd., London (Reprinted in 1983 by International Book Distributors, Dehradun).

Ruby, A. J., Kuttan, G., Dinesh Babu, K., Rajasekharan, K. N. and Kuttan, R. (1995) Antitumor and antioxidant activity of natural curcuminoids. *Cancer Letter* 94 : 79–83.

Sadanandan, A. K. and Hamza, S. (1996) Response of four turmeric (*Curcuma longa* L) varieties to nutrients in an oxisol on yield and curcumin content. *J. Plantn. Crops* 24 (Suppl.): 120-125.

Sadanandan, A. K., Kandiannan, K. and Hamza, S. (1998) Soil nutrients and water management for sustainable spices production. In: Sadanandan, A. K., Krishnamurthy, K. S., Kandiannan, K and Korkanthimath, V. S. (Eds). *Proc. Water and nutrient management for sustainable production and quality of spices.* Indian Society Spices, Calicut, India, pp. 12-20.

Sadanandan, A. K. and Hamza, S. (1998) Effect of organic manures on nutrient uptake, yield and quality of turmeric (*Curcuma longa* L.). *In: Developments in plantation crops research.* Proc. Plantation crops symposium XII (PLACROSYM XII), Mathew, N. M. and Kuruvilla Jacob, C (Eds), Indian Rubber Research Institute, Kottayam., pp. 175-181.

Saha, A. K. (1988) Note on response of turmeric to manure and source of N and P under terrace conditions of mid-altitude Mizoram. *Indian J. Hort.* 45(1-2): 139-140.

Saifudeen, N. (1981). Foliar diagnosis, yield and quality of turmeric (*Curcuma longa* L.) in relation to nitrogen, phosphorus and potassium. M.Sc. (Ag) thesis, KAU, Trichur.

Salvi, N D., George, L and Eapen, S. (2002) Micropropagation and field evaluation of micropropagated plants of turmeric. *Plant Cell Tissue and Organ Culture,* 68(2): 143-151.

Santhanakrishnan, P. and Balashanmugam, S. (1993) Paper presented in All India Tuber Crop Meet at Tamil Nadu Agricultural University, Feb, 1993, Coimbatore.

Sarma, S.S. and Murthi, D.K. (1965) Type of seed rhizomes in relation to time of planting turmeric. *Andhra Agricultural J.* 12: 84-91.

Sarma, U. J. and Chowdhury, D. (2002) Synergestic effect of high density multiple cropping on soil productivity and yield of base crop (coconut). *Indian Coconut J.* 33(8): 18-22.

Sasikumar, B., Ravindran, P.N. and George, J.K. (1994) Breeding ginger and turmeric. *Indian Cocoa Arecanut and Spices J.* 18 : 10-12.

Sekar, K. and Muthuswami, S. (1985). Economics of double intercropping in turmeric. *Indian Cocoa Arecanut and Spices J.* 8(3): 67-69.

Selvaraj, P. K., Krishnamurthi, V. V., Manickasundaram, P., Martin, G. J. and Ayyaswamy, M. (1997) Effect of irrigation schedules and nitrogen levels on the yield of turmeric through drip irrigation. *Madras Agricultural J.* 84: 6, 347-348.

Selvarajan, M. and Chezhiyan, N. (2001) Studies on the influence of *Azospirillum* and different levels of nitrogen on growth and yield of turmeric (*Curcuma longa* L.). *South Indian Hort.* 49: 140-141.

Shankaraiah, V. and Reddy, I. P. (1988) Studies on effect of different manurial and spacing levels with local and high yielding turmeric varieties. *Indian Cocoa. Arecanut and Spices J.* 11(3): 94-95.

Shankaraiah, V., Reddy, I. P. and Rao, R. R. (1987) Studies on intercropping in turmeric with maize, chilly, castor and okra. *Indian Cocoa Arecanut and Spices J.* 11 (2): 50-52.

Shanmugam, P., Vijayakumar, M. and Shanmugasundaram, K. A. (2000) Standardization of planting material with reference to the stages of sprouting in turmeric (*Curcuma longa* L.). *In: Spices and Aromatic Plants: Challenges and Opportunities In The New Century,* Centennial conference on spices and aromatic plants, Calicut, Kerala, India, 20-23 September 2000.pp. 113-116.

Shashidhar, T. R. and Sulikeri, G. S. (1996) Effect of plant density and nitrogen levels on growth and yield of turmeric (*Curcuma longa* L.). *Karnataka J. Agric. Sci.* 9 (3): 483-488.

Sheshagiri, K. S. and Uthaiah, B. C. (1994) Effect of nitrogen, phosphorus and potassium levels on growth and yield of turmeric (*Curcuma longa*) in the hill zone of Karnataka. *J. Spices and Aromatic Crops.* 3(1): 28-32.

Shishodia, S., Sethi, G. and Aggarwal, B. B. (2005) Curcumin: Getting Back to the Roots. In : Natural Products and Molecular Therapy. *Annals of New York Academy Sciences.* 1056: 206–217.

Silva, N. F. da., Sonnenberg, P. E. and Borges, J. D. (2004) Growth and production of turmeric as a result of mineral fertilizer and planting density. *Horticultura Brasileira.* 22(1): 61-65.

Singh, A., Bajrang Singh and Vaishya, R. D. 2002. Integrated weed management in turmeric (*Curcuma longa*) planted under poplar plantation. *Indian J. Weed Sci.* 34(3/4): 329-330.

Singh, A. K. and Edison, S. (2003) Eco-friendly management of leaf spot of turmeric under partial shade. *Indian Phytopathol*. 56(4): 479-480.

Singh, B. P., Maurya, K. R., Sinha, M. K and Singh, R. A. (1986) Reaction of genotypes of turmeric (*Curcuma longa* L.) to Zn stress in a calcareous soil. Proceedings of the *Indian National Science Academy- Biological Sciences.* 52(3): 385-390.

Singh, J. and Randhawa, G. S. (1988) Effect of intercropping and mulch on yield and quality of turmeric (*Curcuma longa* L.). *Acta Horticulturae* 188 A: 183-186.

Singh, R. V. and Rai, M. (2003) Standardization of mango based cropping system for sustainable production. *J. Res. Birsa Agricultural University.* 15(1): 61-63.

Singte, M. B., Yamger, V. T., Kathmale, D. K. and Gaikwad, D. T. (1997) Growth, productivity and water use of turmeric (*Curcuma longa*) under drip irrigation. *Indian J. Agron.* 42: 547-549.

Sivaraman, K. (1992) Studies on the productivity of turmeric x maize and onion intercropping systems under varied populations and nitrogen levels. Ph.D Thesis. Tamil Nadu Agricultural University, Coimbatore.

Sivaraman, K. and Palaniappan, S. P. (1994). Turmeric- maize and onion intercropping systems. I. Yield and land use efficiency. *J. Spices and Aromatic Crops* 3(1):19-27.

Spices Board. (2004) Guidelines for production of organic spices in India. Spices Board. Cochin. p 94.

Sreenivasan, K. G., Ipe, C. V., Haridasan, V. and Mathew, M. (1987) Economics of intercropping in the first three years among new/replanted rubber. Rubber Board Bulletin, India, 23(1): 13-17.

Srinivasan, V., Sadanandan, A. K., Hamza, S. (2000) An IPNM approach in spices with special emphasis on coir compost. International Conference on Managing Natural Resources for Sustainable Agricultural Production in the 21st Century, Feb. 14-18, 2000, IARI, New Delhi. pp. 1363-1365.

Stix., G. (2007) Spice Healer - An ingredient in curry shows promise for treating Alzheimer's, cancer and other diseases. Available on line at *http://www.sciam.com/article.cfm?articleID=131CED4F-E7F2-99DF-3C84BB412D1D3B51&chanID=sa006* (accessed on 12-3-2007).

Subramanian, S., Rajeswari, E., Chezhiyan, N. and Shiva, K. N. (2003) Effect of Azospirillum and graded levels of nitrogenous fertilizers on growth and yield of turmeric (*Curcuma longa* L.). Proceedings of the National seminar on new perspectives in spices medicinal and aromatic plants, 27-29 Nov, Indian Society for Spices, Calicut, pp. 158-160.

Sundararaj, D. D. and Thulasidas, G. (1976) Botony of field crops. The Macmillan company of India Ltd. New Delhi.

Tayde, G. S. and Deshmukh, V. D. (1986) Yield of turmeric as influenced by planting material and nitrogen levels. *PKV Research J.* 10(1): 63-65.

Thomas, A., Swati Barche and Singh, D. B. (2002) Influence of different levels of nitrogen and potassium on growth and yield of turmeric (*Curcuma longa* L.). *J. Spices and Aromatic Crops* 11(1): 74-77.

Umarani, N.K., Patil, R.B.and Pawar, H.K. (1982) Studies on time of harvesting turmeric pp 90-92. In: (.M.K. Nair et al. Eds). *Ginger and Turmeric.* Proc. Natl. Seminar on Ginger and Turmeric. April. 8-9, 1980. Calicut., Central Plantation Crops Research Institute, Kasaragod.

Upadhyay, D. C. and Misra, R. S. (1999) Nutritional study of turmeric (*Curcuma longa* Linn.) cv. Roma under agroclimatic conditions of Eastern Uttar Pradesh. *Progressive Hort.* 31(3/4): 214-218.

Vadiraj, B. A., Siddagangaiah and Poti, N. (1998) Effect of vermicompost on the growth and yield of turmeric. *South Indian Hort.* 46(3/6): 176-179.

Valsala, P. A., Nair, G. S., Nybe, E. V., Koshy, A., Mathew, M. P and Abraham, K. (1998) Optimum spacing for turmeric cultivation. *In: Developments in plantation crops research.* Proc. Plantation crops symposium XII (PLACROSYM XII), Mathew, N M and Kuruvilla Jacob, C (Eds), Indian Rubber Research Institute, Kottayam. pp. 206-208.

Velayudhan, K.C., Muralidharan, V.K., Amalraj, V.A., Gautham, P.L., Mandal, S. and Dineshkumar (1999). *Curcuma.* Genetic Resources, Scientific Monograph No. 4. National Bureau of Plant Genetic Resources, New Delhi, P.149.

Velmurugan, M., Chezhiyan, N. and Jawharlal, M. (2002) Effect of biofertilizers on physiological and biochemical parameters of turmeric cv. BSR 2. *Proceedings of the National seminar on Strategies for increasing production and export of spices,* 24-26 Oct, Indian Society for Spices, Calicut, pp. 138-145.

Venkatesha, J., Khan, M. M., Farooqi, A. A. and Sadanandan, A. K. (1998) Effect of major nutrients (NPK) on growth, yield and quality of turmeric (*Curcuma domestica* Val.) cultivars. *In: Proc. Water and nutrient management for sustainable production and quality of spices.* Sadanandan, A K., Krishnamurthy, K S., Kandiannan, K and Korkanthimath, V S. (Eds). Indian Society for Spices, Calicut, India. pp. 52-58.

Venkateswaran, T.V. (2003) Science of Spice – Turmeric. Dream 2047. 5 (8) : 24-25. (Monthly Newsletter of Vigyan Prasar, Department of Science & Technology, Government of India) available online at h*ttp://www.vigyanprasar.gov.in/dream/may2003/english.pdf* (accessed on 11-3-07).

Vinning, G., Tobgay, S. and Tshomo, K. (2004) World Turmeric Markets. Ministry of Agriculture, Thimphu, Kingdom of Bhutan. Available online at *http://www.trade.gov.bt/ administration/mktbriefs/14.pdf* (accessed on 11-3-07).

WHO. (1999) Rhizoma Curcuma Longae. *In: WHO Monograph on Selected Medicinal Plants Volume 1.* World Health Organization, Geneva, pp. 115-124.

Yamgar, V. T., Kathmale, D. K., Belhekar, P. S., Patil, R. C and Patil, P. S. (2001) Effect of different levels of nitrogen, phosphorus and potassium and split application of N on growth and yield of turmeric (*Curcuma longa*). *Indian J. Agron.* 46(2): 372-374.

12

Vanilla

K. Kandiannan and R. Dinesh

1.0 INTRODUCTION

Vanilla is one of the most popular and widely used flavour materials in the world and it is the most expensive spice traded in the global market after Saffron *(Crocus sativus L).* The major consumers of vanilla beans currently are the developed countries with the US, France and Germany absorbing about 80 per cent of the world imports. Among them, the US imports 50-60 per cent while the latter 10-15 per cent each. They are also main re-exporters of both vanilla beans and processed products. The total area under vanilla cultivation in the world during 2002 was 38,066 ha with the production of around 4956 metric tones and increased to 76,440 ha with a production of 10, 360 tons during 2005 (FAO, 2007). The major vanilla producing countries are Madagascar, Comoros, Indonesia, Mexico and Reunion (Table 1). Area under vanilla in India during 2002-03 was 2545 ha with the production of 92 tonnes. (Spices Board, 2003) and increased to 5370 ha with a production of 196 tonnes during 2004-05.

Vanilla was first introduced to Europeans in 1520, when Diaz, an officer of Cortes, noted its use by the Aztecs in southern Mexico to flavor their *chocolate* (Montezuma was served *chocolate* in golden goblets!). By the end of the 16th century, the Spaniards had established factories to manufacture chocolate with vanilla flavoring, and for many years the Spaniards had control of vanilla production. A plant was successfully grown in England in 1807. From this, plant cuttings were given to botanical gardens in Paris and Antwerp. In 1819, two vanilla plants were sent to Java from Antwerp, and one survived but did not fruit. Finally in 1827, a plant from Paris was taken to Mauritius that did survive and fruit, giving rise to the vanilla industry centered around the Indian Ocean. Madagascar, which received a plant in 1840, now produces 90% of all the natural vanilla extract in the world. Vanilla can be cultivated around the world in the tropics wherever mean temperatures range between 21° and 32°C (http://www.botgard.ucla.edu/html/botanytextbooks/economicbotany/Vanilla/index.html).

1.1 Area and production

Madagascar was holding the prominent position with a cultivated area of 25,900 ha. Of late, with an average production of more than 2000 tones from 1998-2002, Indonesia leads the world production of vanilla (Table 1). These two countries together contribute about 76 per cent of the world average production of 4863 during 1998-2002. Mexico, the native land of vanilla is in third position now after losing its first position to Madagascar in nineteen eighties.

The support side for the commodity is characterized by unstable productions largely due to natural reasons and dominance of market by very small number of countries. There is a decreasing trend in Indonesia's production for the last few years, mainly due to unforeseen agro-climatological factors. During 2003-04 there was a short fall in vanilla production in Madagascar leading to high price rise. This increasing price level acted as catalyst to increase area under crop in many new entrants including India.

Table 1. Country wise production of vanilla

(Metric Tonnes)

Country	1998	1999	2000	2001	2002	Average
Traditional producing and exporting countries						
Madagascar	1650	1650	1815	1815	1518	1690
Indonesia	1900	2102	2102	2102	1800	2001
French Polynesia	30	34	40	40	35	36
Mexico	564	300	550	550	300	453
The Comoros	160	150	180	180	140	162
Reunion	30	30	28	28	35	30
Tonga	100	50	130	130	130	108

Contd...

Table 1. Contd...

Zimbabwe	10	10	10	10	10	10
Cook Islands	2	NA	NA	NA		2
Guadeloupe	8	40	8	8	8	14
Malawi	22	NA	NA	NA	20	21
New entrants						
Kenya	10	NA	NA	NA	8	9
China	550	NA	NA	NA	650	600
Uganda	60	NA	NA	NA	40	50
Turkey	100	NA	NA	NA	170	135
India		10	NA	60	92	54
World	5196	4376	4863	4923	4956	4863

Source: FAO, Rome NA: Not available

Organized cultivation of vanilla in India started in 1990s. Vanilla cultivation is mainly concentrated in Karnataka, Kerala and Tamil Nadu. There are also reports that vanilla cultivation is getting expanded to other states like Maharastra, Andhra Pradesh, Madhya Pradesh, Orissa, West Bengal, Andamans and Northeastern states (Thomas and Rao, 2003).

Since vanilla cultivation started very recently only 50% of area is now under yielding stage. This implies that, the production and productivity is likely to increase substantially in the near future when the entire area under cultivation starts yielding. Table 2 indicates that India has produced 92 Mt of vanilla during 2002-03 with total area of 2545 ha, this is hardly 1.90 per cent of total average world production. For the period ending 2007 the targets have been fixed at 15,000 ha.

Table 2. State-wise area and production of vanilla in India (2002-03)

State	Total Area (ha)	Yielding Area (ha)	Yield (cured) (kgs/ha)	Production (Mt)
Karnataka	1465	545	100	54
Kerala	812	239	78	19
Tamil Nadu	268	130	147	19
Total	2545	914	101	92

Source: Survey Report of Spices Board, India

1.2. Prospects of organic

Aggregate global demand for vanilla is estimated at about 4000-5000 metric tones a year. As per the information available at FAOSTAT (FAO, 2004), 85 countries imported vanilla during 2004. Total world import during 2004 was 4652 Mt valued US$ 395.03 million. The major importers were Singapore (744 Mt), Spain (668 Mt), USA (672 Mt), Netherlands (531 Mt), France (277 Mt), Ireland (203 Mt), Germany (182 Mt), Austria (134 Mt), Canada (133 Mt) and UK (114 Mt). Some of the importing countries are also exporters of vanilla in the form of value added products. In USA, over 95 percent of vanilla beans are processed into extracts while in France at least 20 percent are consumed directly. Madagascar is the main player in organic vanilla trade. World over there is an increasing trend for use of natural flavour and colours and therefore the share of natural vanillin is likely to increase to a great extent.

1.3. Composition and uses

The characteristic and typical flavour develops in the bean after special drying and curing processes. The most abundant and important aroma component formed during curing is vanillin ($C_8H_8O_3$). However this delicate compound alone does not account for the aromatic principals of vanilla extract. Further volatile constituents are responsible for its characteristic aroma with sweet, balsamic, creamy, woody, spicy, fruity, herbaceous, phenolic and cinnamon-like notes. Vanilla essence is largely used in the preparations of ice creams, chocolates, bakery products, puddings, pharmaceuticals, liquors and perfumes.

2.0 BOTANY AND VARIETIES

There are two vanilla types- *Vanilla fragrens* var *planifolia* and *V. tahitensis* (Tahitian Vanilla), it is rarely found outside the pacific region. It has narrower leaves; the pods have lower vanillin content. For *V. fragrens*, the source of biodiversity is thought to be the isthmus of Central America (Guatemala and South Eastern Mexico. Vanilla, *Vanilla planifolia* Andrews (*V. fragrans* (Salisb)) is a climbing orchid. It is indigenous to wet low land forests in South-Eastern Mexico, Guatemala and other parts of Central America. *V. tahitensis* J.W. Moor, the Tahitian vanilla, which is cultivated in the Tahiti and *V. pompona* Scheide, cultivated in some of the South Pacific Islands also yield vanillin, but of inferior quality (Puseglove *et al.* 1981).

Popular, 'breeder's varieties' are not known in vanilla. As vanilla cultivation spread from its country of origin to different growing countries through vegetative cuttings, little variation can be expected. Vanilla cultivars found in Reunion are 'Classique', 'Mexique', 'Megic' and 'Bleue'. Four different local cultivars of vanilla are recognized by growers in Mexico. They are 'mansa' (dura), 'acamaya' (rayada, variegate), 'albo-marginata' and 'oreja de burro (ibid). Mansa is the most widely used to initiate vanilla production in Madagascar, Asia and Africa (Bhat *et al.*, 2006).

3.0 SOIL AND CLIMATE

3.1 Soil

Successful cultivation of vanilla depends on soil quality, soil factors such as texture and pH is more important than soil fertility (Childers *et al.*, 1959). Steep slope land / water logged situation should not be selected for establishing plantation. Under uniform distribution of rainfall, good drainage is essential. Vanilla is a surface feeder. Hence, humus rich forest soil is ideal for growth. In areas where the soils are relatively heavy with high rainfall, the vines planted on mounds or ridges to help alleviate the drainage problem. The best soils for vanilla in Puerto Rico are of lime stone origin, with a pH 6.0 to 7.0 having less incidence of root rot disease compared to Catalina clay soils of pH 5.0 to 5.5 (Childers *et al.*, 1959).

3.2 Climate

Vanilla is a tropical plant requiring a warm, humid climate for proper growth and economic production. Although vanilla thrives best under moist conditions, precipitation should not be very high as it may cause disease problems. The crop can be grown in wide range of temperature but the mean minimum temperature during cooler months should not go below 12° to $15^{o}C$ and the mean maximum temperature should not be above 35 ^{o}C. Potty and Krishnakumar (2003) stated that the optimum is 21 ^{o}-32 ^{o}C, with an average around 27 ^{o}C. The ideal relative humidity is 80 per cent. The rainfall should be moderate and evenly distributed; the quantum required being 2000 to 2500 mm per annum. However, there should be two drier months prior to the flowering season to check vegetative growth and bring the vines to flower. If heavy rain occurs during flowering, vanilla puts forth new growth instead of flowering and consequently it reduced the yield (Ridley, 1912). Vanilla grows well up to 1000 m MSL.

4.0 PLANTING MATERIAL

Vanilla is propagated vegetatively using vines from productive gardens, whereas seed propagation invariably done only for research purposes employing special techniques because the seeds do not germinate readily under natural conditions. Three types of planting materials are used for raising vanilla

Fig 1: A vanilla plant (*Source:* Director, IISR, Calicut)

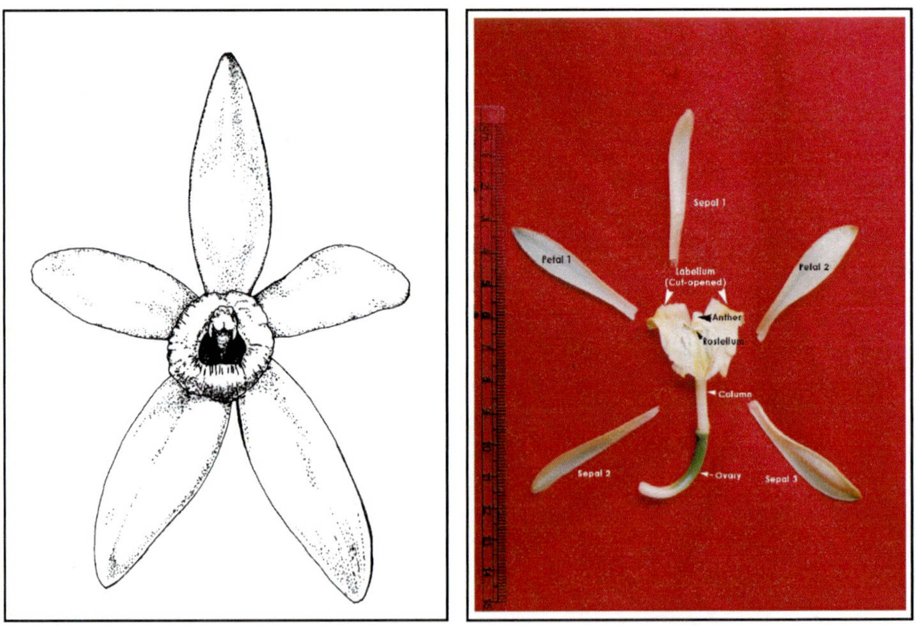

(a) (b)

Fig.2: Vanilla flower (a) and its parts (b) (*Source:* Director, IISR, Calicut)

gardens (Bhat *et al.*, 2006), they are (i) cuttings, (ii) rooted cuttings in bags, (iii) tissue culture derived plants (it is not recommended for organic production).

Vanilla propagated by using cuttings from healthy and vigorously growing plants, which are grown organically. However, to begin with they obtained from conventional plantations. Vanilla is amenable to both sexual and asexual methods of propagation. The stem cuttings are capable of striking roots at nodes when they come in contact with soil or any other rooting medium hence, it is prepared for commercial propagation. Stem cuttings either planted directly in the field, if sufficient material is available, or grown in polythene bags and transplanted.

Only vigorously growing healthy disease and pest free vines selected from yielding plants for collection of cuttings. Any part of the vine with dormant axillary vegetative buds selected for taking stem cuttings. However, vines of current years growth, which are in vegetative phase leaving the tender shoots at the tip, are most ideal. This is because the number of vegetative buds will be less in those portions, which have already, yielded during previous years. Cuttings from already yielded shoots should be avoided for fresh planting as it takes a very long time for sprouting. Such shoots identified by the presence of peduncle stubs that remains after the beans harvested. The cuttings after separation from the mother vines need to be kept under shade for one week to condition them. This will allow hardening of stem, arrest the growth, stimulate the buds and allow healing of the cut ends which will prevent rotting of cuttings when planted.

The length of vine used for planting varies from place to place but it has a profound influence on further growth and time taken to attain maturity. Large cuttings if planted at the beginning of the rainy season maintain a continuous growth and bear flowers and fruits in one or two years while shorter cuttings do not flowers until third or fourth year. Cuttings with 8 to 10 inter nodes (about 1 m long) preferred as they come to flower earlier. Cuttings with less than five to six internodes and 60 cm length not used directly for planting, but when properly rooted in the nursery, they establish well in the field (Sudharshan *et al.*, 2006). Cuttings collected from mother vines in the field or nurseries specially prepared as source of cuttings. The limitation in getting cuttings from field plants is that usually it will be available only during the harvesting season of September - October when pruning is done. In nurseries meant for production of cuttings, they will be available throughout the year. For rooting of stem cuttings, net houses are made with irrigation facility and vines are planted closely in row and the new growth is allowed to trail on plastic netting, coir mats or bamboo supports and is removed when sufficient length is reached. During the period, intensive manuring and irrigation are done to get good vegetative growth. Vines also planted around tall trees and allowed to grow upwards to sufficient length and these vines are also good source of cuttings for planting.

4.1 Nursery

A rapid multiplication nursery raised by forming trenches of 60 cm x 60 cm of any length which is filled with a mixture of soil: sand and cow dung (3:1:1 proportion). Cuttings planted with a spacing of one meter on supports and intensively manured and watered. Shade to an extend of 50% is necessary. This method would result a vine growth of seven meters per year and one hectare of such nursery may produce 40, 000 to 50, 000 cuttings annually.

The planting material for direct planting is not available in adequate quantity; vines are divided into small cuttings (with two nodes and leaves retained) and rooted in poly bags. The basal one node is inserted into a suitable rooting medium in poly bags under nursery condition in a protected shed. The bag size can be 25 cm x 25 cm with bottom holes for drainage. Black or white polythene bags are suitable. Rooting medium is normally a mixture of garden soil, FYM and sand in 3:1:1 proportion. If soil is clayey more sand added to provide drainage. Good top soil alone is sufficient if water drains easily. Solarization of rooting mixture by exposure to sun for 40 days and incorporation of biocontrol agents such as *Trichoderma* protect against root rot diseases. Watering done as and when necessary, excessive watering avoided as it would lead to rotting of stem. The best time for rooting is during August. The cuttings will root within a week and new shoots will emerge from the upper node. The new growth is supported with long bamboo splinter or sticks of about 75-100 cm length and once the vine reaches a maximum of 60 cm (will take about six to seven months), the rooted cuttings can be field planted.

Instead of poly bags, split coconut husk used to root the cuttings. In this case, split coconut husk is filled with a mixture of soil: sand: cow dung (1: 1: 1) and cuttings planted with one node inside the husk. Tying of the husk needed to hold them together. After rooting, the planting done with the husk intact and transplanting shock avoided. The husk placed in horizontal or erect position.

In another method, bamboo splinters of two meter length taken and vines of one meter length tied to this. The lowest three nodes buried under soil in the nursery. The cuttings planted close to each other. Over head shade provided with plastic netting. After rooting, these vines uprooted and planted along with bamboo intact in the field and allowed to lean on the support tree.

Collection of sizable quantities of stem cuttings could lead to arrest of vegetative growth of mother plants. Planting materials production of vanilla has described by Kuruvilla *et al.* (2003). Nurseries established for steady supply of planting materials. Siddagangaiah *et al.* (1996) standardized the rooting medium for propagation of vanilla, mixture of sand and vermicompost is better for rooting. Application of fresh cattle manure to vanilla vines has deleterious effect on growth. August is more suitable month for raising rooted cuttings (Spices Board, 2000). Vanilla also produced through micro-propagation by different labs for commercial planting (Rao *et al.*, 1999), but tissue cultured plants are not used in organic system.

Vanilla to be propagated by using cuttings from healthy and vigorously growing plants, which are grown organically. However, to begin with they can be obtained from conventional plantations. Tissue culture plants should not be used as planting materials in order to keep integrity with the methods of propagation.

5.0 PLANTING AND AFTER CARE

Vanilla is the only member of Orchidaceae which has a real economic value in the food and related industries, owing to its unique flavor and pleasant aroma and vanillin is responsible for it. Vanilla is one of the expensive spices traded in the global market.

5.1 Field preparation

Vanilla can be cultivated either as a pure crop or as an inter-crop with coffee, coconut, arecanut, black pepper etc., under mixed cropping situation all crops grown by following organic methods. An isolation distance of 25m wide provided. A period of three years is required as conversion period for existing plantations. For new planting, if vanilla is cultivated in virgin land or in farm where records are available that no chemical inputs are used, this conversion period relaxed (Spices Board, 1998).

5.1.1 Supports or Standards

Vanilla is a climbing vine, it requires live or non living supports for its successful production. Since it needs shade, living supports cum shade trees provided in pure gardens whereas under mixed cropping, the perennial crops provide the needed shade. An ideal support must be short stature with plenty of branches, are amenable for coppicing, leaves must be small and narrow; the stem bark should be rough. A number of tree species have been recommended. In general, small leaved species, which permit filtered light through the foliage, are useful. Species, which can be easily propagated through long stem cuttings and those that grow faster and produce branches sufficiently low (from 1.5 to 2.0 m from ground) for the vines to hang within easy reach of the workers are found to be the most ideal. The trees selected should be strong enough to support the vines and the beans without breaking down in a strong wind and should never become entirely defoliated during summer months. It is also important for the support tree to have deep penetrating roots so that they do not compete with the shallow rooted vanilla plants for nutrients. The most common tree used is *Gliricidia sepium* world wide. Other trees suitable are *Jatropha curcas* (Physicnut), *Plumaria alba* (Frangipani), *Casuarina* sp., *Erythrina* sp, *Pandanus* sp., *Ficus* sp., and *Morus* sp.

Vanilla also grown on coconut, arecanut, oil palm, cocoa, coffee, rubber or cashew. When grown under coconut or other palms, wires tied above the vanilla rows are necessary to prevent damage to the vines from falling leaves. With cocoa, few branches above vanilla have removed to allow more light to vanilla. Under high-tech horticulture, with plastic net as shade (50%) and concrete or wooden poles, asbestos or PVC pipes, stripes of plastic netting etc are used as supports. Even walls of the compound or house also are utilized as supports for growing vanilla by enterprising homestead gardens. Under the canopy of miscellaneous trees, farmers use even simple wooden poles across temporary supports for training vanilla with good success.

The live supports established at least three months in advance of planting vanilla. Since, glyricidia is commonly used; limb cuttings of 5.0 cm thickness and a length of 1.5 to 2.0 m collected during summer and allowed to stand in a leaning position for rooting. With the onset of rains these stumps planted in pits for establishment. As it starts sprouting a group of branches all around the stem, 1.5 m above ground level developed into an umbrella like fashion to facilitate looping of the vines. Generally it is pruned close to the main stem in June when rains are in full swing while a light pruning done to branches by removing 1/3rd length in September to allow for retention of leaves during summer. If no pruning done the branches will shed the leaves in summer exposing vines to the hot sun. Under agro-forestry or other tall trees shade regulation in rainy season is necessary. The ideal system to get the best yields is yet to standardize in India.

5.1.2 Trellis system

Instead of live support trees, some planters use hard wood / other posts and bars, the bars resting in notches on the top of posts. They put at a height of from 1.25 to 1.75 m from ground and plants brought over them and lopped up as they grow to long. Wire used in some places instead of bars of wood; it has a disadvantage that wire may cut the vine during wind. It noted that coiling or hanging treatment is essential to promote flowering and fruiting of vanilla vine (Muralidharan *et al.*, 1974).

In the telephone system of planting the 'T" shaped pole planted three or four meters apart and wires are tied length wise on both end of the 'T' (at top) and vines allowed to hang on both sides of the pole

5.2 Planting

Vanilla raised as pure crop or as mixed crop with perennials or agro-forestry species or in homestead. In Kerala, India it is common to see as a mixed crop in arecanut, coconut, coffee.

5.2.1 Planting time

Considering the fleshy nature of the vines, it is generally recommended to plant vanilla in the months of August-September when heavy rains are over and good humidity and light prevails. In low rainfall areas, it may be planted at any time with irrigation. So also is true for summer, provided necessary shade is given.

5.2.2 Spacing

Though farmers practice different spacing, too close spacing is not ideal for vanilla because overcrowding may result in severe incidence of diseases and pests. As a pure crop with single supports a spacing of 2.5 to 3.0 m between rows and 2 m between plants in a row is ideal that would accommodate 1600 to 2000 plants per hectare. Few farmers follow a spacing of 2.5m X 1.5m to accommodate 2500 plants per hectare. This is applicable for live standards such as *glyricidia*. In systems where nonliving standards are used, a spacing of 2m X 2m is sufficient. If trailing on walls, a spacing of 1.5 m between plants may be sufficient. In certain high tech systems even one m spacing between plants is adopted. What ever may be the case, adequate aeration, enough light penetration easy management and better disease control should be taken into account w1hile selecting the appropriate spacing. When intercropped with coconut, arecanut or oil palm, only the interspaces should be used for planting vanilla.

5.2.3 Pits

Vanilla being a surface feeder, a small pit of 40 cm x 40cm x 40cm is sufficient for planting. The pit can be filled with good top soil and well decomposed FYM and allowed to settle for a week before planting. The filled earth should form a small mound above the pit to facilitate planting and also to avoid depressions in the pit. Otherwise rain water will accumulate in the pit resulting in rotting of the vine. Bio control agent (*Trichoderma* @50g per pit) can be incorporated to the soil with sufficient moisture when pits are filled to ward off potential diseases.

5.2.4 Planting method

The method of planting depends on the type of planting material used. If cuttings are used, the leaves from the lower three nodes are removed, before planting. The defoliated part of the vine is laid on the loose soil surface and covered with a thin layer of about two to three cm soil. The basal end of the cutting would be kept just above the soil to prevent rotting. The growing end is gently tied to the support for clamping by aerial roots. If the sun is hot the cuttings are shaded with tall dry grass, palm fronds or with other suitable materials. In dry soil, a light sprinkling of water helps for early establishment of cuttings. It takes about four to eight weeks for the cuttings to strike roots and show initial signs of growth. In some areas where diseases are a problem another planting technique is adopted. Here the vine is tied to the support in such a way that the lower cut end of the vine is 10 to 15 cm above the ground level. The vine gradually puts down roots which reach the ground and help establish the plants. In this case spraying with nutrient solutions once in a fortnight may be necessary to stimulate sprouting and root formation. In summer planting, shading of the vines are needed.

When rooted cuttings are planted, a small hole is made in the filled pit to accommodate the root ball, the polybag is removed and planting is done by covering with soil. The growing end is tied to the support.

It is recommended to plant only one cutting per standard though many farmers do plant three or more plants to get quick canopy coverage and high yield. More plants (up to three) may be planted per support if the spacing adopted is wider. When planting is done in coconut, arecanut etc. 3 or 6 or even 10 vines are planted depending on the girth of the palm. Same is the case when non living supports such as asbestos pipes, concrete poles etc. are used.

In homesteads, planting closer to the walls of compounds or house in a large size pots, can be taken up. In all the above cases planting has to be done as discussed above. For pots a good potting mixture of sand, soil and FYM in 3:1:1 proportion may be used.

When long cuttings are used for planting, the basal portion of the stem is to be laid on the ground and coiled around its support, the stem may be covered with a light layer of soil or rotten leaves to save it from injury and hasten its growth. The tip is to be protected from the sun using banana sheath or plaited coconut leaves and the plant must be frequently watered. While planting short cuttings, the base is to be inserted in the ground for about 1.5 m and a stick put close to it to start it climbing or it may be put close to the tree on which it is to climb. At least two nodes of the stem must be above the ground. After one or two weeks, the plant puts out a shoot from one of the leaf axils. This soon lengthens and sends out roots of climbing type and to cling to the support. While planting cuttings by the side of supports, whether tree or trellis, they should in such a position that the aerial roots may come into contact with the support and the cuttings should be tied to it. Vanilla cuttings usually planted at about 3 m apart at the base of supporting trees or poles. A spacing of 1.5 to 3.0 m is adopted. Closer spacing may be inconvenient for field operations. General recommendation is that planting may be done during September to November. It can be done throughout the year under protected cultivation or well managed irrigated condition.

5.3 Mulching

Mulching is very important for vanilla for profuse root development which in turn will induce vigorous growth. Mulching from planting onwards is needed and the base of the vine must be covered with mulches. Dry leaves, rotten wooden logs, coconut husk, coir waste, wood shavings etc. are good for mulching. Sawdust should not be used as mulch, as it thickens like cake preventing aeration and drainage. Some farmers have obtained excellent results with chopped pieces of bamboo (1" or 2") where ramification of roots is excellent. Coconut husk is also very good mulch as it provides potassium and retains moisture.

5.4 Weed management

Weed flora vary with location, cultural methods like digging, manual weeding and hoeing, slash weeding can be adopted. One should not allow wild vines to creep on the supports, if it is not checked it will smother the crop. While weeding, the base of the plant should not be disturbed as much the roots found in top layer of humus (surface feeders).

5.5 Irrigation

Vanilla requires moist climate with frequent but not excessive rains. Under excessive rainfall, there is wide spread of diseases and under drought, the plant may suffer considerable physiological damage and the vines may not recover. In extremely dry years, irrigation should be provided at least once in four to five days. Irrigation is needed during the summer months only after flowering is over. Mulching and irrigating once in two days was found to increase vanilla yield. Thirty minutes of mist irrigation during morning and evening each day was reported to be good. If the vines are irrigated continually after stoppage of rains no flowering will take place. An interval of about a month is needed for flowering. For fruit development irrigation is essential when rains are not enough.

5.7 Shade management

Judicious lopping of branches of the living supports is very important to give shade to vanilla plants. Vanilla leaves show yellowing and sun scald when exposed to direct sunlight. Shade can be provided for vanilla by means of existing tree shade or planting shade trees in between or along the borders. Under high rainfall and high relative humidity, vanilla can withstand more sunshine than under low humidity and drought. Thus, it is important that support trees maintain much of their foliage during dry periods.

5.8 Training and pruning

The way in which the vines are trailed has an effect on flowering. Vanilla is to be trained to keep vines within the limits and to facilitate cultural management. The vines are twisted around the lower branches of the supporting tree or over the lattice of trellis so that they may hang down. Care is required so as not to tear or bruise the leaves, branches or roots. Bending of vines appears to be an important factor in causing it to flower and fruit beyond the bend, which may be due to an accumulation of carbohydrates and possibly other flower inducing substances in the regions of the vine. If vegetative growth of vines arrested by pruning, it turns to reproductive phase. At pruning time special bearing branches are prepared. These are shoots which when 1 - 1.2 m long are bent down round a branch of a support slightly twisted in the process with tip pruned at about 45 cm from the soil. Any shoots appearing on the bearing branches themselves are cut off when 7.5 to 10 cm long but the shoots appearing on the rest of the plant before the bends are allowed to grow. They will constitute the bearing branches of the following year. As a result there is a decreased sap flow towards the bearing branches, which favors flower formation. After the harvesting of beans, the old branches are cut off. Meanwhile, the following year's bearing branches have already been prepared. The vine architecture after three to four years is therefore that of a vine with a number of shoots hanging down over the branches of the live support.

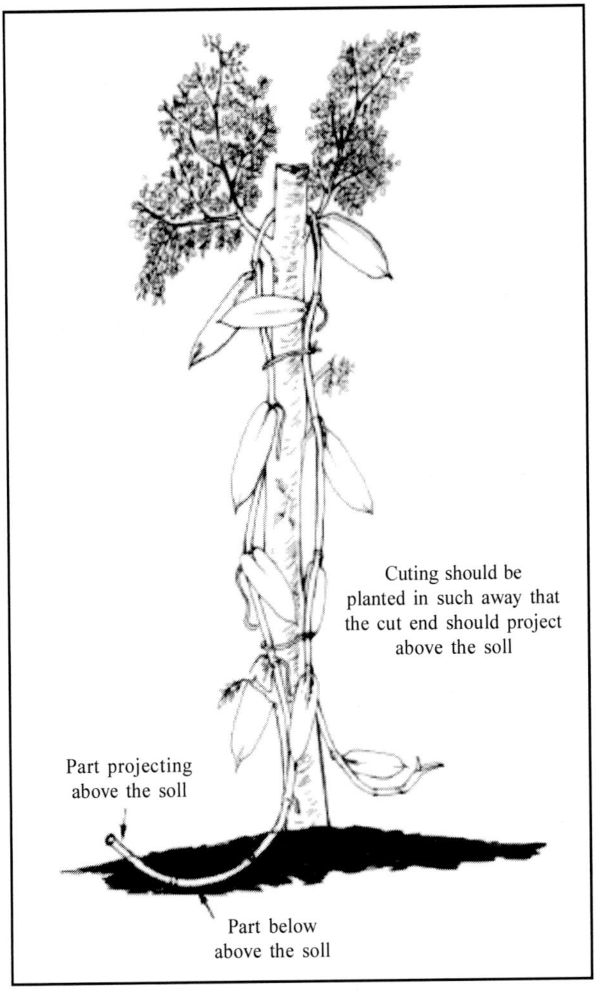

Fig. 3: Training and pruning of vanilla (*Source:* Director, IISR, Calicut)

5.9 Flowering and pollination

Vanilla usually starts flowering in the third year of planting; however, it depends on the size of the original cutting used for planting. Maximum production of flowers occurs during the 7-8th year. Vanilla flowers during December to February and each flower lasts for only a day. Pinching off the top 7.5 to 10.0 cm of the vine, 6-8 months before the flowering season encourages flower production. Similarly, pruning off the older branches (which bore fruits the previous year) also encourages flower production. The flowers are borne in axillary racemes and each inflorescence consists of 15-20 flowers. The flowers are to be artificially pollinated (hand pollination) for fruit set. Since the flowers last only for a day pollination must be done on the same day. The remaining flower buds are nipped off. About 10-12 inflorescences may be pollinated in a vine. In hand pollination method, a pin or needle or small piece of pointed wood or a tooth pick is ideal to apply pollen on the stigma of the flower. The pollen of the vanilla flower is produced in a mass called pollinia, and is covered by hood or anther cap. The stigma is protected with a lip known as 'rosetellum' or 'labellum'. For pollination, the stamen cap is removed by a needle and exposing the pollinia. Then the flap like rostellum is pushed up and the pollinia are brought into contact with the stigma. The ideal time for pollination is 6 am to 1 pm. An efficient worker can pollinate 100-150 flowers a day.

6.0 NUTRIENT MANAGEMENT

Vanilla is not a heavy feeder. Under low soil fertility, proper shade, adequate moisture and mulching are essential. Where the soil is rich in humus of sufficient depth, it is not necessary to add anything in the way of manure. The best known source of nutrients for vanilla is a deep layer of mulch maintained over and around vanilla roots. The source and composition of the mulch is important. Vanilla, being an orchid plant, prefers organic manures. Decomposed organic materials may be applied two to three times a year. Bone meal, well rotten cow-dung, compost or vermicompost @ 4-5 kg/ plant/ year may be used.

7.0 DISEASES AND PEST MANAGEMENT

7.1 Diseases

Some of the disease affecting vanilla is rotting of leaf, stem, root, shoot tip and pods and blight caused by *Fusarium oxysporum* as well as brown spots or anthracnose caused by *Colletotrichum gloeosporioides*. Regular surveillance and plant sanitation are required for managing the diseases. The disease- affected portions are to be removed and burnt. Any injury to roots and other plant parts should be avoided. The plant basins should be kept under mulching. Restricted use of Bordeaux mixture (1%) may be made if necessary.

Occurrence of viral diseases is common in several countries where vanilla is grown on a commercial scale. The crop affected by six viruses (Thomas *et al.*, 2003). Management measures suggested by Bhat *et al.* (2004) are

a) Use of virus - free planting material
b) Regular inspection and removal of infected plants
c) Control of insect vectors like aphids

7.2 Pest

No serious pest attacking vanilla is noticed. However, a few caterpillars, earwigs, snails and slugs damage the tender parts of the plant. They may be controlled mechanically.

Fig.4: Fungal diseases of vanilla. (a) Bean rot induced by *Phytophthora*; (b) *Sclerotium* induced rot on leaves; (c) Premature yellowing and bean shedding; (d) Stem rot induced by *Fusarium* (*Source:* Director, IISR, Calicut)

8.0 HARVEST AND POST HARVEST HANDLING

8.1 Harvest

The beans or pods are ready for harvest 6-9 months after flowering. The beans can be considered as mature when they change from green to pale yellow. At this time, the pods may be 12-25 cm long. It is essential to harvest the pods at the right stage, as immature pods produce an inferior product and over mature pods split during curing. The right picking stage is when the distal end of the pod turns yellow and fine yellow streaks appear on the pods. Daily picking of mature pods is essential. The pods can be harvested by cutting with a knife. A good vanillery yields 300-600 kg of cured beans per hectare per year. About 6 kg of green pods produce 1 kg of cured beans. The yield of vines declines after 12 –14 years.

8.2 Post harvest handling

Due to lack of technical know how, hardly 20 per cent farmers process their produce before sale. Majority farmers do not go for processing just because they are afraid that they may loose the quality because of improper curing i.e. the farmer doesn't have the fool-proof technology for processing, which involves the following five stages:

8.2.1 Washing: (1st. day)

The harvested beans are washed to remove remains *of* Bordeaux mixture spray and other unwanted adhering material. Generally women labourers are engaged in this operation.

8.2.2 Killing: (1st. day)

In this process, 8-10 kg *of* beans in a jute bag, dipped in hot water at 63-65° C for three minutes. In order to carry out this process materials required are 10 jute bags, one container to boil the water and a thermometer. Both men and women labourers were employed to carryout this operation.

8.2.3 Sweating: (1st. day)

In this third process of sweating, three-four kg of killed beans will be put in to a dark blankets and the blankets are arranged in to wooden box. Around 15-16 blankets folded with killed beans will be put in to each wooden box for sweating. For around 8-10 days continuously they will take the beans out and spread under sunlight for about two hours a day. Required, materials for this process are 25 dark blankets, 6 wooden boxes and 18 women labour.

8.2.4 Slow drying: (12th day)

In this process sweated beans are spread on the stand made out of wooden/iron under shade, to allow for complete drying, this processes lasts *for* two months. Materials required for this process are one stand of either wood or iron.

8.2.5 Grading and Packing: (73rdDay)

Last stage in on-farm processing (curing) of beans is grading and packing. In this stage dried beans are segregated based on the grade: First grade having length more than six inches, Second grade having length between 4-6 inches and third grade below four inches. Then the beans are packed grade wise separately in a self stickable polythene covers. Materials required are 20 polythene self-stickable covers, and two women laborers.

Vanilla highly preferred spice for its flavor and aroma in food and confectionary industries. Traditional production centers are not able to meet the global demand and there is a scope to produce vanilla from India by adopting appropriate location specific production packages. Being a new crop vanilla poses a number of constraints for development. There are certain specific bottlenecks such as paucity of good planting material, the requirement of hand pollination and long cumbersome on-farm processing of beans. According to Sudharshan (2002), narrow germplasm base and inadequate research in India for vanilla are constraints.

REFERENCES

Bhat, A.I., Mathew, P.A., Nair, R.R. and Madan, M.S. (2006) Vanilla. pp. 191-227. In : Parthasarathy, V.A. and Rajeev, P. (eds.) *Major Spices – Production and Processing*. Indian Institute of Spices Research, Calicut.

Bhat, A. I., Venugopal, M. N., Bhai, R. S. and Anandaraj, M. (2004) Stem Necrosis - A new virus disease of vanilla in India. *Spice India* 17(7):5-7.

Childers, N. F., Cibes, H. R. and Medina, E. H. (1959) Vanilla - the orchid of commerce. *In: The Orchids. A Scientific Survey.* (Ed) Withner, C.L. Ronald Press & Co. New York, pp 477-508.

FAO Database, (2007) website, www.fao.org.

http://www.botgard.ucla.edu/html/botanytextbooks/economicbotany/Vanilla/index.html (online accessed on 17-5-07).

Kuruvilla, K. M., Rao, Y. S., Mathew, K. M. and Kumar, K. P. (2003) Planting Material Production. *In: Vanilla*: The Prince of Spices. Spices Board, Cochin, pp 10-18.

Muralidharan, A., Nair, E. V.G. and Balakrishnan, S. (1974) Preliminary studies on trailing of vanilla (*Vanilla planifolia*). *Agric. Res. J. Kerala.* 12(2): 174-177.

Potty, S. N. and Krishnakumar, V. (2003) Agrotechniques. *In: Vanilla*: The Prince of Spices. Spices Board, Cochin, pp 19-28.

Purseglove, J.W., Brown, E.G., Green, C.L. and Robbins, S.R.J. (1981) *Vanilla*. In: Spices Vol II. Longman, London, pp 644-735.

Rao, Y.S., Mathew, K.M., Pradip Kumar, K., Madhusoodanan, K.J., Potty, S.N., and George, C.K. (1999) *A manual for micro-propagation of selected spices*. Indian Cardamom Research Institute, Spices Board, Kochi- 25, India.

Ridley, H. N. (1912) Vanilla. *In: Spices*. Mc Millan and Co Ltd. London, pp 23-93.

Spices Board, (1998) *Production of Organic Spices*, Spices Board, Cochin, India.

Spice Board (2000) *Vanilla Status Paper*. Spices Board, Cochin.

Spices Board (2003) *Vanilla Situation in India*. Spice India. 16(6):10-11.

Sudharshan, M. R. (2002) Strategies for increasing production and export of vanilla. *In: Proceedings of National Seminar on Strategies for Increasing Production and Export of Spices*. Indian Society for Spices, Indian Institute of Spices Research, Calicut, pp 18-29.

Sudharshan, M R., Bhatt, S.S., Rao, Y.S., Mathew, K.M., Sivadasan, C.R., Ramesh Babu, N., Prakash, K.V., and Radhika, N.S. (2006) Vanilla. In : Ravindran, P.N., Nirmal Babu, K., Shiva, K.N., Johny, A.K. (eds.). *Advances in Spices Research History and Achievements of Spices Research in India Since Independence*. Agrobios (India). Jodhpur, India, pp. 533 - 569.

Thomas, J., Bhai, R. S. and Gopakumar, B. (2003) Diseases and their management. *In: Vanilla: The Prince of Spices*. Spices Board, Cochin, pp. 29-41.

Thomas, J and Rao, Y S. (2003) *Vanilla: The Prince of Spices*. Spices Board, Cochin.

13

Nutmeg

P. A. Mathew

1.0 INTRODUCTION

Among tree spices, nutmeg (*Myristica fragrans* Houtt) belonging to the family *Myristicaceae* is the most important, providing two spices, i.e. the nutmeg (dried seed) and the mace (dried aril covering the seed). The latter is more important. Nutmeg is cultivated mostly in the tropical areas of Indonesia and West Indies. The world annual production of nutmeg and mace is around 12000 tonnes and 2000 tonnes, respectively (Krishnamoorthy and Rema, 2001). A native of the Moluccas (The Banda Islands of East Indonesia), nutmeg is presently cultivated in several countries such as Grenada, Malaysia, Papua New Guinea, Sri Lanka, Fiji, some Caribbean Islands, Africa and China including India. Of late, Brazil also started its cultivation. The Indonesian nutmeg known as East Indian nutmeg shares 75% and Grenadian known as West Indian nutmeg shares 25% of world market and they are major suppliers in the world. The East Indian nutmeg is highly aromatic which is preferred by the USA. West Indian type is of mild flavour and is preferred in Europe. The major importers are USA, Canada, Japan and European countries. In USA, the oil is generally preferred whereas the Europeans consume it as whole spice.

1.1 Area and production

The area under nutmeg in India is estimated to be 11,070 hectares with a production of 2798 tonnes (Table 1). The spice is grown in Karnataka, Kerala, Tamil Nadu and Andaman & Nicobar Islands. It is also grown in Goa,

Maharashtra, Andhra Pradesh and North Eastern states though the exact area is not known. The area is expanding under the crop due to the prevailing high returns and low risks involved.

Table 1. Area and Production of nutmeg in India

State	Area ('000' ha)	Production ('000' T)
Kerala	10.78	2.7
Karnataka	0.228	0.096
T.Nadu	0.035	-
Andaman & Nicobar Islands	0.029	0.002
Total	**11.07**	**2.798**

Source: Directorate of Arecanut and Spices Development (2007)

India has been a traditional importer of nutmeg though it started exporting nutmeg mainly to USA. India's import in 2005 – 06 was 865 tonnes of nutmeg valued at Rs 1250 lakhs and 525 tonnes of mace valued at Rs 1485 lakhs while it's export was 1250 tonnes of nutmeg and mace valued at Rs.2235 lakhs in 2004-2005 (Directorate of Arecanut and Spices Development, 2007).

Traditionally nutmeg has been cultivated organically especially in the river basins where the rich alluvial soils are very conducive for its growth and productivity. Kalady in Ernakulam district of Kerala is forerunner in nutmeg cultivation in India where huge trees exist even today. The use of fertilizers is of recent origin. Not much research has been done to determine the nutrient and cultural requirements of nutmeg and existing recommendations are only Adhoc (KAU, 2003). Nevertheless, the Spices Board (2001) has come out with an organic package of practices. The organic farming practices discussed herein are mainly based on that publication.

1.2 Perspective of organic farming

Organic farming is gaining momentum globally owing to the increased awareness about health and the environment, pollution, adverse effects of pesticides and chemicals in human welfare, dwindling natural resources, sustainability of agriculture and holistic living. This has resulted in boosting the demand for organic food products i1ncluding spices. It is reported that the organic market is expanding at the rate of 20% per annum (Madan, 2007). Nutmeg is cultivated generally as a homestead crop or intercrop in coconut/

areca nut gardens. Under such situations all the crops are to be cultivated following organic farming principles.

For an existing plantation, a minimum of three years is required as conversion period for organic cultivation. For a newly planted or replanted area raised through organic cultivation practices, the first yield itself can be considered as organic produce. In the case of cultivation on virgin land and for farms where records are available that no chemicals were used previously, the conversion period can be relaxed. It is desirable that organic method of production is followed in the entire farm, but in large scale plantations the transition can be phased out for which a conversion plan is to be prepared.

In order to avoid contamination of organically cultivated plots from neighboring farms, a suitable buffer zone is to be maintained. An isolation distance of at least 25 meters width is to be left from all around the conventional plantation. The produce from this isolation belt shall not be treated as organic. In sloppy lands adequate precaution should be taken to avoid the entry of run off water and drift from the neighboring farms.

All relevant measures should be taken to minimize the contamination from out side and within the farm. Accumulation of heavy metal and other pollutants should be avoided. For protected structure coverings, plastic mulches, fleeces, insect netting and silage wrapping, only products based on polythene and poly propylene or other polycarbonates are allowed. These shall be removed from the soil after use and shall not be burnt on the farm land. The use of polychloride based products is prohibited.

1.3 Composition and uses

The nutmeg has the following composition

Nut oil (%)	7.14
Mace oil (%)	7.13
Oleoresin in nut (%)	2.48
Oleoresin in mace (%)	13.8
Butter in nut (%)	30.9
Myristicin in nut oil (%)	12.48
Myristicin in mace oil (%)	20.03
Elemicin in nut oil (%)	13.65
Elemicin in mace oil (%)	13.65

Nutmeg is used in culinary preparations, bakery products, meat industry, cosmetics etc. Medicinally nutmeg is known for its stimulative and carminative properties; its oil is used in curing urinary track infections, insomnia, skin diseases etc. The fresh rind covering the seed can be used for preparing jellies, jam, syrup, pickles, chutneys or even liquor but seldom used.

2.0 BOTANY AND VARIETIES

2.1 Botany

Botanically nutmeg is known as *Myristica fragrans* Houtt of the family *Myristicaceae* and is described by Purseglove *et al.* (1981). It is an evergreen tree, densely foliated that needs shade for good growth and development. Ridley (1912) reported that in the original habitat in Moluccas, it was associated with the tall pili nut (*Canarium ovatum*) as an upper storey tree. Seedlings start flowering from 4 to 7 years and the population consists of males and females with a low proportion of bisexual trees. (Prestoe, 1884; Flach, 1966). Flowers are born in leaf axils of current season shoots. Male flowers are in clusters of more than 3 whereas female flowers are in clusters of 2 to 3 or single. Fruit is a drupe spherical or slightly ovoid or pear shaped, pale yellow in colour with a longitudinal groove in the centre through which it splits open when ripe exposing the red aril. The tree exhibits orthotropic and plagiotropic growth pattern capable of growing up to 20 m or more and are long lived even more than 200 years. Only the female and the bisexual trees bear fruits but trees are obligatory cross pollinated and hence male trees are essential. Only female trees give good yields. Variation abounds in the population for vegetative, yield and fruit characters with excellent scope for selection (Krishnamoorthy *et al.* 1996). A 15 years old good tree is projected to yield around 2000 fruits.

2.2 Varieties

At present the existing population of nutmeg consists of seedling origin with variation in yield and other attributes. Of late, grafts of high yielding elite lines are preferred and planted. Alternately, high yielding mother plants with desirable nut and mace characters can be used for preparing grafts in view of the shortage of elite planting material. When certified organic planting materials are not available, initially conventional planting materials shall be used. Three improved varieties of nutmeg have been released in India for commercial cultivation (Johny and Ravindran, 2006).

2.2.1 Konkan sugantha

The first variety released by Dr. Balesaheb Sawant Konkan Krishi Vidhya Peeth, (BSKKV) Dapoli, Maharashtra. It is a single plant selection from local seedling population made at the Regional Fruit Research Station Vengurla. It gives an yield of 200 - 526 fruits per tree. Tree canopy is conical, compact and adapted to the Konkan region of Maharashtra.

2.2.2 Konkan Swad

This variety is released by Dr. Balesaheb Sawant Konkan Krishi Vidhya Peeth, (BSKKV) Dapoli, Maharashtra in 2003. It is a selection from seedlings from Ratnagiri district made at the Regional Coconut Research Station, Bhatye, Ratnagiri. The tree yields an average of 761.38 fruits per year. The canopy is erect, conical shape and has 39.8% essential oil in seed and 10.9% in mace. No incidence of pest and disease has been observed. It is recommended for the Konkan Region of Maharastra.

2.2.3 Viswashree

The variety is released by Indian Institute of Spices Research, Calicut, Kerala in 2001. Viswashree is a high yielding variety, recommended for the entire nutmeg growing areas of Kerala (Rema *et al.* 2003). The details of the variety are given in Table 2.

Table 2: Characters of the nutmeg variety IISR - Viswashree

Characters	**Values**
Pedigree	Clonal selection from elite tree
Areas of adoption	All nutmeg growing areas in Kerala
Pre bearing period	6 years for first harvest
Average yield	480kg dry mace and 3122kg dry nuts/ ha at 8th year
Potential yield	4800 kg dry mace and 31220 kg dry nuts/ ha at 25th year
Quality characters	
Nut oil (%)	7.14
Mace oil (%)	7.13
Nut recovery (%)	70
Mace recovery (%)	35
Oleoresin in nut (%)	2.48

Contd...

Table 2. Contd...

Oleoresin in mace (%)	13.8
Butter in nut (%)	30.9
Myristicin in nut oil (%)	12.48
Myristicin in mace oil (%)	20.03
Elemicin in nut oil (%)	13.65
Elemicin in mace oil (%)	13.65
Plant characters	
Height of tree at 9 years (m)	3-5
Width of the canopy (m)	3-3.5
Time taken for flowering	4 years
Colour of mace	Dark red
Colour of seed	Shining black
Size of the seed	Bold
Resistance to major pests and diseases	Low incidence of fruit rot caused by Diplodia spp.
Special characteristics	A high yielding, high quality variety with bushy and compact type (due to grafting with lateral branch).

3.0 CLIMATE AND SOIL

Nutmeg thrives well in the warm humid conditions with an annual rainfall of 150 cm to 250 cm. It grows from sea level up to about 1300 above MSL. Clay loam, sandy loam and laterite soils are ideal for its growth. Dry climate or water logged conditions are not suitable for its cultivation. River banks and sheltered valleys are best suitable for this crop. It is a shade loving crop under warm situations whereas it can thrive well without shade at higher elevations.

4.0 PLANTING MATERIAL

The nutmeg is propagated both by seeds and vegetatively by grafting or budding. Nutmeg crop is a dioecious plant and seed propagation is not in vogue.

4.1 Source and type of planting materials

Fully ripe, tree-burst fruits from organic farms are to be used for raising nurseries. However, initially seeds from conventional farms may be used. Farmers can also use grafted plants wherein the rootstock and scions are collected from organically cultivated area. M*ristica beddomei* and *M.malabarica* are related to nutmeg also can be used as rootstocks besides nutmeg. For raising seedlings for planting or as rootstocks, naturally split healthy fruits are harvested during June-July and the seeds are extracted and sown immediately.

Since the orthotropic and plagiotropic shoots can be used for grafting the resultant plants have a different growth pattern. Orthotropic shoots give rise to erect growing plants whereas plagiotropic (lateral) shoots give rise to bushy spreading plants. However, these may be induced to develop orthotropic shoots by bending above 90 degrees (Mathew, 1985). However, bush grafts are advantageous for high density crops (Chellappan and Roche, 1982).

4.2 Nursery raising of seedlings1

Nursery beds of appropriate size using sand may be prepared in a cool and shady place of convenient length, 1 or 1.5 meters width and 15 cm height. The seeds are sown in the bed and covered with sand to a thickness of 2-3 cm. Regular watering is to be done for good germination. A mixture of garden soil and sand in 3:1 ratio also may be used for preparing beds. The seeds are to be sown immediately after collection. Germination may commence from about 30th day and last up to 90 days after sowing. Approximately 20 day old sprouts are transplanted in pots or biodegradable polythene bags containing a mixture of soil, sand and well decomposed cow dung at 3:3:1 ratio. Retain the seedlings for 18-24 months before planting in the main field.

4.3 Epicotyl grafting

Grafts of elite high yielding trees can be prepared by epicotyle grafting instead of planting seedlings. For this purpose the seed sprouts are extracted from the bed when the first leaf begin to appear and the epicotyle having a diameter of 0.5 cm or more with sufficient length to give a cut of 3cm long and are used as rootstocks. Tender top shoots with two to three leaves, collected from high yielding trees can be used as scion. The stock and scion should have approximately the same diameter. A 'V' shaped cut is made in the stock and the base of the scion is tapered and is fitted carefully into the cut. Bandaging the grafted region may be done with polythene strips. The completed grafts are planted in poly bags of 25 cm x 15 cm size containing potting mixture of soil, sand and cow dung (3:3:1). The scion is covered with a polythene bag and kept in a cool shaded place protected from direct sunlight and well cared for. Those grafts showing sprouting of the scion are opened by removing the polythene bags and kept in shade for further development. The polythene bandage covering the grafted portion can be removed after 3 months. Twelve month old grafts can be planted in the field (Rema *et al.* 2003).

4.4 Top working

Wherever seedlings have been established and trees are poor yielders or predominantly male, top working by budding (Beena and Kurian, 1996) or grafting (NRCS, 1993) can be done to get desired trees. In this case, after ascertaining sex of the seedling, budding can be resorted directly if they are young whereas for grown up trees, these have to be beheaded at the beginning of the rainy season and budding or grafting can be taken up on the coppice shoots during August-September. Instead of keeping 10% male trees, one can graft a few male branches to each female tree to provide necessary pollen.

5.0 PLANTING AND AFTER CARE

5.1 Field preparation and planting

Planting in the main field is done at the beginning of the rainy season. Pits of 0.75 m x 0.75 m x 0.75 m are dug at a spacing of 9 m x 9 m for seedlings and 5 m x 5 m for bush / lateral grafts and filled with organic manure and soil 15 days before planting. If male trees are not available one male graft has to be planted for every 10 female grafts. Alternately, one male branch can be grafted to each female graft to provide enough pollen. The plants should be shaded to protect them from sun scorch during early stages in summer. Permanent shade trees are to be planted when the site is on hilly slopes and when nutmeg is grown as a monocrop. It is grown as mixed crop in homesteads also. As a pure crop it has been found to grow without shade in high elevations such as in Yearcaud (1000 m MSL).

Nutmeg can best be grown as an intercrop in coconut gardens more than 15 years old where shade conditions are ideal especially along river beds and adjoining areas. The study conducted in Rathnagiri showed that the net profit from a coconut plantation intercropped with nutmeg was 177% (Rs 69593) as compared to pure crop of Coconut (Nagawekar *et al.* 2002). Banana may be planted in between nutmeg plants to provide temporary shade and congenial micro climate. Banana and coconut should also be cultivated organically.

The feeder roots are confined to the top 20-30 cm of soil and the tap root goes below two meters in nutmeg. Irrigating once in 15 days with 50 liters per tree during summer helps build a dense root system and promote good growth (Ankegowda, 2006).

5.2 Mulching

No inter-cultivation is done for nutmeg because most of the feeder roots are on the surface. Mulching with dry leaves or other organic wastes may be done to conserve moisture.

5.3 Soil and Water conservation

Soil and water resources should be handled in a sustainable manner and appropriate measures are to be adopted for their conservation. In sloppy land the soil may be disturbed to the minimum in all agricultural operations. Clearing of land through the means of burning organic matter, e.g., slash and burn, straw burning shall be restricted to the minimum and the clearing of primary forest is prohibited. Relevant measures may be followed to prevent salination of soil and water.

6.0 NUTRIENT MANAGEMENT

The organic manurial requirement has not been standardized in nutmeg. However, application of well rotten cattle manure or compost @ 15kg/plant/year during May/June in the initial years is recommended (Spices Board, 2001). The quantity may be increased gradually so that a well grown tree of 15 years and more gets 40 to 50 kg of organic manure. The manure should be free from heavy metals. It is generally experienced that nutmeg is susceptible to potassium deficiency resulting in immature splitting of fruits. In such cases potash deficiency may be replenished by addition of wood ash or granite dust.

7.0 PEST, DISEASE AND WEED MANAGEMENT

Organic farming system should be carried out in such a way so as to minimize the losses from pests, diseases and weeds. Emphasis is given on the use of a balanced nutritional programme, use of crops and varieties which are tolerant to or resistant and well adapted to the environment, fertile soils of high biological activity, crop rotations, companion planting, green manures etc. Weeds can be slashed and applied as mulch.

7.1 Diseases

The following diseases are reported to occur in nutmeg (Rema *et al.* 2003).

i) Die back

The disease is characterized by drying up of mature and immature branches from the tip downwards. *Diplodia* sp. and few other fungi have been isolated from such trees. The infected branches should be cut and removed and the wounds are treated with Bordeaux mixture.

ii) Thread blight

Two types of blights are noticed in nutmeg. The first is a white thread blight wherein fine whitish hyphae aggregate to form fungal threads and traverse along the stem underneath the leaves in a fan shaped or irregular manner causing blighting of affected portions. The dried up leaves with the mycelium forms a major source of inoculum for the spread of the disease. The disease is caused by the fungus *Marasmius pulcherima.*

The second type of blight is called horse hair blight .It is caused by *Corticium equicrinus.*

Fine black silky threads of the fungus form an irregular loose network on the stems and leaves causing blighting of the leaves and stems. However these threads hold up the detached dried leaves on the tree giving the appearance of a birds nest when viewed from a distance.

The above diseases are very severe under heavy shade. These diseases can be managed by adopting phytosanitation and shade regulation. In severe infections Bordeaux mixture (1%) may be sprayed after removal of severely infested branches.

iii) Fruit rot

Immature fruit split, fruit rot and fruit drop are highly prevalent in nutmeg. Immature fruit split and shedding are noticed in some trees without any apparent infection. In the case of fruit rot, the infection starts from the pedicel as dark lesions and gradually spreads to the fruit, causing brownish discolourations of the rind resulting in rotting. In advanced stages, the mace also rots emitting a foul smell. In addition to these, occasional dark sunken lesions, dark scabbing, mostly restricted to the outer layers of the pericarp without affecting the mace have also been noticed. *Phytophthora* sp. and *Diplodia natalensis* have been isolated from affected fruits. However the reasons for fruit rot could be both pathological and physiological. Bordeaux mixture 1% may be sprayed when the fruits are half mature to reduce the incidence of the disease.

iv) Shot hole

The disease is caused by the fungus, *Colletotrichum gloeosporioides.* Necrotic spots develop on the lamina which is restricted by chlorotic halo. In advanced stages the necrotic spots become brittle and fall off resulting in shot holes. A prophylactic spray with Bordeaux mixture 1% is effective against the disease.

The use of Bordeaux mixture in organic farming is restricted to the extent that the total copper applied per hectare per year should not exceed 8 kg per year on rotational basis. Therefore resistant varieties, phytosanitation and preventive measures are crucial in organic farming.

7.2 Pests

The following pests are reported in nutmeg, though not very serious (Rema *et al.* 2003). The black scale (*Saissetia nigra*) infests tender stems and leaves especially in nursery and sometimes on young plants in the field .These are clustered together and are black, oval and dome shaped. They feed on plant sap and severe infestation cause the shoots to wilt and dry. Another pest is the white scale (*Pseudaulacaspis cockerelli*) which is grayish white flat and shaped like a fish scale and occurs clustered together on the lower surface of the leaves especially in nursery seedlings. The pest infestation results in yellow streaks and spots on affected leaves and in severe infestations the leaves wilt and dry. Occasionally Mealy bugs and scale insects (*Saissetia nigra*) infect tender leaves and shoots. The third pest is shield scale (*Protopulvinaria mangiferae*) and is creamy brown and oval and occurs on tender leaves and stems especially in nursery seedlings. The infestation results in wilting of leaves and shoots. This may be controlled using approved botanical pesticides whenever necessary.

8.0 HARVEST AND POST HARVEST HANDLING

Nutmeg tree is slow growing and female tree starts fruiting from the 6th year. The yield increases with age, the most productive period is from 15 to 30 years. Some trees may still be in production at the age of ninety. The grafts start fruiting from the third year of planting but good yields are obtained from 6th year onwards. The fruits are ready to harvest in about nine months after flowering. Flowering and harvesting continue through out the year, but June - August is the peak period. Fruits are ripe and ready for harvest when the peri-carp split open. Harvest is done generally by a knife attached to a stick known as bill hook. It is reported that a large tree yields 3,000 to 10,000 fruits annually. Individual fruits weigh on an average of 60 g of which the seed weight is 6-7 g, 3-5 g mace and the rest is pericarp. The ovary is unilocular and ripens into a pendent, fleshy pyriform / globular fruit with a longitudinal groove in the centre. The outer peri-carp is 1cm thick and split at maturity along the groove into two halves and exposes the nut and scarlet reticulate aril. The dried kernels and arils make the nutmeg and mace of commerce respectively.

8.1 Harvesting

Harvesting should not be carried out in wet (dew or rain) or high humidity conditions, i.e., wherever possible harvesting should be carried out in dry, low humidity conditions. Harvesting equipment should be clean and well maintained. All containers used for primary collection of the produce must be kept free from previously accumulated plant material and when not in use they must be kept in a dry place, free from vermin and inaccessible to farm and domestic animals. Damaged and spoiled produce should be sorted and discarded. Harvested produce should be collected in dry sacks, baskets, trailers or hoppers. It must not be collected on the ground. Mechanical damage, high compaction and storage which promote composting should be avoided for which plastic sacks should not be used during harvest. Sacks must not be overfilled and stacking should be done avoiding compaction. The time between harvest and transport of the produce to the drying site should be kept as short as practicable. The harvested product should be protected from pests and domestic as well as farm animals.

8.2 Processing of mace

The collected fruits are brought to the drying area and spread out to avoid fermentation. The nuts including the mace are carefully removed from the husks. Mace is detached from the seed shell by hand or with a knife, washed and flattened out by hand or between boards and sun dried slowly for 10 to 14 days. During drying mace gradually becomes brittle, turning from scarlet to orange to yellowish brown and it acquires the pungent aroma. Hot air ovens can be used to dry the mace at 50 ^{0}C for retention of better colour and getting a good product. Blanching of mace in hot water at 75 ^{0}C for 2 minutes is advantageous to get a glossy, pleasing product, free from fungal and insect infection with more retention of colour (Dhas *et al.* 2004).The fixed oil content of mace ranges from 20% to 35%.

Central Plantation Crop Research Institute, Kasaragod, Kerala has developed a small dryer suitable for drying mace which has burning cum heat exchanger and drying chamber. Mace is spread in single layer over the wire mesh separating the plenum and drying chamber. The temperature of drying is maintained around 50 ^{0}C. Hot air requires 4 hours to complete the process. The moisture content, volatile oil and oleoresin in dried mace range from 4.85 - 5.05, 11.57 - 12.40 and 21.25 - 22.57 per cent respectively while the lycopene is up to 149.0 - 182.2 mg/100g. If drying is delayed or prolonged even by a day, chances of mould and fungal contamination are very high. Drying to optimum moisture level without losing the inherent qualities especially the colour, is a prerequisite for long storage and better price. Sun drying bleaches the colour and may contaminate the mace with mould. But artificial drying at 50 °C, which takes 28 hours to reduce moisture content from 36 % to 9% showed 'case hardening' effect after drying. Hence, slow drying is preferable. The dried mace can be packed in air tight containers or poly bags which are approved for organic products with protection from sunlight so as to protect the red coloured pigment lycopene which is highly susceptible to oxidation in light.

8.3 Processing of nut

After removal of the mace unshelled nutmeg seeds are dried separately for 4 to 8 weeks in the sun till the kernel rattles in the shell. The nutmegs should be allowed to dry in trays or on clean bamboo mats while sun drying to prevent microbial and other contamination. Shell is then broken with a wooden mallet and nutmegs are separated. In Grenada the nutmegs are allowed to dry slowly below 40°C in large ventilated buildings to get a high quality product. It is said that if the nutmeg is dried above 40°C the fat will melt and cover the kernel which gives a poor quality product. The dried nut will have less than 10% moisture.

8.4 Processing of pericarp

The fleshy pericarp can be used for making jams, jellies and pickles. In Grenada delicious Jams, Jellies, Syrups, Ice creams and liquor are prepared from the pericarp which is very popular there. The mild flavour of the Grenadian nutmeg may be a contributing factor for the excellent quality of these products. Jams and Jellies are also prepared in a limited scale in Indonesia.

8.5 Precautions in processing methods

Processing methods of nutmeg should be based on mechanized physical and biological processes. Care should be taken to maintain the vital quality of an organic ingredient throughout each step of its processing. Processing method is to be selected in such a way that it will limit the number and quantity of additives and processing aids.

8.6 Packaging

Care should be taken to see that organically produced nut/mace is protected from co-mingling with non organic products and they are not stored and transported together except when labeled or physically separated. All products shall be adequately identified through the whole process. The dried nut/mace is to be packed in eco-friendly packing materials such as clean gunny bags, double layered linen, or polythene bags. Use of polythene bags should be minimized to the extent possible. Waste generating packaging materials are to be avoided. Recyclable/reusable packing materials shall be used wherever possible. Care must be taken to avoid materials which might lead to 'sweating' and mould development. The packaging should be such that the maximum weight loss is 10%. Spices must be dried thoroughly prior to shipment.

8.7 Grading and marketing

The grading for organic nutmeg separately is not specified though organic certification through appropriate agencies is essential. However, the grading and quality specifications applicable to conventional nutmegs may be adopted for organic nutmeg also.

8.7.1 Grading

East Indian nutmeg and West Indian nutmeg are the two grades of nutmeg. Good quality has to be maintained for trade of nutmeg and mace. Whole nutmegs are grouped under three broad quality classifications:

- *Sound*: Nutmegs which are used mainly for grinding and to a lesser extent for oleoresin extraction.
- *Substandard*: Nutmegs which are used for grinding, oleoresin extraction and essential oil distillation.
- *Distilling*: Poor quality nutmegs used for essential oil distillation.

In Indonesia, high quality whole nutmeg is traded in grades which refer to their size in numbers of nutmeg per pound: 80s, 110s and 130s or 'ABCD' which is assortment of various sizes. Substandard nutmegs are traded as 'Sound Shriveled', which in general have a higher volatile content than mature sound nutmeg and are used for grinding, oleoresin extraction and oil distillation. 'BWP' (Broken, Wormy and Punkey) are used mainly for grinding, as volatile oil generally does not exceed 8%. Distilling grades of nutmegs are of poor quality .i.e. 'ETEZ' with a volatile oil content of 12 - 13%.

In Grenada, sound nutmegs are sold as unassorted which corresponds to the Indonesian grade 'ABCD'. Substandard nutmegs are classified as floats and defective, the latter is similar to the Indonesian 'BWP' grade but considered of high quality. Distilling grades of nutmegs are primarily exported to the USA and consist of floats.

Mace is classified as whole pale mace, No.1 broken mace, selected, unassorted or siftings (Indonesia) and as whole, broken blades or siftings (Grenada). The international standards applicable for trade in spices of nutmeg and mace are ISO 6577:1990.

During storage and transportation, oils should be protected from light and stored in tightly - packed containers at a temperature not exceeding 25°C. Prolonged storage deteriorates the composition of the oil.

8.7.2 Quality specifications

Though national standards are available for maintaining the quality, the European traders prefer the ASTA cleanliness specifications as they are more stringent than the national standards. The specifications are furnished in Table 3. The quarantine system, plant protection law and the Food Sanitation Act set the quality standards in Japan. The Dutch standard for cleanliness is furnished in Table 4. Aflatoxin and salmonella are the common complaints on the imports on nutmeg. The presence of insects is a major complaint for US importers.

Adulteration is common in nutmeg trade. The essential oil has often been extracted before they are marketed and such nut can be detected by their light weight and are more subjected to insect attack. *M. fragrans* is adulterated with *M. argentea*, *M. malabarica* and *M. otaba* which can be identified by their poor quality. The mace from *M. argentea* is imported as Papuan nutmeg from Papua New Guinea; *M. malabarica* is traded as Bombay nutmeg from India and *M. otaba* as Otaba nutmeg. Trade of wild nutmeg exists and they are marketed as long, female, Papua, Guinea, or Norse nutmeg. All these have been traced to *M. argentea* of New Guinea from where they enter into the market as Madagascar nutmeg.

Table 3. American Spice Trade Association (ASTA) cleanliness specifications

Product	Insect (by count)	Excreta mammalian (mg/lb)	Excreta other (%wt)	Mould (%wt)	Insect infested (%wt)	Foreign matter (%wt)
Nutmeg (broken)	4	5	1.0	*	*	0.5
Nutmeg (whole)	4	0	0.0	*	*	0.0
Mace	4	3	1.0	2.0	1.0	0.5

**Not more than 5% by weight insect defiled and mould infected.*

Table 4. Dutch regulation for cleanliness specifications

Product	Ash content (Max. %)	Sand content (Max. %)	Volatile oil (Min. %)	Others
Nutmeg	3.5	0.5	4.0	NVEE*5.0
Mace	3.5	0.5	4.0	NVEE*4.0

*NVEE - Non - volatile ether extract

REFERENCES

Ankegowda, S. J. (2006) Soil and water conservation measures in spice crops. *Spice India,* 19(6): 2-5.

Beena, S. and Kurian, A (1996) In situ budding to assure femaleness in nutmeg. *J. Plantn. Crops,* 24 (Suppl.): 473- 478.

Chellappan, K.and Roche, J. I. (1982) Studies on clonal propagation of nutmeg (*Myristica fragrans* Houtt.). *South Indian Hort.,* 30: 156-158.

Directorate of Arecanut and Spices Development. (2007) *Arecanut and Spices Database.* Directorate of Arecanut and Spices Development, Calicut, Kerala, India.

Flach, M. (1966) Nutmeg cultivation and sex problem. *Mededelingenvande Landbouwhogeschool,* Wageningen, 66, 1: 87.

Dhas, P. H. A, John Zacharia, T., Rajesh, P.N. and Shinoj Subramannian. (2004) Effect of Blanching and Drying on Quality of Mace (*Myristica fragrans*). *J. Food Sci. Technol.,* 41(3): 306–308.

Johny. A. Kalluparackaral and Ravindran, P. N. (2006) High yielding Spices varieties developed in India. In: P. N. Ravindran, K.Nirmal Babu, K.N. Shiva, A.K. Johny . (eds.) *Advances in Spices Research*: History and achievements of Spices Reseach in India since Independence. Agrobios, Jodhpur, India. pp: 120-121.

Kandiannan. K., Srinivasan, V., Thankamani, C. K. (2003) Organic way of spice production - an overview. In: George. V. Thomas(ed.). *Recent advances in organic farming technologies in plantation crops.* Central Plantation Crops Research Institute, Kasarkode, Kerala, India. pp. 70–75.

Kerala Agricultural University. (KAU). (2003) *Package of Practices: Crops,* Kerala Agricultural University, Vellanikkara, Thrissur, Kerala, India.

Krishnamoorthy, B., Rema, J., Leela, N. K. and Jacob, T. K. (2006) Nutmeg. In: V. A. Parthasarathy and P. Rajiv (eds.) *Major Spices - Production and processing*. Indian Institute of Spices Research, Calicut, Kerala, India. pp. 274 - 288.

Krishnamoorthy, B. and Rema, J. (2001) Nutmeg and mace. In : K.V.Peter (ed.) *Hand book of herbs and spices,* vol.I . Wood Head Publishing Ltd, USA. pp. 238- 248.

Krishnamoorthy, B., Sasikumar, B. and Rema, J. (1996) Genetic variability and segregation of sex in nutmeg (Myristica fragrans Houtt.). *J. Plantn. Crops, (Suppl.):* 468-472.

Madan, M.S. (2007) Organic Spices: Marketing and Economics. Souvenir, *National Seminar on Organic Spice and Aromatic Crops,* 1 - 2 February, Directorate of Arecanut and Spices Development, Calicut, Kerala, India. pp. 83- 85.

Mathew, P.A. (1985) A case of reversible plagiotropy in nutmeg (*Myristica fragrans* Houtt.). *J. Plantn. Crops,* 13: 147-151.

Nagawekar, D.D, Desai, A. G., Joshi, G. D., Magdum, M.B.and Khan,H.H (2002) Performance of Spice crops as intercrops in coconut plantation under Konkan conditions.In: P.Rethinam et.al. (eds.). *Plantation Crops Research and Development in the Millennium*. Proceedings of PLACROSYM - XIV 12- 15 December 2000; Hyderabad, India.Coconut Development Board, Kerala, India. pp. 333- 335.

National Research Center for Spices (NRCS). (1993) *Annual Report 1992- 1993,* Calicut, Kerala, India.

Prestoe. (1884) *Sexes of Nutmeg trees.* Gardeners Chronicle, 1:3.

Purseglove, J.W., Brown, E.G., Green, G.L.and Robbin, S.R.L. (1981) *Spices Vol. I.* Longman Inc., New York.

Rema,J., Krishnamoorthy, B. and Mathew, P. A. (2003) High yielding varieties of Cinnamon and Nutmeg. *Spice India,* 16(10): 7-17

Ridley, H.N.(1912) *Spices.* Macmillan, London.

Spices Board. (2001) *Guidelines for production of organic spices in India.* Spices Board, Cochin, India, 84 – 92.

14

Chilli and Paprika

K.N. Shiva, D. Prasath and V. A. Parthasarathy

1.0 INTRODUCTION

Capsicum or chilli including paprika belonging to family *Solanaceae*, is a native of new world tropics and sub tropics. Within the genus *Capsicum*, mainly five species have been recognized as domesticated namely, *Capsicum annuum, C. baccatum, C. chinense, C. frutescens* and *C. pubescens*. Most of the commercially cultivated chilli cultivars as well as paprika belong to the species *C. annuum*. The cultivated *Capsicum* was introduced to India and Southeastern Asia in the 17th century by the Portugese. The name *Capsicum* is possibly derived from the Greek word 'Kapso' meaning 'to bite' referring to the dominant pungency stimulated by the spice, while 'Paprika' means plants of the genus *Capsicum*, in the Hungarian language. 'Paprika', is defined in the world market as non-pungent (sweet) brilliant red ground *Capsicum* powder, derived from the thick flesh dried red pods, with most of the seeds and veins removed. Both chilli and paprika are widely consumed as vegetable and spice all over the world. Mostly, the cultivars grown in Asian countries are pungent, while less and non-pungent cultivars are more common in European countries. Chilli of commerce is the dried pod, while it was pericarp (outer skin of the dried fruit) for paprika.

1.1 Area, Production and export

India is the largest producer of chillies in the world. Among the various spices produced in India, per capita consumption is the highest for chillies. Chillies contribute to the tune of 3% in total vegetables in world. According to FAO statistics, chillies and peppers (dry) are cultivated in an area of 20.05 lakh hectares with a production of 26.63 lakh tonnes during 2005 (FAO, 2007). India is the only country having diversified varieties of chillies with varying quality aspects. Chilli, a universal spice is being cultivated in almost all the States and Union Territories of the country. During 2004-05, it was grown in an area of 7.37 lakh hectares with an annual production of 11.85 lakh tonnes and recording the highest productivity of 1608 kg ha^{-1}. Though India leads in area, the productivity remains low (1t ha^{-1}), as compared to other countries like USA, South Korea and Taiwan (3-4 tonnes). The major chilli growing states in India are Andhra Pradesh, Maharashtra, Karnataka and Tamil Nadu, together constitute about 75% of the total area. However, the production was more in Andhra Pradesh, Karnataka, Orissa, West Bengal and Tamil Nadu. In terms of productivity, Andhra Pradesh tops the list, followed by Punjab, Arunachal Pradesh and Tripura and Karnataka (Table 1).

Table 1. Area, production and productivity of chillies in India (2004-2005)

State	Area ('000 ha)	Production ('000 t)	Productivity (kg ha^{-1})
Andhra Pradesh	236.5	748.5	3165
Andaman & Nicobar	0.4	1.0	2500
Arunachal Pradesh	1.7	2.6	1529
Assam	14.3	9.2	643
Bihar	3.1	2.2	710
Chhattisgarh	5.6	3.3	589
Gujarat	6.4	5.7	891
Haryana	1.1	1.2	1091
Himachal Pradesh	0.8	0.1	125
Jammu & Kashmir	1.0	1.0	1000
Karnataka	69.9	945	1352
Kerala	0.7	0.7	996
Madhya Pradesh	45.7	39.6	867
Maharashtra	90.0	44.0	489
Manipur	13.4	8.0	597
Meghalaya	1.8	2.1	1167

Contd...

Table 1. Contd...

Mizoram	1.3	0.8	615
Nagaland	0.8	0.9	1125
Orissa	76.1	63.2	830
Punjab	2.6	4.2	1615
Rajasthan	26.4	31.4	1189
Tamil Nadu	67.0	44.6	666
Tripura	1.6	2.3	1438
Uttar Pradesh	16.7	12.8	766
West Bengal	52.2	61.4	1176
Total	737.1	1185.3	1608

Source: Directorate of Economics and Statistics, New Delhi

Though the area under paprika cultivation is very meager, similar types are mainly grown in Karnataka and Andhra Pradesh. Bydagi chilli (paprika type) is traditionally cultivated in Dharwad, Haveri and Gadag districts of Karnataka. During 1998-99, the area under cultivation was as 1,29,000 ha. with production of 86,200 MT, and during 1999-2000, the production was around 38,500 MT (Koshy John, 2000).

India is the highest exporter as well as consumer. Chilli earns the highest amount of foreign exchange among the spices exported from India (113250 tonnes valued at Rs. 403.50 crores during 2005-06). In recent years, value-added products like oleoresin, natural red colour (carotenoids) and pungent principle (capsaicin) are manufactured for use in food and pharmaceutical industry are also exported. Oleoresin of different grades *viz.*, low, medium or high pungency are also exported in large quantities. Chilli powder is another important item of export. Other chilli products include: chilli seed, chilli paste, capsicum oleoresin, paprika oleoresin, etc. Among the various items/products exported from India, dry chilli contributed maximum, followed by chilli powder and paprika oleoresin. India accounts for 60% of the total oleoresins extracted from raw material valued at Rs. 315.48 crores. Of all the oleoresins exported from India, Paprika oleoresins account for 42.13% share valued at Rs. 132.9 crores. The current market for natural colours accounts for about 65% in Europe and is ever increasing. The red and yellow color hues outweigh all other hues. Moreover, usage rate and replacement of synthetic pigments have also increased the demand for paprika in the world market. Hence, India is having a great export potential. In India paprika types *viz.*, Bydagi chilli in

Karnataka and Tomato chilli in Warangal area of Andhra Pradesh are being cultivated by the farmers to meet the huge demand of paprika industry.

1.2 Composition and Uses

Chilli/paprika contain thousands of chemicals including water, fixed (fatty) oils, steam-volatile oil, carotenoids, capsaicinoids, resin, protein, fibre and mineral elements. Chilli is good sources of several vitamins. They produce high concentrations of vitamin C, provitamin A, E, P (citrin), thiamine (B_1), riboflavin (B_2), and niacin (B_3). Paprika contains vitamins - A and C, niacin and minerals like potassium, magnesium, sodium, phosphorous and calcium. The two most important groups of chemicals found in chillies are carotenoids and the capsaicinoids. The carotenoids contribute to chilli colour and its nutritional value, while the capsaicinoids are the alkaloids responsible for their characteristic pungency. Paprika fruit can be portioned into pedicel, seed, placenta and pericarp, in which major portions of pigments occur in the outer pericarp tissues beneath the epidermis. Pigments constituting the color are a complex mixture of carotenoids. Red and yellow components combine to give the total color to paprika. The red color is constituted by two major pigments of capsanthin (30-60%) and capsorubin (6-18%). The inherent characters of the cultivar which affects the stability of the color are: anti-oxidants present in the pericarp and seeds and oxidizing agents. In chillies, 95% of the total pro-vitamin A in green pods and 93% in mature red pods is â-carotene. When red mature pods were measured, the cultivars with the highest and the lowest pro-vitamin A activity were both yellow wax pod types (Bosland and Votava, 2000).

In the world market, three major products traded for use in food processing are: paprika, oleoresin and dried chilli (both in whole and in powdered form). They are widely used in curry powder, paste, pickles, sauces and ketchups for being its characteristics pungency, colour and aroma. Paprika is used primarily for its colour and flavour in processed foods. Paprika oleoresins are used in colouring, seasonings, poultry products, snack foods, fats and oils, pastes and the remaining in all other food categories such as baked food, cheese, condiments, fish and fish products, gravies, juices, meat with gravy, nuts and toppings/ creams/soups. Ground paprika is used in confectionary, meat and meat products, snack and processed foods and vegetables, fish products and bakery, beverages, cereals, dairy, fats and oils.

2.0 BOTANY AND VARIETIES

For organic production of chilli and paprika, land races/traditional (local) varieties which are being cultivated since age-old may be adopted as such.

However, the varieties and hybrids developed by various institutions and agencies need to be studied in detail for their response to organic farming practices.

2.1 Chilli types/landraces

There are several chilli types/landraces/cultivars cultivated traditionally in chilli growing tracts of the country (Table 2) and some popular types/cultivars of chilli are given in Table 3.

Table 2. Chilli types/land races along with its geographical distribution

Chilli type/land race	Geographical distribution
Bird's eye chilli (Dhani)	Assam, Mizoram and some areas of Manipur
Bydagi dabbi / kaddi	Dharwad, Karnataka
Ellachipur Sannam - S4 type	Maharashtra
Guntur Sannam – S4 type	Guntur, Warangal, Khammam districts of Andhra Pradesh
Hindpur-S7	Hindpur in Andhra Pradesh
S7	Hindpur in Andhra Pradesh
Jwala	Kheda, Mehsana in South Gujarat
Kanthari white	Kerala and some parts of Tamil Nadu
Kashmir chilli	Temperate region such as Himachal Pradesh, Jammu & Kashmir and in subtropical regions of North India
Madhya Pradesh – G.T. Sannam	Indore, Malkapur chikli and Elachpur areas of Madhya Pradesh
Madras Pari	Nellore District of Andhra Pradesh
Nagpur and Nalchetti	Nagpur District of Maharashtra
Ramanad Munda	Ramanad and Virudhunagar District of Tamil Nadu
Sangli Sannam – S4 type	Kolhapur District of Maharashtra
Sattur-S4	Dindigal, Sattur, Rajapalayam Sankarankoil and Theni areas of Tamil Nadu
S9 Munda	Ananthapur District of Andhra Pradesh
Tadapally Biglong	Tadapally in Andhra Pradesh
Tomato chilli or Warangal Chappatta	Warangal, Khammaam, East and west Godavari districts of Andhra Pradesh

Source: Johny and Ravindran (2004)

2.2 Varieties

In India, many improved varieties and hybrids of chillies have been released and recommended for cultivation under specific agro-climatic zones (Table 4). However, its suitability for organic production has to be validated.

Table 3. Commercially known chillies that cultivated in different states of India

State	Commercial / local name
Punjab	Serhindi, Sanauri Red, Sanuari yellow, Patna Red, Rajpura long red, Jullundur, Shokati, Longi
Andhra Pradesh	Mundu, Sadasivpet, Shivai, Nallapadu, Warangal, Lanka chilli, Seema Mirapa
Maharashtra	Dondicha, Achalpur, Malkapuri
Tamil Nadu	Salem, Ramanad, Samba Kodakal, Sattur samba Elichup, NP 34
Karnataka	Javari, Javageal, Coilbel danur, Gouridnur, Bydagi type, Sankeshwar type, Chincholi, Kollegal type, Arsikere, Madhugiri
Bihar	Rahuri, sity patna red, Sobour
Gujarat	Gholar, Patta
Assam	Suryamukhi, Latabh, Tiger
Uttar Pradesh	Kalianpur
Kerala	Kandhari

(a) (b) (c) (d)

Fig. 1: Some high yielding varieties of Paprika (a) Arka Abir (b) Bydagi dabbi (c) Kt Pl 19 (d) Vinant of Tomato chilli (*Source:* Director, IISR, Calicut)

Table 4. Improved varieties of chillies/paprika and paprika like chillies and their salient features

S. No	Variety	Source	Yield (kg/ha, dry)	Salient features
Variety				
1	Pusa Jwala	Indian Agriculture Research Institute, New Delhi	610	Plants dwarf, tolerant to thrips, mites and aphids
2	Pusa Sadabahar	"	2000	Plants tall, fruits erect in clusters, resistant to CMV, TMV and leaf curl, has perennial tendency
3	K 1	TNAU, Agriculture Experiment Station, Kovilpatti, Tamil Nadu	1700	210 days duration, suitable for rainfed cultivation in Southern districts of Tamil Nadu, high ly pungent type
4	K2	"	1800	210 days duration, tolerant to thrips
5	Co.1	TNAU, Horticultural College and Research Institute, Coimbatore, Tamil Nadu	2100	Fruits bright red in colour, 210 days duration, high ly pungent type
6	Co. 2	"	2200	200-210 days duration and suitable both for green and dry chillies
7	Co. 3	"	3000-3500	165 days duration, dual purpose chilli, high oleoresin content, suitable for high density planting
8	Co. 4	"	-	Plants dwarf, 165 days duration and suitable for vegetable purpose
9	PKM 1	TNAU, Horticultural College and Research Institute, Periyakulam, Tamil Nadu	2300	Plants dwarf, 180 days duration and suitable for irrigated conditions
10	MDU 1	TNAU, Agricultural College and Research Institute, Madurai	1900	Plants dwarf, 205-215 days duration, cluster bearing and suitable for high density planting,
11	PLR 1	TNAU, Vegetable Research Station, Palur, Tamil Nadu	1800	'Gundu' type, 210 days duration and suitable for green chilli
12	PMK 1	TNAU, Agricultural Research Station, Paramakudi, Tamil Nadu	2300	'Gundu' type, 200 days duration and suitable for rainfed cultivation
13	G3	ANGRAU, RARS, Lam Guntur, Andhra Pradesh	3000	Suitable for both rainfed and irrigated conditions

Contd...

Table 4. Contd...

14	G4 (Bhagyalakshmi)	"	1100 (rainfed) 4000 (irrigated)	Fairly tolerant to pest and disease and wider adaptability
15	G5 (Andhra Jyothi)	"	5000 (irrigated)	Plants tall, tolerant to thrips and wider adaptability
16	Sindhur	"	5000 (irrigated)	Plants tall, dual purpose variety and suitable for summer cultivation
17	Aparna	"	3500	Highly pungent, moderately tolerant to major pests and diseases under field conditions
18	Bhaskar	"	5000 - 6000	Early maturing, highly pungent, tolerant to thrips and mites and suitable for pickles
19	Jwalasakhi	KAU, Vellanikkara, Trichur	-	Plants dwarf, less pungent and suitable for Southern districts of Kerala
20	Jwalamukhi	"	-	Pods dark green, less pungent and suitable for Southern districts of Kerala
21	Ujwala	"	-	Resistant to bacterial wilt, mosaic and leaf roll
22	Anugraha	"	-	Early in flowering and resistant to bacterial blight
23	Phule Jyoti	MPKV, Rahuri, Maharashtra	3000	Dual purpose chilli, tolerant to leaf curl and powdery mildew diseases, resistant to lodging, suitable for both *Kharif* and summer seasons
24	Phule Suryamukhi	"	2800	Tolerant to leaf curl and powdery mildew disease, suitable for both *Kharif* and summer seasons
25	Punjab Lal	PAU, Ludhiana, Punjab	1200	Resistant to TMV, CMV, leaf curl viruses, moderately tolerant to fruit rot and dieback
26	Pant C-1	GBPUAT, Pant Nagar, Uttaranchal	1200	Highly pungent, tolerant to mosaic and leaf curl virus
27	Pant C-2	"	1400	Tolerant to mosaic and leaf curl virus
28	NP 46A	IARI, Pusa, New Delhi	-	Tolerant to thrips but susceptible to viral diseases

Contd...

Table 4. Contd...

Paprika and paprika like chillies				
29	Kt-Pl-19	Indian Agricultural Research Institute - Regional Station, Katrain, Himachal Pradesh	4000-4800	Fruits pendent bearing two-celled, produces about 60 fruits with a yield of 25 – 35 tonnes/ha, colour value - 233 ASTA units , yields 4.0-4.8 tonnes of dry fruit with 60-66% skin recovery and around 2.77 tonnes dry basic product for processing into ground paprika or oleoresin and gives a seed yield of 0.7-0.8 t/ha.
30	Bydagi Chilli	Dharwad, Haveri, Gadag and Hubli Bellary, Raichur and Gulburga Districts of Karnataka	3750-4500	Suitable for intercropping with cotton and cultivation in black soil, color value -157 ASTA units; Capsaicin -negligible
31	Tomato Chilli (Warangal Chappatta)	Warangal, Khammam, East and West Godavari Districts of Andhra Pradesh	-	Fruits oblong and glossy with less seeds, highly red color with very low pungency, very low pesticide usage and the yield also very low, Color value - 125.26 ASTA units; Capsaicin - 0.17;
32	Arka Abhir	Indian Institute of Horticultural Research, Bangalore, Karnataka	2000	Fruits light green, wrinkled turning to deep red on maturity, high color value (1,65,541 cu) and low pungency (0.05% capsaicin), suitable for oleoresin extraction, oleoresin yield (5.78%) without seed.
33	KDC –1	University of Agricultural Sciences, Dharwad, Karnataka	-	Bears semi-wrinkled red fruits with more capsaicin than Byadagi type, tolerant to leaf curl virus, high color and low pungency, suitable for intercropping with cotton and for cultivation in black cotton soil.

Source: Shiva *et al.* (2006); Deshpande (2005)

Apart from this, many private seed companies including Multi National Companies and Indian firms have developed and introduced a number of varieties in India, which are also getting popular among growers.

Chillies are highly prone to pests and diseases. Even though, so far no real resistant variety has been released against biotic stresses, some resistant or tolerant sources have been identified for the benefit of farming community across the country (Table 5). These would form a component in organic production.

Table 5. Resistant/ tolerant sources identified against biotic stresses in chilli

Pest /Disease/ Nematode	Resistant/tolerant source
Leaf curl virus / Mosaic virus	Phule Jyothi, Phule, Suryamukhi, Pusa Jwala, Pusa Sadabahar, PantC-1, Pant-C-2, Punjab Lal, Jawahar-218, HC -1, HC- 15, HC - 22, HC - 28, HC – 44, NP 46 A, Musalwadi
Mosaic, wilt & dieback	Punjab Surkh, LCA – 235, LCA - 305
Leaf curl complex	KDC-1, GPC-82, GPC 69, GPC 80, Jwala, LCA – 235
Virus complex	LCA - 235
TMV	Kranti, Krishna (Hybrids)
Bacterial wilt	Ujwala, Anugraha, Manjari, BC-21-2 (Utkal Rashmi), AAUM-1 & AAUM-2
Bacterial leaf spot	G-5
Die back /Anthracnose	Musalwadi, Pusa Sadabahar, Vardan (Hybrid), HC – 28, HC – 44, Achar
Powdery mildew	Musalwadi, Phule Jyothi, Phule Suryamukhi, Arka Lohit, HC -28
Fruit rot	Jawahar-218, PBC 36, HC – 28, HC -44, K-2
Thrips, mites	Andhra Jyothi, Bhaskar
Thrips	K-2, Chamatkar,Guchedar
Root knot nematode	K 2, G-2
Mites	LCA - 235, LCA - 230, LIC - 13, LIC - 45, Yellow anther mutant
Multiple resistance	Punjab Lal, Lorai, LS - 111,CH - 1

Source: Shiva *et al.* (2006)

3.0 CLIMATE AND SOIL

3.1 Climate

The *Capsicum* (chilli and paprika), a plant of tropical and sub-tropical regions requires a warm humid climate. It is suitable to varying altitudes ranging from sea level (Coastal Karnataka, Kerala and Tamil Nadu) to 2000 m MSL (Himachal Pradesh). Crop can tolerate extremes of temperature but long frost and freezing temperature is not suitable. A soil temperature about 17°C favours normal development. Even though high temperature favours shoot growth, soil temperature above 30°C affects the root development. Fruit coloring is delayed considerably below 15°C. At high temperature, fruit drop and poor fruit set is a problem. Maximum temperature of 20-30°C and minimum temperature not below 10°C is optimum for growth and development. However, performance of paprika types is better under low temperature conditions. It thrives well in areas having a moderate rainfall within the range of 60 - 120 cm. Excessive rainfall associated with high humidity causes defoliation and root rot. It is mostly grown as rainfed crop both for green and dry chilli purpose in Dharwad belt of Karnataka and certain pockets of Andhra Pradesh and Tamil Nadu. Peppers are photoperiod-insensitive (day length does not affect flowering or fruit set).

3.2 Soil

Well-drained, well-aerated loamy soil rich in organic matter with pH range of 6.5-7.5 is ideal for chilli and paprika cultivation. The black cotton soil which retains moisture for long periods is suitable for rainfed cultivation, whereas well-drained red sandy soils are suitable for irrigated crop. To a certain extent, the crop can be raised in coastal alluvial soils. Though chilli comes up well in black cotton soil, sandy loam is best suitable for its growth and yield, if provided with adequate irrigation and manuring. Light loamy or sandy loam soils rich in organic matter are ideal for chilli crop. It is suitable from the Himalayan range in the north to coastal sandy soils of Kerala, Tamil Nadu and Karnataka. Crop is highly sensitive to water logging and thus drainage is important. Not very sensitive to soil acidity but saline and sodic soils are not suitable, as it hinders fruit development. In saline soils, the germination and early vigour of plants are affected. In India, most of the chilli growing area for dry chillies lies in deep black cotton soils (Vertisols) and medium deep soils (Vertic intergrades) occurring in transitional belt of Dharwad (Karnataka), Nagpur, Nanded and Kolhapur (Maharastra), Guntur (Andhra Pradesh) and Coimbaotre (Tamil Nadu). It is grown in soils having pH 5.0 in coastal areas to pH 9.0 in vertisols of the country.

For cultivating chillies organically, a buffer zone of 25 to 50 feet is to be left all around from the conventional farm, depending upon the location of the farm. The produce from this buffer zone belt shall not be treated as organic (www.indianspices.com).

4.0 SEED MATERIAL

4.1 Source of seed materials

Chilli (and paprika) is grown from seeds, which are very light and remain viable for 2-3 years. Successful production of chilli primarily depends on quality of seeds. Prime attributes of good quality seeds are that they should ensure maximum germinability and vigour. Seeds should be collected from healthy plants and healthy fruits. The seeds should be extracted from first picked healthy and fully ripened red color fruits, which may give rise to better growth, early filed emergence and better seedling vigor index. Maximum percentage of viable seeds may be achieved by harvesting the fruits in reddening stage. One kilogram of chilli seeds may contain 123-170 thousand seeds depending upon size and weight (Hosmani, 1993).

For raising nurseries, seeds of high yielding varieties with tolerance to pests and diseases and responsive to organic manures may be used. They should be carefully selected from certified organic farms or from own seed plot which is raised organically. To start with, chemically untreated seeds from local high yielding varieties could also be used, in the absence of organically produced seeds. Seeds should not be treated with any chemical fungicides or pesticides. However, it is always beneficial to adopt indigenous practices for seed treatment, wherever possible. The seeds may be treated with *Trichoderma* @ 10^9 kg^{-1} of seed to prevent incidence of seedling rot in the nursery (www.indianspices.com).

Seed rate: Seed rate of 6.25 kg ha^{-1} is recommended for direct sown crop, while it was 0.5 - 1.0 kg/ha for varieties and 100 - 250 g ha^{-1} for hybrids in transplanting crop.

4.2 Nursery raising

Time of sowing and transplanting depends upon temperature, rainfall and availability of irrigation facilities. Normally, chilli is raised by transplanting of seedlings but in some regions direct sowing is also practiced.

Direct sowing: Direct sowing is recommended for rainfed crop under retentive black soils. Seeds are drilled by the end of July or first week of August. Seed rate is 6.25 kg per hectare. After 30-40 days of sowing, thinning and gap filling is done on a cloudy day. Spacing of 56 x 15 cm is adopted for direct sown crop (www.indianspices.com).

Transplanting: The advantages of transplanting are better care for seedlings, better establishment and growth, optimum plant protection measures compared to direct seeding. Well-drained area near the vicinity of water source is usually selected for raising nursery. Fertile and elevated areas in partial shade are preferred and the selected area is ploughed to a fine tilth by removing the stones and other debris. Raised beds are preferred compared to flat beds. Raised beds of 90 - 100 cm width and of convenient length are prepared to which well-decomposed organic matter at the rate of 20-25 kg and 75 g each of NPK/bed of 10 m^{-2} is incorporated. Application of neem cake 300 -350 g 100 m^{-2} as basal dose before preparation of raised beds is also advisable. The nursery beds should be surrounded by 30 cm width drainage channels. Stagnation of water should not be allowed in nursery beds in order to avoid fungal infection. The nursery area required for planting one hectare is 100 m^2. Around 1,250 g of seeds would be required to raise the seedlings for one hectare. The nursery area earmarked may be subjected to solarization by covering the area with plastic or polythene sheet and sealing the sides so that heat generated kill the pathogen Sowing can be taken up after a gap of 5-6 days.

Treat the seeds with *Trichoderma viride* @ 4-6 g kg^{-1} of seeds and sow thinly 5 cm apart in lines spaced at 10 cm in raised nursery beds and cover with sand. Mulching with paddy straw or dry grass helps to maintain optimum moisture and soil temperature required for better and uniform germination and the mulch material should be removed once germination is noticed. Under normal conditions seeds germinate in 6 – 10 days. Water is given immediately for better establishment of seedlings. Thereafter, regular light watering with rose can in morning and evening is advisable. Drench the nursery with Bordeaux mixture (1%) at 15 days interval against damping off disease, if necessary. The entire nursery area should be protected with 100 mesh nylon net.

Commercially available plastic trays filled with sterilized premixed media can be used to raise quality seedlings of paprika. The advantage includes uniformity in germination to transplant height and growth, ease in management, strong and healthy root system, less risk with soil borne pathogens and ease in transporting to long distances. Trays with cell size of 1.5" X 1.5" are adequate. The trays may be arranged under shade net or green house for proper germination and growth. Seeds take about 10 days to germinate depending upon the availability of favorable temperature and moisture in the soil.

5.0 PLANTING AND AFTER CARE

Successful crop management is primarily based on the soil type, climate, variety/hybrid, availability of adequate nutrients in the soil, good quality seed materials free from seed borne diseases, cultural practices, etc. The cultivation of paprika (paprika like chillies) is almost the same as that of chilli.

5.1 Land preparation and planting /transplanting

Land is ploughed either by wooden plough or iron plough after the receipt of showers for 3-4 times to a fine tilth. Repeated harrowings are done to destroy the weeds. Deep ploughing minimizes the incidence of soil-borne insect-pests of chilli such as cut worms, root grubs, wire worms and fruit borers and soil borne disease such as damping off caused by *Pythium, Fusarium* and *Rhizoctonia* spp. by encouraging drainage (Hosmani, 1993). Tractor ploughing followed by harrowing reduces weed population and increased yield of chilli (Rangaswami, 1984). Farm yard manure is incorporated @ 25-30 tonnes per hectare in the last ploughing. For irrigated crop, ridges and furrows are formed in South India, while bed system is adopted in North India.

5.1.1 Season

Season of cultivation should be in such a way that the crop should receive favorable climatic conditions for growth and development and also to escape from the biotic and abiotic stresses. In South India it is cultivated throughout the year for green fruits under irrigated condition and during *Kharif* season (June - July) for ripe dry fruits. In West Coast of India, chillies are mostly cultivated during *Rabi* season (October - November). In North India, it is mostly grown during *Kharif* and summer since the crop does not withstand low temperature. Sowing and transplanting seasons of different Indian states are given in Table 6.

Soil type influences the time of planting under rainfed cultivation. In light soils, transplanting may be taken up in early part of monsoon than heavy soils, as the moisture retention is low in light soils.

The seedlings are subjected to hardening prior to transplanting so that it can withstand the transplant shock better. This is done by reducing the frequency and quantity of watering during the last week in the nursery. Age of the seedlings for transplanting depends on variety, soil and weather conditions. Early maturing varieties need to be transplanted early compared to late varieties. Young seedlings establish better in cloudy days than in sunny days. Age of the seedling influences the growth and yield of the crop and while planting it should be in active growth stage and young. Generally, 40-45 day old seedlings, when they are about 15 cm height are preferable for transplanting. Optimum number of seedlings per hill is 2 -3 for varieties and in case of hybrids it is 1 -2, which later thinned to one per hill and gap filling is completed 30 days after sowing. To enhance establishment and growth of seedlings, *Azospirillum* @ 2 kg 10 liter^{-1} of water can be used for dipping the root of the seedlings for about 30 minutes.

Table 6. Sowing and transplanting seasons for chilli/paprika in different states

Region/State	Direct sowing	Nursery sowing	Transplanting
Chillies/paprika like chillies			
Andhra Pradesh	June (*Kharif*) September (*Rabi*)	May – July (*Kharif*)	August – September
Karnataka	June – July (*Kharif*)	May - June (*Kharif*) October – November (*Rabi*) January (Summer)	July – August December - January February - March
Tamil Nadu	-	January – February (*Rabi*) June – July (*Kharif*)	March – April (*Rabi*) September - October (*Kharif*)
Maharashtra	-	September – October (*Rabi*) June – July (*Kharif*)	November –December (*Rabi*) August – September (*Kharif*)
Punjab & Rajasthan	-	April – May	June – July
Jammu and Kashmir	-	March – April or January – February (*Rabi*)	May – June or July – August
Kerala	-	April – May (rainfed) August – September (irrigated)	June – July October – November
Gangetic valleys	-	June – July	August – September
North-eastern region / Hill areas	-	March – April	May – June
Paprika			
Paprika types	-	August (Autumn winter crop) November (Spring summer crop)	September – October December - January

Chilli is transplanted mainly in June-July, if grown under rainfed conditions and throughout the year under irrigated conditions in South India. However, main planting seasons for the crop are June-July, November and March. In North India, summer crop is raised by transplanting in February - March under hills, while it is mainly during *Kharif* season under plains. However, crop cannot be raised successfully in *Rabi* season due to low temperature prevalence. The season of sowing and transplanting for different regions/states are given in Table 6. In red soils of Bangalore, Karnataka, transplanting in July is recommended, since late planting attracts more insect-pests and diseases, while it is in August for Hyderabad, Andhra Pradesh.

5.1.2 Spacing

Plant density per unit area is decided by the spacing adopted for the crop. Spacing in chilli crop depends on various factors such as varieties, soil type, soil fertility status, season, stature of the plant, etc. Distance between two ridges depends upon the row spacing of the crop. Healthy, disease free, vigorous and 40-45 days old seedlings are transplanted to the field at proper spacing.

When the crop is raised in light soils, narrow spacing is followed. In black cotton soils, wider spacing facilitates inter-cultivation operations. Wider spacing may be given to the crop receiving higher dose of manures and fertilizers and also grown under irrigated conditions. Increase in plant density results in less lateral branching, which makes the fruits easier to be harvested. In Karnataka and Maharashtra, growing cotton and onion as intercrops along with chilli is a common practice. In these states, under rainfed cultivation, wider spacing (90 x 90 cm) is adopted for widely grown, tall stature and high branching cvs. Bydagi kaddi, Bydagi dabbi, Guntur selections and Sankeshwar type of chillies, while narrow spacing (60 x 60 cm) is adopted for short stature, low branching and short duration type of cvs. Jwala and NP 46 A (Hosmani, 1993). Under rainfed cultivation, paprika variety Kt-Pl-19 gave maximum number of fruits per plant at 60 x 60 cm spacing, while 60 x 30 cm gave highest yield of 12.08 t ha^{-1} (Hosamani, 2000).

In India, different spacings *viz.*, 30 x 30 cm (Selvaraj *et al.*, 1974), 30 x 20 cm (when applied with nitrogen at 120 kg/ha for cv. MDU-1) (Ramachandran and Subbiah, 1981), 45 x 30 cm (cv. Pusa Jwala) (Chougule and Mahajan, 1979), 45 x 20 cm (cv. Dharwad 7-6-6), 45 x 30 cm (cv. NP-46A) (Kumar, 1982), 30 x 30 cm (cv. K-1), 60 x 30 cm (Revanappa *et al.*, 1998), 45 x 60 cm (Patil and Singh, 1979), 90 x 20 cm (cv. Byadagi in rainfed with 125 kg N ha^{-1} (Kurdikeri, 1982) are being adopted to obtain maximum yield in chilli. In Karnataka and Maharasthra, spacing of 75 x 75 cm or 90 x 90 cm is generally practiced, while in Andhra Pradesh and Tamil Nadu, narrow spacing of 45 x 45 cm or even closer is followed. A Spacing of 30 x 15 cm in flat beds suits to cv. CO-3 at Coimbatore, Tamil Nadu. However, the general spacing of 60 x 45 cm for varieties and 50 x 45 cm for hybrids is recommended by AICRP on Vegetables.

5.2 Inter-cultivation

Inter-cultivation aims at to keep the field free from weeds and to facilitate soil aeration and proper root development, to conserve soil moisture and to prevent evaporation. Chilli requires frequent inter-cultivation. In direct sown crop, blade harrow is used starting from 30th day of sowing. Four inter-cultivation operations are needed at 10 days interval alternated with blade harrow and tyned harrow or junior hoe. Final inter-cultivation is given by the country plough. For an irrigated crop, inter-cultivation is given either by junior hoe or light plough after each irrigation. Intercultivation is followed by hand weeding to check the weed growth.

5.2.1 Weed Management

Weeds which attract pests should be allowed to grow in the field to act as trap and removed before flowering. Generally, weed intensity is higher in red soils (oxisols) than in black cotton soil; higher weed intensity is noticed in *Kharif* season than in *Rabi* and summer seasons. In chilli, some of the weeds act as potential alternate hosts for viral diseases and its vectors. Weeds such as *Solanum nigrum, Acanthospermum hispidum, Blainvillea rhumboidea, Flaveria australacia* and *Ageratum conyzoides* against pathogens causing leaf curl disease and insect-pest, white fly (*Bemisia tabaci*) (Mariappan and Narayanaswamy, 1977; Alegbejo, 1987) and *Gynura angustosa* and a common weed of Compositae family (Kalimpong area of W.B.) against aphid (*Aphis gossypii*), a polyphagous pest (Agarwala and Raychoudhari, 1979). In Andhra Pradesh, Karnataka, Tamil Nadu and Maharashtra, perennial weeds are problematic in clayey soils and annual weeds in red soils. Common perennial weeds associated with chilli crop are *Cynadon dactylon, Cyperus rotundus* and *Convolvulus arvensis.* Among annual grasses, *Dianebra retroflexa, Panicum isachne, Digetaria marginata* and *Dactyloctinum aegyptium* are common. Other dicot weeds commonly observed are *Commelina benghalensis, Cynotis* spp, *Phyllanthus amarus, Sida* spp. Weeds associated with chilli in Tamil Nadu are *Trianthema portulacostrum, Gynandropsis phentophylla, Cynadon dactylon* and *Cyperus rotundus.* In north India regions, *Cyperus rotundus, Tribulus terretris, Celosia argentia, Digera arvensis, Digiteria sanguinalis, Eragrostis* spp, *Eleusine aegyptiacum, Eleusine indica* and *Amaranthus viridis* are common (Hosmani, 1993).

Manual weeding is a common practice in chilli. Weeding is done by hand hoeing. It is 3-4 for irrigated crop and 2-3 for rainfed crop. Cover crops (*Stylosanthus hamata*, lucerne, *Centrosoma* and *Calopogonium*) sown during the second fortnight of July with 90 x 60 cm spacing reduced weed dry weight at all stages of growth. Among the cover crops, *Calopogonium* resulted in lower weed dry weight than the other cover crops at 45 days after transplanting. Also 1:2 row proportion of chilli: cover crop and 45 days cutting interval gave good weed control. (Basavanagouda *et al.,* 2000).

5.2.2 Mulching

Crop residues, straw, leaves, plastic films, aluminum foil, gravel, saw dust and dry soil are used as mulch materials. Mulch checks soil erosion, conserves moisture, reduces temperature and suppresses weeds. Plastic films used as mulch and tunnels increased growth and yield of chilli by 30 to 50 per cent (Lee and Yoon, 1975) by increasing temperature and reducing evaporation losses (Kapitancy, 1971). Similarly, transparent polyethylene mulch improved the germination and induced early flowering (Albregts and Howard, 1973). Mulch materials such as decomposition of organic materials, maize stalks, crop residues (sorghum - 10t/ha, maize - 5 & 10 t/ha and cotton - 10t/ha) induced early flowering, suppressed weeds, conserved soil moisture and ultimately increased the yield (Hosmani, 1993). Aluminum foil, saw dust, straw dust and wood chips act as repellents to insects. Aluminum foil as mulch repellent to aphids thereby reduces the mosaic incidence (Black and Rolston, 1972). Likewise, straw dust and wood chips reduced the incidence of mosaic disease (Kemp, 1978). Mulching may be practiced with black polythene film with rice husk in order to conserve soil moisture and to add organic matter to the soil over a period. In cv. CO 2 of chilli, mulching with 25μ black linear low density polyethylene film significantly increased the yield parameters and net returns (Lourduraj, 2003). Similarly, plastic mulching (black LLDPE) and irrigation at 0.8 IW/CPE resulted in the highest yield and net returns (Rs. 74,600/ha) (Nagalakshmi *et al.*, 2002).

5.2.3 Earthing up

Earthing up is done 30 days after transplanting. Sometimes, it is performed along with hoeing to drain the surplus rain water under rainfed cultivation, which also helps in soil aeration, checking weeds and mixing the manures with soil.

5.2.4 Crop rotation, mixed and inter-cropping

Chilli can be cultivated organically as an inter or mixed crop provided all the other crops are grown following organic methods. It is desirable to include a leguminous crop in rotation with chilli in order to improve/maintain soil fertility status.

Chilli is rotated with jowar, ragi, cotton, groundnut and castor under rainfed conditions and with sugar cane, turmeric, beans, maize or with vegetables avoiding brinjal and potato exclusively. In general, chilli is grown as a mixed crop under rainfed conditions, while it is grown as pure crop under irrigated condition (Hosmani, 1993). Even, chillies are also grown as mixed crop under irrigated conditions in some pockets. In Tamil Nadu, onion, brinjal and coriander are grown as intercrops under rainfed conditions. Barrier crops such as castor and agathi are grown on bunds in chilli field. In Karnataka and Maharashtra, it is a common practice to grow cotton and onion as intercrops along with chilli. These crops provide additional income to the farmers. Moreover, mixed cropping is generally practiced with dry chilli. In Dharwad District of Karnataka, short-duration crops like garlic, onion or coriander are also cultivated in black soils, whereas horsegram and castor crops are grown as mixed crop in sandy soils (Hosmani, 1993).

Under chilli-spinach cropping system at Parbhani, Maharashtra, application of soil conditioner (vermicompost & other organics) @ 2.5 t ha^{-1} in combination with organic booster @ 1 litre m^{-2} improved the physical condition of soil by reducing bulk density, increasing porosity, water holding capacity and infiltration rate and also the yields of chilli and spinach as compared to recommended dose of NPK (Hangarge *et al.*, 2002a). Chilli intercropped with garlic/onion consistently recorded with lesser infestations of whitefly, thrips, aphids, jassids and pod borers with higher yield, as compared to chilli as sole crop in Davanagere (Karnataka) during *kharif* and summer (Aswathanarayanareddy *et al.*, 2006).

5.3 Irrigation

Normally chilli is grown under rain-fed condition. However, under irrigated condition, care should be taken to avoid using water contaminated with fertilizers, pesticides and fungicides. Irrigation should be done judiciously. Stagnation of water should not be allowed in fields in order to avoid fungal infection. Chilli cannot withstand heavy moisture. Hence, over irrigation should be given only when necessary. Frequent and heavy irrigations induce lanky vegetative growth and cause flower shedding. The number of irrigations and interval between irrigation depend upon soil and climatic conditions. If the plants show drooping of leaves at 4 PM it is an indication that irrigation is needed.

Chilli crop is raised as a rainfed crop, where the amount of annual rainfall received is around 80-100 cm and well distributed. In northern parts of Karnataka and southern parts of Maharashtra, where the rainfall received is around 80 cm, it is extensively cultivated as a rainfed crop during *Kharif* season. Two protective irrigations at 40 and 60 days after transplanting produced higher fruit set and yield of dry chilli (Bydagi) in sandy loam soil in the transitional belt of Dharwad (Hulamani, 1988). In Andhra Pradesh and Tamil Nadu, the crop is raised under irrigation for riped dry fruit, where the amount of rainfall received is not sufficient for the crop (Hosmani, 1993). Irrigation reduces the fruit yield in the first picking but increases in the second and subsequent pickings (Berenyi, 1968). First irrigation is given just after transplanting and a life saving (light) irrigation on third day of transplanting and at weekly interval thereafter. Gap filling may be done after 10 days of transplanting during the second irrigation. In India, generally 8 to 9 irrigations are given depending on the rainfall, soil type, humidity and prevailing climatic conditions.

Studies conducted in field conditions at Indian Institute of Horticultural Research, Bangalore indicated that alternative and widely spaced furrow system requires about 40 percent less irrigation water, as compared to every furrow irrigation with 40 mm depth of water (Hegde, 1989). Irrigation to chilli at 0.8 IW/CPE produces highest fruit yield with a water consumption of 542 mm. Water use efficiency was higher in the farmyard manure and coirpith treatments (3.88 and 3.95 kg mm^{-1} $ha^{-1,}$ compared with 3.32 kg mm^{-1} ha^{-1}in the control) (Selvaraj *et al.*, 1998). Under semi-arid conditions of South India, scheduling irrigation at 40 mm depth and at 0.75 IW/CPE ratio gave the highest dry chilli yield (Prabhakar and Hebbar, 1998). Highest yield of dry chilli could be obtained in sandy clay loam of pH 7.1 by irrigating all furrows and at 0.90 IW/CPE ratio (Subramanian *et al.*, 1998). Chillies could be grown near pitcher for effective utilization of saline and high residue sodium carbonate water in sandy loam soil of arid and semi-arid regions of Rajasthan (Chauhan *et al.*, 1999). Sprinkler irrigation in the field and mist irrigation in plastic green houses reduces *Leveillula taurica* (Mones and Welvart, 1981) and drip irrigation reduces the *Phytophthora* foot rot incidence (Xie *et al.*, 1999). Use of drip irrigation increased the yield of chilli cv. K-1 by 25 per cent and reduced weed intensity by 50 per cent, besides saving water by 62 per cent in Coimbatore (Tamil Nadu) conditions (Sivanappan *et al.*, 1978; Sivanappan, 1979). For the same location, Lourduraj (2003) reported significant increase in yield parameters of chilli by irrigation at IW/CPE ratio of 0.80.

5.4 Cover crops

In chilli, cover crops were sown during the second fortnight of July with 90 x 60 cm spacing. Among the cover crops, lucerne cover crop produced higher dry biomass (11.038 t/ha) and N addition (269.315 kg ha^{-1}) and highest soil organic carbon (0.740%). Highest N uptake (66.69 kg ha^{-1}) by chilli was obtained with *Stylosanthes hamata*. Row proportion of 1:1 of chilli: cover crop (60.84 kg ha^{-1}) and 30 days cutting interval (60.85 kg ha^{-1}) resulted in higher N uptake. Sole chilli recorded higher N uptake (63.27 kg ha^{-1}) than chilli with cover crops (except *S. hamata*). A similar trend was observed with P uptake by chilli (Basavanagouda *et al.*, 2000).

6.0 NUTRIENT MANAGEMENT

There are a number of organic sources available in contributing nutrition to the crop. Among them, organic and green manures, biofertilizers are important. Chilli crop responds well to application of manures and fertilizers both under rainfed and irrigated conditions.

6.1 Nutrient uptake

Being long duration crop, chilli/paprika needs judicious application of manures and fertilizers. Sub-optimal levels of major nutrients often lead to deficiency symptoms in chilli, as detailed below:

Nitrogen: Stunted appearance leaves pale green, older leaves are small and uniformly pale green bleaching from margin towards leads to bleaching of entire leaf and to pale white

Phosphorus: Leaves small and bluish green in the beginning, which turns to dirty grayish green, later

Potassium: Retarded plant growth to a lesser extent, leaves normal green but smaller in size with crinkled surface. *Foliar symptoms*: appearance of very small whitish necrotic spots dispersed over the entire lamina in older leaves.

Nitrogen deficiency causes pale yellow colour on older leaves, phosphorus by bluish green colour on older leaves, potassium by marginal necrosis, and subsequently marginal scorching, magnesium by interveinal chlorosis of matured leaves and zinc by smaller leaves in shorter internodes. Nitrogen reduces plant growth and fruit yield considerably but the effect was low in calcium deficiency (Balakrishnan, 1999). Chilli crop takes up about 100 kg of nitrogen from one hectare (Kotur, 2005).

In chillies, nutrient requirement varies based on soil type, type of cultivation (rainfed/irrigated), climatic conditions and variety to be grown. Chilli crop responds well to nitrogen and potash application rather than phosphorus. In chilli, extractable color (red, yellow and total carotenoids; red to yellow carotenoids ratio) increased rapidly during fruit maturation, especially towards the end by potassium application. In general, about 240:300:350 kg ha^{-1} NPK is recommended along with FYM @ 25 t ha^{-1}. However, for organic production, the required quantity of nutrients should be applied through organic source with restricted use of mineral/inorganic fertilizers.

6.2 Organic manures

Organic manures such as farmyard manure is applied @ 4 to 5 t ha^{-1}. It is always the best to use on-farm compost/farmyard manure rather than procuring it from outside. Sheep manure @ 3-5 quintals and neem cake @ 3-4 quintals per hectare may be given at the time of land preparation. Restricted use of permitted mineral fertilizers under organic system can be done depending on requirement, on the basis of soil analysis. Application of soil amendments such as tank silt, basic slag and gypsum is also allowed in a limited manner (www.indianspices.com). Organic manures including compost and oil cakes are the traditional source of nutrients for the crop.

In general, chilli fruit yields are fairly high, when fertilizers are applied with organic manures rather than either fertilizers or organic manures alone.

Fertile soils with adequate humus are most desirable for the crop. Chilli crop demands heavy application of organic manures under irrigated conditions. In general, FYM 25-30 t ha^{-1} is incorporated in soil before transplanting. Under rainfed conditions, spot manuring is practiced before transplanting. The amount of organic manure applied under rainfed conditions varies from 5 to 10 t ha^{-1}. Sheep penning is a common practice in transitional belt of Karnataka, Maharashtra and Rajasthan. The best keeping quality and highest ascorbic acid content in chilli was recorded for 100% poultry manure, which was found superior among the organic fertilizers tested (Sharu and Meerabai, 2001).

In chilli, phosphate rich organic manure (PROM), a value-added product by co-composting different organic wastes with high-grade rock phosphate in fine size can be used as a useful and effective alternative to costly chemical fertilizers (Shaktawat *et al.*, 2004). Sreekumar *et al.* (2004) obtained significant increase in the yield of chilli with the application of high grade rock phosphate (PR, 34/74) along with FYM, vermicompost, green manure, VAM and Microphos inoculant.

Vermicompost is reported to increase the yield parameters in chilli (Yadav and Vijayakumari, 2003). Application of vermicompost at 5 t ha^{-1} + organic booster at 1 litre m^{-2}, and soil conditioner (Tera care) at 2.5 t ha^{-1} + organic booster at 1 litre m^{-2} enhanced the availability of N, P, K and organic C content in soil (Vertisol) under chilli-spinach cropping system in Parbhani, Maharashtra (Hangarge *et al.*, 2004). In a pot culture experiment at Coimbatore, Tamil Nadu, vermicomposted vegetable waste, alone and in combination with different organic manures (FYM and neem cake) increased the biochemical characters and nutrient content in chilli (Yadav and Vijayakumari, 2004).

In a pot culture experiment with chilli cv. Suryamukhi in Keonjhar, Orissa, chilli yielded appreciably high when vermicastings based upon karanj and niger cakes were applied thrice to the potted plants. Besides, use of these oil cakes as additives in vermiculture units resulted in qualitative vermicastings with pest suppressing abilities against jassids and thrips (Subhasmita *et al.*, 2005).

Application of soil conditioner – coir pith compost (Tera care) @ 2.5 t/ha in combination with organic booster @ 1 litre m^{-2} improved the physical condition of soil by reducing bulk density, increasing porosity, water holding capacity and infiltration rate and also the yields of chilli and spinach as compared to recommended dose of NPK under chilli-spinach cropping system at Parbhani, Maharashtra (Hangarge *et al.*, 2002a). They also obtained highest yield (105.67 q ha^{-1}) and N (51.10 kg ha^{-1}), P (5.39 kg ha^{-1}) and K

(49.34 kg ha^{-1}) uptake of chilli with the Application of coir pith compost (2.5 t ha^{-1}) and organic booster (1 litre m^{-2}) (Hangarge *et al.*, 2002b).

Compost prepared in sunlight from water hyacinth + basic slag (Maurya and Dhar, 1983) and Water hyacinth at 25 t ha^{-1} with 100 kg of nitrogen as urea and 100 kg P_2O_5 as rock phosphate gave the highest N content, yield and nutrient content (protein, minerals and vitamins A, B_1 and C contents) of chillies (Maurya and Dhar, 1984).

Application of vermicompost (2.0 t ha^{-1}) had beneficial effect on yield and quality parameters in paprika variety 'Kt-Pl-19'under rainfed cultivation in Karnataka (Savanur, 1999) (Table 7). Application of Zinc sulphate @ 25 kg per hectare is done, if the soil is really deficit in Zinc (www.indainspices.com).

Table 7. Effect of manures and biofertilizers on yield and quality attributes of paprika variety 'Kt-Pl-19"

Treatment	Wt. of single fruit (g)	Wt. of pericarp/fruit (g)	Yield (t/ha)	Total extractable color (ASTA units)
Vermicompost	28.20	25.45	25.95	198.40
Farm Yard Manure	28.07	25.41	25.49	198.05
Azospirillum	28.28	25.51	26.01	198.59
Without *Azospirillum*	28.02	25.34	25.42	197.86
0 kg ha^{-1}	25.77	22.99	23.11	173.90
50 kg N ha^{-1}	28.08	25.39	24.47	197.23
100 kg N ha^{-1}	28.65	25.97	262.45	208.43
150 kg N ha^{-1}	30.10	27.35	29.04	213.35

Source: Savanur (1999)

6.2.1 Green manures and green leaf manures

They are the main source of nitrogen, carbon and improve the physical condition of the soil and population of beneficial microbes. In Ghana, incorporation of *Canavalia ensiformis* as green manure is equally as effective as inorganic fertilizer in increasing yield of chilli. *C. ensiformis* as mulch was not as effective as incorporated manure (Boateng and Peprah, 2005).

6.3 Biofertilizers

Biofertilizers are fungal and bacterial cultures which are capable of fixing atmospheric nitrogen and making unavailable soil phosphorus into available form to the plants. Use of bio-fertilizers can also be resorted to the combination with organic inputs. Application of biofertilizers (100g each of *Azotobacter*, *Azospirillum*, phosphate solubilizing bacteria and 200g vesicular arbuscular mycorrhiza mixed with 1 kg of FYM) with reduced level of chemical fertilizers has beneficial effect on Bydagi Dabbi chilli (Mahantesh *et al.*, 2002).

Azospirillum: Azospirillum is an effective biofertilizer, which can be applied as a seed treatment, seedling treatment as well as direct application to the soil. It is more effective as seed treatment. For seed treatment, 200g of *Azospirillum* should be mixed with 200 ml of boiled and cooled rice water. About 500 g seeds of chilli can be soaked in the solution and dried in shade for 30 minutes and sown. *Azospirillum* @ 40 g in 2 litres of water is recommended for root dipping of seedlings (for 15 minutes) to boost the growth and vigour of the plant (Shinde, 2005). Dipping the seedlings of Bydagi Chilli in *Azospirillum* culture 'ACD-15' reduced the requirement of nitrogen by chilli crop (Hosamani *et al.*, 1996). Treat the seedlings with *Azospirillum* gave significantly high yield and quality parameters in paprika variety 'Kt-Pl-19' under rainfed cultivation (Savanur, 1999) (Table 7). For soil application, *Azospirillum* @ 2 kg/ha, mixed with 20 kg FYM/compost as carrier is recommended.

Vesicular-arbuscular mycorrhiza (VAM): In nursery, inoculation of VAM cultures like *Glomus microcarpum, G. vesiculatum* are found effective (Hosamani, 2000). Organic amendments increased the proliferation of VAM fungi. Inoculation of *Glomus macrocarpum* in conjunction with worm casts with a narrow C:N ratio was found as optimum for improving the growth and yield of chilli (Sreenivasa, 1994). More bacteria were present in the endorhizosphere of chilli plants inoculated with *Glomus fasciculatum* or *G. macrocarpum* than in the rhizosphere of uninoculated plants, with the number of bacteria increasing with percentage root infection by mycorrhizal fungi. Increased colonization by bacteria on plant roots with VAM fungi, with a corresponding increase in the percentage P in shoots and in shoot biomass due to the increased rhizosphere bacteria which liberate organic acids and solubilize otherwise insoluble P in the rhizosphere, which is then taken up and transferred to the plant by the VAM fungi (Krishnaraj and Sreenivasa, 1992). Shrihari and Sreenivasa (1998) recommended VAM (*Glomus macrocarpum*) @ 50 g per10 kg of black clayey and 60 g per 10 kg of red sandy loam soils as optimum for better P-nutrition, growth and yield for Karnataka.

6.4 Micro nutrients

Chilli/paprika crop responds well to foliar application of nutrients. In the Zinc deficient soil, Zinc sulphate can also be sprayed @ 2 gm per litre of water using 1250 gm Zinc sulphate per hectare (www.indianspices.com). Spraying of Zn, Fe and B @0.1 % either alone or in combination at 30, 50 and 70 days after transplanting increased the yield and quality of chillies (Bharambe *et al.*, 2004). In cv. K-2 of chilli in Annamalai, Tamil Nadu, six foliar sprays of the individual amino acid as at 0.50 and 0.75% given at 20-day intervals commencing from the 30th day after transplanting improved the micronutrients, Fe, Zn, Mn and Cu. However, Fe and Zn uptake was influenced by humic acid application (Maheswari *et al.*, 2003).

7.0 DISEASES AND PEST MANAGEMENT

Chilli/paprika crop is attacked by a number of diseases caused by fungi, bacteria and viruses. The loss due to diseases was estimated to be in the range of 15-50 % depending upon location and season. Occurrence of these diseases is endemic in certain chilli- growing tracts except mosaic, which is universal in nature. Some of the important diseases infecting chilli are damping off, fruit rot, leaf spot (fungal and bacterial), bacterial wilt, powdery mildew and viral diseases (mosaic and leaf curl) (Kumar, 2005; Bajpai, 2005 and Muthukrishnan *et al.*, 2002). Wind breaks (Jowar) and alternate crops like marigold and castor may be grown to avoid pest and disease (bacterial) occurrence in main field.

7.1 Diseases

7.1.1 Damping off

Damping off is a common disease of seedlings in solanaceous vegetable crops. It occurs in almost all chilli growing tracts of India. The disease is caused by *Pythium aphanidermatum* (Edson) Fitzp and *Rhizoctonia solani* Kuhn.

Both the pre emergence and post emergence symptoms are noticed. Seeds may rot or the seedlings may be killed before they emerge from the soil. Water soaked lesion and shriveling of stem may occur in the young seedlings after emergence, which may lead to fall over and death of seedlings. The disease may start in patches in the nursery and in the course of two to four days, the entire lot of seedlings may be destroyed. The disease severity is more in moist soils with poor drainage conditions. The fungus is mainly soil-borne and is capable of living for many years in the soil.

The disease can be controlled by seed treatment, soil treatment, maintaining optimum seedling population in the nursery, raised beds with good drainage and light irrigation at frequent intervals. Soil solarization is an effective method to control soil born pathogens (Raj and Kappor, 1993). Partial sterilization of soil by burning trash in the surface helps in checking the disease. Steam sterilization of nursery soil is commonly practiced in Western countries. Raj *et al.* (1997) reported the soil solarization for 40 days with transparent polythene mulch (25 mm) was effective to control pre and post emergence damping off pathogens of chilli. Providing better drainage by improving soil texture with addition of sand and by forming the raised beds helps to avoid the disease. Appropriate cropping pattern/rotation may be followed to reduce the disease problems. Chilli may be planted every 3-4 years in the same soil.

Biocontrol agents *viz., Trichoderma viride* and *Pseudomonas fluorescens* are also found effective in controlling the damping off in chilli (Sharifi-Tehrani and Omati, 1999; Manoranjitham *et al.*, 2000). The pre emergence and post emergence damping off disease percent was lowest with combined seed treatment of *Trichoderma viride* (@ 4g/kg) and *Pseudomonas fluorescens* (@ 5g/kg).

7.1.2 Anthracnose and fruit rot

Anthracnose is a common disease in chilli, caused by *Colletotrichum capsici* (Syd.) Butler and Bisby (Muthumary and Jayachandhra, 1994).The disease symptoms appear mostly on ripened fruits and hence the disease is also called as ripe fruit rot. The spots are usually circular and sunken with black margins. The sunken spot is covered with pinkish mass of fungal spores. As the disease advances the spots spread forming concentric markings with dark fructifications representing the fungal acervuli. The fruits with many spots drop off prematurely resulting in heavy yield loss. The fungus also infects the fruit stalk and spread along the stem causing die-back symptoms. The fungus is externally seed borne and the secondary spread is through air borne inoculum. The disease spreads rapidly through windy rains. The fungus survives for long period on the dead twigs stored under dry conditions (Smith and Crossan, 1958).

To avoid source of inoculums in the field, destruction of diseased material is important. Among the fungal and bacterial antagonists to check the mycelial growth of *Colletotrichum capsici, Saccharomyces cervisiae* exhibited the maximum reduction, followed by *Trichiderma viride* (Jeyalakshmi *et al.*, 1998). Pure culture of *Pesudomonas fluorescens* was more effective in reducing *C. capsici* infection followed by *T. harzianum* than the fungicides treated and untreated seeds. Formulations of *P. fluorescens* (5 g/kg) were also effective in reducing *C. capsici* infection and in increasing the seed germination, vigour index and field emergence, followed by *T. harzianum* (5 g/kg) (Srinivas *et al.*, 2006). Highest dry chilli yield per hectare and maximum cost benefit ratio were obtained under two sprays of *P. fluorescens* (0.2%) at 105 and 120 days after sowing as compared to control (Paramasivam *et al.*, 2005). Cultivating resistant/tolerant varieties such as Pant C-1 (fruit rot), HC-44 (fruit rot and die-back), HC-28 (fruit rot and die-back) and K-2 (fruit rot) help to over come the disease problem.

Hot water seed treatment is recommended to destroy seed borne fungi. Soak seed at 52^{0}C for 30 minutes. Following treatment, plunge the hot seeds into cold water, dry on paper. Freshly harvested seeds withstand heat treatment better than one or two year old seeds (AVRDC, 2004).

Fig. 2: Fruits affected by anthracnose disease (*Source:* Director, IISR, Calicut)

7.1.3. Cercospora leaf spot

The disease is caused by *Cercospora capsici* Haeld and Wolf, which is a seed borne nature but does not live in the soil. Leaf spot disease occurs both in chilli and bell pepper. The disease is characterized by chlorotic lesions with angular to irregular in shape, which in later stage turns to grayish brown with profuse sporulation at the centre of the spot. The severely infected leaves drop off prematurely resulting in yield reduction. The main source of survival of the fungus is in diseased plant tissues.

Crop rotation and destruction of infected plant material in the field minimize infection.

7.1.4 Fusarium wilt

This disease is caused by *Fusarium annuum* and is characterized by wilting of the plant with upward and inward rolling of leaves. The leaves turn yellow and die. The fungus can live in the soil for several years.

Restricted use of *Pseudomonas fluorescens* 15 days after transplanting and three months after first application @ 2.5 kg/ha mixed with 50 kg farmyard manure or compost is beneficial against wilt disease (Anon., 1998).

7.1.5 Powdery mildew

It is caused by *Oidiopsis taurica* (syn. *Leveillula taurica* (Lev) Arn.). The disease has been reported from India, Mauritius, Hungary, Libya and elsewhere. The leaves and stems are covered by white powdery dust resulting in rapid defoliation of leaves and wilting of the plant, ultimately.

Growing powdery mildew resistant/tolerant varieties *viz*., HC-44 and HC-28.

7.1.6 Bacterial leaf spot

This disease is caused by *Xanthomonas vesicatoria*. It can cause damage to both leaves and fruits. On young leaves, the spots are yellowish green and on older leaves, they are dark and water soaked. Later, they appear with straw-coloured center and dark margins. On the fruits, small blister-like spots occur at early stages and later become warty appearance.

The bacteria are seed as well as soil-borne and hence proper crop rotation is to be followed. Provide appropriate wind breaks to avoid bacterial diseases. The screening of genotypes against bacterial leaf spot resulted in identification of resistant (G-4 and X 206) and highly resistant (G-5, CA 960, CAP (63), Co.1 and X 197) types.

7.1.7 Viral diseases

Chilli is vulnerable to several biotic stresses caused by viruses and phytoplasmas (Green and Kim, 2001) and as many as 36 viruses have been reported to infect chilli. These include, Leaf curl virus; Chilli mosaic caused by Tobacco Mosaic Virus (TMV); Cucumber mosaic virus (CMV) is one of the devastating viruses in pepper all over the world (Greenleaf, 1956); the poty virus naturally occurring in chilli are potato virus Y (PVY), potato virus X (PVX), tobacco etch virus (TEV), pepper veinal mottle virus (PVMV) and pepper vein banding virus (PVBV) (Bidari and Reddy, 1981). TMV, CMV, PVY, PVX, leaf curl virus and chilli mosaic virus are the most important viruses of chilli throughout India (Ramakrishnan, 1959; Jeyarajan and Ramakrishnan, 1969). Leaf curl is the most serious disease, among the viral diseases affecting the chilli crop. If the virus infection takes place at the seedling stage or in the early stage of plant growth, the reduction in yield is enormous and as high as 70-80%. The disease incidence due to potyvirus in Uttar Pradesh ranged from 5 - 75 per cent (Prakash *et al.*, 2002).

Insect transmitted viral diseases in chillies

Virus	Disease	Vector
Potyvirus	Potato virus Y, Chilli veinal mottle virus	Aphids
	Tomato spotted wilt virus	Thrips
Cucumovirus	Cucumber mosaic virus	Aphids
Geminivirus	Chilli leaf curl virus	White fly
Tobamovirus	Tobacco mosaic virus, tomato mosaic virus	Mechanically transmitted

Management

Use seeds from healthy plants.

Grow seedlings in an insect proof net-house to prevent early infection, which leads to severe damage. Cultivars reported to be resistant/tolerant to leaf curl virus are Pusa Jwala, HC-28, HC-44, Musalwadi, Pant C-1 & C-2 and NP 46-A (Bhalla *et al.*, 1983).

Use of yellow sticky traps to monitor and to reduce aphid population. Also, the use of yellow water troughs, which attract aphids away from the crop and kill them by drowning is effective in reducing aphid numbers and incidence of virus infection (AVRDC, 2004).

Avoid sequential planting because thrips can continue to emerge from the soil for 2-3 weeks after crop residues are destroyed. Eliminate all sources of thrips at the end of each crop to avoid harboring a small population ready to infest an incoming crop (AVRDC, 2004).

Planting of barrier crops such as sesamum for summer crop and wheat for winter crop, 20 days before transplantation of main chilli crop were found more effective in checking viral diseases and enhancing yield of chilli (Dhawan and Rishi, 1999).

The infected plants should be uprooted and burnt or buried immediately to avoid further spread. The incidence and intensity of the disease may be delayed /reduced by controlling of insect vectors. Mineral oil spray @ 1.5 - 3.0% at weekly intervals was effective in checking vector-borne viruses.

Removal of weed hosts around chilli crop also helps to control the vector multiplication (Singh, 1993). Growing non-host barrier trap crops such as sesame and wheat for summer and winter crops, respectively. The border crops like maize, jowar or bajra should be sown around the field at least 50-60 days in advance.

7.2 Pests

There are about 50 species of insects and pests, causing damage to chillies in India (Reddy and Puttaswamy, 1983). Among these pests, thrips, mites, aphids and fruit borers are of serious in nature (Butani, 1976; Reddy and Puttaswamy, 1980). They are major limiting factors for profitable cultivation of chilli. Ahmed *et al.*, (1987) reported an overall yield reduction of 34 - 77 per cent due to pest infestation.

7.2.1 Sucking pests

7.2.1.1 Thrips

Thrips (*Scirtothrips dorsalis* Hood, *Thrips palmae*) is a polyphagus pest having wide range of host plants. Both nymphs and adults damage the crop. Adults are tiny, slender, fragile and yellowish straw colour and have grey-coloured heavily fringed wings. Thrips and mites cause leaf curl complex of chilli. The extent of yield losses due to leaf curl complex ranged from 25-80 per cent (Tatagar *et al.*, 2001).

Thrips are wide spread throughout the world with pre ponderence of tropical species. It is a polyphagus pest having wide range of host plants. The life cycle on chilli ranges from 16 days in summer to 19 days in winter. The duration of different stages of *S. dorsalis* is 4-6, 7-8, 0.75-1 and 10-15 days for egg, larval, pupal and adult stages respectively. The insect reproduces sexually as well as parthenogenetically. It occurs throughout the year (Borah, 1987) and does not under go hibernation (Raizada, 1965).

Both nymphs and adults damage the crop. Adults are tiny, slender, fragile and yellowish straw colour and have grey-coloured heavily fringed wings. They lacerate the leaf tissue and suck the sap. The infested leaves develop crinkles and curl upwards. The severely infested plants develop bronze colour. If the plants are infected at early stage, plants remain stunted in growth and flower production and pod set are arrested causing severe yield loss (Ayyar, *et al.*, 1935; Reddy and Puttaswamy, 1983; Borah, 1987). Pest will become very serious during dry months and is a vector for leaf curl virus (Ananthkrishnan, 1969). High temperature, low relative humidity and rainfall causes drastic decline in population (Velayudhan *et al.*, 1985).

Fig. 3: Thrips infestation of paprika (upward curling of leaves)
(*Source:* Director, IISR, Calicut)

7.2.1.2 Mites

Mites are tiny accarids found in large numbers on the ventral surface of the leaves under protective cover of fine webs. Mites of different genera have been found feeding on leaves. The eggs are minute and oval in shape and are laid on the ventral surface of the young leaves or leaf buds. Sexual and parthenogenetic (arrhenotokus) reproduction has been observed in mites. It is a polyphagous pest on wide range of weeds and cultivated plants. The pest occurs throughout the year under field conditions. However, two peak periods of infestation are October - November and February - March (Mote, 1976). The favourable temperature and humidity requirements for the mites are 18^{0}C - 25^{0}C and 80 - 90 per cent respectively. Fewer incidences were reported under the dry weather combined with high temperature and low relative humidity (Kabir, 1979).

Both nymphs and adults suck the cell sap (Butani, 1976). The mites first appear in the terminal or axillary tender shoots of plants. Nymphs and adults feed exclusively on the lower surface of the leaves. The affected leaves curl downward along the margins of the leaf and attains on inverted boat shape. The pedicles of the leaves get elongated and the young leaves at the tip of the branch cluster. The affected plants develop dark green colour. In the affected plants, the vegetative growth is inhibited and flower production is ceased and yields are considerably reduced.

7.2.1.3 Aphids

The aphid (*Aphis gossypii*) and green peach aphid (*Myzus persicae* Sulz.) are serious pests on chilli at Maharastra (Deshpande, 1933), Karnataka (Khan *et al.*, 1978), Tamil Nadu (Saivaraj *et al.*, 1979) and Andhra Pradesh (Reddy *et al.*, 1981).

Nymphs of aphids are small, ovate, soft bodied and adults appear on the tender shoots, leaves and on the lower surface of the old leaves. They suck the sap and reduce the vigour of the plant. They secrete sweet substance, which attract ants and develop sooty moulds. The pods that develop black colour due to sooty moulds loose quality and fetch low price. The yield levels are also reduced by aphid infestation directly and more through the spread of virus diseases (mosaic) as vectors (Nariani and Sastry, 1958; Basha and Balasubramanian, 1980).

Organic management of sucking pests

Application of neem seed kernel extract (NSKE 3%) can be done for control of thrips, aphids and mites. Release of larvae of *Chrysoperla carnea*, a biocontrol agent, once in 15 days is also helpful in controlling thrips and mites (Anon., 1998).

Installation of yellow sticky traps to attract and kill insects especially thrips. Incase of severe mite infestation, clipping off the affected shoots in the early stages of the crop is recommended.

7.2.2 Pod borers

The following two species bore into the tender and matured fruits and feed the seeds resulting in hallowing and dropping of fruits.

Helicoverpa armigera Hb:	Larvae are yellowish green in colour and adults are grey in colour
Spodoptera litura F.:	Larvae are brownish with three strips and adults are grayish brown.

Pod bores are polyphagaous and appear on both in vegetative and reproductive phase (at the time of pod formation). Borers enter chilli pods by second and third instars by making a hole near calyx and feed on chilli seed. The affected pods drop off or develop white colour on drying. The fully-grown caterpillars enter the soil for pupation.

Restricted installation of pheromone traps (5 acre^{-1}) in the field helps to monitor the adult moths. Ten days after spotting the moths in the traps, spraying with Nuclear Polyhedrosis Virus (NPV) @ 500 ml ha^{-1} 4-5 rounds is beneficial to control early larval stage of pod borers. The egg masses of *Spodoptera* can be mechanically collected and destroyed. *Trichogramma* may be released two days after appearance of moths. Spraying of neem products like neem oil or neem seed kernel extract and restricted use of *Bacillus thuringensis* @ 1 kg ha^{-1} are beneficial. All the shed fruits and inflorescence parts should be collected and destroyed at regular intervals (Anon, 1998). Grow castor as trap crop along the field bunds.

7.2.3 Root grubs

Roots grubs (*Anomala bengalensis, Holotrichia consanguinea, Holotrichia reynaudi*) damage the plants by feeding the root system. The affected plants die and can be easily identified. Its damage can be seen in the field up to the end of October.

To avoid infestation of root grub, only well rotten farmyard manure should be applied in the field. Application of 250 kg neem cake ha^{-1} along with farmyard manure reported to be effective in root grub control (Anon., 1998). Summer ploughing is beneficial as it exposes grubs and adults. Installation of light traps between 6 and 9 am is also beneficial to attract adult beetles during monsoon showers.

7.2.4 Nematodes

In India, the occurrence of root knot nematode in chilli was reported in 1960. Among the three species of *Meloidogyne*, *Meloidogyne arneria* is most harmful (Rajagopalan *et al.*, 1971). Root-knot nematodes are polyphagus pests with a wide range of host plants. In general, nematodes infest the roots and produce tiny galls and affected plants show symptoms of withering and wilting of leaves. The growth of the plant is stunted and adversely affects fruiting.

Exposure of soil by repeated ploughing reduces the infestation of nematodes. Application of neem cake @ 500 g m^{-2} and crop rotation with cereals or intercropping with marigold were reported as effective methods to overcome nematode problems. Use of solarized soil is also effective against nematodes. Most plant parasitic nematodes are killed between 44 and 48^{0}C (AVRDC, 2004).

Resistant/tolerant sources identified against pest and diseases in chilli

Variety/hybrid	Institute	Characteristic features
Pusa Jwala	IARI	Fairly tolerant to thrips, mites and aphids
Pusa Sadabahar	IARI	Resistant to CMV, TMV, Leaf curl virus
NP 46A	IARI	Tolerant to thrips, susceptible to viral diseases
Arka Lohit	IIHR	Tolerant to powdery mildew
Pant C-1, C-2	GBPUAT	Tolerant to mosaic and leaf curl
Punjab Lal	PAU	Resistant to TMV, CMV, leaf curl virus Moderately tolerant to fruit rot and dieback
CH 1, CH 3	PAU	Tolerant to wilt, fruit rot, mosaic and virus
Phule Suryamukhi	MPKV	Tolerant to leaf curl and powdery mildew
Jawahar 218	JNKV	Tolerant to leaf curl and fruit rot
Manjari, Ujwala	KAU	Resistant to bacterial wilt
LCA-304, LCA-301, LCA-312 & B-1-1	UAS-D	Resistant to thrips and mites

8.0 HARVEST AND POSTHARVEST HANDLING

8.1 Harvesting and yield

In chilli, flowering begins 1-2 months after transplanting depending upon the variety, climate, nutritional status of soil and it takes another month of green fruits. Harvesting should be done at the right stage of maturity. Ripe fruits are harvested at an interval of 1-2 weeks and it continues over a period of about 3 months with 6-10 pickings depending upon the season, cultivar and cultural practices adopted. The ripe chillies are generally dried under sun for 8-15 days depending upon weather conditions, while commercially dried at about 54.4°C in 2-3 days. This imparts deep red colour and glossy texture to the fruits. Fruits are spread on drying yards in layers of 8-10 cm. From the fifth day onwards, the produce is inverted on alternate days so that the pods in the lower layers are brought up to ensure quick and uniform drying. To avoid microbial activity and aflatoxin production, moisture in the dried pods should be brought down to 10%. Solar drier and tray drier can be used. While drying, the produce can be covered with polythene sheets during night time to avoid dew deposition and resultant colour fading.

About 100 kg of fresh ripe fruits yield 25-40 kg dry chilli depending upon the variety and thickness of inner wall (Nath, 1976). Normally, chilli crop yields 7.5 - 10.0 tonnes green chilli and 2.0 - 2.5 tonnes dry chilli from a hectare. Under rainfed conditions, commercial chilli varieties yield 1.0 - 1.5 t/ha dry pods, while it was 2.0 - 2.5 t/ha under irrigated conditions. Fully ripened fruits of paprika/paprika like chillies could be harvested at 60 days after flowering. Usually 4-5 pickings are done. Paprika (cv. Kt-Pl-19) gives average yield of 25-30 t/ha with a dry yield of 4.0 - 4.8 t/ha with a skin recovery of 66% and dried under shade to a safer moisture level (8-12%). Stages of harvesting of chilli/paprika for various uses / purposes are given in Table 8.

Table 8. Commercial cultivation of chilli in India: fruits of various market types

Stage of harvesting	Consumptio n pattern	Cultivars types	Species	Preferred fruit type/size	Degree of pungency
Red ripe fruits	Intact fruits or grinded (power)	Landraces, improved populations and hybrids	*C. annuum, C. chinese, C. frutescens*	Cayenne fruits 10-12 x 2-3 cm	High colour retention and highly pungent
Green fruits	Intact fruits or sauce preparation	Landraces, improved populations and hybrids	*C. annuum, C. chinese, C. frutescens*	Cayenne fruits 6-8 x 2-3 cm	Mild pungent to highly pungent
Red ripe fruits (paprika)	Oleoresin extraction	Landraces (e.g. Tomato, Chilli, Bayadgi Kaddi etc.)	*C. annuum*	Cayenne fruits with very less capsaicin and high oleoresin	High oleoresin with no pungency
Red ripe fruits	Pickle formation	Landraces	*C. annuum*	Jalapeno type fruits, but with thin pericarp	Very mild pungent
Red ripe fruits	Specific flavour	Landraces (e.g. Dello of North-East region)	*C. baccatum*	Typical bell shaped with distinct flavour	Regional preferences

Source: Rai (2005)

Grading is to be done to remove defective and discoloured pods. Packing is done in gunny bags, or jute boras. Chillies should be properly stored to avoid infestation of pests. It is preferable to store dried chilli in refrigerated condition (cold storage) to retain colour. The temperature in a cold store is maintained in the range of 5-8°C and relative humidity is kept at 55-60 %. There were great loss of oleoresin and capsanthin when it stored ambient temperature for 20 days after removed from the cold store. Therefore, the chilli produce should be used immediately after termination from cold store for extraction of colour and oleoresin and other products.

8.2 Good processing practices (GPP)

1. Selective harvesting
2. Ideal temperature for drying (above 37°C)
3. Drying surface
   Clean drying yards, cemented/kadapakal/terracotta tiles/HDPE sheets
   Mechanical driers - solar driers
4. Chilli thickness for drying should be 10 cm
5. After 2 days turning required (once in two days)
6. Time required 12-15 days for drying
7. Avoid sprinkling of water (and also impure water) over dry chillies to gain weight and to prevent aflatoxin growth
8. Grading - remove fungus effected/ discoloured/ insect infested pods and other extraneous matter
9. Packing
   Use clean dry gunny bags of 30 kg; for export - bags of 25 kg
   Packing may be in morning hours
10. Avoid synthetic colour on gunny bags
11. Pack value added chilli by sealed plastic bags or glass containers
12. Storage of grounded chilli below 13°C reduces the risk of Aflatoxin production
13. Redrying chilli powder after grounding and pack with moisture proof plastic bags
14. Transportation by load/tractor/lorry

8.2.1 Processing of chilli

Harvesting
↓
Washing and Sorting (Remove dirt, debris and diseased fruits)
↓
Storage
(Storage at refrigerator temperature below 7°C)
↓
Drying
(Optimum moisture level of 8%) - Sun drying – 10 to 15 days - Hot drier- 40-60°C
↓
Grounding (Stone mill) (Hammer mill)
↓
Storage & Re-drying (Moisture content <14%) (Storage temperature <13°C)

8.2.2 Processing of paprika

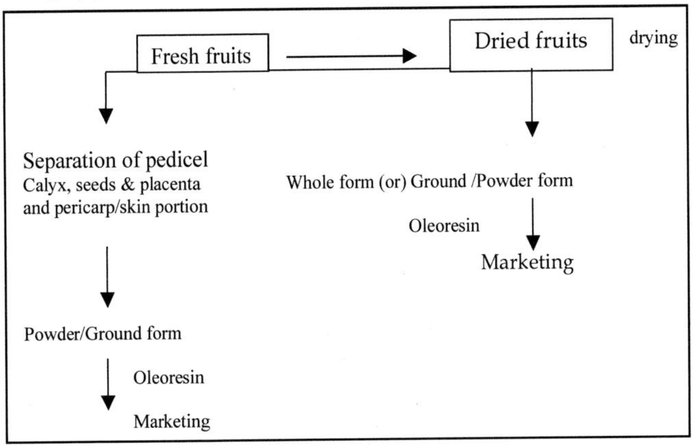

8.3 Marketing

Indian chilli and its products are imported by a number of countries. Important among them are: Sri Lanka, Bangladesh, South Korea, and USA for dry chilli; USA, Germany, Japan, UK and France for oleoresin. China and Pakistan are main competitors for India, who offer chillies at low prices in international markets. The varieties of chillies having foreign demand include Bydagi Guntur, Paprika, Sannam - 4, Indian wrinkled chillies, Wonder hot, Bird's eye chilli, Kashmiri chilli, and Warangal Chappatta.

Marketing of chilli remains a major constraint. There are several types marketing channels are prevailing for chillies depending upon the end use as produce [Producer → wholesaler → retailer → consumer or Producer → commission agent → wholesaler → processing factory → spice trader → selling agents (within country and abroad) → consumer]. For products/value-added products the chain is; [Producer → commission agent → wholesaler → processing factory → spice trader → selling agents (within country and abroad) → consumer or Producer → processor → commission agents (abroad) → retailer → consumer]. The commission agents still take a major share of consumer's price. The major marketing centers located in various states are: Nasik, Ahmed Nagar, Sholapur, Aurangabad, Nanded, Amaravathi, Lasalgaon in Maharashtra; Guntur, Warangal, Hyderabad, Guddapah, Vijayawada, Rajamundri, Nellore in Andhra Pradesh; Dharwad, Mysore, Hassan, Bangalore, Bellary, Ranibennur, Hubli, Bydagi in Karnataka; Pollachi, Ramanad, Madurai, Trichi, Theni, Dindigul, Virudunagar, Sattur in Tamil Nadu (Shiva *et al.*, 2006).

The market for chillies is expanding at faster rate. According to International Trade centre, world trade in capsicum (which includes chillies, pimento and paprika) grew on average by 15% annually, both in volume and value terms. China and India are the principal producers and suppliers of chillies to international markets. The total world imports of capsicum reached 371,000 tons valued at US$ 590 million, of which China and India's exports contributed US$ 140 million and US$ 94 million, respectively, for quantities exceeding 85,000 tons each. The United States, Malaysia and Germany are among the major markets for capsicum. The United States accounted for 24% of the total value of capsicum imports, followed by Malaysia 8% and Germany 7% during 2004.

REFERENCES

Agarwala, B.K. and Raychoudhuri, D.N. (1979) Biotic potential of weeds in respect of *Aphis gossypii* Glover infesting some economic plants in Kalimpong, West Bengal. *Indian Agric.* 23, 25-29.

Ahmed, M.K., Shivpuji, P.R. and Jadhav, G.D. (1987) Residues of synthetic pyrethroids and acephate in/on chilli fruits. *Pestology*, 9, 16-18.

Albregts, E.E. and Howard, C.M. (1973) Effect of fertilization and mulching with bio-degradable polyethylene coated paper on response to okra and pepper. *Hort. Sci.* 8, 36-38.

Alegbejo, M.D. (1987) Identification of a weed host of pepper veinal mottle virus in northern Nigeria, Samaru. *J. Agric. Res.* 5, 65-70.

Aswathanarayanareddy, N., Ashok Kumar, C.T. and Gowdar, S.B. (2006) Effect of inter cropping on population dynamics of major pests of chilli (*Capsicum annuum* L.) under irrigated conditions. *Indian J. Agric. Res.* 40, 294-297.

Ananthkrishnan, T.N. (1969) Indian Thysanopetera Monograph No. 1., Publication and Information Directorate, New Delhi, pp. 171.

Anonymous (1998) Production of Organic spices - Concepts, Principles, Basic standards, Production guidelines, Documentation, Inspection and Certification. Spices Board, Cochin, Kerala.

AVRDC (2004) Pepper diseases-Fact sheets, Asian Vegetable Research and Development Center, Taiwan.

Ayyar, T.V.R., Subbaya, M.S. and Krishnamurti, P.S. (1935) The leaf curl disease of chilli due to thrips. *Madras Agric. J.* 23, 408 - 411.

Bajpai, N.K. (2005) Chilli nursery and plantation practices. Spice India, 18, 11-13.

Balakrishnan, K. (1999). Studies on nutrients deficiency symptoms in chilli (*Capsicum annuum* L.). *Indian J. Plant Physiol.* 4, 229-231.

Basavanagouda, K.H., Halikatti, S. I. and Palled, Y.B. (2000). Effect of cover crops on weed control, soil fertility and nutrient uptake in chilli. Karnataka *J. Agrl. Sci.* 13, 540-545.

Basha, A.A. and Balasubramanian, M. (1980) Newer chemicals for the control of green peach aphid (*Myzus persicae* Sulz.) on chillies. *Pesticides*, 17, 12-13.

Bhalla, P.L., Bagehel, B.S. and Krishna, A. (1983) Screening of chilli cultivars/lines against leaf curl and anthracnose. *Proc. Nat. Seminar Tomato and Chillies*. Tamil Nadu Agric. Univ., Coimbatore, pp. 147.

Bharambe, P.R., Ghodpage, R.M., Patil, M.N., Tamgadge, D.B., Ingle, S.N. and Gabane, V.V. (2004) *Glimpses of Recent Soil Science Research*. Dr. PDKV, Akola, Maharashtra. pp. 39-67.

Berenyi, M. (1968) The effect of irrigation on the quality and quantity of red pepper crop in two extremely rainy yars. *Hort*. Abst. 39, 4947.

Bidari, V.B. and Reddy, H.R. (1981) "Etiological agents in chilli murda in Karnataka". In: *Proc. III International Symp. Pl. Path.* 14-18, December, 1981, IARI, New Delhi.

Black, L.L. and Ralston, L.H. (1972) Aluminium foil mulch reduces virus infection of pepper. Lousiana Agri., 15, 6-7.

Boateng, S.A., Peprah, S.A. (2005) Response of pepper (*Capsicum frutescens*) to Canavalia ensiformis green manure in the rain forest zone of Ghana. *Agricultural and Food Science J.* Ghana, 4, 317-328.

Borah, D.C. (1987) Bio-ecology of Polyphagotarsonemus latus and Scithrips dorsalis infesting chilli and their natural enemies. *PhD thesis*, Hisar Agri Univ., Hissar.

Bosland, P.W. and Votava, E.J. (2000) Pepper: In: *Vegetable and Spice Capsicums*. CABI publishers, UK.

Butani, D.K. (1976) Pests and diseases of chillies and their control. Pesticides, 10, 38-41.

Chauhan, V., Singhania, R. A., Singh, A. K. and Kumar, A. (1999) *Indian J. Agrl. Res.* 33, 62-66.

Chougule, A. B. and Mahajan, P. R. (1979) Effect of varying levels of plant population, nitrogen, phosphorus and potash on growth and yield of chilli (*Capsicum annum* L.). *Veg. Sci.* 6, 73-80.

Deshpande, A.A. (2005). "Paprika improvement for exports and value-addition". *In: Souvenir of National Symposium on Current trends in onion, garlic, chillies and seed spices* - production, marketing and utilization, 25-27 November 2005, NRC for Onion and Garlic, Rajgurunagar, Pune (M.S.), pp. 122-125.

Deshpande, R.B. (1933) South Indian Chillies 3. The inheritance of some characters in Capsicum annuum L. *Indian J. Agric. Sci.* 3, 219-300.

Dhawan, P. and Rishi, N. (1999). Eco-friendly management of viral diseases of chilli by non-host barrier trap crops. *Capsicum and Eggplant Newsletter*, 18, 52-55.

FAO (2007). *FAO Database*. FAO, Rome

Green, S.K. and Kim, J.S. (2001) Characteristics and control of viruses infecting peppers: a literature review. *AVRDC Tech. Bulletin*. No.18, 60 pp.

Greenleaf, W.H. (1956) Inheritance of resistance to tobacco etch virus in *Capsicum frutescens* and *Capsicum annuum. Phytopathol*. 6, 371-375.

Hangarge, D.S., Raut, R.S., More, S.D., Birajdar, R.R. and Pholane, L.P. (2002a). Effect of vermicompost and soil conditioner on some physical properties of soil under chilli-spinach system. *J. Soils and Crops,* 12, 41-45.

Hangarge, D.S., Raut, R.S., Malewar, G.U., More, S.D., and Keshbhat, S. S. (2002b). Yield attributes and nutrients uptake by chilli due to organics and inorganics on vertisol. *J. Maharashtra Agril. Univ.* 27, 109-110

Hangarge, D.S., Raut, R.S., Gaikwad, G. K., Adsul, P.B. and Dixit, R.S. (2004) Influence of vermicompost and other organics on fertility and productivity of soil under chilli-spinach cropping system. *J. Soils and Crops,* 14, 181-186.

Hegde, D.M. (1989) Effect of method and volume of irrigation on yield and water use of sweet pepper (*Capsicum annuum* L.). *Indian J. Hort.* 46, 225-229.

Hosamani, R.M., Madalageri, B.B., Ganiger, V.M. and Hadli, J.M. (1996) "Effect of Azospirillum - biofertilizer cultures and nitrogen levels on rainfed green chilli". *In: Souvenir of National Symposium on Horticultural Biotechnology*, held at Bangalore, 28-30 October 1996, pp.58-59.

Hosamani, R.M. (2000) Paprika in North Karnataka – Status, Research and Development. *Indian J. Arecanut, Spices and Medicinal Plants*, 2, 50-57.

Hosmani, M.M. (1993) *Chilli crop*. Directorate of Extension, UAS, Dharwad, Karnataka.P.232.

Hulamani, K.H. (1988) Effect of growth regulators and nutrients on growth and yield of Bydagi chilli (*Capsicum annuum* L. var. *acuminatum*) under rainfed conditions of Dharwad tract. *M.Sc. (Agri.) thesis,* Univ. agrl. Sciences, Dharwad.

Jeyalakshmi, C., Durairaj, P., Seetharaman, K. and Sivaprakasam, K. (1998) Biocontrol of fruit rot and die-back in chilli using antagonistic microorganisms. *Indian Phytopath.* 51, 180-183.

Jeyarajan, R. and Ramakrishnan, K. (1969) Potato virus Y in chilli (*Capsicum annuum*) in Tamil Nadu. *Madras Agric. J.* 56, 761-766.

Johny, A.K. and Ravindran, P.N. (2004) Chilli varieties for higher yield. *Spice India,* 17, 2-8

Kabir, A.K.M.F. (1979) Bioecology and behaviour of yellow jute mite. In: Rodriguez, J.G. (ed.) *Recent Advances in Knowledge of the Phytoseiidae*, Univ. California.

Kalloo, G., Pandey, S. K., Chakrabati, S. K. and Kaushik, S. K. (eds.) (2005) "Vegetables". *In: Technologies for production of quality of seed and planting material in Horticultural*

Crops. (Technical Bulletin No. 71). Central Potato Research Institute, Shimla (H.P.), pp. 01-41.

Kapitancy, J. (1971) Investigations on the effect of plastic mulching in *Capsicums. Zoldegtermesztesi Kutat Intezet Bulletinje*, 6, 97-106.

Kemp, W.G. (1978) Mulches protect peppers from viruses. *Canada Agri.* 24, 22-24.

Khan, M.M., Rajagopal, D. and Hanumanna, P. (1978) Plant protection practices and problems of chilli growers of Kolar district. *Mysore J. Agric. Sci.* 12, 159-162.

Krishnaraj, P.U. and Sreenivasa, M.N. (1992) Increased root colonization by bacteria due to inoculation of vesicular-arbuscular mycorrhizal fungus in chilli (*Capsicum annuum). Zentralblatt fur Mikrobiologie*, 147, 131-133.

Koshy John (2000) Promotional activities of spices Board on paprika. *Indian J. Arecanut, Spices and Medicinal Plants,* 2, 47-49.

Kotur, K. C. (2005) "Refinement of fertilizer management of onion and chilli using isotope-aided investigations". *In: Souvenir of National Symposium on current trends in onion, garlic, chillies and seed spices* - Production, marketing and utilization, 25-27 November 2005, Rajgurunagar, Pune (M.S.), pp. 82-87.

Kumar, A. (2005) Prospects for successful integrated management of yield reducing disease of chillies. *Spice India*, 18, 28-31.

Kumar, V. (1982) Influence of spacing on growth and yield of chillies on dry land cultivation. *M. Sc. Thesis*, UAS, Bangalore.

Kurdikeri, C. B. (1982) Quoted by Hosmani (1982) Lit. Cit.

Lee, B.Y. and Yoon, J.Y. (1975) The effect of polyethylene film mulches on soil temperature and growth and yield of red pepper. *J. Korean Hort. Sci.* 16, 185-191.

Lourduraj, A. C. (2003) Effect of mulching and irrigation regimes on chillies (*Capsicum annuum* L.). Agricultural Science Digest. 23, 208-210.

Mahantesh, S. K., Gowda, K.K., Kumar, S.N. and Sreeramu, B.S. (2002) Effect of biofertilizers on growth and yield of chilli (*Capsicum annuum* L.) cv. Bydagi Dabbi at different levels of nitrogen and phosphorus. *J. Spices and Aromatic Crops,* 11, 58-61.

Maheswari, T.U., Haripriya, K. and Kamalakannan, S. (2003) Impact of foliar organic nutrients on nutrient uptake pattern of chilli (*Capsicum annuum* L.) cv. k2. *South Indian Hort.* 51, 168-172.

Manoranjitham, S.K, Prakasam, V., Rajappan, K. and Amutha, G. (2000) Control of chilli damping off using bioagents. *J. Mycol. Pl. Pathol.,* 30, 225-228.

Mariappan, V. and Narayanaswamy, P. (1977) Characterization of virus affecting weeds. II. Leaf curl diseases. *Madras Agric. J.* 64, 740-744.

Maurya, K.R. and Dhar, N.R. (1983) Effect of different composts on yield and composition of chilli (*Capsicum annuum* L.). *Anales-de-Edafologia-y-Agrobiologia*, 42, 183-191.

Maurya, K.R. and Dhar, N.R. (1984) Effect of nitrogen, phosphorus and crude organic matter on the yield and composition of potato and chilli. *Anales-de-Edafologia-y-Agrobiologia*, 43, 827-834.

Mote, U.N. (1976) Seasonal fluctuation in population and chemical control of chilli mites (*Polyphagotarsonemus latus* Banks). Veg. Sci. 3, 54-61.

Muthukrishnan, C.R., Thangaraj, T., Chatterjee, R. and Maity, T.K. (2002) "Chilli and Capsicum". In: Bose, T.K., Kabir, J., Maity, T.K., Parthasarathy, V.A. and Som, M.G. (eds.). *Vegetable Crops* (Vol. I), Naya Prokash, Kolkata, India, pp. 155-264.

Muthumary, J. and Jayachandra, J.A. (1994) Histopathological and ultrastructural studies on *C. capsici* causing anthracnose in chilli. *Indian Phytopath*. 47, 292.

Nagalakshmi, S., Palanisamy, D., Eswaran, S. and Sreenarayanan, V.V. (2002) Influence of plastic mulching on chilli yield and economics. *South Indian Hort*. 50, 262-265

Nariani, T.K. and Sastry, K.B.M. (1958) Two additional vectors of chilli mosaic virus. *Indian Phytopath*. 11, 193-194.

Nath, P. (1976) *Vegetables for the Tropical Region*. ICAR, New Delhi, pp. 21.

Paramasivam, M., T. Kalaimani., J.R. Kannan Babu., R. Ramjegadesh and Manivannan. (2005) "Current trends in chilli fruit rot disease management and increasing productivity in Southern Tamil Nadu". In: Abstracts, *National symposium on current trends in onion, garlic, chillies and seed spices* – production, marketing and utilization, 25-27, November, 2005, Rajgurunagar, Pune.

Patil, P. B. and Singh, V. (1979) Yield performance of chilli varieties under different plant spacing. *Res. Bull. Maharashtra Agric. Univ*. 3, 160-161.

Prakash, S., Singh, S.J., Singh, R.K. and Upadhyaya, P.P. (2002) Distribution, incidence and detection of a Potyvirus on chilli from eastern Uttar Pradesh. *Indian phytopath*. 55, 294-298.

Prabhakar, M. and Hebbar, S. S. (1998) Water management in chilli under semiarid conditions. In: Sadanandan, A.K., Krishnamurthy, K.S., Kandiannan, K. and Korikanthimath, V.S. (eds.) *Water and nutrient management for sustainable production and quality of spices*. Proceedings of the National Seminar, 5-6 October 1997, Madikeri, Karnataka, India, pp. 131-135.

Rai, M. (2005) "Chilli improvement in India for value addition and export". *In: Souvenir of National Symposium on current trends in onion, garlic, chillies and seed spices* – Production, marketing and utilization, 25-27 November 2005, Rajgurunagar, Pune (M.S.), pp. 12-18.

Raizada, U. (1965) The life history of S. dorsalis with detail external morphology of its immature stages. *Bull. Entomol*. 6, 30-49.

Raj, H. and Kapoor, I.J. (1993) Soil solarization for the control of tomato wilt pathogen (*Fusarium oxysporum* Schl.). *Z. Pflkrankh. Pflschutz*. 100, 652-661.

Raj. H., Bhardwaj, M.L. and Sharma, N. K. (1997) Soil solarization for the control of damping off of different vegetable crops in the nursery. *Indian Phytopath.* 50, 524-528.

Rajagopalan, P.M. Seshadri, A.R. and Muthukrishnan, T.S. (1971) Preliminary observations on the varietal resistance of chillies to root knot nematodes. *Madras Agric. J.* 58, 604.

Ramachandran, S. and Subbiah, K. K. (1981) Studies on the effect of plant density and graded level of N on yield and yield components of chillies (*Capsicum annuum* L.). South Indian Hort. 29, 178-181.

Ramakrishnan, K. (1959) Potato virus Y on chillies (*Capsicum* sp.). *South Indian Hort.* 7, 41-52.

Rangaswami, R. (1984) Effects of different tillage methods in weed infestation under different crops – sorghum, chilli, groundnut and cowpea. *M.Sc., thesis,* University of Agricultural Sciences, Bangalore, Karnataka.

Reddy, D.N.R. and Puttaswamy. (1980) Pests infesting chilli in the nursery. *Mysore J. Agric. Sci.* 17, 246-251.

Reddy, D.N.R. and Puttaswamy. (1983) Pests infesting chilli in the transplanted crop. *Mysore J. Agric. Sci.* 17, 122-125.

Reddy, P.S., Prasada Rao, V.L.V. and Lakshiminarayana, K. (1981) Control of chilli aphids in Andhra Pradesh. *Pesticides,* 15, 38-39.

Revanappa, Nalawadi, U. G. and Sheelavantar, M. N. (1998) Influence of plant density on growth and yield of green chilli varieties. *Karnataka J. Agric. Sci.* 11, 543-545.

Saivaraj, K., Kumaraswamy, T. and Jayaraj, S. (1979) Evaluation of certain newer insecticides for control of green peach aphid (*Myzus persicae* Sulz.) on chillies. *Pesticides,* 13, 20-21.

Savanur, V. (1999) Influence of inorganic nitrogen levels, *Azospirillum* species and organics on yield and quality of paprika. *M.Sc. (Agri.) thesis,* University of Agricultural Sciences, Dharwad, Karnataka.

Selvaraj, K. V., Subramanian, S. and Rajasekharan, R. (1974) Influence of spacing on chilli varieties. *Madras Agric. J.* 61, 889.

Selvaraj, K.V., Iqbal, I.M., Dawood, M.S., Krishnasamy, S.M., Muralidharan, V. and Hariharan, M.S. (1998) Irrigation management under differential moisture regime for chilli crop (*Capsicum annuum* L.) In: Sadanandan, A.K., Krishnamurthy, K.S., Kandiannan, K. and Korikanthimath, V.S. (eds.) *Water and nutrient management for sustainable production and quality of spices.* Proceedings of the National Seminar, 5-6 October 1997, Madikeri, Karnataka, India, pp. 161-163.

Shaktawat, M.S., Singh, M., Aery, N.C., Swami, B.N., Katewa, M.K. (2004) Phosphate rich organic manure: an alternate to phosphatic fertilizers. viii + 194 pp.

Sharifi-Tehrani, A. and Omati, F. (1999) Biocontrol of *Phytophthora capsici* the causal agent of pepper damping-off by antagonistic bacteria. Mededelinger *Faculteit Larabouvkudigeen Toegepaste Biologische Wetenschapten Universitei*, Genet, 64, 419-423.

Sharu, S.R and Meerabai, M. (2001) Effect of integrated nutrient management on yield and quality in chilli (*Capsicum annuum* L.). *Vegetable Science*, 28, 184-185

Shinde, P.B. (2005) "Micronutrient management in onion, garlic and chilli". *In: Souvenir of National Symposium on current trends in onion, garlic, chillies and seed spices* – Production, marketing and utilization, 25-27 November 2005, Rajgurunagar, Pune (M.S.), pp. 71-76.

Shiva, K.N., Prasath, D., Zachariah, T.J. and Parthasarathy, V.A. (2006) "Chilli and paprika". In: Parthasarathy, V.A. and Rajeev, P. (eds.). *Major spices – production and processing*. Indian Institute of Spices Research, Calicut, Kerala, pp. 228-273.

Shrihari, P.C. and Sreenivasa, M.N. (1998) Optimum dose of arbuscular mycorrhizal inoculum for chilli in two types of soil. *Karnataka J. Agric. Sci.* 11, 696-700.

Singh, J. (1993) "Improvement of chillies". In: Chadha, K.L. and Kallo, G. (eds). Advances in Horticulture Vol. 6 – *Vegetable crops*: Part 2. Malhotra Publishing House, New Delhi, pp 763-827.

Sivanappan, R.K. (1979) Drip irrigation for vegetable crops. *Punjab Hort. J.* 19, 83-85.

Sivanappan, R.K., Rajagopal, A. and Palaniswamy, D. (1978) Response of chillies to drip irrigation. *Madras Agric. J.* 65, 576-579.

Smith, R.W. and Crossan, D.F. (1958) The taxonomy, etiology and control of *Colletotrichum piperatum* (E&E) and *C. capsici*. *Plant Dis. Reptr.* 42, 1099-1103.

Sreekumar, M.R., Satyakeerthi, M.R.P. and Kumari, P.M.V. (2004) Studies on the use of high grade rock phosphate as direct fertilizer along with integrated plant nutrient system (IPNS) components. *Phosphate rich organic manure: an alternate to phosphatic fertilizers*, 70-87.

Sreenivasa, M.N. (1994) VAM in conjunction with organic amendments improve growth and yield of chilli. *Environment and Ecology*, 12, 312-314.

Srinivas, C., Niranjana, S.R., Praveen Kumar, L., Chandra Nayaka, S. and Shetty, H.S. (2006) Effect of fungicides and bioagents against *C. capsici* on seed quality in chilli. *Indian phytopath.* 59, 62-67.

Subhasmita, P., Samal, M. K. and Patnaik, H.P. (2005). Effect of oil cake based vermicompost on the incidence of sucking insect pests and fruit yield in chilli. *J. Applied Zoological Researches*, 16, 184-185

Tatagar, M.H., Prabhu, S.T. and Jagadeesha, R.C. (2001) Screening chilli genotypes for resistance to thrips, *Scirtothrips dorsalis* Hood and mite, *Polyphagotarsonemus latus* (Banks). *Pest Management in Hortl. ecosystems*, 7, 113-116.

Velayudhan, R., Gopinath, K. and Kthavasalam, M. (1985) Pollination potential, population dynamics and dispersal of thrips species infesting flowers of *Dolichus lab-lab*. Proc. Indian Natl. Sci. Acad (Part B)., 51, 574-580.

www.indianspices.com (Assessed on 3rd January 2007)

Xie, J., Cardenas, E. S., Sammis, T. W., Wall, M. M., Lindsey, D. L. and Murray, L. W. (1999) Effects of irrigation method on chile pepper yield and *Phytophthora* root rot incidence. *Agricultural Water Management*, 42, 127-142.

Yadav, H. and Vijayakumari, B. (2003) Influence of vermicompost with organic and inorganic manures on biometric and yield parameters of chilli [*Capsicum annuum* (L.) var. Plri]. *Crop Research*, Hisar, 25, 236-243

Yadav, R.H. and Vijayakumari, B. (2004) Impact of vermicompost on biochemical characters of chilli (*Capsicum annuum*). *Journal of Ecotoxicology and Environmental Monitoring*, 14, 51-56.

15

Coriander

S.K. Malhotra and B.B. Vashishtha

1.0 INTRODUCTION

Coriander is a native of southern Europe and Asia Minor. In a treatise, Farrell (1999) has described coriander as indigenous to the Near East and Mediterranean region of North Africa and Southern Europe. According to early Sanskrit writings, coriander was known as far back as 5000 BC for its cultivation in India. Persia grew coriander 3000 years ago for adding fragrance to hanging garden of Babylon and seeds were found in Egyptian tombs of 960 BC. It has been used as an ingredient to traditional and folk medicine in the Middle East and Asia for more than 2000 years. There is archaeological evidence of coriander cultivation from Early Bronze Age in Macedonia (Fragiska, 2005).

1.1 Area and Production

Coriander (*Coriandrum sativum* L.) belonging to family *Apiaceae* is a herbaceous plant widely grown in most of the states of India as a spices crop. The stems, leaves and fruits have a pleasant aromatic flavour and thus used in flavouring different kind of foods and beverages worldwide. Coriander is commercially cultivated in India, Morocco, Russia, Hungary, Poland, Romania, Bulgaria, Czechoslovakia, Guatemala, Mexico and USA. In India, it is cultivated largely in Rajasthan, Madhya Pradesh, Gujarat, Tamil Nadu and Andhra Pradesh. During the year 2004-2005, 2.60 lakh ton of coriander seeds was produced from 3.82 lakh ha area which contributed and 46 % of the total seed spices produced in India (Table 1). The foreign exchange of Rs.

64 crores was earned by exporting seed of 22750 metric tonnes during 2005-2006. The value-added products of coriander such as coriander powder, essential oil and oleoresins were exported in 19581, 3168, 2.2 and 39 metric tonnes and earned export value of Rs. 5085, 1324, 58.3 and 466.7 lakh, respectively (Table 2). The major importing countries of Indian coriander are USA, Saudi Arabia, Germany, Canada, Bahrain, Indonesia, South Africa, Singapore etc.

Table 1. Area, production & Productivity of Seed Spices in India (2004-05)

S No.	Seed Spices	Area (ha)	Production (t)	Productivity($kgha^{-1}$)
1	Coriander	381710	259480	738
2	Cumin	367680	176070	479
3	Fennel	39730	51038	1285
4	Fenugreek	50445	62798	1245
5	Others	40715	15334	377
	Total	**850280**	**564720**	**664**

Table 2. Export of raw and value added products in coriander

	Value (Rs lakhs)			Quantity (ton)		
	2003-4	2004-5	2005-6	2003-4	2004-5	2005-6
Seed	5979.8	6982.9	5085.2	18344.2	30518.1	19581.9
Powder	1222.1	1283.1	1324.8	2673.4	3231.9	3168.1
Seed oil	78.4	28.7	58.3	1.4	1.2	2.2
Oleoresin	82.9	99.9	466.7	12.3	10.1	39.1

1.2 Chemical composition and Uses

1.2.1 Chemical composition

Coriander is quite popular for its peculiar sweet fragrance in leaves and fruits and is recognized well as good source of vitamins and minerals (Table 3). The composition of compounds varies with different stages of crop plant part in use. Coriander leaves contain 87.9% moisture, 3.3% protein, 0.6% fat, 6.5% carbohydrates and 1.7% mineral matter whereas, the seeds possess 6.3-8.0% moisture, 1.3% protein, 0.3-1.7% volatile oil, 19.6% non-volatile oil, 31.5% ether extract, 24.0% carbohydrates, 5.3% mineral matter and vitamin A 175 IU per 100 g (Sharma and Sharma, 2004). The coriander leaves are very rich in vitamin A, containing 5200 to 12000 I.U. per 100 g as mentioned by various authors (Shanmugavelu *et al.*, 2002). The coriander is known to have little nutritive value, the green leaves and seeds are mostly used for adding flavour and taste to the foods and various culinary preparations.

Table 3. Proximate principles and mineral composition of coriander (per 100 g dry leaves)

Constituent		Content	Constituent		Content
Moisture (%)		87.9	Fat (%)		0.6%
Total ash (%)	:	1.7	Vitamin B_2 (Riboflavin,mg/100g)	:	60.0
Carbohydrates (%)	:	6.5	Vitamin C(mg/100g)	:	135
Protein (%)		3.3	Vitamin A (IU/100g)	:	10000
Phosphorus (%)	:	0.06	Niacin (mg/100g)	:	0.8
Iron (%)	:	0.01	Energy (kcal)	:	28
Calcium (%)	:	0.14			

The aroma in coriander seeds is contributed by essential oil and its quantity in fruits vary with the variety and source (Table 4). European and Russian coriander are reported to be rich in essential oil (1.4-2.0%), whereas Indian varieties contain little quantity (0.1-1.0 %). High volatile oil is found in Norweigian coriander (1.4-1.7%) followed by Bulgarian (0.1-0.5%) and Indian coriander (0.1-0.4) varieties (Agarwal and Sharma, 1990). Volatile oil is made up of hydrocarbons and oxygenated compounds. Hydrocarbons accounts for about 20 per cent of oil but the major oxygenated compounds present are

Table 4. Proximate composition of coriander seed (100 g edible portion)

Water	8.9 g	Iron	16 mg
Food energy	298 kcal	Magnesium	330 mg
Protein	12.4 g	Phosphorus	409 mg
Fat	17.8 g	Potassium	1267 mg
Total carbohydrate	55.0 g	Sodium	35 mg
Fibre	29.1 g	Zinc	5 mg
Ash	6.0 g	Riboflavin	2 mg
Calcium	709 mg	Other vitamins	Insignificant

Source: Farrell (1999)

d-linalool or coriandrol (45-70%), Figueredo *et al.* (2004) have reported 77.48 per cent of linalool in coriander oil from Brazil. Besides essential oil, fruits are known to contain 19-21 per cent of fatty oils. A little quantity of herb oil (0.1-0.95%) can be extracted from stems and leaves. Herb oil has a pronounced odour of decylaldehyde or other higher fatty aldehyde.

1.2.2 Uses

The tender leaves stem and fruits of coriander have a pleasant aromatic flavour and thus is indispensable food adjuncts in Indian cookery. Coriander is popular as a garnish for salads, sandwiches and cooked dishes, as flavouring to soups, chutneys, sauces and curries, and as a salad ingredient in small amounts because of its charming flavour. The seeds (botanically fruits) are used as condiment in preparation of curry powder, pickling soups, sausages and seasonings. In the food industry, coriander has big use in flavouring pastry, cakes, biscuits, chewing gums tobacco products and beverages. Coriander oil and oleoresin are primarily used in seasoning for sausage and other meat products. Coriander is also employed for flavouring liquor, particularly gin, pork, frankfurter, meat, fish, salad, soda and syrup, gelatin dessert, pudding and candy preserve, etc. The fresh leaves and seeds are an essential ingredient in many European, Indian, Chinese, Vietnamese, Mexican and Portuguese foods.

The medicinal properties of coriander are well form and it is used in Indian system of medicinal *viz.,* Ayurvedic and Unani medicinal preparations. Coriander seeds possess immense medicinal values and are considered to be carminative, diuretic, tonic, anti-stomachache and refrigerant. They are also reported to lessen the intoxicating effects of spirituous liquors. An infusion of the seeds in combination with cardamom and caraway seeds is useful against flatulence, indigestion, vomiting and intestinal disorders. The seeds are chewed to correct foul breath; coriander water is useful in indigestion and other bowl complaints. Decoction of milk and sugar is beneficial during bleeding piles. The paste of coriander, nutmeg, clove, honey, lemon peel and rose water lightens and tones up the skin. Coriander leaves mixed with juice of apple, milk, multani mitti tones skin during summer and for similar action a paste of apricot, coriander leaves is also used. As a cosmeceutical application, coriander and tender coconut mixed together are popularly used in India to cure pimples.

2.0 BOTANY AND VARIETIES

2.1 Botany

Coriander is an annual aromatic herb, 30-80 cm in height with erect and branching stem. Leaves are alternate, compound and pinnatified as they reach upper extremities. Inflorescence is compound terminal umbels and the flowers are zygomorphic and hermaphrodite. Each umbel consists of several umbellets. Flowers at periphery are bisexual whereas the inner ones are unisexual white or rose coloured with five calyx teeth and a short conical style. Calyx is tubular, adnate to ovate at base with 5 lobes and corolla are 5, spreading, stamens 5 in number and ovary is bilocular. Fruit is schizocarpic round to globular, 2 to 3.5 mm in size, yellowish brown at maturity, green when young, ribbed composed of two concavo-convex mericarps with inner face of carpel having two vittae. Purseglove *et al.* (1981) have given a detailed botanical description of the plant.

2.2 Varieties

Varieties selected should have adaptation to the soil and climatic conditions and preferably have resistance/tolerance to pests and diseases. Efforts should be made to use organic seed of recommended variety for organic farming. When certified organic seed of coriander is not available, initially conventional seeds can be used (Anon., 2001). The description of some of the important cultivated varieties recommended for different states are given in Table 5.

Table 5. Recommended varieties of coriander for different states

Variety	Salient characters
Rajasthan	
RCr- 41	• Suitable for irrigated areas. • Highly resistant to stem gall and wilt but moderately tolerant to powdery mildew. • Matures in 130-140 days and average yield is 9.2 q/ha.
RCr-20	• Suitable for cultivation as rainfed crops or limited moisture conditions. • Moderately tolerant to powdery mildew, wilt as well as stem gall. • Matures in 120 days and average yield is 900 kg/ha.
RCr-435	• Adapted to irrigated conditions under high fertility soil. • Moderately resistant to root knot nematode and powdery mildew.

Contd...

Table 5. Contd...

RCr-436	• An average yield of 1000 kg/ha can be obtained in 135 days. • Suitable for cultivation in heavy soils of Rajasthan as rainfed crop. • Moderately resistant to root knot nematode and less susceptible to aphid.
RCr-684	• Matures in 90-100 days and average yield is 1100 kg/ha • A variety from medium maturity group with bold grains • It is resistant to stem gall and is suitable for irrigated cropping system.
RCr-446	• Average yield - 990 kg/ha in crop duration of 130 days. • Medium in maturity bears bold grains. • Suitable for irrigated cropping system • Moderate resistance to stem gall and wilt. • An average yield of 1200 kg/ha can be obtained in crop duration of 130 days.
NRCSS-ACr-1	• It is a long duration variety for irrigated conditions. • Dual purpose variety suitable for leaves and seeds production, • The grains are medium to small, having essential oil content up to 0.5% • Average yield 1250 kg seeds per hectare. • Resistant to stem gall and have tolerance to powdery mildew. • Suitable for growing as greens during off-season also.
Gujarat	
GCr-1	• Suitable for cultivation as early crop under irrigated conditions. • It has moderate tolerance to wilt and powdery mildew. • Matures in 112 days with an average seed yield of 1100 kg/ha.
GCr-2	• Suitable for early sowing under rainfed areas. • It is moderately tolerant to wilt and powdery mildew. • It matures in 120 days with an average yield of 1400 kg/ha.
Tamil Nadu	
CO 1	• Suitable for leaf and seed production. • Has less water requirement and suited for rainfed areas of Tamil Nadu.

Contd...

Table 5. Contd...

CO2	• It matures in 100-120 days and average yield is 400 kg/ha. • It is a dual purpose variety and is suitable for saline, alkaline and drought prone areas of Tamil Nadu.
CO3	• The variety matures in 90-100 days with an average grain yield of 5.2 q/ha & green yield of 1000 kg/ha in 40 days after sowing. • Plants are dwarf with medium-bold, oblong, brownish-yellow grains. • Suitable for both *kharif* and *rabi* season cultivation in Tamil Nadu • Tolerant to wilt, powdery mildew and grain mould. • Matures in 86-104 days and average yield is 650 kg/ha.
CS 287	• Early maturing variety is suitable for cultivation under both rainfed and irrigated conditions of Tamil Nadu. • Field tolerance to wilt and grain mould. • Plants are early-maturing with medium-sized, oblong, straw grains • Matures in 78-97 days and average yield is 600 kg/ha.
Bihar	
Rajendra Swathi	• An early variety and semi erect type is suitable for intercropping. • Resistant to stem gall and moderately resistant to wilt, aphids and weevil. • It matures in 100 days and average yield is 910 kg/ha.
Rajendra Sonia	• It is a medium tall, high yielding variety. • Resistant to stem gall, moderately resistant to wilt, aphid, weevil and tolerant to fruit fly. • Average yield is 1200 kg/ha in crop duration of 110 days.
Andhra Pradesh	
Sadhna	• A semi-erect variety is early maturing and dual purpose type. • Suitable for cultivation under rainfed conditions of A.P. • Field tolerance to white fly and mites. • Matures in 100 days and average yield is 1030 kg/ha
Swathi (CS 6)	• Semi-erect early maturing variety and is suitable for rainfed condition and late sown crop.

Contd...

Table 5. Contd...

	• It has field tolerance to fruit fly.
	• The seed yield is 855 kg/ha in 95-105 days crop duration.
Sindhu	• Suitable for intercropping and rainfed farming system.
	• Tolerant to wilt, powdery mildew, and aphids.
	• Average seed yield of 1000 kg/ha in 102 days cropping period.
Haryana	
Hisar Sugandh	• Medium maturity group.
	• Suitable for dual purpose cultivation both as leaves and seeds.
	• This variety possesses resistance to stem gall. The average yield of 1400 kg/ha
Hisar Anand	• It is a dual purpose high yielding variety has wider adaptability to different soil conditions
	• Average yield 1400 kg/ha
Uttar Pradesh and Uttaranchal	
Pant Haritama	• A dual purpose variety, late maturing and high yield potential (1500 kg/ha)
	• It is tolerant to stem gall disease.
Azad Dhania- 1	• Tolerant to moisture stress
	• Tolerant to powdery mildew and aphids
	• Maturity in 120-125 days
Madhya Pradesh and Chhattisgarh	
Gwalior No. 5365	• A medium seed sized variety from medium maturity group
	• This variety produces 800-1000 kg/ha seed in 100-113 days cropping period.

3.0 CLIMATE AND SOIL

3.1 Climate

Coriander a tropical crop requires a cool and comparatively dry and frost free climate. It is cultivated successfully during winter season in frost free weather in plains and as summer crop in temperate areas. Moderately cool and dry weather conditions favour high fruit yield and quality but occurrence of frost may prove injurious particularly during flowering and fruit developmental phases. Germination of coriander is severely reduced at temperature above 30^{0}C and below 10^{0}C. Heavy rains are harmful for the crop and continuous cloudy weather invites diseases and aphids. High temperature and high wind velocity during anthesis and seed formation accelerates sterility and reduces yield. Relatively cool weather during early stage favours better vegetative growth, while relatively high temperature favours both yield as well as quality of seed. For green leaf, it can be grown throughout the year provided moisture is made available. It is also successfully grown during *kharif* season for the production of greens. However, it does not grow well in spring summer season for greens, because it switches over within short time form vegetative growth to reproductive phase as soon as temperature raises beyond 20^{0}C. Coriander is cultivated either as a pure crop or mixed crop under irrigated as well as rainfed conditions.

3.2 Soil

Coriander is grown in a wide range of soils from medium to heavy loam in nature. Deep and fertile loamy soils are best suited for quality production of greens and fruits under the irrigated production system. But owing to high moisture retention, the heavy soils are ideally suited for rainfed cultivation of coriander and is thus grown successfully in areas under black cotton soils in southern and northern parts of India, as rain fed crop with conserved soil moisture. It can be grown successfully in soil with a pH of 6-7 but very light soils are not suitable for quality greens production of coriander hence Sandy, alkaline and saline soils should be avoided. However, coriander as an irrigated crop can be cultivated in all type of soils containing sufficient organic matter.

4.0 SEED MATERIAL

Coriander is propagated through seeds. Being bilocular each schizocarp has two mericarps and thus before sowing fruits are rubbed to split into two halves. The seed rate required is 10-12 kg per hectare for irrigated and 20 kg for rainfed cropping system.

4.1 Seed treatement

The use of bioinoculant *Azospirillum or Azotobactor* as seed treatment and soil treatment in combination with 10 t sheep manure/ha could give seed yield of 9 q/ha and thereby reducing the incidence of stem gall to 8% under semi-arid conditions (Malhotra *et al.* 2006). The Fusarium wilt incidence in coriander can be better managed by seed pelleting with *Trichoderma viride* (CFU10^6 @ 4g/kg of seed) plus neem cake application (150 kg/ha) reduces wilt incidence (Ravindran *et al.* 2001).

5.0 PLANTING AND AFTER CARE

5.1 Requisite for Organic farming

A minimum of two years is required as conversion period for organic cultivation and undergoing the process of certification by the accredited agency. The general guidelines for selection of farm land, buffer zone, certification etc. are required to be followed (Anon., 2001; Malhotra and Vashishtha, 2004). The procurement management which includes operationalization of activities for production of organic seed spices crops and creating conducive environment for required kind of quality and quantity of produce by way of adopting pre and post-harvest practices and deciding marketing system suitably. It shall ultimately be guideline to the grower/ processor/ exporter in planning the production/supply of organic seed spices for export (Malhotra and Vashishtha, 2007) is given here under:

1. Selection of site/ area in terms of soil and crop conditions, irrigation facility and time suitability for procurement
2. Selection of varieties in terms of
   a) Requirement of overseas markets
   b) Resistance to biotic and abiotic stresses
   c) In terms of ultimate use of the produce (herbs or seeds)
   d) Local infrastructure availability
3. Adoption of fields/farmers for better agronomic practices
   a) Use of organic manures/ composts/ biofertilizers/ other permittable nutrient sources etc.
   b) Optimum to low irrigation intensity.
   c) Adoption of fields as per crop rotation and suitable crop mixing.

5.2 Preparation of land

Soil is brought to fine tilth by 2-3 ploughing with harrow or *desi* plough. Stubbles of the previous crop should be collected and removed from the field. Clods should be broken and field should be levelled with the help of plank. Beds of convenient size with provision of irrigation channels should be prepared before sowing of seed. Well rotten farm yard manure or compost is to be incorporated into the soil preferably 3 weeks before sowing.

5.3 Sowing time

Coriander is grown throughout the year for leaves except sever summer months. The main crop of coriander for grains/leaves as spices is grown during October-November and as *rabi* season crop in north India (Singh and Gangwar, 1991). In southern states Andhra Pradesh, Tamil Nadu and part of Karnataka coriander is cultivated both in *kharif* and *rabi* season. The *kharif* crop is sown during June-July while the *rabi* crop in October-November. The rainfed crop of coriander is also grown in Northern Rajasthan, Madhya Pradesh, Tamil Nadu and Andhra Pradesh during mid September to October. Bhati *et al.* (1989) found that November sown crop of coriander produced maximum number of umbels and seed yield than in December. He further reported that crop sown around 31 October produced highest yield in Rajasthan (Bhati and Shaktawat, 1994). Sowing of coriander on 20 October yielded highest coriander seeds (965 kg/ha) and good quality essential oil (Venkateswarlu *et al.* 1992) and highest biomass production was reported when crop sown on 20th October (Reddy *et al.* 1993) in Andhra Pradesh. At Hisar and Delhi highest seed yield of coriander was obtained when the crop was sown on 30th and 31st October (Baswana *et al.* 1989; Sharma and Israel, 1991). Coriander sown on 1st November gave maximum plant height, number of branches, umbels per plant and seed yield but yield decreased gradually as sowing was delayed (Chaudhari *et al.*, 1995; Das *et al.*, 1993) in Cooch Bihar (West Bengal).

5.4 Sowing method

The crop is usually sown by broadcasting of seeds but line sowing is better. Soil depth should not exceed 2.5 cm. A line to line spacing of 20-30 cm and plant to plant spacing of 15 cm should be maintained. Soaking seeds in water for 8-12 hours before sowing hastens germination (Malhotra, 2005). The field should be prepared well in order to produce fine seed bed. For rainfed crop, the field should be prepared before the onset of monsoon so as to make soil loose to absorb maximum of moisture to depth. Coriander seeds initially require sufficient moisture to give good germination. The seeds take about 10-15 days to germination depending upon the climate. Pareek and Sethi (1985) obtained the highest seed yield of coriander when the plants were grown in 45 cm apart rows. Coriander line sowing at 30 cm gave the maximum plant spread and the highest number of secondary branches, umbels per plant and seed yield (Sukhadia *et al.* 1986). It was reported that branches per plant increased significantly with an increase in row spacing form 20 to 40 cm whereas, maximum seed yield of coriander was obtained at 30 cm row spacing in Rajasthan (Bhati 1988). While, Baswana *et al.*, (1989) found that row spacing had no effect on growth characters but maximum seed yield of coriander was obtained at 20 x 20 cm spacing in Haryana whereas, Nehra *et al.,* (1996) recorded the maximum plant height of coriander at closer spacing of 20 x 20 cm, while branches, umbels per plant and seed used were higher at 30 x 20 cm. A spacing of 30 cm from line to line and 10 cm from plant to plant has been found suitable for getting the higher seed yield of coriander (Bhati and Shaktawat, 1994; Venkateswarlu *et al.,* 1992).

5.5 Irrigation

Coriander is cultivated both as rainfed and irrigated crop but in the later production system, about 4-6 irrigations are required after germination, depending upon the soil and climate. Initial moisture is essential for good germination and crop stand. The *kharif* crop being rainfed, require no irrigation but the *rabi* crops should be irrigated at 12-15 days interval. Accruing to Sharma and Sharma (2004), the crop requires 4-5 irrigations after germination. The first irrigation should be given 30-35 days after sowing, the second at 60-70 days, the third at 80-90 days, the fourth at 100-105 days and the fifth at 110-150 days, respectively. An overall requirement of 490 mm irrigation water in the cropping season is required for growing coriander. Safe irrigation water should be used for irrigation; use of sewage water is strictly prohibited. Under Chhattisgarh conditions, Lakpale *et al.,* (2004), obtained maximum seed yield of coriander when five irrigations at 0.8 IW/CPE were applied.

5.6 Intercultural operations

Coriander is slow growing plant initially and therefore thinning, first hoeing and weeding should be done 30 days after sowing. Two weedings followed by hoeing should be practiced *viz.,* 30 and 60 days after sowing, if required, third weeding can be done. During first weeding the plants should also be thinned to the desired distance.

5.7 Cropping system

Coriander can be grown as mixed or intercrop provided all the other crops are grown following organic farming methods. It is desirable to include leguminous crops or other suitable crops exhibiting lesser competition. Likewise rotation crops should also be grown by adopting organic farming methods.

Bimbraw (2006) suggested following crop rotations under the Punjab conditions:

(i) Maize - Potato - Coriander
(ii) Summer moong - Coriander
(iii) Rice - Coriander

Some of the other cropping systems suggested for seed spices growing area can be (Mehta and Malhotra, 2007):

(i) Green gram/black gram- Coriander

(ii) Cluster bean/cowpea-Coriander

(iii) Cluster bean-Coriander-Summer Maize

5.8 Frost protection

Frost causes damage to coriander crop at flowering and fruit setting stage. During severe winter months when there is danger of frost, the crop should be irrigated and smoke environment should be created before sunset in the field.

6.0 NUTRIENT MANAGEMENT

The manurial recommendations have been made from different states. In general the well rotten farm yard manure of 10-15 t/ha for heavy soils and 15-20 t/ha for light soils with low organic matter, is added while preparing the field. The application of vermicompost at 10 to 15 t/ha significantly increased the concentration of trace elements (Fe, Mn) in leaves and Zn content was also highest with 5t /ha quantity of vermicompost (Reddy *et al.*, 1993). The good herbage yield of 6067 kg /ha at 60 days after sowing was obtained, when 15t/ha of vermicompost was applied whereas highest seed yield of 1314 kg/ha was obtained with 20t/ha vermicompost (Vadiraj *et al.*, 1998).

7.0 DISEASES AND PEST MANAGEMENT

Coriander is prone to attack by some of the serious diseases and insect-pests. The plant protection measures under organic farming system should involve emphasis on crop management practices time of sowing, balanced nutrition, crop rotation, green manuring etc. for reducing the incidence of diseases and pests, selection of resistant varieties and adoption of bio control measures.

7.1 Diseases

7.1.1 Coriander wilt *(Fusarium oxysporum* f. sp. *corianderii)*

The disease is more prevalent in light soils and affected plants show yellowing, drooping of leaves and terminal branches followed by the withering and drying of the plant. In the severe incidence of disease, plant dies and can be pulled out easily of the ground.

Control

- Pathogen being coriander-specific can be managed through crop rotation.
- The occurrence of disease can be reduced by the application of oilcakes in the field with high soil pH 8.2
- Solarization of soil during summer and use of disease free seed minimizes the disease incidence.
- The antagonistic fungi *Trichoderma viride* 4 g / kg seed treatment is effective in reducing the wilt incidence to 5.3 % (Muthulakshmi *et al.*, 2002).
- Seed treatment with *Pseudomonas fluorescens* @ 10 g/kg seed + soil application of neem cake @ 150 kg/ha is effective biocontrol method (Rajeshwari *et al.*, 2004)
- Extreme dry climate condition should be avoided as it favours the disease.
- Adopt field sanitation measures
- Locate wilt resistant variety suiting to the production area

7.1.2 Powdery mildew *(Erysiphe polygoni D.C.)*

The symptoms appear as white powdery mass on the leaves, twigs and fruits of the plant. The incidence is more during February- March and on severe attack the leaves become yellowish and dry up. The disease makes it appearance when the atmosphere gets damp, particularly during the flowering stage of the crop. The powdery mildew can completely ruin the crop, if adequate control measures are not taken in time.

Control

- The application of Sulphur dust, wettable sulphur suspension (0.6%) and sulphur spray (Patel *et al.*, 2006) and the restricted use of sulphur is permitted and described in guidelines from Spices Board (Anon., 2001)
- Foliar spray of onion extract (5%) is effective in reducing the severity.
- Grow resistant/tolerant varieties.
- The incidence is more during February-March, thus sowing of coriander should be completed in first week of October, as late sown crop has more severe damage.
- Uprooting of all affected plants followed by burning or burying.
- Early planting manages the disease well (Kalra *et al.*, 2000).

7.1.3 Stem gall *(Protomyces macrosporus Unger)*

The symptoms appear as galls on the stem, leaf stalk and peduncles. The seeds are deformed and the yield is reduced. The leaf yield and quality is badly affected.

Control

- Uproot and burn all the infected plants completely out of the field.
- Besides, solarization of soil and use of resistant variety RCr 41 can also be followed.
- Wet growing conditions should be avoided
- Use of healthy seed is best remedy
- Follow long term crop rotation

7.1.4 Blight *(Alernaria poonensis)*

The disease appears in the form of dark brown spots on the stem and the leaves. The disease mostly appears under cloudy weather. The quality of green leaves is affected severely. The disease is of minor importance and occurrence is sporadic.

Control

- Avoiding excessive irrigation and high humidity
- Using disease free seed
- Adopting field sanitation measures
- Using resistant variety

7.2 Pests

7.2.1 Pest monitoring

Pest monitoring is an important part for organic cultivation of coriander and should be done once in seven days to work out economic threshold level (ETL). It will help to keep close watch on the appearance of insect pests, diseases and bio-control fauna.

For sucking pests: Population should be counted on three leaves i.e. top, middle and lower per plant.

For cut worms and defoliators: Per cent damage assessment can be made by counting total number of plants and affected plants

Management strategy

1. Set up yellow pan/sticky traps for monitoring of aphids/whitefly @ 10 yellow pans/sticky traps per hectare. Locally available empty yellow palmolive tins coated with grease/Vaseline/castor oil on outer surface may also be used.
2. Collect the larvae of cutworm and leaf defoliators

7.2.2 Aphids (Hyadaphis coriandri Das.)

Aphid incidence is generally greater on late flowering varieties than on the early flowering. The aphids are found in colonies sometimes very serious in patches on tender leaves, stem and inflorescence. Both nymphs and adults suck sap from the tender leaves thus reducing marketability. The severe infestation affects the yield and quality of leaves badly. Moreover secretions from aphids invite attack of sooty mould. The crop become chlorotic and the fruits get shriveled.

Control

- Clip off the heavily infested plant parts.
- Spray the crop with pressurized water.
- Application of fish oil, resin soap or *neem* seed kernel-extract (3%), *neem* oil (2%) or tobacco decoction (0.05%) is effective measures (Ravindran *et al.*, 2001).
- Spray of neem oil (1%, 5 litre/ha) or NKE (3%, 15 litre cow urine + 1.5 kg neem kernel extract/ha) are effective in controlling aphids (*Hyadaphis coriandri* Das) when sprayed three times at 15 days interval (Gupta and Rai, 2004).

7.2.3 Mites (Petrobia latens)

The mites frequently attack the coriander crop and whole plant becomes whitish yellow and appears sickly. The insect mostly feeds on young leaves and the infestation is more severe on young inflorescence. Small mites are seen on lower side of leaves and when serious, cause webbing and feed within the web. Plants get stunted at severe infestation.

Control

- Spray neem oil (1%)

8.0 HARVEST AND POST HARVEST HANDLING

8.1 Harvesting and Yield

Coriander crop matures in about 100-150 days of sowing depending upon varieties, climate and geographical location. The green leaves are normally harvested 60-75 days after sowing and left for flowering and seed formation, whereas coriander crop grown exclusively for green leaves is harvested after 35-40 days of sowing. The coriander compound leaves with stem are prepared in bunches of desirable size for marketing. The stage of harvest depends upon the market requirement. For green coriander seeds, the crop is harvested when seeds are green and have attained full size. The green seed crop is dried under shade for retaining the green colour. Keeping in view, the market requirement, the crop can be harvested at yellow or brown colour of seeds. Delay in harvesting should be avoided, otherwise shattering and splitting of seeds may occur. Harvesting should be done by cutting the whole plant when 60 per cent of seeds in main umbels attain desired size and colour and piled into small stacks in partial shade to dry for 2-3 days. Beating plants on threshing floor separates the seed and thereafter it is cleaned, winnowed carefully by winding up almost all chaffs, dust and dirt. Clean seeds should be allowed to dry on tarpaulin for 4-5 days or so till moisture content of seed reaches to 9 per cent before storing (Singh and Malhotra, 2007). Solar drier or electric drier may also be used for early drying if available.

An average seed yield of 4.5 to 6.0 q/ha can be obtained from rainfed crop, whereas average seed yield of 8-12 qha^{-1} along with additional marketable leaf yield of about 5-7 q/ha can be obtained from irrigated crop while the genetic potential of 20-22 q/ha seeds of this crop has been reported. The crop grown exclusively for leaf purpose yielded green leaves of about 48.41 qha^{-1} when three cuttings were taken (Thakral *et al.*, 1992). In a cutting management trial, Singh *et al* (2000) obtained green leaf yield of 76 q/ha from two cuttings with additional seed yield of 8.36 q/ha and the reduction in seed yield due to cutting effect was recorded to 27%.

8.2 Post harvest management

Coriander is mostly grown for green leaves as vegetable, dehydrated leaves and seed. The post harvest operations are followed as per the type of commodity grown. The fresh greens consisting edible portion are the tender leaves and stem, which are cut to length of about 5-7 cm above ground. The yellow, diseased, and damaged leaves are trimmed off and weed plants and straw is culled during cleaning and dressing. Healthy and disease free leaves are tied into small bunches for the convenience in handling, transportation and marketing. The leaves of coriander are sun-dried or dehydrated in a suitable dehydrator, for further use in off-season. The dried coarse powder or rubbed leaf powder is mostly used for flavouring by sprinkling on the various food items.

8.2.1 Processing

The mature dried seed of coriander are distilled to obtain the essential oil. Hydro or steam distillation method is generally used for extraction of essential oil. On an average, the dried seeds of coriander yield 0.1 to 1.7% volatile oil depending upon variety and location. Good quality oleoresin extracted from coriander seed has great demand in importing countries.

Coriander is sold as whole coriander, ground coriander, essential oil, oleoresin, coriander herb oil, coriander dhana dal. The volatile oil is used chiefly as a flavouring agent for liquors, cocoa and chocolate industry and it is also valuable ingredient in perfumes. The oleoresin is used for flavoring foods, beverages, pickles, sweet and sausages. The other important product is *dhania dal* which is good mouth freshener and is as a major adjunct in *supari* and *pan masala*.

8.2.2 Storage

The fresh coriander leaves are highly perishable and deteriorates rapidly at temperature above 5^0C. Therefore the fresh leaves bunched should be marketed soon after harvesting. The leaves can be sorted only for about 24-36 hours after harvesting under ambient condition, however in cold store at 0^0C temperature and 90% relative humidity, the storage period can be extended for one week.

The dried seed is filled in plastic bags or in gunny bags lined with plastic sheet. Each bag is sealed and stored under clean, dry and ventilated place.

8.3 Quality specifications for export

8.3.1 Whole seed

The quality specification for coriander can be classified into three categories *viz.*, commercial requirements, cleanliness and health specifications and have been detailed by Thamil Selvan, (2007) and Malhotra, (2005).

Commercial specifications

These requirements vary from country to country and crop to crop. It depends on many factors like customer needs, acceptability and country specifications. The commercial specifications of seed spices crops are colour, appearance, taste, pungency, texture, shape, volatile oil and packaging. The general commercial and general requirements for coriander seed are yellow-brown or green colour, small seeded, large seeded, free flowing, high volatile oil (min. 0.25%).

Cleanliness specifications

The cleanliness specifications as per ASTA, ESA and ISO are given below

Crop	Whole insect dead (by count)	Excrets mammalian (mg/lb)	Excreta other (mg/lb)	Mould (% wt)	Insect defiled / infected (%/wt)	Extraneous foreign matter (%/wt)
Coriander	4	3	10	1	1	0.5

Other quality standards as per ESA and ISO

Crop	Ash level % w/w (min.)	Acid insoluble ash % w/w (max.)	Moisture content % (max)	Volatile oil % (Min.)
Coriander	7	1.5	12	0.3

Health specifications

The micro contaminants such as pesticide residues, microbial counts, aflatoxins, heavy metals etc. are the health specifications. These contaminants make the food unsafe. The importing countries have limited the maximum level of these contaminants and limits are given here.

Maximum Residue Level = 0.05 to 0.2%

Microbial counts = 1×10^2 / g to 1×10^6 /g (*Salmonella, E. coli*, yeast, moulds)

Aflatoxin = 5 ppb to 10 ppb (Max.) Aflatoxin, mycotoxins

Heavy metals = Arsenic (5 mg/kg), copper (20 mg/kg) lead (10 mg/kg), Zinc (50 mg/kg)

8.3.2 Essential oil

The specifications for coriander oil are (Singhal *et al*, 1997):

- Colourless or pale yellow
- Specific gravity (20ºC, 0.870 to 0.885
- Optical rotation (20ºC), +8 to+13

- Refractive index, 1.463 to 1.471
- Solubility (70% ethanol), 2-3 vols
- Aldehydes (Linalool) 45 to 70%

8.3.3 Oleoresin

The coriander oleoresin should be prepared with recommended organic solvents followed by the subsequent removal of the solvent as per specifications of importing countries.

8.3.4 Powder

The coriander powdered hygienically through permittable devices, should at least 95 % of the ground product pass through a U.S. Standard No. 30 sieve (Farrell, 1999).

8.4 Marketing system

The export of coriander mainly involves indirect exports through links with agents or export management firms which will eventually lead to the handling of own exports. The other entry options are licensing, contract production (supply), management contracting and joint ventures. The details can be sought from Spices Board at Cochin.

- Export via independent representative (agent)
- Establishment of overseas sales subsidiary own or with foreign party collaboratly.

Marketing channels involved are:

- Direct channel - Foreign distributor, foreign retailer, state controlled trading companies and end users.
- Indirect channel - Export brokers, export agents, export management organizations and co-operative exporters.

REFERENCES

Agarwal, S. and Sharma, R. K. (1990) Variability in quality aspect of seed spices and future strategy. *Indian Cocoa, Arecanut and Spices J.* 13:127-129.

Ali, S. A., Pathak, R. K. and Saraf, R. K. (1998) Fungicides seed treatment in coriander, *Advances in Plant Sciences,* 11(1):303-304.

Anonymous (2001) Guidelines for production of organic spices in India. Spice Board, Cochin, pp 64-67.

Baswana, K. S., Pandita, M. L. and Sharma S. S. (1989) Response of coriander to dates of planting and row spacing. *Indian J. Agron.* 34 (3):355-357.

Bhati,D. S. (1988) Effect of nitrogen application and two spacing on coriander production under irrigated condition in semi-arid Rajasthan. *Indian J. Agric. Sci.* 58 (7): 568-569.

Bhati, D. S. Aggarwal, H. R. and Sharma, R. K. (1989) The grain yield and yield attributes of coriander varieties as influenced by sowing dates. *Indian Cocoa, Arecanut and Spices J.* 13(2):58-60.

Bhati, D. S. and Shaktawat M. S. (1994) Effect of sowing date, row spacing and nitrogen on quality parameters of coriander. *Progressive Hort.* 26(1/2):14-18.

Bimbraw, A. S. (2006) *Agrotechniques for umbelliferrous Medicinal and aromatic plants.* Inernational Book Distributing Co. Luknow.pp 96-137.

Chaudhari, S. M. Kharchae, S. M. and Desai, U. T. (1995) Effect of sowing dates on seed production in coriander. *J. Maharashta Agric Univ.* 20(1):139.

Das, A. K. Sadhu, M. K. Som M. G. and Bose, T. K. (1993) Effect of sowing dates on growth and yield of coriander. *Indian Cocoa Arecanut and Spices J.* 17(1&2) :26-27.

Farrell, K. T. (1999) *Spices, condiments and seasonings.* Aspen Pub. Inc. Gaitherberg, Maryland, USA. pp 78-81

Figueiredo, R. D., Marques, M., Nakagawa, J., Figueiredo, R. O. and Ming, L. C. (2004) Composition of coriander essential oil from Brazil. *Acta Horti.* 629:135-137.

Fragiska, M. (2005) Wild and cultivated vegetables, herbs and spices in Greek Antiquity. *Environmental Archaeology,* 10 (1): 73-82.

Gupta, M. R. and Rai, H. S. (2004) Bioefficacy of neem products against the incidence of coriander aphid. In Extended Summaries, *National Seminar on Opportunities and Potentials for Spices for Crop Diversification* held at JNKVV, Jabalpur from 19-21, January, 2004, pp 77-79.

Kalra, A., Parameshwaram, T. N., Ravindra, N. S., Rao, M. G., Sushil Kumar and Kumar, S. (2000) Effects of planting date and dinocap application on control of powdery mildew and yields of seed and seed oil in coriander. *J. Agric. Sci.* 135:193-197.

Lakpale, R., Shrivastave G. K. and Tripathi R. S. (2004) Effect of irrigation schedules on yield and economics of different seed spices crops under Chhattisgarh condition. *Proc. National Seminar on New Perspectives in Spices, Medicinal and Aromatic Plants* held at ICAR Research Complex Goa, 27-29 Nov., 2003 pp329-331.

Malhotra S. K. (2005) 'Quality Standard of Seed Spices' Lead paper in *Proc.National Seminar on Spices* held at KNK College of Horticulture, Mansor, from 20-21 March, 2005.

Malhotra S. K. and Vashishtha B. B. (2004) Organic production of seed spices in India-Status and Strategy. *Seed Spices Newsletter*, 4(1):1-4.

Malhotra S. K. and Vashishtha B. B. (2007) Production technology for seed spices crops. NRCSS, Ajmer, pp 1-80.

Malhotra, S. K., Vashishtha, B. B. and Apparao, V. V. (2006) Influence of nitrogen, Azospirillum sp. and farmyard manure on growth, yield and incidence of stem gall disease in coriander (*Coriandrum sativum* L.). *J. Spices and Aromatic Crops*, 15 (2): 115-117.

Mehta, R. S. and Malhotra, S. K. (2007) Seed spices based cropping system. In *Production, Development, Quality and Export of Seed Spices*. Eds. Malhotra, S.K. and Vashishtha, B.B. NRCSS, Ajmer, Rajasthan, India, pp 181-189.

Muthulakshmi P., Chezhiyan N., Muthukrishnan K. and Doraisamy S. (2002) Management of coriander wilt using biocontrol agents. *J. Spices and Aromatic Crops*. 11 (2) : 138-140

Nehra, B. K., Rana, S. C., Malik, J. S., Negi, A. and Singh, A. (1996) Effect of varieties, spacings and fertility levels on growth and seed yield of coriander. *Proc. Silver Jubilee National Symp. on Arid Hort.*, held from Dec.5-6,1996 at H.A.U,Hisar,: 363-365.

Pareek, S. K. and Sethi, K. L (1985) Response to irrigation and fertilization in coriander. *Indian Perfumer* 29(3-4):225-228.

Patel, N. R., Patel, K. D., Agalodiya, A.V. and Patel, P. K. (2006) Abst. in *SYMSAC II, Crop and product diversification* held at Lal Bagh, Bangalore from 8-10, November, 2006, pp 56.

Purseglove, J. W., Brown, F. G., Green, C. L and Robins, S. R. J. (1981) *Spices*, Vol II. Longman, London and New York, pp 736-88.

Rajeshwari, E., Subramanian, S., Balakrishnamoorthy, G. and Shiva, K. N. (2004) Abst. In *SYMSAC I, Commercialization of Spices, Medicinal and Aromatic Crops*, held at IISR Calicut from November, 1-2, 2004, pp26.

Ravindran P. N., Kallupurackal, J. A and Sivaraman K. (2001) All India coordinated Research project on spices-Status of seed spices research. *Indian J. Arecanut, spices & Medicinal Plant*, 3(3): 97-110.

Reddy, S. E. Venkateswarlu, K. and Reddy, Y. N. (1993) Effect of certain coriander cultivars in biomass production. *Indian Cocoa, Arecanut and spcies J*. 16(3-4):110-112.

Shanmugavelu, K. G., Kumar, N. and Peter, K. V. (2002) *Production Techniques of Spices and Plantation Crops*. Agrobios (India), Jodhpur, pp. 121-130.

Sharma, M. M. and Sharma R. K. (2004) Coriander. In *Handbook of Herbs and spices*. ed. K.V. Peter, Woodhead publisher U.K.pp 145-161.

Sharma, R.N. and Israel, S. (1991) Effect of date of sowing and levels of nitrogen and phosphorus on growth and seed yield of coriander. *Indian J. Agron.* 36:181-184

Singh D. and Gangwar, B. (1991) Management practices for the higher productivity of coriander in India-a-review. *Agric Rev.* 12(1):15-21.

Singh, J. Malik, Y. S., Thakral, K. K. and Mehla, C. P. (2000) Effect of sowing time, nitrogen levels and leaf cuttings on green & seed yield of coriander. *Haryana J. Hort Sci.* 29(3-4):225-8.

Singh, R. and Malhotra, S. K. (2007) Harvesting and maturity indices in seed spices crops. In *Production, Development, Quality and Export of Seed Spices.* Eds. Malhotra, S.K. and Vashishtha, B.B. NRCSS, Ajmer, Rajasthan, India, pp 195-200.

Singhal, R. S., Kulkarni, P. R. and Rege, D. V. (1997) In *Handbook of Indices of Food quality and Authenticity,* Woodhead Publishing Limited, England, pp. 386-456.

Sukhadia, H.L. , Sinha , M.P., Malavia, D.D., Kaneria, B.B. and Patel, J.C. (1986) Response of coriander to planting patterns and seed rates. *Indian J. Agron.* 31(3):296-297.

Tamil Selvan, M. (2007) Quality management in seed spices. In *Production, Development, Quality and Export of seed spices.* Malhotra,S.K. and Vashishtha,B.B. (Eds.), NRCSS, Ajmer, pp 23-38.

Thakral K. K. Singh G. R., Pandey U. S. and Srivastava U. K. (1992) Effect of N levels and cuttings on production of green leaves and seed yield of coriander cv Narnaul Selection. *HAU J. Res.* 22(1): 35-39.

Vadiraj, B. A., Siddagangalah, and Potty, S. N. (1998) Response of coriander cultivars to graded levels of vermicompost. *J. Spices and Aromatic Crops.* 7(2):141-3.

Venkateswarlu, K. Reddy, S. E. and Reddy, Y. N. (1992) Nitrogen uptake by coriander cultivars sown at different dates. *Indian Cocoa, Arecanut and Spices J.* 16(2):72-73.

16

Fenugreek

K. N. Shiva and S.K. Malhotra

1.0 INTRODUCTION

Fenugreek is one of the oldest cultivated spice crops of the world and grown for its medicinal value and forage in India, Western Asia and Nile valley since remote antiquity. In Egypt, it has been cultivated since 1000 BC and is part of the Indian diet for over 3000 years. Fenugreek is used as an important component of holy smoke in Egyptian incense. It is found growing wild in parts of northern India and cultivated all over the subcontinent for its green leaves and seeds. Fenugreek is a native of south-eastern Europe and Western Asia. India is also said to be a native of fenugreek (Shanmugavelu *et al.*, 2002) and found growing wild in Kashmir, Punjab and upper Gangetic plains. Fenugreek, the multi-use and commercially important spice crop, is extensively grown almost in every part of the country during winter season for its seeds, tender shoots, and fresh leaves. The genus *Trigonella* is reported to have 50 species (most of them originated in the Iranian Indian region and oriental origin). Among these, eleven species is reported occur in India, out of which *Trigonella- foenum-graecum* L. (Fenugreek) and *Trigonella corniculata* L. (Kasuri methi) are grown as spices in India (Singhania *et al.*, 2006).

1.1 Area, production and export

The major fenugreek producing countries are India, Argentina, Egypt, Southern France, Morocco, Spain, Turkey and China but India enjoys the status of largest producer in the world. In India, its production is concentrated mainly in Rajasthan, Madhya Pradesh, Maharashtra, Haryana, Punjab, Gujarat and Uttar Pradesh states. Major fenugreek growing districts in Rajasthan are Nagaur, Jaipur, Jodhpur, Ajmer, Kota, Bhilwara, Nimbahera and Pratapgarh ; whereas in Uttar Pradesh, it is grown in Aligarh, Bulandshar, Meerut and Bareily and in Madhya Pradesh the districts are Jawra, Neemuch and Mandsaur. The crop is grown in the Mediterranean and some tropical countries, for cattle fodder and also to improve soil health and fertility.

Among the seed spices, fenugreek occupies the third position, next to coriander and cumin in terms of area and production, covering an area of 0.50 lakh hectares with a production of 0.64 lakh tonnes and a productivity of 1269 $kgha^{-1}$ (2003-04) (DASD, 2007). India is one of the major producers and exporters of fenugreek. In India, Rajasthan has maximum area and production contributing 90% to the country's production, followed by Gujarat, Uttaranchal and Uttar Pradesh. However, maximum productivity was in Uttaranchal, followed by Rajasthan and Gujarat (Table 1). It is also cultivated in the states of Gujarat, Andhra Pradesh, Tamil Nadu and Punjab. During 2005-06, India exported 12250 tonnes of fenugreek valued at Rs. 2400 lakhs (Spices Board, Cochin). The value added products of fenugreek such as fenugreek seed, fenugreek powder, and oleoresins exported in quantity is 11582, 640 and 23.1 metric tones and earned export value of Rs. 2012, 350.3 and 105.8 lakh, respectively during 2005-06 (Table 2). The major importing countries for Indian fenugreek are France, Germany, Canada, Israel, Japan, Jordan, Korea, Kuwait, Malaysia, Netherlands, Saudi Arabia, Singapore, South Africa, Switzerland, UAE, UK and USA.

Table 1. Area, production and productivity of fenugreek in India (2003-04)

State	Area (ha)	Production (tonnes)	Yield ($kgha^{-1}$)
Rajasthan	43856	56175	1281
Uttar Pradesh	221	156	706
Gujarat	6258	7494	1198
Uttaranchal	261	396	1517
Total	50596	64221	1269

Source: DASD (2007)

Table 2. Export of raw and value added products from fenugreek

Items	Value (Rs. lakhs)			Quantity (tonnes)		
	2003-04	2004-05	2005-06	2003-04	2004-05	2005-06
Seed	1287	2263	2012	6376	13027	11582
Kasuri seed	22	48	38.2	26	49.4	28.2
Powder	245.9	349.2	350.3	530.5	673.4	640
Seed oil	-	-	-	-	-	-
Oleoresin	90.7	47.7	105.8	23.9	21.1	23.1

Source: Spices Board

1.2 Composition and Uses

The nutritional significance of fenugreek is well-recognized world over as a vital source of essential minerals, vitamins and dietary fibers. The fenugreek green leaves supply 35 calories and contain moisture 86.1%, protein 4.4%, fat 0.9%, fibre 1.1%, other carbohydrates 6.0%, and ash 1.5%. In addition, the fenugreek green leaves are rich in vitamins and contain carotene content 2.34 mg, thiamine 0.04 mg, riboflavin 0.31 mg, nicotinic acid 0.8 mg, and vitamin C 52.0 mg/100 g edible portion (Duke, 2003). Fenugreek leaves also contain about 296 and 59 µg/g dry weights of aliphatic and aromatic secondary amines, respectively (Shah and Bhinde, 1985). The green leaves are also known to contain sulphur containing amino acids. Seeds are rich in protein, minerals, especially iron, calcium, vitamins (A, B_2 and C) and contain substances like volatile oil, fixed oil, cellulose, starch, sugars, alkaloids and enzymes. The volatile oil content of seed is low (0.02%) but it is extremely odoriferous due to a compound called 3-hydroxy-4, 5- dimethyl-oxolane-2-furanone. The fixed oil content averages 7.6% and has a very bitter taste. The seeds of fenugreek are also known to have steroidal compounds fenugreekine, trigonelline, diosgenin, choline and malonic acid. The bitter taste in seeds as well as in leaves is due to the presence of alkaloids such as trigonelline and choline. Diosgenin has been identified as the major steroid sapogenin in the leaves of *Trigonella foenum-graecum,* along with smaller amounts of tigogenin and gitogenin (Varshney and Sood, 1971). The proximate principles and mineral composition of fenugreek seeds are given in Table 3.

Every part of this multipurpose crop is useful and utilized in one or the other form as food, fodder, medicine, and cosmetics. Its green fresh leaves and tender immature pods are used as green cooked vegetable, and sun dried leaves, which are having aromatic qualities, are used as spice for seasoning a variety of foods in off-season. Being odoriferous, the dry seeds and their powder are used as condiment/flavoring agent and for medicinal purposes. The seeds are also used for making dye and extraction of alkaloids or steroids. In certain parts, the green herbage as well as the seeds of fenugreek along with other green or dry fodder is usually fed to the cattle. It is recognized well as commercial uses in syrups, pickles, baked foods, condiments, chewing gums, icings, and cooked food seasonings. The fenugreek leaves and seeds have use in cosmetics and hair conditioning. In Punjab, dried plants are added to store grains as insect repellent. Fixed oil finds its use in perfumery.

Table 3. Proximate composition of fenugreek seed (100 g edible portion)

Water	8.8 g	Magnesium	191 mg
Food energy	323 kcal	Phosphorus	296 mg
Protein	23.0 g	Potassium	770 mg
Fat	6.4 g	Sodium	67 mg
Total carbohydrate	58.4 g	Zinc	3 mg
Fibre	10.1 g	Ascorbic acid	3 mg
Ash	3.4 g	Riboflavin	2 mg
Calcium	176 mg	Other vitamins	Insignificant
Iron	34 mg		

Source: Farrell (1999)

Fenugreek is one of the few spices that are used extensively for medicinal purposes owing to its hypoglycemic and hypocholestremic properties. The seeds are carminative, tonic and aphrodisiac and customarily used for the treatment of colic, flatulence, dysentery, diarrhea, dyspepsia with loss of appetite, chronic cough, dropsy, enlargement of liver and spleen, rickets, gout and diabetes, and are good for the elimination of bad breath and body odor. It is believed that regular intake of fenugreek seed reduces the level of glucose, total cholesterol, and triglycerides in blood. Fenugreek is known to possess galactagogue properties, and thus, Indian women consume the seeds for its power to promote lactation. Fenugreek seeds substantially contain diosgenin (0.40-1.26%), which is used as a starting material in the synthesis of sex hormones as stimulant and contraceptive (Thangaraj and Vijayakumaran, 2000). Fenugreek also helps to combat dandruff and is a cure for baldness.

The ground powder mixed with cotton seed is fed to cows and buffalos for increasing the flow of milk. It is used in cattle feeds and in veterinary medicines.

2.0 BOTANY AND VARIETIES

2.1 Botany

Fenugreek (Common methi): It is a quick growing herb producing erect shoots up to a height of one meter. The leaves are dark green leaves with or without pink margin. It produces 2-3 small white flowers at the base of each leaf. The pods are slender and straw coloured when ripe, beak shaped and are about 8-10 cm long with 8-15 yellowish brown coloured and smooth surfaced seeds.

Kasuri methi: It is a slow grower and remains in a rosette condition during most of its vegetative growth period, which produces bright orange to yellow flowers. It produces sickle shaped pods smaller than the common methi (Som and Maiti, 1986).

2.2 Varieties

Varieties selected should have adaptation to the soil and climatic conditions and preferably have resistance/tolerance to pests and diseases. Efforts should be made to use organic seed of recommended variety for organic farming. When certified organic seed of coriander is not available, initially conventional seeds can be used (Anon., 2001). The description of some of the important cultivated varieties recommended for different states is given in Table 4. Of these, some of the varieties are having resistance/tolerance to various biotic and abiotic stresses (Table 5), coupled with high yield and quality parameters.

3.0 SOIL AND CLIMATE

3.1 Soil

Fenugreek is well adapted to a wide range of soils but grow best on well-drained loamy soils. Organic matter rich clay-loam soils may also be used provided adequate drainage facilities are available. However, the crop does not thrive best in sandy or gravely soils. In rain fed areas, black cotton soils are best suited for its successful cultivation. Although the crop is tolerant to salinity (Habib *et al.*, 1971) but always gives higher yield with better quality of leaves in neutral soils having a pH range from 6.0 to 7.0 (Ramaswamy and Ananthan, 1999) hence, its cultivation should usually be avoided in problematic soils, i.e., saline, alkaline and acidic one.

Table 4. Improved varieties in fenugreek

Variety	Institution	Av. Yield (dry) kg ha^{-1}	Salient features
CO.1	Horticulture College & Research Institute (TNAU), Coimbatore, Tamil Nadu	680	Dual-purpose variety, early maturing variety
Rajendra Kranti	Tirhut College of Agriculture (RAU), Dholi, Bihar	1300	Suitable for intercropping
RMt-1	SKN College of Agriculture (RAJAU), Jobner, Rajasthan	1500	Vigorous growth habit, moderately resistant to root knot nematode and powdery mildew
Lam sel.-1	Regional Agrl. Res. Station (ANGRAU), Guntur, AP	740	Dual-purpose varieties, early maturing, bushy type
Hisar Sonali	Department of Vegetable Crops (CCS HAU), Hisar, Haryana	1700	Dual-purpose variety, high yielding, vigorous growing habit
CO.2	Dept. of Spices & Plant. Crops, HC & RI (TNAU) , Coimbatore, Tamil Nadu	482	Short duration dual purpose variety (85-90 days), suitable for both *kharif* and *rabi* season
RMt-303	SKN College of Agriculture (RAJAU), Jobner, Rajasthan	1900	Medium maturity variety seeds bold with typical yellow color
Guj. Methi-1	Main Spices Research Station (SDAU), Jagudan, Gujarat	1864	The first variety from Gujarat
RMt-143	SKN College of Agriculture (RAJAU), Jobner, Rajasthan	1650	Moderately resistant to powdery mildew
RMt-305	SKN College of Agriculture (RAJAU), Jobner, Rajathan	1800	First determinate type
Hisar Suvarna	Department of Vegetable Crops (CCS HAU), Hisar, Haryana	1900	A quick growing dual purpose, medium maturity variety suitable for cultivation throughout the country, moderately resistant to powdery mildew resistant
Hisar Madhavi	Department of Vegetable Crops (CCS HAU), Hisar, Haryana	2000	Moderately resistant to downy mildew, quick growing dual purpose medium maturity variety

Contd...

Table 4. Contd...

Hisar Mukta	Department of Vegetable Crops (CCS HAU), Hisar, Haryana	2000	A quick growing seed type, medium maturity, moderately resistant to powdery mildew
Pant Ragini	GBPUAT, Pantnagar, Uttaranchal	1500	A dual purpose type
RMt-351	SKN College of Agriculture (RAJAU), Jobner, Rajathan	1840	High yield; bold attracive seed
Gujarat Methi-2	Main Spices Research Station (SDAU) Gujarat	1920	High yield; bolder lustrous grain of uniform in size; medium maturity (120 days)
NRCSS-AM 1	National Research Centre for Seed Spices (NRCSS), Ajmer, Rajasthan	2070	Exclusively for leaf purpose gives a yield of 76 q/ha from 3 cuttings; seeds less bitterness with bold in large pods of 8 mm size; crop duration - 137 days
NRCSS-AM 2	National Research Centre for Seed Spices (NRCSS), Ajmer, Rajasthan	1810	Exclusively for leaf purpose gives a yield of 72 q/ha from 3 cuttings; crop duration - 138 days
Pusa Early Bunching	Indian Agricultural Research Institute (IARI), PUSA, New Delhi	1200-1500	Suitable for taking 2-3 leaf cuttings; high yielder with bold seeds; crop duration - 156 days
Pusa Kasuri (Other kasuri type local varieties: Kasuri methi, Nagauri Methi, Marwari methi and Champa methi	Indian Agricultural Research Institute (IARI), PUSA, New Delhi	600-700	Exclusively for greens with a yield of about 80 q/ha; heavy yielder of green leaves with special fragrance; rich in vitamin C

Source: Parthasarathy *et al.*, 2005; Anandaraj *et al.*, 2006; NRCSS, Ajmer

3.2 Climate

Fenugreek performs well in province of low to moderate rainfall but cannot withstand heavy rainfall. Continuous moist and cloudy weather invites insect-pests and a number of diseases. The crop requires cool climate for its better growth and is capable of tolerating 10-15°C frost and freezing weather for a short period. A cool growing season without extremes of temperature is favorable for best development. Dry weather during crop maturity is essential for harvesting better seed yield. Having wider adaptability, the crop can be grown successfully both in tropical and temperate regions up to an altitude of 2000 m above mean sea level.

Table 5. Resistant/tolerant sources identified against biotic and abiotic stresses in fenugreek under AICRPS

Resistant/ tolerant source	Reaction	Developed/released by
CO.1	Tolerant to root rot	Coimbatore (TNAU)
CO.2	Field tolerant to *Rhizoctonia* root rot	Coimbatore (TNAU)
Rajendra Kranti	Suitable for intercropping; field tolerant to *Cercospora* leaf spot	Dholi (RAU)
Lam Sel.1	Field tolerant to major pests and diseases	Lam, Guntur (ANGRAU)
RMt-303	Less susceptible to powdery mildew	Jobner (RAJAU)
RMt-305	Resistant to powdery mildew and root-knot nematode),	Jobner (RAJAU)
RMt-351	Resistant to powdery mildew and root-knot nematode, suitable for crop rotation with wheat and barley	Jobner (RAJAU)
Pant Ragini	Resistant to downy mildew and root rot	Pantnagar (GBPUAT)
Hisar Suvarna	Resistant to *Cerospora* leaf spot	Hisar (CCS HAU)
Hisar Madhavi	Suitable for irrigated as well as rainfed conditions; resistant to powdery mildew	Hisar (CCS HAU)
Hisar Mukta	Resistant to downy mildew	Hisar (CCS HAU)
Guj. Methi-2	Tolerant to powdery mildew, resistant to root-knot nematode & downy mildew	Jagudan (GAU)

4.0 SEED MATERIAL

4.1 Source of seed material and selection

For organic production, seeds should be carefully selected from certified organic farms or from own seed plot which is raised and maintained organically. However, for initiating organic production, the seeds may be collected from the plots which are adopting conventional system of cultivation. For raising organic crop, the seed is the prime requirement in deciding yield and quality. Hence, chemically untreated seeds of high yielding varieties with tolerance to pests and diseases and responsive to low organic inputs may be used. Seeds should not be treated with any chemical fungicides or pesticides. However, it is always beneficial to adopt indigenous practices for seed treatment, wherever possible (Anon., 1998). The seeds may be treated with *Trichoderma* @ 10^9/kg of seed to prevent incidence of seedling rot in the nursery (www.indianspices.com).

4.2 Seed treatment

The seeds germinate in 7-10 days, depending on sowing depth and soil temperature, but under a condition of favorable temperature and adequate moisture, the seeds germinate in 5-7 days in common fenugreek whereas kasuri fenugreek germinates late by another 3-5 days (Malhotra and Vashishtha 2007). Germination is usually delayed, if the temperature is too low, soil moisture is inadequate, sowing depth is more than 2 cm, or there is any obstacle in way of newly sprouts emerging out of the soil, i.e., formation of hard soil crust due to unexpected rain follows sowing may impede early emergence.

Fenugreek, being legume crop, fixes nitrogen about 283 kg per hectare per year from the atmosphere into the soil. The role of *Rhizobium* in fenugreek production is well established, and inoculation of seed before sowing has proved beneficial in getting higher seed yield (Chaudhary, 1999; Yadav and Kumawat, 2003) since field inoculation with *Rhizobium melilotii* bacterium is very useful in absorbing nitrogen from the atmosphere and fixing it into the soil. Seeds should also be inoculated with *Rhizobium* culture prior to sowing, especially when the crop is sown in new field as the symbiotic nitrogen fixation takes place more effectively if an interrelationship is established between a particular legume and a specific *Rhizobium* strain (Dart *et al.*, 1976). Crop growth and yield improvement in legume crops due to *Rhizobium* inoculation have been reported by Bajpai *et al.* (1974) and Wange (1990).

4.3 Seed rate

The quantity of seed required for the sowing of unit area depends on the purpose for which the crop is sown. In order to sow one-hectare area, about 20-25 kg seed of *common fenugreek* and 10-12 kg of *kasuri fenugreek* type is required. Randhawa *et al.* (1996) recommended a seed rate of 30 kg/ha for receiving good seed yield, however, Mavai *et al.* (2000) suggested a seed rate of 20 kg/ha for getting higher seed yield. The initial soil moisture should be adequate to ensure satisfactory germination.

5.0 PLANTING AND AFTER CARE

A minimum of two years is required as conversion period for organic cultivation and undergoing the process of certification by the accredited agency. The general guidelines for selection of farm land, buffer zone, certification etc. are required to be followed (Anon., 2001; Malhotra and Vashishtha, 2004). The procurement management which includes operationalization of activities for production of organic seed spices crops and creating conducive environment for required kind of quality and quantity of produce by way of adopting pre and post-harvest practices and deciding marketing system suitably. It shall ultimately be guideline to the grower/ processor/ exporter in planning the production/supply of organic seed spices for export as detailed in coriander chapter in this treatise are applicable to this crop also (Malhotra and Vashishtha, 2007).

5.1 Land preparation and planting

For organic cultivation, the land should be given minimum tillage by plough or harrowed two times. The stubbles of previous crop are removed and the clots are broken. Then, the land is leveled by planking. Well-rotten cattle manure or compost @ 4-5 t/ha has to be incorporated into the soil at the time of sowing (Anon., 1998).

5.2 Season

Fenugreek, being cool season crop, is sown in the month of October to November in northern plains, whereas, in hilly tracts, it is sown from March to May, depending on altitude. In areas with mild climate, fenugreek for fresh greens may be grown round the year except extreme hot months of summer and rainy season, hence, in southern states of India, particularly Karnataka, Andhra Pradesh and Tamil Nadu, fenugreek is sown twice, once in *rabi* (September-December) and again in *kharif* season (June-July). *Kasuri* type varieties require preferably extra cool weather for longer duration, is thus raised more successfully in northern states during winters than the southern states of India. For a continuous supply of leaves, crop can be sown at weekly intervals. For seed crop, the sowing time is so adjusted that the pod development and seed maturity phase may coincide with a dry and rain free period.

5.3 Sowing time

According to Singh and Nand (1984), the optimum time for the sowing of fenugreek is between mid of October to the mid of November under Uttar Pradesh conditions, however, sowing on 15th October (Rathore and Manohar,

1989) and 7th November (Bhati, 1988) gave good growth and the highest seed yield, which was adversely affected with the delay in sowing in Rajasthan. In Haryana, Baswana and Pandita (1989) compared different dates of sowing and found that the crop sown on 5th October gave maximum plant growth and seed yield. In another report from Hisar maximum seed yield was obtained from the crop sown on December, 1 followed by November, 16 and December, 16 sown crops (Sheoran *et al.*, 2000), whereas Lal *et al.* (2003) reported to obtain maximum returns with three cuttings by sowing the crop by 15th November, in Haryana. While, Kaswan *et al.* (1994) observed maximum returns from variety Pusa Early Bunching when grown with two leaf cuttings taken at 60 and 80 days after sowing. In Punjab, the highest seed yield was obtained when the crop was sown between last week of October and first week of November (Randhawa *et al.*, 1996). In the mid hills of Himachal Pardesh, Korla and Saini (2003) found 21st September as the best time for getting the higher leafy yields. Under conditions of Karnataka, Halesh *et al.* (1998) found that crop sown between 1st and 15th July recorded higher seed yield of 16-23 q/ha during *kharif* season.

5.4 Method of sowing and spacing

Fenugreek is sexually propagated crop and sown either in lines or by broadcasting seeds in well-prepared flat seedbeds and raking the bed surface prudently, however, sowing in lines is comparatively better than the broadcasting method since it facilitates the intercultural operations, like hoeing and weeding, and also makes the cutting of leaves very easy. Sometimes, line sowing becomes too essential when moisture in soil is scarce, while broadcasting method of sowing seeds is successful only when the field is having sufficient soil moisture at the time of sowing. For fresh greens, seeds are usually sown sparsely in 15-20 cm apart shallow furrows and for seed crop 25-30 cm apart and later the plants are thinned to a spacing of 10-15 cm within lines. Although the depth of sowing seeds depends on soil type and soil moisture at the time of sowing, but being small size, the seeds of *common fenugreek* are usually sown at a depth of 2-3 cm and *kasuri fenugreek* at 1.0-1.5 cm.

Seed yield of fenugreek increased when sown at closer row spacing of 15 cm as reported by Singh and Nand (1984), whereas Bhati (1988) reported that plant height and primary branches in fenugreek were not influenced by row spacing but closer row spacing of 20 and 30 cm gave significantly higher seed yield than 40 cm row spacing. Baswana and Pandita (1989) reported that the row spacing failed to influence the plant height, while fenugreek grown in rows 20 cm apart gave higher yield than 30 or 40 cm row spacing under Haryana agro-climatic conditions. Pandita and Randhawa (1996) obtained highest fenugreek seed yield at a plant spacing of 20 cm and without taking greens cutting. Randhawa *et al.* (1996) recommended a row-to-row spacing of 22.5 cm for receiving good seed yield. In a study of leaf cutting management in Gujarat, Mehta *et al.* (1995) obtained the highest seed yield (16.8 q/ha) with no greens cutting followed by 10.42 q/ha seed yield with one leaf cutting at row to row spacing of 20 cm. The spacings recommended for various states is given in Table 6.

Table 6. Spacing adopted in various states for fenugreek

State	Spacing
Rajasthan	25 cm x 10 cm
Gujarat	15 x 10 and 30 x 10 cm
Bihar	20 x 10 cm
Uttar Pradesh	30 x 10 cm
Tamil Nadu	15 x 10 cm or 22.5 x 10 cm
Haryana	30 x 10 cm
Andhra Pradesh	-

Source: Parthasarathy *et al.* (2005)

5.5 Inter-cultivation

Fenugreek takes 6 to 8 days for germination. The slow growing nature of the crop in the initial stage of establishment makes conducive atmosphere for weeds growth. The growth of the crop becomes vigorous after four to five weeks, during which the crop does not allow weeds to compete with it. Hence, weeding at the early stage of the crop is very much essential to combat the weeds growth (Saini, 2005). In fenugreek cv. Pusa Early Bunching, Tripathi and Singh (1993) estimated 14.2% yield reduction due to weed competition during first 30 days after sowing (DAS) and 69% during entire cropping season. Weeds emerging 30 DAS caused a very little reduction in yield and there was no significant gain in increasing weed free period beyond 30 days, thus, first 30 days were identified as a critical period with respect to crop weed competition in fenugreek.

Two hoeings and weeding are good enough to keep the crop well aerated and weed-free. First hoeing and weeding should be done at the time of thinning about 25-30 DAS and the second at 50-60 DAS. Thinning is practiced in broadcasted crop to maintain the plant population by adopting the plant distance of about 8-10 cm.

5.6 Thinning

Usually, thinning is not practiced in fenugreek crop exclusively grown for fresh green leaves. However, for seed crop it is necessary to maintain proper spacing between lines and plants within rows. In line-sown seed crop, the plants 20-30 days after sowing as and when they are large enough to handle, are thinned to a spacing of 10-15 cm within the row. Similarly, the plants should be thinned to a suitable spacing to avoid competition in broadcasted crop. Retaining one to two plants per hill the surplus plants are uprooted by pulling out. This operation is soon followed by top dressing with a small quantity of nitrogen and a light irrigation to set the disturbed soil and seedling roots. In crop grown for fresh green leaves the young shoots are nipped off 10-30 days after sowing at a height of 4-5 cm from the ground level to increase the number of branches.

5.7 Cropping system

Fenugreek can be grown as mixed or intercrop provided all the other crops are grown following organic farming methods. Being leguminous crops it fits well as component crop for most of intercropping systems involving fennel, coriander, ajowan, dill and winter vegetable crops. Likewise rotation crops should also be grown following organic farming methods.

Some of the other cropping systems suggested are (Mehta and Malhotra, 2007):

(i) Maize/Pearl millet- Fenugreek
(ii) Sesame- Fenugreek
(iii) Sesame- Fenugreek -Summer Maize

5.8 Irrigation

Fenugreek, being primarily an irrigated crop, requires light irrigation at frequent interval for its quick growth but can also be cultivated under rain fed conditions in certain parts of the country. Usually the crop is sown when the field is having plenty of soil moisture so irrigation is not applied unless the seedlings attain 2-4 true leaves, however, if the initial moisture in the field at sowing is inadequate, a light irrigation should be applied very soon after sowing and should be followed by another light irrigation on third day to facilitate rapid and uniform germination. Subsequent irrigations are given at 7 to 10 days interval, depending on soil type, season, rainfall, and other temporary weather conditions. In general, frequent and light irrigations are essential for quick foliage growth and as a thumb rule, each cutting should be followed by a light irrigation. The early growth period and seed setting are the critical stages for irrigation requirement (Malhotra and Vashishtha, 2007). For fenugreek it is believed that irrigation can double the yield of a crop than non-irrigated one, while a good variety can increase the yield at least by 30%. Too much irrigation is also as harmful as the scarcity of moisture, since excessive moisture in any form and at any stage increases the incidence of root rot and powdery mildew. Maintenance of optimal moisture is essential to prevent blossom and pod drop. Under Chhattisgarh conditions, Lakpale *et al.* (2004) obtained maximum seed yield from fenugreek when five irrigations were applied at IW/CPE ratio of 0.8, however, under AICRP application of seven irrigations at 1.0 IW/CPE ratio was found suitable as a general recommendation for getting higher production. Sprinkler irrigation with IW/CPE ratio of 0.75 was found optimum for Gujarat (Ravindran *et al.*, 2001). Care should be taken to avoid water stress at pod and seed development stages.

6.0 NUTRIENT MANAGEMENT

6.1 Nutrient Uptake

Fenugreek is commonly grown on residual fertility of the previous crops. The crop removes 34.9 kg N, 8.2 kg P_2O_5, 13.9 kg K_2O, 4.3 kg Ca, 1.8 kg Mg, 0.17 kg Fe, 0.01 kg Mn, 0.04 kg Zn and 0.01 kg Cu from soil to produce one tonne seed yield (Peter *et al.*, 2000). The highest Fe, Mn and Zn uptake by fenugreek were recorded for plants grown on alluvial and the lowest on calcareous soil (El-Koumey *et al.*, 2000). For organic production, the required quantity of nutrients should be supplied through organic source with restricted use of mineral/inorganic fertilizers. At the time of field preparation about 10-15 tonnes of FYM / ha (Sharma and Bhati, 1984) or compost should be incorporated into the soil. Application of FYM (15 t/ha) had a beneficial effect on the enhancement of vegetative growth and resulted in higher dry matter production of fenugreek (Khiriya *et al.*, 2001). Application of one tonne per hectare of neem cake has also proved beneficial. It is essential to know the nutrient requirement of fenugreek. A basal dose of 40 kg N, 40 kg P_2O_5 per hectare should be applied in the soil at the time of sowing. If soil fertility is high, the dose of nitrogen maybe reduced. Fertilizer application did not influence the diosgenin content at Slovenia (Zupancic *et al.*, 2001). The recommended fertilizer schedule for different states of India is given in Table 7.

6.2 Biofertilizers

Being leguminous crop, fenugreek is having advantage of fixing atmospheric nitrogen. It has been estimated to fix about 28 kg nitrogen per

Table 7. Fertilizer recommendation for fenugreek

State	Recommendation				
	FYM* (t ha^{-1})	N (kg ha^{-1})	P_2O_5 (kg ha^{-1})	K_2O (kg ha^{-1})	Remarks
Rajasthan	10.0	40.0	40.0	-	20 kg N & 40 kg P_2O_5 / ha at the time of sowing
Gujarat	10.0 -15.0	40.0	40.0	-	40 kg N, 25 kg P_2O_5 in two equal splits as basal and top
Bihar	10.0 -15.0	40.0	60.0	20.0	-
Uttar Pradesh	10.0 -15.0	60.0	50.0	10.0	-
Tamil Nadu	10.0 -15.0	50.0	25.0	40.0	-
Haryana	10.0 -15.0	-	40.0	-	-

* Applied at the time of field preparation
Source: Parthasarathy *et al.* (2005)

hectare (Saini, 2005). The microbial inoculation with *Azospirillum*, Azotobactor and *Rhizobium* has been reported to show suitable means for organic cultivation of fenugreek. The *Azospirillum* in combination with organic manure 10 t/ha yield high organic seed yield of 13.9 q/ha, which was 25% less than the yield obtained from combinations applied with chemical nitrogen with reduced incidence of root rot (8.4%) (Malhotra *et al.*, 2003). In addition, fenugreek responds well to the field inoculation of *Rhizobium melilotii* and produced higher yield with delayed maturity. *Rhizobium* inoculation increased the number of branches per plant, pod length, number of pods per plant, number of seeds per plant, seed and straw yield, test weight and yield (Shivran *et al.*, 1995; Chaudhary, 1999). Kumawat *et al.*, (1998) achieved higher yield by seed inoculation of *Rhizobium* and adopting 30 cm spacing with the application of 40 kg P_2O_5. Spraying suspension of N_2-fixing strain KUPOs of *Klebsielea* (isolated from leaves of rice) on foliage of fenugreek 4 times at weekly interval improved yield components (Mukhopadhyay and Sen, 1997).

Rhizobium inoculation along with 40 kg N and 40 kg P_2O_5 ha^{-1} increased the yield contributing traits, and seed yield by 60.0, 29.6 and 31.3%, respectively over control. Incorporation of *Tephrosia purpurea* (4 t/ha) increased the number of pods per plant by 14.8%. *Rhizobium* inoculation along with 20 or 40 kg N + 40 kg P_2O_5 ha^{-1} resulted in the highest number of branches per plant (7.6 and 8.0, respectively), number of pods per plant (25.7 and 26.3), and number of seeds per pod (20.3 and 21.0). *Rhizobium* inoculation 40 kg N + 40 kg P_2O_5 ha^{-1} increased the seed yield by 95.9% and the biological yield by 21.7% over the control. Seed and straw yields increased by 37.0 and 16.1%, respectively (Yadav and Kumawat, 2003)

Among the 31 fenugreek genotypes (9 of the sub. sp. *indica*, 9 of the sub. sp. *foenum-graecum* and 13 intersubspecific hybrids) analyzed, the *indica* genotypes significantly surpassed the *foenum-graecum* genotypes in fresh weight, seed yield, nodulation ability, resistance to root rot and duration of the inter-stage and growing periods. There were no observed differences between the subspecies leafiness and biochemical composition of seeds. The most promising genotypes yielded 1650-1938 g/m^2 of fresh weight, 98.2-116.1 g/m^2 of seeds and accumulated 1.45-1.64% diosgenin in their seeds. High positive correlations were found between fresh weight, seed yield, plant leafiness and height. Inoculation of Nakhichevanskaya Shambala with *Rhizobium meliloti* strains 851 and 852 increased its fresh weight by 94.5 and 93.6% and its seed yield by 83.1 and 60.5%, respectively. Significant correlation obtained for nodule number with the fresh weight and seed yield indicated the symbiotic activity, which may be useful for improving fenugreek productivity (Provorove *et al.*, 1996). Soil inoculation with *Rhizobium* strain ENRRI or TAL 380 increased fat, fibre and protein contents. Mo increased 1000 seed weight and protein content of the seeds. However, N had no significant effect on quality of the seeds. Hence, inoculation with compatible and effective *Rhizobium* strain could improve the seed composition and quality of fenugreek (Abdelgani *et al.*, 1999).

A field experiment was conducted at Egypt to evaluate the effect of biofertilization (*Rhizobium leguminosarum*), sheep dung, two levels of P-fertilization (15 and 30 kg P_2O_5/f. [1 feddan=0.42 ha]) on microbiological changes and the yield of fenugreek in saline calcareous soil (52% CaCO3) irrigated with saline groundwater (6400 ppm). The results indicated that bacterial population, aerobic cellulose decomposers, fungi, actinomycetes and asymbiotic nitrogen fixers, (*Azotobacter* and *Azospirillum*) increased with increasing rate of P application, using sheep dung and biofertilization. Meanwhile, the production of the crop was also increased by 237% over the control (El-Ghany *et al.*, 1999).

Fenugreek was inoculated with vesicular-arbuscular mycorrhiza (VAM; *Glomus, Gigaspora, Scutellospora*) fungi and grown in nutrient-deficient, sandy-loam soils amended with organic matter (compost made from *Albizzia* [Albizia] and *Populus* leaves) in India. Under field conditions, shoot and root dry weights and total uptake of P and N were significantly increased in all inoculated plants. About 60% VAM colonization and greater number of propagules were produced in fenugreek. The extent of green yield increase following VAM inoculation was greater in plots amended with leaf compost at a soil: compost ratio of 1:1 than in plots amended at a ratio of 2:1. High levels of VAM colonization and large numbers of infection propagules demonstrated the potential of this crop as substrates for inoculum production and VAM also increased crop yield (Gaur *et al.*, 2000).

7.0 DISEASES AND PEST MANAGEMENT

Coriander is prone to attack by some of the serious diseases and insect-pests. The plant protection measures under organic farming system should involve emphasis on crop management practices time of sowing, balanced nutrition, crop rotation, green manuring, etc. for reducing the incidence of diseases and pests, selection of resistant varieties and adoption of bio control measures. Some of the important diseases are discussed in detail here.

7.1 Diseases

7.1.1 Powdery mildew

The disease is reported to occur almost in all fenugreek growing areas of the country. The disease is caused by *Erysiphe polygoni* DC and *Leveillula taurica* (Lev.) Am. Mildew caused by *Erysiphe polygoni* is characterised by white flowery patches on both surfaces of leaf and aerial parts of the plants, where as pathogen, *Leveillula taurica* produces mycelial patches on the leaves. The pathogen is an obligate parasite which perennates through the cleistothecia present in the debris of the previous crop (Agrawal, 2001; Chattopadhyay and Maiti, 1990). These cleistothecia appear as black specks in the later stage. The disease reduces the yield as well as grain size besides affecting the quality. The disease development occurs in the favorable cloudy and humid weather (Singh *et al.*, 2001). The disease becomes serious at temperatures between 15°-25°C and at a relative humidity of 60-70%. This disease deteriorates the quality of leaves. The following measures can be adopted for the management disease:

- Restricted use of sulphur dusting (Anon., 1998)
- Early sowing of crop (October 20 and 30) instead of late sowing (November 20 and 30) (Sharma and Sharma, 1999)

- Growing resistant/tolerant varieties/lines like RMt-1, RMt-16, RMt-143, RMt-303, RMt-305, RMt-351, Hisar Madhavi, Gujarat Methi-2 and Lam selection -1 (Parthasarathy *et al.*, 2005; Anandaraj *et al.*, 2006).

7.1.2 *Downy mildew*

In fenugreek, downy mildew is caused by fungal pathogen, *Peronospora trigonellae* Gaum. This disease was first reported from Maharashtra (Uppal *et al.*, 1934) and later on from Punjab (Thind, 1942). The symptoms of the disease are yellow patches on the upper surface of the leaves and cottony grayish-violet mycelium on the corresponding lower surface. The disease is spread through oospores. On advanced stages, the leaves turn yellow and fall off, ultimately the growth of the plant is stunted (Agrawal, 2001; Singh *et al.*, 2001). The disease can be managed by adopting the following practices:

- Give hot water treatment to seed before sowing
- Cultivating resistant/tolerant varieties such as Pant Ragini, Hisar Mukta and Gujarat Methi - 2 (Parthasarathy *et al.*, 2005; Anandaraj *et al.*, 2006)

7.1.3 *Root rot*

This is a soil-borne disease. The causal organism of the disease is *Alternaria alternata* (Fr.) Kessler (Dwivedi *et al.*, 1982). It is a serious disease in Tamil Nadu and some parts of Rajasthan. The incidence of the disease varies from 20-40%. Young ones are more susceptible to this disease recording up to 75% rot at seedling stage. Severe incidence of disease was noticed in heavy waterlogged conditions. The disease leads to under development or complete destruction of finer roots (Agrawal, 2001) results in poor development of root system. In severe cases, the infected plants show stunted appearance and the plants are easily detached and pulled out from the soil (Chattopadhyay and Maiti, 1990). The following measures can be adopted for the management disease:

- Application of neem cake @ 150 kg/ha combined with seed pelleting with biocontrol agents like *Trichoderma viride, T. harzianum* @ 4g/kg of seeds (Parthasarathy *et al.*, 2005).
- Deep summer ploughing
- Adopting suitable crop rotation
- Use of resistant/tolerant varieties like CO1, RMt-1, RMt-143
- Use of bioinoculants *Azospirillum* or *Azotobactor* plays significant role in reducing incidence of root rot to 8.2% (Malhotra *et al.*, 2003).

7.1.4 Damping off

Damping off, also known as collar rot is caused by *Rhizoctonia solani*. It was reported to be serious disease in fenugreek growing tracts of Bangalore (Hiremath *et al.*, 1978). The pathogen causes lesions mostly confined to collar region. Water soaked lesions appear on the collar region of the plant results in damping off by rotting the epidermal or vertical tissues of the collar area. On advanced stage of the disease, the stem becomes constricted at the base and the plants collapse rapidly. The infected soil is the predisposal media for the spread of the disease. The sclerotia of the fungus are having the ability to survive in the soil for several months depending upon the temperature and moisture conditions of the soil. In the infected tissues, the mycelia grow in all directions and produce sclerotia. High humid and warm climatic conditions favour the disease development at faster rate and become destructive (Agrawal, 2001). The disease can be managed by adopting the following measures:

- Removal of diseased plants immediately to check the further spread of the disease
- Use of soil amendments like *neem* cake @ 10q/ha and farmyard manure @10-25t/ha is beneficial.
- Adoption of proper crop rotation and deep summer ploughing
- The antagonistic fungi *Trichoderma viride* and soil application of *neem* cake effectively control the disease
- Application of 1% Bordeaux mixture as prophylactic spray (Anon., 1998)

7.1.5 Leaf spot

This is a minor disease in India. The disease is caused by *Cercospora traversiana* Sacc. The disease is reported to occur commonly in India and also in other parts of the World (Lepik, 1959).

The pathogen is reported to be internally seed-borne (Lepik, 1959; Khare *et al.*, 1981; Richardson, 1990) and attacks all parts of the plant starting from leaves to seeds and also the seedlings. It produces purple discoloration on seed coat, followed by heavy yield loss (Rastogi *et al.*, 1998a). They also recorded an incidence of 0.5 - 81%. The first symptom usually appears on leaves as large spots during and after anthesis, when the disease spreads. Immediately the young pods get infection and spread to immature seeds and over winters on them (Agrawal, 2001). The mycelium is seen in all seed parts. Even though the inoculum does not affect the germination of seeds, it causes heavy post-emergence losses and transmits the inoculum from seed to seedlings (Rastogi *et al.*, 1998a). The following measures can be adopted for the management disease:

- Follow at least 2-year crop rotation in infested areas.
- Healthy seed must be used.
- Growing resistant/tolerant varieties such as Rajendra Kranti, Hisar Suvarna

7.1.6 Rust

The rust is caused by *Uromyces anthyllidis*. The yellow spots bearing aecia in clusters are earlier symptoms. The disease is observed in the form of minute brown pustules on both the surfaces of leaves. This disease is of minor importance. Under cool and wet climate in northern India entire plant may be killed.

The disease can be managed by adopting good cultural practices and field sanitary measures.

7.1.7 Fenugreek mosaic wilt

This is a virus disease. Symptoms include vein clearing, severe mottling, curling at the edges, leaf size reduction, stunting of the plant and ultimately wilting. Young developing leaves are completely distorted, filiform and malformed. The disease causes severe losses on the part of leaf yield and quality.

The disease can be checked by the removal of all the infected plants and weeds and by the use disease-free seeds.

7.1.8 Soil-borne Mycoflora

In fenugreek, mycoflora like *Alternaria alternata* (an ecto- and endophytically seed borne fungi) (Jain and Jain, 1995) and *Aspergillus flavus, A. fumigatus, A. niger* and *Pencillium* spp. (mycotoxigenic fungi) were found associated with seeds. In fenugreek seeds having brown discoloration, Rastogi *et al.* (1998b) recorded an incidence of 2.0 -99.0% by the fungus, *A. alternata* in seed lots. The fungus expressed both symptom less and symptomatic infection. Mostly the infection was confined to seed coat and occasionally to endosperm and embryo. Severe infection resulted in partial or complete disintegration of mucilaginous endosperm. The fungus was observed to be pathogenic causing blight symptoms in seedlings raised from artificially inoculated seeds.

7.2 Pests

Fenugreek is attacked by several insect-pests such as sucking pests including mites, leaf eating caterpillars, etc. Among them, sucking pests cause considerable damage to the crop, which is detailed below:

7.2.1 Sucking pests

Fenugreek is commonly attacked by aphids (*Acrythosiphon pisum, Aphis craccivora, Hydaphis coriandri* and *Myzus persicae*) and occasionally by mites (*Petrobia latens*). Aphids suck sap from the tender parts and flowers. The infected plants turn yellow and give shriveling appearance of grains, ultimately reduction in yield and quality aspects (Singhania *et al.*, 2006), while mites cause severe infestation on young leaves (Singh *et al.*, 2001). The following measures can be adopted to check the pests:

- Regular monitoring and surveillance in the field
- Clip off the heavily infested plant parts.
- Spray the crop with pressurized water.
- Application of fish oil rosin soap or neem seed kernel extract (NSKE-3%), neem oil or tobacco decoction (0.05%) (Ravindran *et al.*, 2001)
- Utilization biocontrol agents such as *Beauveria*, *Metarhizium* and predators and parasites such as *Trichogramma*, lady bird beetle, *Chrysoperla*, spiders, etc. (Jangir and Rathore, 2007)
- Use of pheromone traps, sticky traps, etc.
- Use of tolerant/resistant varieties like Rajendra Kranti and Lam selection-1

7.2.2 Leaf eating caterpillar

This pest (*Spilarcitia oblique* and *Spodoptera litura* F.) attacks the crop and damages the leaf and pod. The eggs are laid in clusters and young larvae gregariously feed on leaves. They scrab out the green matter from the leaves and gives the appearance of papery white structure. The last instar larvae feed voraciously causing defoliation of plants and thereby causing considerable losses in yield and quality of greens. The pest can be managed by adopting the following measures:

- Destroy the egg mass and gregarious young larvae
- Hand picking and killing of large larvae

- Spray *neem* seed kernel-extract (4%) in the early stage of larvae.
- Spray Ha NPV @ 250 to 400 LE/ha during evening hours (Jasvir Singh and Tamil Selvan, 2004).

7.2.3 Pod borer

The pest (*Helicoverpa armigera* Hubner) is found to feed on leaves, flowers and pods. Eggs are laid on young leaves, which are later damaged by young larvae. The first instar larvae bore the fruits and damage them, causing 10-90% losses, if left unprotected. Control measures are same as suggested for leaf eating caterpillar.

8.0 HARVESTING AND POST HARVEST HANDLING

8.1 Harvesting and yield

The common fenugreek becomes ready for cutting fresh green leaves and young shoots in about 20 days after sowing while *kasuri* type fenugreek is ready in 25-30 days after sowing and subsequent cuttings may be taken at an interval of 15-20 days. The crop when grown for dual-purposes after taking one cutting, which does not affect the seed yield, is left for seed production. The crop after harvest is bunched and marketed. The cutting is usually done with sharp knife by leaving stubs 3-4 cm above ground level and after taking 4-5 cuttings the crop grown exclusively for green leaves is uprooted. Common fenugreek can be harvested by clipping the young plants from the base and the clipped plants are allowed to grow further and their tops are nipped periodically until flowering. If fenugreek is harvested late, its leaves develop a bitter taste.

Depending upon variety and season of growing the seed crop takes about 80-165 days from sowing to harvesting. The entire plant is either pulled out or cut from the base with sickle when 70% of the pods turn yellow, and made into small bundles for drying them in sun. Seeds are separated by beating with stick and winnowing. The seed is cleaned and sun dried before packing in Jute bags containing polyethylene lining.

The fenugreek yield depends upon the variety used, location and season of growing. Usually, the cultivars yielding higher grain yield tend to produce low fresh green leaf yield. Under irrigated conditions, the common type fenugreek varieties normally give a fresh green leaf and seed yield of 70-80 and 15-20 q/ha, respectively, and *kasuri* type 80-100 q green leaves per hectare. However, higher yield can further be expected under good management practices. The *kasuri* type fenugreek varieties give higher yield due to more cuttings in comparison to common fenugreek.

8.2 Post harvest management

Keeping in view the requirement of targeted market, the post harvest operations are advised to follow suitably for green leaves as vegetable, dehydrated leaves and seeds. In case of fresh greens, the edible portion is the tender leaves and stem, which are cut to length of about 7-10 cm. After harvesting, the yellow, diseased and damaged leaves are trimmed off and thereby healthy and disease free leaves are tied into small bunches for the convenience in handling and marketing. Since the dried leaves can be stored for one year for further use in off-season, the fenugreek leaves are sun-dried or dehydrated in a suitable dehydrator. However, during dehydration, chlorophyll is oxidized and ascorbic acid is lost, so as to retain green color of leaves, blanching in boiling water (80°C) for 3-6 minutes can be practiced (Chaudhary and Rao, 1979). The palatability of fenugreek leaves improve with steaming and with addition of seasoning rather than with boiling or frying (Rowlands *et al* 1966). The average vitamin C value in fenugreek leaves is 43.10 mg/100g and after boiling in water or steaming and then frying, the leaves lose 10.8% and 7.4% of this vitamin, respectively.

Green leaves are very perishable in nature; therefore, they are marketed soon after harvesting. However, well-dried leaves can be stored for about 10-12 months. The leaves can be stored only for about 24 hours after harvesting under ambient conditions, however, in cold stores at 0°C temperature and 90-95% relative humidity the storage period can be extended up to10 days. Seeds are stored in gunny bags lined with polythene film. Vacuum gravity separator is used for cleaning fenugreek seeds. The properly cleaned fenugreek seeds are stored with an initial moisture level of 7-8% and at an equilibrium relative humidity of 40%. Fenugreek seeds well packed is stored in ventilated dry and cool place under ordinary conditions till sowing of next season crop.

8.3 Quality specifications for export

8.3.1 Whole seed

The quality specification for fenugreek can be classified into three categories *viz.* commercial requirements, cleanliness and health specifications and have been detailed by Tamil Selvan (2007) and Malhotra (2005).

Commercial specifications :

These requirements vary from country to country and crop to crop. It depends on many factors like customer needs, acceptability and country specifications. The commercial specifications of seed spices crops are colour, appearance, taste, pungency, texture, shape, volatile oil and packaging. The general commercial and general requirements for fenugreek seeds are :

Yellow seeded or green seeded, small or large seed size, free flowing, volatile oil (min. 0.02%)

Cleanliness specifications :

The permit table cleanliness specifications as per ASTA, ESA and ISO are given below :

ASTA cleanliness specifications for fenugreek

Whole insect dead (by count)	Excreta mammalian (mg/lb)	Excreta other (mg/lb)	Mould (% wt)	Insect defiled / infected (%/wt)	Extraneous foreign matter (%/wt)
4	3	10	1	1	0.5

Other quality standards of fenugreek as per ESA and ISO

Ash level % w/w (min.)	Acid insoluble ash % w/w (max.)	Moisture content % (max)	Volatile oil % (Min.)
7	1.5	12	0.3

Health specifications

The micro contaminants such as pesticide residues, microbial counts, aflatoxins, heavy metals etc. are the health specifications. These contaminants make the food unsafe. The importing countries have limited the maximum level of these contaminants and limits are given here:

Maximum Residue Level = 0.05 to 0.2%

Microbial counts = 1 X 10^2 / g to 1 X 10^6 /g (*Salmonella, E. coli*, yeast, moulds)

Aflatoxin = 5 ppb to 10 ppb (Max.) Aflatoxin, mycotoxins

Heavy metals = Arsenic (5 mg/kg), copper (20 mg/kg) lead (10 mg/kg), Zinc (50 mg/kg)

8.3.2 Essential oil

Fenugreek seeds contain little quantity of essential oil of 0.02% and the specifications for fenugreek oil are not available. The seeds should contain 0.78- 1.9% diosgenin.

8.3.3 Fenugreek oleoresin

The fenugreek oleoresin should be prepared with recommended organic solvents, followed by the subsequent removal of the solvent as per specifications of importing countries.

8.3.4 Fenugreek powder

The fenugreek powdered hygienically through permitted devices, should at least 95 % of the ground product pass through a U.S. Standard No. 25 sieve (Farrell, 1999).

8.4 Marketing

The major trading centers for fenugreek seeds are: Unjha and Visnagar in Gujarat, Indore, Khjner in Madhya Pradesh, Malerkotla in Punjab, Varanasi, Hapur in Uttar Pradesh and Jaipur, Jodhpur, Ramganj, Bhavani, Baran in Rajsathan.

REFERENCES

Abdelgani, M.E., Flsheikh, E.A.E. and Mukhtar, N.O. (1999) The effect of *Rhizobium* inoculation and chemical fertilization on seed quality of fenugreek. *Food Chemistry*, 64, 289-293.

Agrawal, K. (2001) "Disease status of seed spices". In: Agrawal S., Sastry, E.V.D. and Sharma, R.K. (eds.) *Seed Spices -Production, Quality, Export.* Pointer publishers, Jaipur. pp. 59-76.

Anandaraj, M., Johny A. Kallupurackal and Shiva, K.N. (2006) New spices varieties identified. *Spice India*, 19(8), 25-28.

Anonymous (1998) *Guidelines for production of organic spices in India.* Spices Board, Cochin, Kerala.

Anonymous (2001) *Guidelines for production of organic spices in India.* Spices Board, Cochin, pp. 64-67.

Bajpai, P.D., Lehri, L.K. and Pathak, A.N. (1974) "Effect of seed Inoculation with *Rhizobium* strains on the yield of leguminous crops". In: *Proc. Indian Nat. Sci. Acad.* 40, 571-575.

Baswana, K.S. and Pandita, M.L. (1989) Effect of time of sowing and row spacing on seed yield of fenugreek. *Seed Res.* 17, 109-112.

Bhati, D.S. (1988) Fenugreek (*Trigonella foenum-graecum* L.) response to sowing dates and spacing. Indian J. Agric. Sci. 56, 437-439.

Chattopadhyay, S.B. and Maiti, S. (1990) *Diseases of Betel vine and Spices*, Indian Council of Agricultural Research: New Delhi.

Chaudhary, G.R. (1999) Response of fenugreek to N, P and *Rhizobium* inoculation. *Indian J. Agronomy*, 44, 424-426.

Chaudhary, A.T. and Rao, B.Y. (1979) Potential of chlorophyll and ascorbic acid in dried fenugreek. *Indian Food Packer*, 33, 35.

Dart, P.J., Day, J.M., Islam, R. and Doberiener, J. (1976) "Symbiosis in tropical grain legumes: Some effects of temperature and composition of rooting medium". In: Nutman, P.S. (ed.), *Symbiotic Nitrogen Fixation in Plants* IBP, Cambridge Univ. Press., 7, 361-384.

DASD (2007) *Arecanut and Spices Database*. Directorate of Arecanut and Spices Development, Calicut, Kerala, pp.94-95.

Duke, J.A. (2003) *Handbook of Legumes of World Economic Importance*. Scientific Publisher (India), Jodhpur, pp. 268-271.

Dwivedi, D.K.,Shukla, D.N. and Bhargava, S.N. (1982) Two new root rot diseases of spices. *Curr. Sci*. 51, 243-244.

El-Ghany, B.F.A., Salem, M.O. and El-Sibaie, M.A.F. (1999) Effect of bacterial inoculation, P and organic fertilization on microbiological properties, growth and production of fenugreek plant under saline calcareous soil conditions. *Desert Institute Bulletin*, Egypt, 47, 1-13.

El-Koumey, B.Y., Bayoumi, N.A. and Tantawy, M.F. (2000) Relationship between Fe, Mn and Zn-uptake by plant and their extractable amounts from soils. *Egyptian J. Soil Sci*. 40, 109-129.

Farrell, K.T. (1999) *Spices, condiments and seasonings*. Aspen Pub. Inc. Gaitherberg, Maryland, USA, pp. 78-81.

Gaur, A., Adholeya, A. and Mukerji, K.G. (2000) On-farm production of VAM inoculum and vegetable crops in marginal soil amended with organic matter. *Tropical Agric*. 77, 21-26.

Habib, A.F., Swami, R.T. and Mansirkai, S.W. (1971) Salt tolerance in methi at germination stage. *Curr. Sci*. 40, 67.

Halesh, D.P., Gowda, C., Farooqui, A.A., and Vasundhara, M. (1998) "Response of fenugreek to nitrogen and phosphorus". In: *Proc. Nat. Sem. Water and Nutrient Mgt. for Sustainable Production and Quality of Spices*, held at IISR, Madikeri from 5-6, October 1997, pp. 76-79.

Hiremath, P.C., Ponnappa, K.M., Janardhan, A. and Sundaresh, H.N. (1978) Chemical control of collar rot of fenugreek. *Pesticides*, 12, 30-31.

Jain, M.P. and Jain, S.C. (1995). Seed borne fungi of seed spices. *J. Spices and Aromatic Crops*, 4, 78-89.

Jangir, R.P. and Rathore, D.M.S. (2007) "Present status, problems and prospects of organic farming of seed spices in Rajasthan". In: Tamil Selvan, M., Homey Cheriyan,

Sadanandan, A.,K. and Hameed Khan, H. (eds.) *Souvenir of National Seminar on Organic Spices and Aromatic Crops,* 1-2 February 2007, Calicut, Kerala, pp. 77- 82.

Jasvir Singh and Tamil Selvan, M. (2004) Integrated pest management package for fennel. *Indian J. Arecanut, Spices and Medicinal Plants,* 6, 89-92.

Kaswan, M.P., Singh, G.R., Thakral, K.K. and Mangal, J.L. (1994) Effect of nitrogen and leaf cutting on seed quality and economics of fenugreek cultivars. *Haryana Agrl. Univ. J. Res.* 24, 123-125.

Khare, M.N., Agarwal, S.C. and Sharma, N.D. (1981) A seed borne fungus of *Trigonella foenum – graecum* . *Indian Phytopath.* 34, 71-77.

Khiriya, K.D., Sheoran, R.S. and Singh, B.P. (2001) Growth analysis of fenugreek (*Trigonella foenum-graecum* L.) under various levels of farmyard manure and phosphorus. *J. Spices and Aromatic Crops,* 10, 105:110.

Korla, B.N. and Saini, A. (2003) Effect of dates of sowing and cutting on seed yield of fenugreek. *Haryana J. Hort. Sci.* 32(1&2), 120-122.

Kumawat, S.K., Agrawal, H.R. and Pareek, R.G. (1998) Effect of row spacing and phosphorus levels on fenugreek with and without *Rhizobium* inoculation. *Annals of Biology-Ludhiana.,* 14 (2) : 161-164.

Lakpale, R., Shrivastave, G.K. and Tripathi, R.S. (2004) "Effect of irrigation schedules on yield and economics of different seed spices crops under Chhattisgarh condition". In: *Proc. Nat. Sem. New Perspectives in Spices, Medicinal, and Aromatic Plants,* held at ICAR Research Complex Goa, 27-29 November 2003, pp. 329-331.

Lal Sant, Rana, M.K. and Partap, P.S. (2003) Economic returns from various fenugreek genotypes as influenced by date of sowing and greens-cutting levels. *Haryana J. Hort. Sci.* 32(1&2), 117-120.

Lepik, E.E. (1959) World distribution of *Cercospora* traversiana. *F.A.O. Pl. Prot. Bull.* 8, 19-21.

Malhotra, S.K. (2005) "Quality Standard of Seed Spices". In: *Proc. National Seminar on Spices,* held at KNK College of Horticulture, Mansor, from 20-21 March, 2005.

Malhotra, S.K. and Vashishtha, B.B. (2004) Organic production of seed spices in India-Status and Strategy. *Seed Spices Newsletter,* 4(1), 1-4.

Malhotra, S.K. and Vashishtha, B.B. (2007) *Production technology for seed spices crops.* National Research Centre for Seed Spices, Ajmer, Rajasthan, pp. 1-80.

Mavai, D., Lal, S., Baswana, K.S., Singh, A. and Singh, N. (2000) Response of fenugreek to seed rate nitrogen and phosphorus fertilizer. *Haryana J. Hort. Sci.* 29, 244-246.

Mehta, R.S. and Malhotra, S.K. (2007) "Seed spices based cropping system". In: Malhotra, S.K. and Vashishtha, B.B. (eds.) *Production, Development, Quality and Export of Seed Spices.* National Research Centre for Seed Spices, Ajmer, Rajasthan, pp. 181-189.

Mehta, K.G., Patel, P.G., Kachhadia, B.T. and Edison, S. (1995) Varietal response to leaf cutting management on seed and green leaf yields of fenugreek (*Trigonella foenum-graecum* L.). *Indian Cocoa, Arecanut and Spices J.* 19(3), 91-93.

Mukhopadhyay, D. and Sen, S.P. (1997) Augmentation of growth variables and yield components of plants yielding spices by foliar application of diazotrophic bacteria. *Indian J. Agric. Res.* 31, 1-9.

Pandita, V.K. and Randhawa, K.S. (1996) Row spacing and leaf cutting in relation to seed production of fenugreek (*Trigonella corniculata* L. cv. Pusa Kasuri). Seed Res. 22, 127-129.

Parthasarathy, V.A., Shiva, K.N. and Johny, A. K. (2005) "Research and Development in Major Seed Spices under AICRPS". In: Malhotra, S.K. (ed.) *A Course Manual for Winter School on Advances in Seed Spices Production*, 2-22 September 2005, National Research Centre for Seed Spices, Ajmer (Raj.), pp. 255-271.

Peter, K.V., Srinivasan, V. and Hamza, S. (2000) Nutrient management in spices. *Fertilizer News*, 45, 13-18,21-25 & 27-28.

Provorove, N.A.. Soskov, Y.D., Lutova, L.A. Sokolova, O.A. and Bairamov, S.S. (1996) Investigation of fenugreek genotypes for fresh weight, seed productivity, symbiotic activity, callus formation and accumulation of steroids. *Euphytica*, 88, 129-138.

Ramaswamy, N. and Ananthan, M. (1999) Cultivating fenugreek scientifically. *Spices India*, 12(12), 18.

Randhawa, G.S., Gill, B.S., Saini, S.S. and Singh, P. (1996) Agronomic technology for production of fenugreek seed. *J. Herbs, Spices & Medicinal Plants*, 4(3), 43-49.

Rastogi, A., Singh, T. and Singh, D. (1998a) Incidence, location and transmission of natural seed-borne inoculum of *Cercospora* traversiana in fenugreek. *J. Mycol. Pl. Pathol.* 28, 218-222.

Rastogi, A., Singh, T. and Singh, D. (1998b) Colonisation and phytopathological effects of *Alternaria alternata* in seeds of fenugreek. *J. Phytological Res.* 11, 137-139.

Rathore, P.S. and Manohar, S.S. (1989) Effect of date of sowing, level of nitrogen and phosphorus on growth and yield of fenugreek. *Madras Agric.* J. 76, 647-648.

Ravindran, P.N., Kallupurackal, J.A. and Sivaraman, K. (2001) All India Coordinated Research Project on Spices-Status of Seed Spices Research. *Indian J. Arecanut, Spices & Medicinal Plants*, 3(3), 97-110.

Richardson, M.J. (1990) An Annotated List of Seed-Borne Diseases (IV edn.). *Proc. Int. Seed Test. Ass :* Wageningen, Netherlands.

Rowlands, R., Arondekar, U.S. and Devdas, R.P. (1966) Effect of different methods of cooking on the ascorbic acid and papatability of fenugreek leaves (*Trigonella foenum-graecum* L.). *J. Nutr. Dietet.* 3(1), 15.

Saini, G.S. (2005) Scientific method to grow fenugreek. *Spice India*, 18, 39-42.

Shah, A.S., and Bhide, S.V. (1985) Content of nitrate and secondary amines in vegetables, legumes and marine foods. *J.Food Sci.Technol.* 22(4), 266.

Shanmugavelu, K.G., Kumar, N. and Peter, K.V. (2002) *Production Techniques of Spices and Plantation Crops*. Agrobios (India), Jodhpur, Rajasthan, pp. 109-130.

Sharma, R.K. and Bhati, D.S. (1984) Prabha-A promising variety of fenugreek for Rajasthan and Gujarat. *Indian Cocoa, Arecanut and Spices J*. 8, 14-15.

Sharma and Sharma, S. (1999) Effect of sowing date on the powdery mildew of fenugreek. *J. Mycology and Plant Pathology*, 29, 144-145.

Sheoran, R.S., Sharma, H.C., Panuu R.K. and Niwas Ram, (2000) Influence of sowing time and phosphorus on phenology, thermal requirement and yield of fenugreek (*Trigonella foenum-graecum* L.) genotypes. *J. Spices and Aromatic Crops*, 9 (1), 43-46.

Shivran, P.L., Choudhary, G.R., Gora, D.R. and Shivran, A.C. (1995) Response of fenugreek to nitrogen and *Rhizobium* inoculations. *Indian J. Agron*. 40, 720-721.

Singh, J.N. and Nand, K. (1984) Effect of dates of sowing row spacing and nitrogen levels on the grain yield of fenugreek. *Indian Drugs*, 21(2), 536-540.

Singh, D., Chaudhary, G.R., Shyam Singh and Singhania, D.L. (2001) Production technology for seed spices. In: Agrawal S., Sastry, E.V.D. and Sharma, R.K. (eds.) *Seed Spices -Production, Quality, Export*. Pointer publishers, Jaipur. pp.89-114.

Singhania, D.L., Raje, R.S., Dhirendra Singh and Rajput, S.S. (2006) "Fenugreek". In: Ravindran, P.N., Nirmal Babu, K., Shiva, K.N. and Johny A. Kallupurackal (eds.) *Advances in Spices Research – History and Achievements of Spices Research in India Since Independence*. Published by Agrobios, Jodhpur (Raj.), pp. 757-783.

Som, M.G. and Maiti, T.K. (1986) "Fenugreek". In: Bose, T.K. and Som, M.G. (eds.), *Vegetable Crops in India*. Naya Prakash, Calcutta.

Tamil Selvan, T. (2007) "Quality management in seed spices". In: Malhotra, S.K. and Vashishtha, B.B. (eds.) *Production, Development, Quality and Export of Seed Spices*. National Research Centre for Seed Spices, Ajmer, Rajasthan, pp. 23-38.

Thangaraj, T and Vijayakumaran, M. (2000) "Fenugreek". In: Thamburaj, S. and Singh, N. (eds.) *Vegetables, tuber crops and spices*. Indian Council of Agricultural Research, New Delhi, pp. 364-371.

Thind, K.S. (1942) The genus *Peronospora* in the Punjab. *J. Indian Bot. Sci.*, 21, 197-215.

Tripathi, S.S. and Singh, G. (1993) "Crop weed competition studies in fenugreek". In: *Proceedings of an Indian Society of Weed Science International symposium*, Hissar 18-20 November 1993, Indian Society for Weed Science, vol. II, pp. 41-43.

Uppal, B.N., Patel, M.K. and Kamat, M.N. (1934) The fungi of Bombay. Bull. Dept. Ld. Rec. Agric., Bombay.

Varshney, I.P. and Sood, A.R. (1971) Sapogenins from *Trigonella foenum-graecum* stem and leaves and *T. corniculata* leaves and flowers. *Indian J. Appl. Chem.* 34(5), 208.

Wange, S.S. (1990) "Effects of seed Inoculation with *Rhizobium* strains on the yield of different pea varieties". In: *Proc. 8th South. Reg. Conf. Microbial Inoculants,* held at College of Agriculture, Pune-5 on 5-6 February, pp. 27-28.

www. Indianspices.com (Assessed on 3rd January 2007).

Yadav, G.L. and Kumawat, P.D. (2003) Effect of organic, inorganic fertilizer and *Rhizobium* inoculation on the yield and yield attributes of fenugreek (*Trigonella foenum-graecum* L.). *Haryana J. Hort. Sci.* 32, 147-148.

Zupancic, A., Baricevic, D., Umek, A. and Kristl, A. (2001) The impact of fertilizing on fenugreek yield (*Trigonella foenum-graecum* L.) and diosgenin content in the plant drug. Rostlinna, Vyroba, 47, 218-224.

17
Fennel

S.K. Malhotra

1.0 INTRODUCTION

The fennel (*Foeniculum vulgare* Mill.) is important seed spices belonging to family *Apiaceae,* and a native of Southern Europe and Mediterranean area. Later it spread to the far East and far north in Europe. To the ancient Greeks, fennel represented success and so it was called "Marathon" in reference to the battle, fought in fennel field where the Greeks defeated the Persians in 490 B.C. Fennel was also a symbol of success to the Romans. During the thirteenth century in England, fennel was considered as a royal spice and was served to kings with fruits, bread and pickled fish seasoned with fennel seeds. Fennel was even described by Longfellow in his poem *The Goblet of Life.* According to early Sanskrit writings, fennel has been described as *Madhurika* and was known as far back as 2000 BC for its cultivation in India.

1.1 Area and Distribution

It is cultivated on a large scale in Romania, Russia, Germany, France, Italy, India, Argentina and USA. Besides, the other growing countries are Bulgaria, China, Denmark, Egypt, Syria, Morroco and Japan. The major fennel producing states in India are Gujarat, Rajasthan and Uttar Pradesh. In many other states Punjab, Tamil Nadu, Bihar, Karnataka, Maharashtra and J & K grown at small scale are Khera, Mehasana, Banaskantha, Baroda, Ahmedabad in Gujarat; Tonk, Sirohi, Jodhpur, Ajmer, Udaipur and Pali in Rajasthan; Belgaum and Dharwad in Karnataka; Jallandhar, Amritsar, Ludhiana, Ferozepur in Punjab are the important districts of fennel cultivation. During the year 2004-2005, 51038 tonnes of fennel seed was produced from 39730 ha area, which contributes to 9.0 % and 4.7% of total seed spices production and area in the country. The current productivity of fennel is 1245 $kgha^{-1}$. The foreign exchange of Rs. 64 crores was earned by exporting 22750 metric tonnes of seeds during 2005-2006.

1.2 Chemical composition and Uses

1.2.1 Chemical composition

The physicochemical properties of fennel depend upon the species, variety and climatic conditions of the area where the crop is raised. The fennel seed has been reported to contain 1 to 6% volatile oil and 10 to 20% fixed oil. Besides volatile oil, seeds contain about 9.5% protein, 10% fat and 13.4% minerals (calcium, potassium, magnesium, phosphorus, sodium) and vitamin A and niacin. The bitter fennel contains 50% trans anethole, 10-20% fenchone (contributes to bitter flavour), 10-30% limonene, 3-11% α-phellandrene, 12-16% α- pinene, with α-thujene, β- pinene, estragol (methyl chavicol), myrcene, and 1,8-cineole. The sweeter variety has 50-80% anethole, little (5%) or no fenchone, slightly higher levels of limonene with estragole, safrole and pinene (Raghavan, 2007) (Table 1). Singh *et al.* (1990) reported 20 compounds in fennel essential oil and 18 constituted 96.04% of the total oil and the major components were anethole (68%), limonene (11%), fenchone (3.7%) and few others. Turkish bitter fennel is rich in methyl chavicol (47.09%), besides other constituents such as limonene (29.07%), fenchone (13.43%), α- terpinene

Table 1. Proximate composition of coriander seed (100 g edible portion)

Water	8.8 g	Magnesium	385 mg
Food energy	345 kcal	Phosphorus	487 mg
Protein	15.8 g	Potassium	1694 mg
Fat	14.9 g	Sodium	88 mg
Total carbohydrate	52.3 g	Zinc	4 mg
Fibre	15.7 g	Niacin	6 mg
Ash	8.2 g	Vitamin A	135 IU
Calcium	1196 mg	Other vitamins	insignificant
Iron	19 mg		

Source: Farrell (1999)

(2.5%), fenchyl acetate (exo) (1.95%) and cis- β- ocimene (Ozcan and Akgul, 2001). The presence of high anethole (60-70%) makes the fennel oil highly odorous and of good quality. He further reported that sweet or Roman fennel contains anethole, phellandrene and d-limonene and very little or no fenchone. The fixed oil primarily contains petroselinic acid (60%), oleic acid (22%), linoleic acid (14%) and palmitic acid (4%) (Singh *et al.*, 1990).

1.2.2 Uses

Fennel dried fruits are aromatic and mostly used for flavouring soups, meat dishes and sauses, bread rolls, pastries and confectionery. The seeds also have use for flavouring liquors and in the preparations of various type of pickles. The leaves are used in fish sauce and for garnishing, white leaf stalks are used both as vegetable and salad. Chinese, Indians and Egyptians have used it as a condiment. The thickened leaf stalks of Florence fennel are branched and used as vegetable. In India, fennel seeds are used as masticatory or for chewing alone or in pans. Farrell (1999) reported use of fennel seeds in English-style soups, German breads, Polish borscht, spaghetti, salads, sweet pickles, and vegetable dishes. The essential oil is used as flavouring agent in various food items, pickles, perfumes, soaps, liquors, cough drops and liquorice candy.

Fennel seeds, leaves and roots all possess medicinal uses. The leaves are used as diuretic, fruits are aromatic, stimulant and carminative, and the roots as purgative. Fennel volatile oil relieves abdominal air, stomach pain and bloating, stimulates appetite, diuretic and anti-inflammatory. It is used as ingredient in gripe water for curing infantile colic and flatulence. It checks gripping in purgatives and is considered a good vermicide against hookworm. The residue left after extraction of essential oil from the fruits is used as a feed for cattle. It contains 14-22% proteins and 12-18.5% fat (Guenther, 1950). Singh (2007) have cited that infusion prepared from fennel seeds, is taken as a gargle for sore throat and as a mild expectorant and increased breast milk of mothers. The fennel seeds aid to weight loss and longevity. Chinese, Indians and Egyptians use fennel seeds as a remedy for snakebites and scorpion stings. The essential oil of fennel has been reported to possess oestrogen stimulation properties which makes it useful for treating postmenopausal syndrome (PMS) and regulates menstrual period. Owing to antimicrobial properties it is used in cosmetic creams, body lotions and moisturizers. They are official in the pharmacopeia of many of the countries and are considered useful in treating diseases of the chest, spleen and kidney.

1.3 Prospects for Organic Fennel

The value-added products of fennel such as fennel seed, fennel powder, essential oil and oleoresins were exported in 1876, 143.4, 1.1 and 5.4 metric tonnes and earned Rs. 3956, 84.7, 11.5 and 46.5 lakh, respectively (Table 2). The major importing countries of Indian fennel are Canada, Germany, Hong Kong, Indonesia, Israel, Mauritius, Netherlands, US, Saudi Arabia, Bahrain, South Africa, Singapore, UK, etc. The leading states recently came up are Egypt and China.

Table 2. Export of raw and value-added products of fennel

Item	Value (Rs. or lakhs)			Quantity (in ton)		
	2003-4	2004-5	2005-6	2003-4	2004-5	2005-6
Seed	2153	2445	2956	4923	6956	1876
Powder	58.5	83.9	84.2	83.2	143.5	143.4
Seed oil	0.2	8.8	11.3	0.05	0.40	1.1
Oleoresin	9.0	36.6	46.5	0.9	4.3	5.4

2.0 BOTANY AND VARIETIES

2.1 Botany

The plants are annual or biennial or perennial aromatic herb. It grows to a height of 150-180 cm and has erect, shiny bright green stems, much branched bearing compound leaves. Leaves are stalked with long petioles; blades are much divided of the lower leaves tripinnate, dark dull green, upper leaves smaller and much less divided. Small yellow flowers with fine petals are borne on umbellets in umbels, bisexual, petals five, yellow, ovary inferior, bicarpellary, fruits oblong with persistent style. Umbels long stalked, large, slender, curved, often unequal rays, quite without involucers. Flowers are protandous and hence highly cross pollinated. The seeds (botanically fruits) are 6-8 mm long, straight or slightly curved, oval in shape, mesocarp is 5 ridged and contain agreeable, aromatic and sweet aroma.

2.2 Varieties

Varieties selected should have adaptation to the soil and climatic conditions and preferably resistance/tolerance to pests and diseases. Efforts should be made to use organic seed of recommended variety for organic farming. When certified organic seed of fennel is not available, initially conventional seeds can be used (Anon., 2001). The description of some of the important cultivated varieties recommended for different states are given in Table 3.

Table 3. Recommended varieties of fennel for different states

Variety	Salient characters
Rajasthan	
NRCSS-AF-1	• Suitable for growing both for early sowing and as *rabi* crop. • The seeds are attractive, bold, medium sized, fragrant with volatile oil content upto 1.6%. • It is from late maturity group and produces average yield of 19.5 qha^{-1} during *rabi* season and 25.1 qha^{-1} when grown as an early crop. • The variety has tolerance to *Ramularia* and *Alternaria* blight.
RF-101	• It bears large umbel with long and bold grain. • It matures in 150-160 days and gives an average yield of 15.5 qha^{-1}.
RF-125	• The plants are early, short stature with compact umbels Long bold grains with17.3 quintal of seed yield $hectare^{-1}$.
RF-143	• Very suitable for the loamy and black cotton soils. • Medium duration type, medium tall, having an average yield of 1200 kg ha^{-1} and oil content 1.87%.
Gujarat	
PF-35	• The plants are tall and spreading with medium sized hairless and green seeds. • It is moderately tolerant to sugary disease, leaf spot and leaf blight. • It matures in 215-225 days and gives an average yield of 12.89 q ha^{-1}.
Gujarat Fennel-1	• The plants are tall, bushy type suited for *kharif*, *rabi* and drought.

Contd...

Table 3. Contd...

	• The plants bear large sized umbel and more number of umblets. The seeds are bold, long dark green colour and contain 1.6% essential oil.
Gujarat Fennel 2 (Gf-2)	• It is tall and long duration variety. • Average yield is 18 q ha^{-1}.
Gujarat Fennel 11	• This variety is suitable for rabi season growing under irrigated conditions. • It produces an yield of 24 q ha^{-1}.
Tamil Nadu	
Co-1	• The plants are medium with profuse branching. • It matures in 210-220 days and gives an average yield of 5.67 q ha^{-1}. • It is suited for intercropping, saline and alkaline soil and hilly area, drought prone and water logged area.
Haryana	
Hisar Swarup	• Has a yield potential of 17 q ha^{-1}. • Plant has an average height of 132 cm, spreading, late in maturity. • Resistant to lodging and shattering of grains. • Fruits are long bold, oil content 1.6%.

3.0 CLIMATE AND SOIL

3.1 Climate

Fennel is a cool season crop, mainly grown during winter season in north India. A dry and cold weather favours higher seed production. A temperature of 15-20^0 C is the optimum for growth and high temperature results premature flowering and very low seed yield. Crop is susceptible to frost injury at flowering stage. The optimum temperature for seed germination is 20-29^{0}C. During maturity, crop requires warm climate. High winds at the time of maturity can cause shattering and very hot winds at flowering reduce seed setting. Cloudy weather and high humidity at the time of flowering and seed formation increases the chances of aphid infestation and attack of *Ramularia* blight disease (Malhotra and Vashishtha, 2007a).

3.2 Soil

Fennel thrives well on drained loamy soil rich in plant nutrients. However, heavy soils rich in organic matter with better water holding capacity are more desirable than light soils. A neutral to slightly alkaline soil in the pH range of 6.5-8.0 is highly preferable. Fennel is low accumulator of heavy metals and thus can be grown on polluted soil. Fennel is poor host of root-knot nematodes, thus in soils having problem of root-knot nematode, fennel can give reasonably good yield. Large amount of Na removed by stover indicate the capacity of plant to endure moderate degree of sodic tolerance. Soils containing lime with proper drainage have also been considered suitable (Malhotra and Vashishtha, 2007a).

4.0 SEED MATERIAL

Fennel is propagated through seeds. Seeds may be sown directly in the field or seedlings raised in nursery first and then planted in main field. About 2.5 kg seed is required for raising seedling in nursery for one hectare area but for main season crop 8-10 kg seed is required for direct seeding in one hectare land. Seed soaking for 8-12 hours before sowing helps to improve germination.

4.1 Raising nursery

For raising seedlings in the nursery, select fertile land having good drainage. About 2.5 kg seed is sown in a nursery area of 1000 m^{-2} for raising seedling for one hectare land. In the well prepared seedbeds, seeds are broadcasted or sown in rows. The beds are covered with straw or stalk mulch to protect against high temperature and strong sunlight and are irrigated. The seeds germinate within 10 to 12 days. The mulch is removed after germination. Irrigation is given as per requirement. The seedlings are ready for transplanting after about six weeks (Malhotra and Vashishtha, 2007a).

5.0 PLANTING AND AFTER CARE

5.1 Pre-requisite for Organic Production

A minimum of two years is required as conversion period for organic cultivation and undergoing the process of certification by the accredited agency. The general guidelines for selection of farm land, buffer zone, certification etc. are required to be followed (Anon., 2001; Malhotra and Vashishtha, 2004). The procurement management which includes operationalization of activities for production of organic seed spices crops and creating conducive environment for required kind of quality and quantity of produce by way of adopting pre- and post-harvest practices and deciding marketing system suitably. It shall ultimately be guidelines to the grower/

processor/ exporter in planning the production/supply of organic seed spices for export, as detailed in coriander chapter for this crop also (Malhotra and Vashishtha, 2007b).

The cultural requirements including practices as per procurement management for organic fennel production are described here:

5.2 Preparation of land

Field preparation should start with early rain. The first ploughing should be done by soil turning plough and afterward 2-3 ploughing should be done by *desi* plough or harrow. The ploughed field should be made fine and levelled by planking. There should be good initial moisture level in soil for better germination. In order to manage problem of soil pests summer ploughing should be done.

5.3 Sowing time

Sowing of fennel may be done in both late *kharif* (early) and *rabi* season. For early fennel, seed is sown is nursery in the month of July and about 45 days old seedlings should be planted in the main field. Last week of September to first week of October time is ideal for direct sowing of the fennel as main season crop. Delayed sowing results in low yield levels. Sowing of fennel on 21 October in Punjab has been reported to give highest yield (Randhawa *et al.,* 1981). Bhati and Agarwal (1986) reported first week of October as the best time of sowing for Rajasthan. It was reported that oil content in seed and herb did not vary when fennel was sown on different dates. However the earliest sown crop produces the highest oil yield (Randhawa *et al.,* 1981; El-Gengaihi and Abdallah, 1979). The high oil content was recorded when it was harvested at full bloom stage.

5.3 Sowing method

In a popular farmer's practice, seeds are broadcasted in the beds after the seedlings of chillies or cole crops are established in the field preferably at the time of first weeding and hoeing. Yadav *et al.* (1997) reported that sowing in the third week of October gave the highest seed yield and biological yield in Haryana.

5.3.1 Sowing/ Transplanting

The transplanting of seedling should be done at line-to-line spacing of 45 to 60 cm and plant-to-plant distance of 20 cm in the evening hours otherwise plants may die in hot sunny days. Irrigation should be given immediately after transplanting. For direct seeding the sowing is done either by broadcasting method or in line. The line sowing of seed is more convenient to facilitate proper weeding and other intercultural operations. Randhawa and Mahey (1985) found that closer row spacing of 30 cm gave significantly higher seed yield than wider row spacing of 45, 60, 75 and 90 cm. Ahmed *et al.* (1988) reported that plant spacing of 45 x 30 cm gave the higher seed yield. El-Gengaihi and Abdallah, 1979 recorded highest oil yield at 30 cm as compared to 20 and 40 cm spacing.

5.4 Irrigation

Being long duration crop, fennel requires more irrigations than other seed spices. Depending upon temperature and soil type, 6-10 irrigations during growth period are needed. The irrigation interval should be kept 10-20 days. In direct seeded crop, light irrigation is given immediately after sowing, if initial moisture is low. The flowering and seed formation stages are critical stages for irrigation. Water stress during flowering and seed formation stages may adversely affect the seed formation and the seed yield. Seed yield and biological yields were highest with 3 irrigations, followed by 2 irrigations (one at crown stage and second at main development stage. Patel *et al.* (2007) revealed that higher seed yield was recorded, when irrigation applied at IW/CPE ratio of 0.90.

5.5 Intercultural operation

The crop growth is slow initially and has to face severe weed crop competitions. Moreover, wider row spacing and frequent use of water favours growth of weeds. To keep the field weed free, about 3 to 4 weedings and hoeings are required. Earthing up of plants at the time of last hoeing and weeding proves beneficial and prevents the plants from lodging.

5.6 Protection against frost

Fennel crop is susceptible to frost at flowering and seed formation stage. Irrigate the crop to protect it from frost. During the period of frost danger, an arrangement of smoke should also be made in the field. Wind breaks should be planted set to protect crop from northern cool winds.

5.7 Cropping system

Fennel can be grown as mixed or intercrop provided all the other crops are grown following organic farming methods. It is desirable to include leguminous crops or other suitable crops exhibiting lesser competition. Since fennel is long duration crop and have slow initial growth. The slow growth rate of fennel for first 50 to 60 days has been very conducive for successful intercropping with any of the vegetable crop such as chilli, potato, spinach, carrot, cole crops, coriander and fenugreek. According to Singh *et al.* (1999), sugarcane + fennel proved a good intercrop combination on the basis of net return and benefit cost ratio. Inter cropping with garlic was also reported (Maurya *et al.*, 1997) to have least depressive effect on fennel and reduced the aphid infestation. Likewise rotation crops should also be grown adopting organic farming methods.

Bimbraw (2006) suggested that fennel can be included in the following crop rotations under the Punjab conditions:

(i) Paddy - Fennel
(ii) Maize - Fennel
(iii) Moong - Fennel
(iv) Mash - Fennel
(v) *Kharif* fodder - Fennel

Some of the other cropping systems suggested for seed spices growing area can be (Mehta and Malhotra, 2007):

(i) Green gram/black gram- Fennel
(ii) Cluster bean/cowpea-Fennel

6.0 NUTRIENT MANAGEMENT

6.1 Manuring

Manurial requirement varies from region to region due to type and fertility status of the soil. Therefore organic manure should be applied based on soil testing report. For good production of fennel about 15 t/ha well decomposed FYM should be applied at least 3-4 weeks before sowing. The biofertilizer, *Azospirillum* applied both as seed inoculation (500 g/ha seed) and soil application (2.5 kg ha^{-1} along with application of 10 t well rotten farm yard manure proved best and produced highest fennel yield under organic manurial studies (Malhotra *et al.*, 2004).

7.0 DISEASES AND PEST MANAGEMENT

Fennel is prone to attack by some of the serious diseases and insect-pests. The plant protection measures under organic farming system should involve emphasis on crop management practices *viz.*, time of sowing, balanced nutrition, crop rotation, green manuring etc. for reducing the incidence of diseases and pests, selection of resistant varieties and adoption of bio-control measures.

7.1 Diseases

7.1.1 *Alternaria* blight

First symptoms are seen on lower buds which turn yellow or brownish. The infected plants show dark brown blighty appearance on the leaves and stem resulting in withering off affected plants.

7.1.2 *Ramularia* blight

Infection causes minute angular brown necrotic areas on the lower surface of the older leaves. Later on whitish eruptions are seen. Infection spread on all plant parts and seed setting is adversely affected.

Control measures for *Alternaria* and *Ramularia* blight

- Avoiding excessive irrigation and high humidity
- Use disease-free seed
- Adopting field sanitation measures
- Using resistant variety

7.1.3 *Powdery mildew*

White powdery mass appears on the leaves and twigs of the plant in the initial stages. At later stage, whole plant is covered with the whitish powder.

Control

- Spray can be repeated after 15 days if required
- The application of Sulphur dust, wettable sulphur suspension (0.6%) and sulphur spray (0.1%) and the restricted use of sulphur is permitted as described in the guidelines of Spices Board (Anon., 2001)
- Foliar spray of onion extract (5%) is effective in reducing the severity.
- Grow resistant/tolerant varieties.
- Uprooting of all affected plants, followed by burning or burying.
- Early planting manages the disease well.

7.1.4 *Nematode (Meloidogune javanica)*

Fennel, being poor host to root knot nematode, general precautionary measures can take of care the crop as given below:

Control

- Using disease-free seed
- Adopting field sanitation measures
- Using resistant variety
- Application of neem cake 1000 kg per hectare
- Cover nursery beds with 25 ì LDPE film for solarization and rabbing with castor husk @ 7 kg per m^2 (Patel *et al.*, 2005)

7.2 Pests

7.2.1 Pest monitoring

Pest monitoring is an important part for organic cultivation of fennel and should be done once in seven days to work out economic threshold level (ETL). It will help to keep close watch on the appearance of insect-pests, diseases and bio-control fauna (Singh and Tamil Selvan, 2004).

For sucking pests: Population should be counted on three leaves i.e. top, middle and lower per plant.

For cut worms and defoliators: Per cent damage assessment can be made by counting total number of plants and affected plants

Management strategy

1. Set up yellow pan/sticky traps for monitoring of aphids/whitefly @ 10 yellow pans/sticky traps per hectare. Locally available empty yellow palmoline tins coated with grease/Vaseline/castor oil on outer surface may also be used.
2. Collect the larvae of cutworm and leaf defoliators

7.2.2 Aphids

Aphids attack fennel crop at flowering stage and cause serious damage. Three types of aphids-Hyadaphis, Cavariella and Aphis have been reported to damage the fennel crop. Both nymphs and adults suck the sap from leaves, flowers and fruits.

Control

- Clip off the heavily infested plant parts.
- The botanical insecticide, nemarin (1%) and seed extract (1%) of neem (*Azadirachta indica),* Karanj (*Pongamia* sp.) have been reported to reduce aphid population by more than 50% within 7 days.

- Conserve predators Coccinellids, Chrysopids and Syrphids
- Spray NSKE (5%) or Azadirachtin (0.03%) @ 1 litter ha^{-1} (0.5%)
- Avoid excessive irrigation under cloudy and high dew conditions

7.2.3 Thrips

Both nymph and adults congregate in between the leaf sheath and stem causes drying of leaves. The infestation is commonly observed during March-April.

Control

As recommended for aphids

7.2.4 Midge

The midge is phytophagous and female insert its ovipositor through the fruit wall to lay eggs between the pericarp and ovule in the developing fruits in field conditions. Larvae feed upon the embryo and destroy than.

Control

- Avoid delay in sowing, October first week sowing exhibit less infestation (4%) than delayed sowing (32%) (Kashyap *et al.,* 1994)
- Use clean packaging material
- Maintaining suitable temperature and humidity during storage

8.0 HARVEST AND POST HARVEST HANDLING

8.1 Harvesting and yield

Time of harvesting depends upon the type of the products marketed. Mostly the crop is harvested before the fruits are fully ripe. For green *saunf* used for chewing purpose, umbels are harvested about 30-40 days after flowering when these are still green and have attained half-length size. Since not all the fruits mature at a time, harvesting of umbels has to be done 4-5 times as and when they become ready. Umbels are plucked when fruits have green colour and fully mature. Harvested umbles should be dried in shade under well aerated conditions particularly for green fennel. Umbels should never be piled as it may deteriorate the quality (Singh and Malhotra, 2007). The dried umbel are separated and cleaned by winnowing for removing chaffs, dust and dirt. Under scientific management of the crop, a yield of 20-25 qha^{-1} can be harvested.

8.2 Processing

The fennel seeds are dried in partial shade in the sun and the moisture content of seed should be kept to nine per cent. High seed moisture content may lead to chances of storage contamination by fungus. The dried, cleaned and graded produce is packed in the standard sized packs/containers and appropriately labelled. The dried seed is filled in gunny bags lined with degradable environment-friendly plastic film. Waste generating packaging material should be avoided. Each bag is sealed and stored under clean, dry and ventilated place.

Processing methods of fennel should be based on mechanised, physical and biological processes. Care should be taken to maintain the vital quality of an organic ingredient throughout each step of its processing. Processing method is to be selected in such a way that it will limit the number and quantity of additives and processing aids. The mature dried seeds are distilled to obtain the essential oil. The mature dried seeds of fennel are distilled to obtain the essential oil. Hydro or steam distillation method is generally used for extraction of essential oil. On an average, the dried seeds yield 0.7-2.0% essential oil. The percentage of essential oil varies depending upon variety and type of fennel. The volatile oil is lowest in Indian fennel (0.7-1.2%) and highest in European fennel. The oleoresins prepared from fennel have good demand in the international market. The processed products include cumin essential oil, cumin powder, cumin oleoresin and fixed oil. Volatile oil should be kept in well sealed bottles or aluminium containers. The essential oil of fennel is used for scenting soap and as flavouring material for cakes. The by product *ark saunf* after extraction of essential oil possess good medicinal properties for curing indigestion problems at home level remedies.

8.3 Quality specifications for export

8.3.1 Whole seed

The quality specification for fennel can be classified into three categories *viz.*, commercial requirements, cleanliness and health specifications as detailed by Tamil Selvan (2007) and Malhotra (2005).

Commercial specifications :

These requirements vary from country to country and crop to crop. It depends on many factors like customer needs, acceptability and country specifications. The commercial specifications of seed spices crops are colour, appearance, taste, pungency, texture, shape, volatile oil and packaging. The general commercial and general requirements for fennel seed are : light green and light brownish green, bitter or sweet , small or large seeds, aromatic, free flowing, high volatile oil (min. 1 %- Germany, 3%- Netherlands, 2% - UK).

Cleanliness specifications :

The permit cleanliness specifications as per ASTA, ESA and ISO are given below :

ASTA cleanliness specifications for fennel

Whole insect dead (by count)	Excrets mammalian (mg/lb)	Excreta other (mg/lb)	Mould (% wt)	Insect defiled / infected (%/wt)	Extraneous foreign matter (%/wt)
2	2	2	2	1	0.5

Other quality standards of fennel as per ESA and ISO

Ash level % w/w (min.)	Acid insoluble ash % w/w (max.)	Moisture content % (max)	Volatile oil % (Min.)
9	2	12	1.5

Health specifications :

The micro contaminants such as pesticide residues, microbial counts, aflatoxins, heavy metals etc. are the health specifications. These contaminants make the food unsafe. The importing countries have limited the maximum level of these contaminants and limits are:

Maximum Residue Level = 0.05 to 0.2%

Microbial counts = 1 X 10^2 / g to 1 X 10^6 /g (*Salmonella, E. coli*, yeast, moulds)

Aflatoxin = 5 ppb to 10 ppb (Max.) Aflatoxin, mycotoxins

Heavy metals = Arsenic (5 mg/kg), copper (20 mg/kg) lead (10 mg/kg), Zinc (50 mg/kg)

8.3.2 Essential oil

The specifications for fennel oil are:

- Colourless or pale yellow
- Specific gravity (20° C, 0.965 to 0.977

- Optical rotation (20° C), +11 to+24
- Refractive index, 1.528 to 1.539
- Solubility (80% ethanol), 5-8 vols
- Congealing point, not below 5° and as high as 10° in good oils
- Anethol 50 to 80%(Singhal *et al.*, 1997)

8.3.3 Fennel oleoresin

The fennel oleoresin should be prepared with recommended organic solvents followed by the subsequent removal of the solvent as per specifications of importing countries.

8.3.4 Fennel powder

The fennel powdered hygienically through permitted devices, should at least 95 % of the ground product pass through a U.S. Standard No. 30 sieve (Farrell, 1999).

REFERENCES

Ahmed, A., Farooqui, A. A. and Bojappa, K. M. (1988) Effect of nutrient and spacings on growth, yield and essential oil content in fennel. *Indian Perfumer*. 32(4):301-305.

Anonymous, (2001) *Guidelines for production of organic spices in India.* Spice Board, Cochin, pp 64-67.

Bhati, D. S. and Aggarwal, H. R. (1986) Performance of fennel varieties under different sowing dates. *Haryana J.Agron*. 2:21-25.

Bimbraw, A. S. (2006) *Agrotechniques for umbelliferrous Medicinal and aromatic plants*. Inernational Book Distributing Co. Luknow. pp 96-137.

El-Gengaihi, S and Abdallah, N. (1979) The effect of date of sowing and plant spacing on the yield of seed and volatile oil of fennel. *Pharmazie*. 33(9): 605-606.

Farrell, K. T. (1999) *Spices, condiments and seasonings.* Aspen Pub. Inc. Gaitherberg, Maryland, USA. pp 83-87.

Guenther, E. (1950) Oil of fennel. In *The Essential Oil.* D.Van Nostrand Co. Inc. New York., 4:634.

Kashyap, R. K., Choudhary, O. P. and Thakraal, K. K. (1994) Influence of management practices on incidence of Systole albipennis Walker in fennel seeds. *J. Spices and Aromatic Crops*. 3(2):124-129.

Malhotra, S. K. (2005) Quality Standards and strategies for enhancing export of seed spices, Lead paper published in *Proc. National Seminar on Spices* held at KNK College of Horticulture, Mansor, from 20-21 March, 2005, pp 109-112.

Malhotra, S. K., Vashishtha, B. B. and Rao, V. V. A. (2004) Effect of *Azospirillum*, nitrogen sources and their combinations in fennel. In : *Organic farming in Horticulture*. Eds. Pathak,R.K. Kishun, R., Khan, R M and Ram,RA. Proc. National Symposium on Organic farming in Hort. for sustainable Production at CISH, Lucknow form 29-30 Aug.,2003 pp. 266-269.

Malhotra, S. K. and Vashishtha, B. B. (2004) Organic production of seed spices in India-Status and Strategy. *Seed Spices Newsletter*. 4(1):1-4.

Malhotra, S. K. and Vashishtha, B. B. (2007a) *Production technology for seed spices crops.* NRCSS, Ajmer, pp 1-80.

Malhotra, S. K. and Vashishtha, B. B. (2007b) Procurement management and marketing of seed spices. In : *Production, Development Quality and Export of Seed Spices*. Eds. Malhotra, S.K. and Vashishtha, B.B. NRCSS, Ajmer, Rajasthan, India, pp 234-235.

Maurya, K. R., Singh,V. K., Navin Kumar and Sing, A. B. (1997) Economics of parallel cropping in fennel. *J. Applied Biology*. 7(1-2):52-54.

Mehta, R. S. and Malhotra, S. K. (2007) Seed spices based cropping system. In : *Production, Development, Quality and Export of Seed Spices*. Eds. Malhotra, S.K. and Vashishtha, B.B. NRCSS, Ajmer, Rajasthan, India, pp 181-189.

Ozcan, M and Akgul, A. (2001) Chemical composition of the essential oil of bitter fennel. *J. Spices and Aromatic Crops* 10(1):49-50.

Patel, B. S., Joshi, M. K., Patel, J. C. and Patel, N. B. (2007) Influence of irrigation based on IW:CPE ratio and nitrogen levels on yield and water use efficiency of fennel. In : *Production, Development, Quality and Export of Seed Spices*. Eds. Malhotra, S.K. and Vashishtha, B.B. NRCSS, Ajmer, Rajasthan, India, pp 262-266.

Patel, S. K., Patel, H. V., Patel, A. D. and Patel, D. J. (2005) Integrated management of root knot nematode in fennel. *J. Spices and Aromatic Crops*. 14 (2): 152-154.

Raghavan, Sushila. (2007) *Handbook of Spices, Seasoning and Flavouring*. CRC Press, Boca Raton, Florida, USA, pp 101-103.

Randhawa, G. S., Bains, D. S. and Mahajan, V. P. (1981) Seed and oil yield of fennel as influenced by sowing time and levels of nitrogen. *J. Res. Punjab Agric. Univ*. 18(2):244-247.

Randhawa, G. S. and Mahey, R. K. (1985) Research and Development of essential oils in the Punjab. *Indian Perfumer*, 29(1-2):99-104.

Ravindran, P. N., Kallupurackal, J. A and Sivaraman, K. (2001) All India coordinated Research project on spices-Status of seed spices research. *Indian J. Arecanut, spices Medicinal Plant* 3(3): 97-110.

Singh, G., Upadhyay, R., Narayanan, C. S. and Padinkumari, K. P. (1990) Chemical investigation of the essential oil of fennel. *Indian Perfumer*, 34(1-4): 247-249

Singh, J. and Tamil Selvan, M. (2004) Integrated Pest Management package for fennel. *Indian J. Arecanut, Spices and Medicinal Plants*, 6(3):89-92.

Singh, R. and Malhotra, S. K. (2007) Harvesting and maturity indices in seed spices crops. In : *Production, Development, Quality and Export of Seed Spices*. Eds. Malhotra, S.K. and Vashishtha, B.B. NRCSS, Ajmer, Rajasthan, India, pp 195-200.

Singh, S. P. (2007) *Production Management of Spices*. Agri Hortica Publications, Junagarh, Gujarat. pp. 129-139.

Singh, S. N., Shukla, J. P., Agarwal, M. L. and Singh, G. P. (1999) Spices intercropping with autumn planted sugarcane. *Indian J. Agronomy*. 44(1):64-67.

Singhal, R. S., Kulkarni, P. R. and Rege, D. V. (1997) In : *Handbook of Indices of Food quality and Authenticity*, Woodhead Publishing Limited, England, pp. 386-456.

Tamil Selvan, M. (2007) Quality management in seed spices. In : *Production, Development, Quality and Export of seed spices*. Malhotra, S. K. and Vashishtha, B. B. (Eds.), NRCSS, Ajmer, pp 23-38.

Yadav, A. C., Batra, B. R., Malik, Y. S. and Sadanandan, A. K. (1997) Irrigation requirement of fennel (*Foeniculum vulgare* Mill.) : Proc. of National Seminar, Madikeri, Karnataka held from 5-6, Oct. 1997. pp 136 -140.

18

Cumin

B.B. Vashishtha and S.K. Malhotra

1.0 INTRODUCTION

The cumin (*Cuminum cyminum* Linn.) belonging to family *Apiaceace* is an important seed spice believed to be indigenous to northern Egypt, East Mediterranean region, Iran, Turkey and India (Amin, 2003; Raghavan, 2007). It is popularly known as safed jeera and is one of the most common spices of daily use in the Indian sub-continent. Cumin is well known to ancient civilization. It is one of the oldest spices known since biblical times, valued mostly for their digestive properties and for flaovouring bread and other dishes during the periods of ceremonial fasting. As early as 5000 BC, Egyptians preserved the bodies of their kings by mummifying them with cumin, anise and marjoram. Cumin is mentioned in Isaiah xxvi. 25 and 27, and Matthew xxii 23, and in the works of Hippocrates and Dioscorides. As cited by Amin (2003), Pliny described that Romans took ground seed medicinally with bread, water or wine and in the thirteenth and fourteenth centuries, it was much in use as a culinary spice. Excavation in the Indus valley civilization revealed that cumin had been used even before about 1000 BC, cumin seed is one of the ancient spices, mentioned a number of times in the old testament and once called the best of all condiments by the Roman naturalist, Pliny. The Middle East, North Africa, Mexico, China and India on the other hand continued to be loyal to cumin and made it an important secret of many of their classic dishes. It was used as a condiment in England, since 13th century and by 1419 cumin was one of the taxable imports in London.

1.1 Area and Production

It is mainly grown in India, Egypt, Turkey, Iran, Pakistan, Syria, and Italy. In India it is extensively cultivated in Rajasthan, Gujarat and in some parts of Madhya Pradesh and Uttar Pradesh as *rabi* crop. Cumin is one of the important cash crops for the Rajasthan and Gujarat farmers, hold prime position amongst seed spices and contribute more than 80% in total area and production in India. In Rajasthan, it is cultivated in Jalore, Barmer, Pali, Ajmer, Nagaur, Jaipur, Tonk and Bhilwara districts, whereas in Gujarat, the cultivation of cumin is concentrated in North Gujarat covering Mehsana, Banaskantha, Sabarkantha, Palanpur, Ahmedabad, Saurashtra and Bhuj. In India, the total acreage under this crop is about 3.68 lakh hectares with an annual production of 1.76 lakh tonnes of cumin seeds, which contributed 43 and 31 % of total seed spices area and production, respectively (Table 1). During 2005-06, India exported 13750 tonnes valued at Rs. 10190 lakh. The value-added products of cumin such as cumin powder, essential oil and oleoresins were exported in 10023, 1448, 5 and 11.8 metric tonnes and earned export value of Rs. 7313, 1248, 56 and 108 lakh, respectively (Table 2).

Table 1. Area, production and productivity of seed spices in India (2004-05)

Seed Spices	Area (ha)	Production (t)	Productivity($kgha^{-1}$)
Coriander	381710	259480	738
Cumin	367680	176070	479
Fennel	39730	51038	1285
Fenugreek	50445	62798	1245
Others	40715	15334	377
Total	850280	564720	664

Table 2. Export of raw and value cumin added products from cumin

	Value Rs lakhs			Quantity 1000kg		
Item	2003-4	2004-5	2005-6	2003-4	2004-5	2005-6
Seed	4708.1	9030.0	7313.2	6419.2	11992.0	10023.3
Powder	1008.0	848.2	1248.3	1184.5	1007.4	1448.5
Seed Oil	4.7	33.8	56.1	0.1	1.0	5.0
Oleoresin	85.6	118.8	107.6	9.7	12.4	11.8

1.2 Composition and Uses

1.2.1 Composition

Cumin seed contains 2-4.5% of essential oil depending upon age and regional variations (Table 3). The peculiar odour and flavour is principally due to presence of aldehydes. The major constituents of essential oil are cuminic aldehyde (33%), β-pinene (13%), terpinene (29.5%), ρ-cymene (8.5%), ρ-mentha-1,3-dien-7-al (5.6%), cuminyl alcohol (2.8%), β-farnesene (1.1%), β-bisabolene (0.9%), ρ-menth-3-en-7-ol (0.7%), α-pinene (0.5%), limonene (0.5%), myrcene (0.3%), β-phellandrene (0.3%), 1-8-cineole (0.2%) with smaller quantities of α-phellandrene, α-terpinene, *cis* and *trans* sabinene, myrtenol, α-terpineol and phellandral. It also contains 10% fixed oil with strong aromatic flavour.

The proximate composition of cumin seed is given bellow:

Table 3. Proximate composition of cumin seed (100 g edible portion)

Water	8.1 g	Phosphorus	499 mg
Food energy	375 kcal	Potassium	1788 mg
Protein	17.8 g	Sodium	168 mg
Fat	22.3 g	Zinc	5 mg
Total carbohydrate	44.2 g	Ascorbic acid	8 mg
Fibre	10.5 g	Thiamine	1 mg
Ash	7.6 g	Niacin	5 mg
Calcium	931 mg	Vitamin A	1270 IU
Iron	66 mg	Other vitamins	insignificant
Magnesium	366 mg		

Source: Farrell (1999)

1.2.2 Uses

Cumin as condiment is an essential ingredient in all mixed spices, curry powders and for flavouring soups, sausages, pickles etc. Cumin is also a popular ingredient in Mexican, Southeast Asian, Cuban and Tex-Mex cooking. Seeds are also used for seasoning bakery products such as bread and cakes. It is widely used in Indian, Iranian, Egyptian and Turkish dishes. Whole cumin is used to make various pickles in India, Pakistan and Iran. Cumin is a common flavour in confectionery, meat, sausage and bread manufacturing and as a preservative in food processing. The essential oil is used in perfumery, cosmetics and flavouring beverages.

Cumin is known to possess carminative, stimulant, stomachic, antimicrobial and astringent properties. Cumin is widely used as traditional medicine to treat flatulence, digestive disorders, diarrhoea, insomnia, colds, and fevers and in the treatment of wounds. In India, it is popularly used as a carminative in veterinary practice. Consumption of cumin seeds increase secretion of mother milk after birth of child. Cumin powdered seed mixed with honey, salt and butter is applied externally for scorpion bite. The by product *Ark Jeera* after extraction of essential oil possess good medicinal properties for curing indigestion problems at home level.

2.0 BOTANY AND VARIETIES

2.1 Botany

Cumin is a slender herbaceous annual plant growing to a height of 30-45 cm. The plant has branched stem having alternate, dark green, dissected leaves with filiform segments, angular, sparsely hairy, bluish green and petioles sheathing the stem at base. It bears white or pinkish flowers in terminal compound umbels. The flowers are bisexual or polygamous, sepals five free, petals five free; stamens five, anthers yellowish brown; ovary inferior, fruits greyish brown, small and oblong. The aromatic seed like fruit, commonly known as seed is elongated, oval, elliptical, deeply furrowed, approximately 6 mm long, light yellowish brown. The cumin seed is a mixture of untied and separated mericarp. The surface has five primary ridges alternating with four less distinct secondary ridges bearing numerous short hairs. Some fruits have short attached stalk.

2.2 Varieties

Varieties selected should be adapted to the soil and climatic conditions and should preferably have resistance/tolerance to pests and diseases. Efforts should be made to use organic seed of recommended variety for organic farming from certified organic farms or own seed rose organically. When certified organic seed of cumin is not available, initially conventional seeds can be used (Anon., 2001). The description of some of the important cultivated varieties developed for different states with information on salient characters are given in Table 4.

Table 4. Recommended varieties of cumin for different states

Variety	Salient characters
Rajasthan	
RZ 19	• The plants are erect in growth behaviour bear pink coloured flowers and bold pubescent grains. • Takes 120 –140 days to mature and average yield is 5-6 q/ha.
RZ 209	• The variety has shown tolerance to wilt. • The crop matures in 140-150 days and average seed yield is 6.5 q/ha.
RZ 223	• The variety possesses resistance to wilt. • The essential oil content is 3.23 % and gives seed yield of 6.0 q/ha.
Gujarat	
GC 1	• The plants are erect with pink flowers and bold, linear, oblong ash brown colour grains. • The variety is tolerant to wilt disease. • It matures in 105-110 days and average yield is 7.0 q/ha.
GC 2	• The plants are bushy with good branching habit, attractive grains. • Matures in 100 days and average seed yield is 7.0 q/ha.
GC 3	• The variety is resistant to wilt. • It matures in 100 days and average yield is 7.0 q/ha • The essential oil content is 3.5%.
GC 4	• It gives an average yield of 8.75 q/ha • It is resistant to Fusarium wilt and tolerant to powdery mildew and blight

3.0 CLIMATE AND SOIL

3.1 Climate

Cumin is a tropical crop and is successfully cultivated in moderately cool and dry climate. It does not prefer humidity in atmosphere during flowering and seed setting stage and is prone to attack severely by diseases under such climatic conditions. It is also susceptible to frost injury during flowering and early fruit setting stage. Therefore, the cultivation of the cumin is restricted during *rabi* season in the area in which atmospheric humidity is low and winters are not severe. At the time of maturity cumin needs dry climate. The frequent rains during flowering or early fruit setting period are highly undesirable. High humidity associated with cloudy weather, is harmful and increases incidence of powdery mildew and blight. Frost is the most dangerous and limited factor for cumin cultivation.

3.2 Soil

The cumin can be cultivated in wide range of soils, however sandy loam to medium heavy soils having plenty of organic material with better fertility status are most suitable The soil should have better drainage facility because stagnated water and excessive moisture both are very harmful to cumin crop. In sandy soils, incidence of wilt disease has been reported more. The soil should be free from stones, clods, stumps and weed plants.

4.0 SEED MATERIAL

According to Bhati (1992), seed rate of 14 kg/ha row spacing of 22.5 cm resulted in highest seed yield in Rajasthan. Cumin seeds are small in size therefore; sand or screened friable soil is mixed with seed just before sowing so that uniformity is maintained when seeds are dropped in shallow opened furrow. Since germination is a problem in cumin, therefore soaking of seed in water for 8 hours before sowing improves germination. The soaked seed should be dried in shade before sowing.

5.0 PLANTING AND AFTER CARE

5.1 Requisite for Organic farming

A minimum of two years is required as conversion period for organic cultivation and also required to be certified. The general guidelines for selection of farm land, buffer zone, certification etc. as detailed by Anon., (2001); Malhotra and Vashishtha, (2004) are required to be followed. The procurement management which includes operationalization of activities for production of organic seed spices crops and creating conducive environment for required kind of quality and quantity of produce by way of adopting pre and post-harvest practices and deciding marketing system suitably. It shall ultimately be guideline to the grower/processor/ exporter in planning the production/ supply of organic seed spices for export (Malhotra and Vashishtha, 2007).

The cultural requirements including practices as per procurement management for organic cumin production are described here.

5.2 Preparation of land

Soil is brought to fine tilth by 2-3 ploughing with harrow or *desi* plough. Stubbles of the previous crop should be collected and removed from the field. Clods should be broken and field should be levelled with the help of plank. Beds of convenient size with provision of irrigation channels should be prepared before sowing of seed. Well rotten farm yard manure or compost is to be incorporated into the soil preferably 3 weeks before sowing.

5.3 Sowing time

The time of sowing may vary a little according to place. However the optimum time of sowing is mid October to mid November. Singh and Joshi (1964) reported that cumin can be sown in Pali district of Rajasthan throughout October but the second week of October is the optimum time of sowing to get higher seed yield. Whereas, Chandola *et al.* (1970) recommended that the seed should be sown in November-December to get the highest seed yield.

5.4 Sowing method

The quality seed of 12-15 kg is required for sowing of cumin crop in one hectare area. Sowing of cumin is done by two methods i.e. line sowing and broad-casting. Traditionally farmers sow cumin by broadcasting method but sowing done in lines facilitates intercultural operations. The line to line distance should be kept at 25 cm. The seed should not be sown deeper than 1.5 cm and the same should be covered with a thin layer of soil.

5.5 Irrigation

The water requirement for raising cumin crop is comparatively less than many spices crops. Irrigation should be given judiciously using uncontaminated water. After sowing a light irrigation should be applied and initial moisture helps in getting good germination. The germination of the seed will be visible after 10-12 days of sowing. The crop is irrigated at an interval of about 30 days depending upon the weather condition and soil type. It is however important to keep in mind that irrigation should be avoided when the crop is at maturity stage, as this may affect seed quality adversely. Niazi and Anees (1972) reported that a single irrigation just before flowering increased cumin yields by an average of 21.6 % and an additional irrigation after flowering raised the increase to 31.8%. Patel *et al.* (1992) found that use of water increased significantly with each increment in IW:CPE from 0.30 to 0.60. The mean maximum water use was under IW:CPE of 0.30, 0.45 and 0.60 being 92.23, 107.13 and 126.05 mm, respectively. The higher water use under higher rates may be attributed to increased availability of water during the crop growth.

5.6 Intercultural operations

The crop should be kept free from weeds for proper growth and development of plants. Generally 2-3 hand weeding are required to keep the crop weed free and reducing the crop-weed competition for growth resources like nutrients, moisture, light and others (Gora *et al.*, 1996). The first hoeing and weeding is done at the time when the plants are 5 cm tall. During this operation remove all the weeds, off type plants and those showing poor growth by maintaining optimum plant-to-plant spacing of 10-12 cm. The subsequent intercultural operations should be done on appearance of weeds. The weed like Zeeri (*Plantago pumila* L) is a serious weed in cumin and is difficult to distinguish at early growth phase but can be identified easily at flowering stage. Mehriya *et al.* (2007) in his studies for deciding the critical period of crop-weed competition suggested to keep cumin field free from weeds till 60 days after sowing for lowering weed intensity and realizing higher seed yields.

5.7 Cropping system

Cumin is not recommended to be grown as mixed or intercrop. However the rotation crops required to be grown should also be based on organic farming methods. Some of the suggested cropping systems for cumin growing areas could be (Mehta and Malhotra, 2007):

(i) Dhaincha- Cumin

(ii) Dhaincha- Cumin-Green gram

(iii) Maize- Cumin- Summer Moong

(iv) Pearl Millet- Cumin

5.8 Rotation against frost

Cumin crop is susceptible to frost during flowering and seed formation stage. During the frost danger period to protect the crop it is necessary to make provision of smoke in the field.

6.0 NUTRIENT MANAGEMENT

In Rajasthan, Misra (1963) suggested that 6-8 tonnes of FYM should be applied at the time of seedbed preparation. The requirement of fertilizer depends upon the fertility status of the soil. Therefore, fertilizer should be applied based on soil testing report. In order to maintain good soil structure and biological property apply 10 t / ha FYM or 5 t/ha compost three weeks before sowing of the crop. Inoculation with VAM fungi, *Gigaspora calospora,*

Glomus fasciculatum, Glomus mosseae and *Acaulospora laevis*) not only reduced the incidence of wilt in cumin but also enhanced nutrient uptake (Champawat, 1992). The use of bioinoculant *Azospirillum or Azotobacter* as seed treatment and soil treatment in combination with 5 t sheep manure/ha could give higher seed yield under semi-arid conditions (Malhotra *et al.,* 2004).

7.0 DISEASES AND PEST MANAGEMENT

Cumin is prone to attack by some of the serious diseases and insect-pests. The plant protection measures under organic farming system should involve emphasis on crop management practices time of sowing, balanced nutrition, crop rotation, green manuring etc. for reducing the incidence of diseases and pests, selection of resistant varieties and adoption of biocontrol measures.

7.1 Diseases

7.1.1 Blight

The infected plants show dark brown blighty appearance on the leaves and stem resulting in withering off in affected plants. It is a serious disease in cumin cultivated areas and infection is favoured by humid conditions.

Control

- Avoiding excessive irrigation and high humidity
- Using disease free seed
- Adopting field sanitation measures
- No resistant variety is available

7.1.2 Wilt

The infected plants exhibit drooping of leaves and epinasty later lead to death of the plant. The wilt infection may occur in patches at any stage of crop growth. After occurrence of the wilt in cumin, it becomes very difficult to reduce the damage. The disease can be managed effectively by adopting integrated approach as given below:

- Summer ploughing should be done
- Crop rotation of at least three year should be followed with non-host crops preferably mustard or pearl millet
- Pathogen being cumin-specific can be managed through crop rotation.
- The occurrence of disease can be reduced by the application of oilcakes in the field with high soil pH 8.2

- Use of disease free seed minimizes the disease incidence.
- The antagonistic fungi *Trichoderma viride/ T. harzianum* 4 g / kg seed treatment is effective in reducing the wilt incidence (Aghnoom *et al.*, 1999).
- Extreme dry climate condition should be avoided as it favours the disease.
- Adopt field sanitation measures
- Locate wilt resistant variety suiting to the production area (GC 3 and GC 4)
- Inoculation with VAM fungi reduced the incidence of wilt in cumin (Champawat, 1992).
- Amendment of soil with a combination of mustard residues and mustard oil-cakes (2.5 + 0.5 t ha^{-1}) was effective in control of *Fusarium* population in soil and incidence of wilt on cumin (Israel *et al.*, 2005; Lodha and Singh, 2007).

7.1.3 Powdery mildew

White powdery mass appears on the leaves and twigs of the plant in the initial stages. Later on whole plant is covered with this whitish powder.

Control

- The application of Sulphur dust, wettable sulphur suspension (0.6%) and sulphur spray and the restricted use of sulphur is permitted and described in guidelines from Spices Board (Anon., 2001)
- Foliar spray of onion extract (5%) is effective in reducing the severity.
- Grow resistant/tolerant varieties.
- The incidence is more during February-March, thus sowing of coriander should be completed in first week of October, as late sown crop has more severe damage.
- Uprooting of all affected plants followed by burning or burying.

7.2 Pests

7.2.1 Pest monitoring

Pest monitoring is an important part for organic cultivation of cumin and should be done once in seven days to work out economic threshold level (ETL). It will help to keep close watch on the appearance of insect pests, diseases and bio-control fauna.

For sucking pests: Population should be counted on three leaves i.e. top, middle and lower per plant.

For cut worms and defoliators: Per cent damage assessment can be made by counting total number of plants and affected plants.

Management strategy

1. Set up yellow pan/sticky traps for monitoring of aphids/whitefly @ 10 yellow pans/sticky traps per hectare. Locally available empty yellow palmoline tins coated with grease/Vaseline/castor oil on outer surface may also be used.
2. Collect the larvae of cutworm and leaf defoliators.

7.2.2 *Aphids*

Aphids attack cumin crop at flowering stage causing serious damage.

Control

- Clip off the heavily infested plant parts.
- Spray the crop with pressurized water.
- Application of fish oil, resin soap or *neem* seed kernel-extract (3%), *neem* oil (2%) or tobacco decoction (0.05%) is effective measures (Ravindran *et al.*, 2001).
- Spray of neem oil (1%, 5 litre/ha) or NKE (3%, 15 litre cow urine + 1.5 kg neem kernel extract/ha) are effective in controlling aphids (*Hyadaphis coriandri* Das) when sprayed three times at 15 days interval.

7.2.3 *Mites (Petrobia latens)*

The mites frequently attack the coriander crop and whole plant becomes whitish yellow and appears sickly. The insect mostly feeds on young leaves and the infestation is more severe on young inflorescence. Small mites are seen on lower side of leaves and when serious, cause webbing and feed within the web. Plants get stunted at severe infestation.

Control

- Spray neem oil (1%)

8.0 HARVEST AND POST HARVEST HANDLING

8.1 Harvesting and yield

Generally, cumin crop takes about 90- 120 days to reach maturity. Harvesting should be done when plants turn yellowish brown preferably early in the morning by cutting with sickles and not by uprooting the whole plants. The harvested plants should be stacked for few days and dried in the partial sun in the threshing yard. There after the crop is beaten and trampled on a clean threshing floor (Singh and Malhotra, 2007). Under the scientific management condition the seed yield of 8-10 qha^{-1} of cumin of improved varieties can be obtained Mechanical thresher may be followed in large scale cultivation followed by winnowing, cleaning and grading through mechanical devices.

8.2 Processing

The cumin seeds are dried in partial shade in the sun and the moisture content of seed should be kept to 9 per cent. High seed moisture content may lead to chances of storage contamination by fungus. The dried, cleaned, graded produce is packed in the standard sized packs/containers and appropriately labelled. The dried seed is filled in gunny bags lined with degradable environment friendly plastic film. Waste generating packaging material should be avoided. Each bag is sealed and stored under clean, dry and ventilated place.

Processing methods of cumin should be based on mechanised, physical and biological processes. Care should be taken to maintain the vital quality of an organic ingredient throughout each step of its processing. Processing method is to be selected in such a way that it will limit the number and quantity of additives and processing aids. The mature dried seeds are distilled to obtain the essential oil. Hydro or steam distillation method is generally used for extraction of essential oil. The dried seeds are crushed or as such are steam distilled to yield 2.5 to 4.5 % of valuable volatile oil depending upon variety and location. The oleoresin from cumin has good demand in the international market. The processed products include cumin essential oil, cumin powder, cumin oleoresin and fixed oil. Volatile oil should be kept in well sealed bottles or aluminium containers.

8.3 Quality specifications for export

8.3.1 Whole seed

The quality specifications for cumin whole seed can be classified into three categories *viz.,* commercial requirements, cleanliness and health specifications and have been detailed by Tamil Selvan (2007) and Malhotra (2005).

Commercial specifications :

These requirements vary from country to country and crop to crop. It depends on many factors like customer needs, acceptability and country specifications. The commercial specifications of seed spices crops are colour, appearance, taste, pungency, texture, shape, volatile oil and packaging. The general commercial and general requirements for coriander seed are light brown, large seeded, non-splitting, free flowing, aromatic, high volatile oil (min. 2.2 %).

Cleanliness specifications :

The permit cleanliness specifications as per ASTA, ESA and ISO are given in below :

ASTA

Whole insect dead (by count)	Excretes mammalian (mg/lb)	Excreta other (mg/lb)	Mould (% wt)	Insect defiled / infected (%/wt)	Extraneous foreign matter (%/wt)
4	3	5	1	1	0.5

Other quality standards for cumin as per ESA and ISO

Ash level % w/w (min.)	Acid insoluble ash % w/w (max.)	Moisture content % (max)	Volatile oil % (Min.)
14	3.0	13	1.5

Health specifications :

The micro contaminants such as pesticide residues, microbial counts, aflatoxins, heavy metals etc. comprise the health specifications. These contaminants make the food unsafe. The importing countries have limited the maximum level of these contaminants and limits are given here.

Maximum Residue Level = 0.05 to 0.2%

Microbial counts = 1 X 10^2 / g to 1 X 10^6 /g (*Salmonella, E. coli*, yeast, moulds)

Aflatoxin = 5 ppb to 10 ppb (Max.) Aflatoxin, mycotoxins

Heavy metals = Arsenic (5 mg/kg), copper (20 mg/kg) lead (10 mg/kg), Zinc (50 mg/kg)

8.3.2 Essential oil

The specifications for cumin oil are (Singhal *et al.*, 1997):

- Colourless or pale yellow
- Specific gravity (25°/ 25°C, 0.905 to 0.925)
- Optical rotation (20°C), +3 to+8
- Refractive index, 1.501 to 1.506
- Solubility (80% ethanol), 8 vol
- Aldehydes (cuminic aldehyde) 40 to 52%

8.3.3 Oleoresin

The cumin oleoresin should be prepared with recommended organic solvents followed by the subsequent removal of the solvent as per specifications of importing countries.

8.3.4 Cumin powder

The cumin powdered hygienically through permittable devices should at least 95 % of the ground product pass through a U.S. Standard No. 30 sieve (Farrell, 1999)

REFERENCES

Agarwal, S. and Sharma, R. K. (1990) Variability in quality aspect of seed spices and future strategy. *Indian Cocoa, Arecanut and Spices J.* 13:127-129.

Aghnoom, R., Falahati-Rastegar, M. and Jafarpour, B. (1999) Comparison of chemical and biological control of cumin wilt (*Fusarium oxysporum* f. sp. *cumini*) in laboratory and greenhouse conditions. Iranian Journal of agricultural Sciences 30 : 619-630.

Amin, G. H. (2003) Cumin. In : *Handbook of Herbs and Spices.* Vol 1 Ed. K.V. Peter, Wood head publisher U.K. pp 164-167.

Anonymous, (2001) *Guidelines for production of organic spices in India.* Spice Board, Cochin, pp 64-67.

Bhati, D. S. (1992) Effect of crop geometry and seed rate on yield of cumin. *Indian Perfumer.* 36 (1-4):144-146.

Chandola, R. P., Mathur, S. C. and Srivastava, V. K. (1970) Cumin cultivation in Rajasthan. *Indian Farming.* 20(4):13-16.

Champawat, R. S. (1992) Interaction between vesicular arbuscular mycorrhizal fungi and *Fusarium oxysporum* f.sp. *cumini* - their effects on cumin. *Proceedings of the Indian National Sciences.* 57 (1):59-62.

Farrell, K. T. (1999) *Spices, condiments and seasonings*. Aspen Pub. Inc. Gaitherberg, Maryland, USA. pp 83-87.

Gora, D. R., Meena, N. L. and Shivran, P. L. (1996) Effect of weed control and time of nitrogen application in cumin. *Indian J. Agron*. 41(3):50-51.

Israel, S., Mawar, R. and Lodha, S. (2005) Soil solarization, amendments and bio-control agents for the control of *Macrophomina phaseolina* and *Fusarium oxysporum* f. sp. *cumini* in aridisols. *Annals of Applied Biology* 146:481-491.

Lodha, Satish and Singh, V. (2007) Bio-control of Aspergillus versicolor in managing cumin wilt pathogen. In : *Production, Development, Quality and Export of Seed Spices*. Eds. Malhotra, S.K. and Vashishtha, B.B. NRCSS, Ajmer, Rajasthan, India, pp 308-315.

Malhotra, S. K. (2005) 'Quality Standard of Seed Spices' Lead paper in *Proc. National Seminar on Spices* held at KNK College of Horticulture, Mansor, from 20-21 March,2005.

Malhotra S. K. Mehta, R. S. and Vashishtha B. B. (2004) Effect of biofertlizer on the growth and seed yield of cumin. Extended Summary in *National Seminar on Opportunities and potential of spices for crop diversification*, at JNKVV, Jabalpur from January 19-21, 2004: 121-122.

Malhotra S. K. and Vashishtha B. B. (2004) Organic production of seed spices in India-Status and Strategy. *Seed Spices Newsletter*. 4(1):1-4.

Malhotra S. K. and Vashishtha B. B. (2007) *Production technology for seed spices crops*. NRCSS, Ajmer, pp 1-80.

Malhotra, S. K. and Vashishtha, B. B. (2007) Procurement management and marketing of seed spices. In : *Production, Development, Quality and Export of Seed Spices*. Eds. Malhotra, S.K. and Vashishtha, B.B. NRCSS, Ajmer, Rajasthan, India, pp 234-235.

Mehriya, M. L., Yadav, R. S., Jangir, R. P. and Poonia, B. L. (2007) Study of critical period of crop-weed competition and weed control measures on yield of cumin (*Cuminum cyminum* L.). In : *Production, Development, Quality and Export of Seed Spices*. Eds. Malhotra, S.K. and Vashishtha, B.B. NRCSS, Ajmer, Rajasthan, India, pp 289-297.

Mehta, R. S. and Malhotra, S. K. (2007) Seed spices based cropping system. In : *Production, Development, Quality and Export of Seed Spices*. Eds. Malhotra, S.K. and Vashishtha, B.B. NRCSS, Ajmer, Rajasthan, India, pp 181-189.

Misra, D. K. (1963) Jeera production in Rajasthan. *Indian Farming*.12:12-13.

Niazi, M. H. and Anees, S. M. (1972) Irrigation-cum-fertilizer trials on white jeera (Cuminum cyminum). *J.Agric. Res*. 9(2):124-127.

Patel, K. S., Patel, J. C., Patel, B. S. and Sadaria, S. G. (1992) Influence of irrigation, N and P on consumptive use of water, water use and water expense efficiency of cumin. *India J. Agron*. 37(1):209-211.

Raghavan, Sushila. (2007) *Handbook of Spices, Seasoning and Flavouring*. CRC Press, Boca Raton, Florida, USA, pp 101-103.

Ravindran P. N., Kallupurackal, J. A. and Sivaraman K. (2001) All India coordinated Research project on spices-Status of seed spices research. *Indian J. Arecanut, spices & Medicinal Plant* 3(3): 97-110.

Singh, M and Joshi, M. S. (1964) Jeera is as good crop as any. *Indian Farming*. 14:25-40

Singh, R. and Malhotra, S. K. (2007) Harvesting and maturity indices in seed spices crops. In : *Production, Development, Quality and Export of Seed Spices*. Eds. Malhotra, S.K. and Vashishtha, B.B. NRCSS, Ajmer, Rajasthan, India, pp 195-200.

Singhal, R. S., Kulkarni, P. R. and Rege, D. V. (1997) In : *Handbook of Indices of Food quality and Authenticity*, Woodhead Publishing Limited, England, pp. 386-456.

Tamil Selvan, M. (2007) Quality management in seed spices. In : *Production, Development, Quality and Export of seed spices*. Malhotra,S.K. and Vashishtha,B.B. (Eds.), NRCSS, Ajmer, pp 23-38.

Chapter - 19

联 合 国
粮 食 及
农业组织

FOOD AND AGRICULTURE ORGANIZATION OF THE UNITED NATIONS

ORGANISATION DES NATIONS UNIES POUR L'ALIMENTATION ET L'AGRICULTURE

ORGANIZACION DE LAS NACIONES UNIDAS PARA LA AGRICULTURA Y LA ALIMENTACION

منظمة الأغذية والزراعة للأمم المتحدة

Viale delle Terme di Caracalla, 00153 Rome, Italy | Cables FOODAGRI ROME | Telex: 625852 FAO I 625853 FAO I / 610181 FAO I | Facsimile 57053360 | Telephone 57051

Our Ref A65/2007 Your ref

AUTHORIZATION TO REPRODUCE FAO COPYRIGHTED MATERIAL
AUTORISATION POUR REPRODUIRE DU MATERIAL COPYRIGHT DE LA FAO
AUTORIZACION PARA REPRODUCIR MATERIAL DE LA FAO PROTEGIDO POR DERECHOS

For publication in / *pour sa publication dans* / para su publicación en:

Organic Spices

Addressee / *Destinataire* / Destinario:

Mr. K. Kandiannan
Senior Scientist
Indian Institute of Spices Research (ICAR)
Calicut 673 012
INDIA

Specific FAO copyrighted material to be used / *Matériel copyright utilisé* / Material de la FAO protegido por derechos de propiedad intelectual que va a usarse:

Use chapter on "Suggested Good Agricultural Practices for selected agricultural components" from the "Framework of Good Agricultural Practices" FAO (COAG/2003/6)

Published by / *Publié par* / Publicado por:

FAO

Conditions / *Conditions* / Condiciones.

1. **Due acknowledgement shall be made to the Food and Agriculture Organization of the United Nations (spelled out) and the source document shall be cited.** / *Il sera indiqué que le copyright en est la propriété de l'Organisation des Nations Unies pour l'alimentation et l'agriculture (en toutes lettres) et le document original sera cité.* / Se indicará que los derechos de propiedad intelectual corresponden a la Organización de las Naciones Unidas para la Agricultura y la Alimentación (cuyo nombre se hará constar) y se citará el documento del que procede el material

2. **Fee or royalty to be paid to FAO** / *Redevance à verser à la FAO* / Derechos que se pagarán a la FAO:

   NONE

3. **Two complimentary copy(ies) shall be forwarded to the address shown, citing the number of this authorization.** / Envoyer *spécimen(s) gratuit(s) à l'adresse suivante. Prière d'indiquer le numero de la présente autorisation.* / ejemplar(es) de obsequio se enviará(n) a la dirección que se indica, citando el número de la presente autorización.

   Chief, Electronic Publishing Policy and Support Branch, Communication Division, Food and Agriculture Organization of the United Nations, Viale delle Terme di Caracalla, 00153 Rome, Italy.

4. **Other** / *Autres* / Otras:

Authorization granted / *Autorisation accordée* / Autorización concedida

Date: 25 September 2007

Stephen A. Dembner

Chief, Electronic Publishing Policy and Support Branch / *Chargé du sous-division des politiques et de l'appui en matière de publications électronique*/ Oficial encargado del Subdirección de Políticas y Apoyo en Materia de Publicación Electrónica

19

Good Agricultural Practices

V.A.Parthasarathy, K. Kandiannan, V. Srinivasan

1.0 INTRODUCTION

The crop production is as old as human civilization, it is an art, science and business. In the world of 'Fast Food,' food safety is a watch word. To accomplish safety food to consumers and fair price to producers through appropriate market channels, the concept of Good Agricultural Practices (GAP) has evolved in recent years. The stakeholders include governments, food processing and retailing industries, farmers, and consumers, who seek to meet specific objectives of food security, food quality, production efficiency, livelihoods and environmental benefits in both the medium and long term. GAP offers a means to help reach those objectives.

Food safety described as the guarantee that food has no negative effects on the health of the consumer when the product is processed, cooked and eaten. Food safety is part of the quality of food. Quality means that the product meets consumer expectations. Food safety considered a minimum standard, which is not negotiable. Flavour, smell, colour and taste are other quality of the produce (Asandhi *et al.* 2006).

The steps have been taken to establish Codex Alimentarius by FAO and WHO to develop food standards, guidelines and related texts such as codes of practice. The Codex Alimentarius Commission was created in 1963. The purpose was to protect the health of consumers, ensuring fair trade practices in the food trade and promoting co-ordination of all food standards of work undertaken by international governmental and NGO's. According to Codex (1997a), people have the right to expect the food they eat to be safe and suitable for consumption. Eating habits have undergone major change in many countries over the last two decades and new food production, preparation and distribution techniques have developed to reflect this. Effective hygiene control, therefore, is vital to avoid the adverse human health and economic consequences of food borne illness, food borne injury, and food spoilage. Everyone, including farmers and growers, manufacturers and processors, food handlers and consumers, has a responsibility to assure that food is safe and suitable for consumption.

The CODEX general principles of food hygiene (Codex, 1997a):

- Identify the *essential* principles of food hygiene applicable *throughout the food chain* (including primary production through to the final consumer), to achieve the goal of ensuring that food is safe and suitable for human consumption;
- Recommend a HACCP-based approach as a means to enhance food safety;
- Indicate *how* to implement those principles; and
- Provide a *guidance* for specific codes which may be needed for - sectors of the food chain; processes; or commodities; to amplify the hygiene requirements specific to those areas.

Primary production should be managed in a way that ensures that food is safe and suitable for its intended use. Where necessary, this will include:

- Avoiding the use of areas where the environment poses a threat to the safety of food;
- Controlling contaminants, pests and diseases of animals and
- Plants in such a way as not to pose a threat to food safety;
- Adopting practices and measures to ensure food is produced under appropriately hygienic conditions.

1.1 Definition and scope of GAP

FAO (2003) defines GAP as application of available knowledge to addressing environmental, economic, and social sustainability for on-farm production and post-production processes resulting in safe and healthy food and non-food agricultural products. Sustainable agricultural methods such as integrated pest management, integrated nutrient management, and conservation agriculture are GAP.

At present, GAP is formally recognized in the international regulatory framework for reducing risks associated with the use of pesticides, taking into account public and occupational health, environmental, and safety considerations. The use of GAP is also being promoted increasingly by the private sector through informal codes of practice and indicators developed by food processors and retailers in response to emerging consumer demand for sustainably produced and wholesome food. This trend may create incentives for the adoption of GAP by farmers by opening new market opportunities, provided they have the capacity to respond.

The relationships of concept and tools in food safety described by Jongen and Meulenberg (1998) (Fig.1). Following are widely used private quality and food safety control systems, although more systems and standards exist (de Bode, 2006).

- HACCP - Hazard Analysis of Critical Control Points
- ISO-International Standardization Organization
- SQF - Safe Quality Food
- EUREP-GAP - Euro-Retailer Produce Working Group (EUREP)-Good Agricultural Practices (GAP)

1.2 Approach to GAP and applications

Agriculture is expected to assure food security in a range of settings. Consumer concern is growing in all parts of the world over the environmental, economic and social sustainability, public health implications and safety of agricultural practices and products. Processors and retailers must match the anticipated market demands with the available supply of food in a lengthening food chain. Farmers need to have the capacity to make new farming and technology choices to meet demands for a safe and healthy diet in response to new regulations and standards, changing global consumption patterns, improved market access (through provision of safe food) and potential value-added opportunities. Governments provide the enabling policy and regulatory framework particularly concerning food safety, agricultural production and trade, while seeking to meet food security objectives (FAO, 2003).

Agriculture depends on viable communities and local food systems that provide the mechanism for farmers and consumers to benefit from a closer relationship between production and market, empowering local communities by creating and keeping financial and human resources within the community. The specific issues and constraints facing small-scale producers in developing countries need to be taken into consideration when formulating policies and programmes to develop and promote GAP. The food chain approach to food safety and quality implies that GAP should be extended along the food chain to put greater emphasis on *primary production practices*.

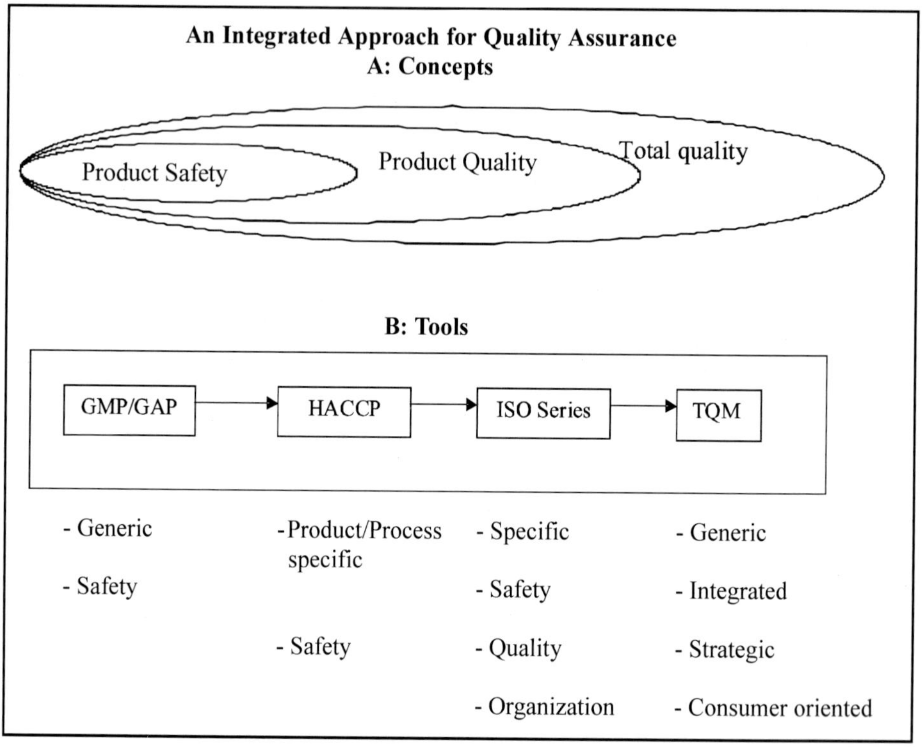

Fig. 1. Realtionships between commonly used food safety definitions and tools (Source: de Bode, 2006 and Jongen and Meulenberg, 1998) (GMP- Good Management Practice; GAP-Good Agricultural Practice; HACCP-Hazard Analysis of Critical Control Points; ISO-International Standardization Organization; TQM- Total Quality Management)

Governments, international agencies and NGOs promote sustainable agricultural methods such as integrated pest management, integrated nutrient management and conservation agriculture, among others, aimed at mitigating specific environmental and societal risks in a range of production and farming systems. The private sector, in particular industrial processors and retailers, uses GAP with a view to attaining quality assurance, consumer satisfaction and profit in the production of safe and high quality food along the food chain. These efforts increasingly incorporate sustainability criteria in response to consumer demand. The promotion of GAP by food processors and retailers can facilitate the adoption of sustainable agricultural practices by creating incentives through potential value-added opportunities for farmers (FAO, 2003). A concrete example of the GAP approach is precise standards for organic production aimed at achieving optimal sustainable agro-ecosystems.

According to FAO (2003), a specific version of GAP is applied within established codes of practice for food safety, under Codex Alimentarius, to minimize or prevent contamination of food. The Codex Alimentarius Commission develops and adopts standards, guidelines and related texts on all aspects of food safety and quality reflecting consensus at the international level. Codex standards are reference points for developing and harmonizing national standards. Codex defines GAP in the use of pesticides to include "...nationally authorized safe uses of pesticides under actual conditions necessary for effective and reliable pest control"(Codex, 1997b). The actual conditions include any stage of the production, storage, transport, distribution and processing of food commodities and animal feed. GAP in this context is used to set maximum residue levels for pesticides and is also recognized in the International Code of Conduct on the Distribution and Use of Pesticides.

The good practices from a broad range of approaches and systems, many of which combine traditional agriculture based on local knowledge with modern agriculture; empowering producers and strengthening farmers' organizations to begin to adopt GAP; recognizing the importance of fair returns to farmers for investments in environmental sustainability; and focusing efforts on humanely produced, safe and high quality foods that address food security.

The GAP principles, indicators and practices may provide a reference point to guide debate on national policies and actions. It may also ensure that stakeholders at all levels of development benefit from the application of GAP in on-farm agricultural production and post-production systems. Such a framework would also provide transparency among all actors along the food chain, and promote harmonization of approaches and their indicators of achievement.

2.0 GAP PRINCIPLES AND FRAMEWORK

In the context of international goals to reduce hunger and promote food security, four principles of GAP apply to all scales of farming (FAO, 2003) :

- Economically and efficiently produce sufficient, safe and nutritious food;
- Sustain and enhance the natural resource base;

- Maintain viable farming enterprises and contribute to sustainable livelihoods;
- Meet the cultural and social demands of society.

GAP provides a means to assess and decide on farming practices at each step in the process of production. For any given agricultural production system, a sound and comprehensive management strategy must be in place providing for the capability for tactical adjustments in response to changes in circumstances. Implementing such a management strategy requires knowing, understanding, planning, measuring, monitoring, and record-keeping, with the aim of achieving production, safety and sustainability goals. Successful implementation depends upon developing the skill and knowledge bases, on continuous monitoring and analysis of performance, and the use of expert advice as required (FAO, 2003).

Accordingly, the proposed process of developing and supporting the adoption of GAP is to:

- Formulate a set of generic practices and indicators from which guidelines for good agricultural practices for on-farm production post-production systems can be developed, collaboratively by the public and private sectors and civil society.
- Focus existing knowledge, options, and solutions into effective food safety and environmental risk analysis guidelines available for use as policy instruments.
- Review existing codes of practice.
- Translate codes of practice into management guidelines for crop and livestock systems in specific agro-ecozones.
- Engage in discussion with governments on their strategies, priorities and instruments to move towards sustainable agriculture and rural development practices.

The generic indicators and practices of GAP include aspects related to soil and water management, crop and fodder production, crop protection, animal production and health, harvesting and on-farm processing and storage, on-farm energy and waste management, human welfare, health and safety, and wildlife and landscape.

The implementation of GAP is generally through a process of assessing the critical management choices that are made sequentially throughout the production of crops and livestock. At each decision control point the implications for GAP are assessed in the context of the indicators. This process may be used to prepare codes of practice for major agricultural production systems, and detailed management guidelines for individual production systems within specific agro-ecozones. By definition, GAP should be explicitly linked to a farming systems categorization in order to apply indicators and practices within a defined domain.

The value of the framework itself will depend upon the adoption of practices by farmers, involvement of the food industry, demand by consumers and support from governments through enabling policies and extension services. Meanwhile, as consumers are increasingly demanding sustainable agricultural practices, application of GAP can, in some cases, result in higher production, processing and marketing costs, which informed consumers might be prepared to absorb. This will create further incentives for the adoption of GAP and its promotion by the private sector. In parallel to the development of GAP, new codes of conduct for agriculture are under discussion in the emerging areas of biotechnology and biosafety, and their relationship to GAP will need to be considered.

2.1 GAP guidelines

The GAP guidelines for on-farm production and post-production systems, involving the participation of farmers, and bringing together scientific and technical expertise, and civil society have to be developed. The technical resources of FAO provide policy advice and technical support to address current developments concerning GAP in Member Nations in collaboration with other International Agencies, NGOs/CSOs and the private sector. This has been interpreted in the Medium Term Plan 2002-07 under the Integrated Production Systems Priority Area for Interdisciplinary Action as:

- Maintain a common information base on promising integrated production techniques for major agroecological zones, by collecting, analysing and disseminating information (data, summary reports) on good practices in relevant geographical contexts.
- Pilot projects to verify the conditions of application of selected promising techniques in systems that have potential for sustainable production increases.
- Decision-support systems promoted in countries based on knowledge accumulated through the above activities.

3.0 GAP FOR SELECTED AGRICULTURAL COMPONENTS (FAO, 2003)

This information is taken from Food and Agricultural Organisation of the United Nations Material on "Framework of Good Agricultural Practices" FAO (COAG/2003/6) with permission from FAO A65/2007.

3.1 Soil

- Soil is a base for crop production. The properties and functions, organic matter and biological activity of the soil are fundamental to sustaining agricultural production and determine, in their complexity, soil fertility and productivity. Appropriate soil management aims to maintain and improve soil productivity by improving the availability and plant uptake of water and nutrients through enhancing soil biological activity, replenishing soil organic matter and soil moisture, and minimizing losses of soil, nutrients, and agrochemicals through erosion, runoff and leaching into surface or ground water. Though soil management is generally undertaken at field/farm level, it affects the surrounding area or catchment due to off-site impacts on runoff, sediments, nutrients movement, and mobility of livestock and associated species including predators, pests and biocontrol agents.
- Good practices related to soil include maintaining or improving soil organic matter through the use of soil carbon-build up by appropriate crop rotations, manure application, pasture management and other land use practices, rational mechanical and/or conservation tillage practices; maintaining soil cover to provide a conducive habitat for soil biota, minimizing erosion losses by wind and/or water; and application of organic and mineral fertilizers and other agro-chemicals in amounts and timing and by methods appropriate to agronomic, environmental and human health requirements.

3.2 Water

- War for water is predicted in future. Water management is essential for sustainable agriculture production. Careful management of water resources and efficient use of water for rainfed crop and pasture production, for irrigation where applicable, and for livestock, are criteria for GAP. Efficient irrigation technologies and management will minimize waste and will avoid excessive leaching and salinization. Water tables should be managed to prevent excessive rise or fall.
- Good practices related to water will include those that maximize water infiltration and minimize unproductive efflux of surface waters from

watersheds; manage ground and soil water by proper use, or avoidance of drainage where required; improve soil structure and increase soil organic matter content; apply production inputs, including waste or recycled products of organic, inorganic and synthetic nature by practices that avoid contamination of water resources; adopt techniques to monitor crop and soil water status, accurately schedule irrigation, and prevent soil salinization by adopting water-saving measures and re-cycling where possible; enhance the functioning of the water cycle by establishing permanent cover, or maintaining or restoring wetlands as needed; manage water tables to prevent excessive extraction or accumulation; and provide adequate, safe, clean watering points for livestock.

3.3 Crop Production

- Seed to seed involves various crop production practices such as the selection of annual and perennial crops, their cultivars and varieties, to meet local consumer and market needs according to their suitability to the site and their role within the crop rotation for the management of soil fertility, pests and diseases, and their response to available inputs. Perennial crops are used to provide long-term production options and opportunities for intercropping. Annual crops are grown in sequences, including those with pasture, to maximize the biological benefits of interactions between species and to maintain productivity. Harvesting of all crop and animal products removes their nutrient content from the site and must ultimately be replaced to maintain long-term productivity.

- Good practices related to crop production will include those that select cultivars and varieties on an understanding of their characteristics, including response to sowing or planting time, productivity, quality, market acceptability and nutritional value, disease and stress resistance, edaphic and climatic adaptability, and response to fertilizers and agrochemicals; devise crop sequences to optimize use of labour and equipment and maximize the biological benefits of weed control by competition, mechanical, biological and herbicide options, provision of non-host crops to minimize disease and, where appropriate, inclusion of legumes to provide a biological source of nitrogen; apply fertilizers, organic and inorganic, in a balanced fashion, with appropriate methods and equipment and at adequate intervals to replace nutrients extracted by harvest or lost during production; maximize the benefits to soil and nutrient stability by re-cycling crop and other organic residues; integrate livestock into crop rotations and utilize the nutrient cycling provided by grazing or housed livestock to benefit the fertility of the entire farm; rotate livestock on pastures to allow for healthy re-growth of pasture;

and adhere to safety regulations and observe established safety standards for the operation of equipment and machinery for crop and fodder production.

3.4 Crop Protection

- Maintenance of crop health is essential for successful farming for both yield and quality of produce. This requires long-term strategies to manage risks by the use of disease- and pest-resistant crops, crop and pasture rotations, disease breaks for susceptible crops, and the judicious use of agrochemicals to control weeds, pests, and diseases following the principles of Integrated Pest Management. Any measure for crop protection, but particularly those involving substances that are harmful for humans or the environment, must only be carried out with consideration for potential negative impacts and with full knowledge and appropriate equipment.
- Good practices related to crop protection will include those that use resistant cultivars and varieties, crop sequences, associations, and cultural practices that maximize biological prevention of pests and diseases; maintain regular and quantitative assessment of the balance status between pests and diseases and beneficial organisms of all crops; adopt organic control practices where and when applicable; apply pest and disease forecasting techniques where available; determine interventions following consideration of all possible methods and their short- and long-term effects on farm productivity and environmental implications in order to minimize the use of agrochemicals, in particular to promote integrated pest management (IPM); store and use agrochemicals according to legal requirements of registration for individual crops, rates, timings, and pre-harvest intervals; ensure that agrochemicals are only applied by specially trained and knowledgeable persons; ensure that equipment used for the handling and application of agrochemicals complies with established safety and maintenance standards; and maintain accurate records of agrochemical use.

3.5 Animal Production

- Good practices related to animal production will include those that site livestock units appropriately to avoid negative effects on the landscape, environment, and animal welfare; avoid biological, chemical, and physical contamination of pasture, feed, water, and the atmosphere; frequently monitor the condition of stock and adjust stocking rates, feeding, and water supply accordingly; design, construct, choose, use and maintain equipment, structures, and handling facilities to avoid injury and loss; prevent residues from veterinary medications and other chemicals given in feeds from entering the food chain; minimize the non-therapeutic use of antibiotics; integrate livestock and agriculture to avoid problems of waste removal, nutrient loss, and greenhouse gas emissions by efficient recycling of nutrients; adhere to safety regulations and observe established safety standards for the operation of installations, equipment, and machinery for animal production; and maintain records of stock acquisitions, breeding, losses, and sales, and of feeding plans, feed acquisitions, and sales.

3.6 Animal Health and Welfare

- Good practices related to animal health and welfare will include those that minimize risk of infection and disease by good pasture management, safe feeding, appropriate stocking rates and good housing conditions; keep livestock, buildings and feed facilities clean and provide adequate, clean bedding where livestock is housed; ensure staff are properly trained in the handling and treatment of animals; seek appropriate veterinary advice to avoid disease and health problems; ensure good hygiene standards in housing by proper cleansing and disinfection; treat sick or injured animals promptly in consultation with a veterinarian; purchase, store and use only approved veterinary products in accordance with regulations and directions, including withholding periods; provide adequate and appropriate feed and clean water at all times; avoid non-therapeutic mutilations, surgical or invasive procedures, such as tail docking and debeaking; minimise transport of live animals (by foot, rail or road); handle animals with appropriate care and avoid the use of instruments such as electric goads; maintain animals in appropriate social groupings where possible; discourage isolation of animals (such as veal crates and sow stalls) except when animals are injured or sick; and conform to minimum space allowances and maximum stocking densities.

3.7 Harvest and On-farm Processing and Storage

- Good practices related to harvest and on-farm processing and storage will include those that harvest food products following relevant pre-harvest intervals and withholding periods; provide for clean and safe handling for on-farm processing of products. For washing, use recommended detergents and clean water; store food products under hygienic and appropriate environmental conditions; pack food produce for transport from the farm in clean and appropriate containers; and use methods of pre-slaughter handling and slaughter that are humane and appropriate for each species, with attention to supervision, training of staff and proper maintenance of equipment.

3.8 Energy and Waste Management

- Good practices related to energy and waste management will include those that establish input-output plans for farm energy, nutrients, and agrochemicals to ensure efficient use and safe disposal; adopt energy saving practices in building design, machinery size, maintenance, and use; investigate alternative energy sources to fossil fuels (wind, solar, biofuels) and adopt them where feasible; recycle organic wastes and inorganic materials, where possible; minimize non-usable wastes and dispose of them responsibly; store fertilizers and agrochemicals securely and in accordance with legislation; establish emergency action procedures to minimize the risk of pollution from accidents; and maintain accurate records of energy use, storage, and disposal.

3.9 Human Welfare, Health and Safety

- Good practices related to human welfare, health and safety will include those that direct all farming practices to achieve an optimum balance between economic, environmental, and social goals; provide adequate household income and food security; adhere to safe work procedures with acceptable working hours and allowance for rest periods; instruct workers in the safe and efficient use of tools and machinery; pay reasonable wages and not exploit workers, especially women and children; and purchase inputs and other services from local merchants if possible.

Many countries started developing their own GAP for generic and crop specific. The categories of standards adopted in INDON-GAP (Indonesian Good Agricultural Practices) (Table 1.) are; recommended practices (R), highly or strongly recommended (SR) practice, and the must (M).

Farm applying for INDON-GAP certification can be approved based on three categories or schems (Table 2).

- Prima One (P I) is granted to farmers that applies agricultural practices covering safety, quality and environment friendly production process

Table 1. Summary of Category of standards

S. No.	Group of controlling item	Application Status of Regulation		
		Must (M)	Strongly Recommended (SR)	Recommended(R)
1	Land and Location Selection	3	14	7
2	Seed and variety	-	3	1
3	Planting	-	3	1
4	Fertilization	2	10	8
5	Pest Control	5	30	10
6	Irrigation	-	3	7
7	Crop Management	-	3	4
8	Harvest	-	5	4
9	Post Harvest Handling	-	19	8
10	Machinery Utilization	-	1	2
11	Environmental Sustainability	-	1	1
12	Working Safety	1	12	3
13	Cleaners and Sanitary	-	3	-
14	Recording controlling	-	3	8
Total Controlling Items		**11**	**110**	**66**

Source : APEC, 2006

Table 2. Indo GAP Prima

P I	P 2	P 3
Apply all 100% (M) regulations	Apply all 100% (M) regulations	Apply all 100% (M) regulations
Apply ≥(90%) (SR) regulations	Apply ≥(70%) (SR) regulations	Apply ≥(60%) (SR) regulations
Apply ≥ (60%) (R) regulations	Apply ≥ (40%) (R) regulations	Apply ≥ (20%) (R) regulations

- Prima Two (P 2) is granted for those farms that adhere to safety and quality practices.
- Prima Three (P 3) is granted to farms that only apply practices to ensure food safety

Spices are grown in varied agro-climatic situations ranging from tropical plains to humid mountain. There are 105 spices are recognized by ISO and their habitat vary. It may not be possible to have common code of conduct for all these crops. However, based on set principles and framework of GAP, a country or an organization can make separate GAP for each crop or for same crop in different production centers. One such attempt is under way from International Pepper Community, Jakarta, Indonesia to provide broad guidelines and framework to formulate GAP for black pepper. Similar attempt may be formulated for other spices at major spice production centers like India and other tropical countries.

REFERENCES

FAO. (2003) *Development of a Framework for Good Agricultural Practices*, Committee on agriculture, Seventeenth Session, Rome, 31 March-4 April 2003, Rome. (available online accessed on 15-5-07) (http://www.fao.org/DOCREP/MEETING/006/Y8704e.HTM)

Codex. (1997a) Basic text on food hygiene. Codex Alimentarius, FAO, Rome (www.haccphelp.com/Documents/Codex.pdf) (online accessed on 15-5-2007)

Codex. (1997b) *Codex Alimentarius Commission Procedural Manual*. Tenth edition, 1997. p. 43

Asandhi, A.A., Schoorelmmer, H., Adiyoga, W., Dibyantoro, L., van der Voort, M. Nurhartuti and Sulastirini, I. (2006) Development of a good agricultural practice to improve food safety and quality in Indonesian vegetable production. Plant Research International, Wageningen, The Netherlands. (available online accessed on 15-5-07 (http://documents.plant.wur.nl/ppo/agv/hortin.pdf)

APEC. (2006) Proceedings of the capacity building seminar on good agricultural practices for APEC developing economies. The Asia Pacific Economic Cooperation, Singapore). http://www.apec.org/apec/enewsletter/mar_vol12/publication.primarycontentparagraph.0002) (online acessed on 15-5-07).

de Bode (2002) Developing good agricultural practices for vegetable production in Hanoi: a chain-oriented approach. http://www.lei.dlo.nl/vegsys/files/30a656760d4a810adc55fe19e9f86c09.pdf (online accessed on 15-5-07)

Jongen, W.M.F., Meulenberg, M.T.G. (1998) *Innovation of food production systems: Product quality and consumer acceptance*. Wageningen Pers, Wageningen.

20

Organic Certification and Quality Standards

V. Srinivasan, T. J. Zachariah and B. Chempakam

1.0 INTRODUCTION

The most significant factors distinguishing organic farming from eco-farming are the existence of production standards and certification procedures. Certification of organic farming is the procedure by which a written assurance is given by the certification agency that a clearly identified production or processing system has been methodically assessed and conforms to the specified requirements. It is a process certification rather than a product certification. Certification and labelling is usually done by an independent body to provide a guarantee that the production standards are met. In case of organic products, certification is an acknowledgement that the products are produced according to organic production standards of certification bodies or associations or of the country. Organic production system is a land use system in accordance with set standards to the location specific conditions by integrating practices that foster cycling of resources, promoting ecological balance and conserve biodiversity. The prime principles of organic farming are,

- Production of sufficient quantities of food with nutritional quality
- Work in harmony with the natural ecosystem instead of attempting to dominate it
- Promotion and fortification of biological cycles within the farming system, including micro-organisms, soil fauna and flora, plants and animals

- Preservation and promotion of the long-term soil fertility, e.g. by manifold crop rotation instead of monocultures
- Avoidance of burdens being put on the environment e.g. by the use of renewable resources
- Creation of the best possible closed operation cycles concerning organic matter and nutrients, e.g. by perceptive and careful use of farmyard manure
- Combating causes rather than symptoms in protecting crops
- Active promotion of landscape elements with ecological and recreational value
- Creation of satisfying life and work conditions

2.0 ORGANIC FARMING STANDARDS

Standards in different countries are developed by certification bodies or private associations or by the state itself. Universally accepted established guidelines for the production, processing, labeling and marketing of organically produced foods have been prepared by Codex Alimentarius, a commission established by FAO and WHO in 1963, as a first step towards harmonization of the guidelines for organic production requirements. The International Federation of Organic Agriculture Movements (IFOAM), a global network of ecological farming movements founded in 1972 provides a platform for global exchange and co-operation in organic agriculture. The organic standards set by IFOAM are recognized by the Codex Alimentarius and the most widely accepted voluntary norms for organic agriculture and food processing. These IFOAM organic standards are applied as a world-wide standard for the evaluation of ecological farming systems equivalent to Codex organic guidelines. IFOAM also established the accreditation criteria for certifying bodies in 1992 to meet the certification norms. The basic organic standards emphasis the following conditions for environment management under plant production system (IFOAM, 2004).

- The whole farm, including livestock, should be converted according to the standards over a period of three years ensuring a clear separation between organic and conventional production
- Any contamination from outside has to be prevented
- Minimum burning of the organic residue (only by permission of the controlling body)

- Use of measures to prevent erosion and saving energy
- Use of crops and varieties which are adapted to soil and climatic conditions
- Use of seed and planting material from recognized ecological production (as far as available), and prohibiting the use of genetically modified organisms (GMOs)
- Maintaining the genetic diversity
- Choice of an appropriate, wide-ranging, most diverse crop rotation with sufficient legumes, temporary pastures with legumes and/or deep-rooting plants for green manure
- Purpose of the fertilization should be maintaining and increasing the soil fertility and the biological activity of the soil
    - The use of artificially synthesized fertilizers is not permitted
    - The basis for fertilizers should be organic material from ecological working farms e.g. farmyard manure, composts derived from animals and plants, as well as slurry and dung water
    - Other organic or mineral fertilizers like wastes from slaughter houses and textile industry or ground rock serve only as an addition
    - Depending on the local conditions, maximum stocking rates per acre are to be fixed by the agencies
    - The heavy metal concentration must not increase in the long term by the application of fertilizers
- Weed regulation only by the use of preventative cultivation methods and by mechanical measures and use of herbicides of any kind is not allowed
- Plant protection has to be by the promotion of natural enemies of pests through the creation of suitable environmental conditions and use of artificial chemical pesticides is not allowed

Processing, storage and transport

- The storage and the transport has to be carried out in such a way that no mixing or mix up with conventionally produced goods can occur
- Storage rooms have to be free of pests and suitable for the product

However, because there are locally different production conditions, regional certification standards are developed and adapted in different countries. The basic standards of IFAOM or Codex give the statutory certification norm framework within which the national organizations can have their own voluntary certification regulations. In US, National Organic Programe (NOP) regulates the standards for producing/selling crops organically through United States Department of Agriculture (USDA) standards for organic foods. Like wise European Union has EEC Regulation No. 2092/91 in EU member states and Japan has Japanese Agricultural Standards (JAS) for organic products. Other developed countries also have their set of standards for organic production and certification like UK Register of Organic Food Standards (UKROFS) and Australian Quarantine and Inspection Service (AQIS) etc (Yadav *et al.*, 2007). In India, the guidelines for organic cultivation of crops at national level has been set under the National Programme for Organic Production (NPOP) set by the national steering committee for organic agriculture under Ministry of Commerce, Govt. of India (www.apeda.com).

National Programme for Organic Production (NPOP) having standards for products and labelling as 'India Organic' was started in May 2000. This programme for organic production proposes to provide an institutional mechanism for the implementation of National Standards for Organic Production, through a National Accreditation Policy and Programme (NPOP, 2005). The aims of the National Programme for Organic Production includes,

- To provide the means of evaluation of certification programmes for organic agriculture and products as per the approved national standards
- To accredit certification programmes
- To encourage the development of organic farming and organic processing

This programme is being implemented by the Government of India through its Ministry of Commerce as the apex body. The National Steering Committee for National Programme for Organic Production has formulated a National Accreditation Policy and Programme (NAPP) and draw up National Standards for Organic Products, including standards for organic production and processes as well as the regulations for use of the National Organic Certification Mark.

Under NPOP, national standards for organic production and processing, accreditation criteria and procedures for inspection and certification have been prepared and approved. It is mandatory that all certification bodies, whether already engaged or proposed to engage in inspection and certification of organic produces should be accredited by an accreditation agency. At present in our country six authorized accreditation agencies namely, Agricultural and Processed Food Products Export Development Authority (APEDA), Coffee Board, Tea Board, Spices Board, Coconut Development Board and Directorate of Cashew and Cocoa Development have been approved by Ministry of Commerce, Govt. of India. Govt. has taken steps to have indigenous certification system to help small and marginal growers and to issue valid organic certificates through certifying agencies accredited by APEDA/ Coffee Board/ Tea Board/ Spices Board. Private (Foreign or Indian) certification agencies accredited by APEDA/ Spices Board such as

- IMO Control Pvt. Ltd., Bangalore (Institute of Marketecologie, Switzerlands)
- Skal International (The Netherlands) India, Bangalore
- SGS (Societe Generale de Surveillance, Switzerlands), India Pvt. Ltd., Gurgaon
- ECOCERT (Ecological Certification, Germany)
- Laon Gmbh (Germany) Aluva, Kerala and
- INDOCERT, Aluva, Kerala are in operation.

Every certifier will implement certification programme and a programme cannot be accredited without accrediting the certifier. Eligible certification agencies are issued with the accreditation after proper evaluation by the accreditation agency with the help of an evaluating agency. The certifying agencies operating or intending to operate in India may apply for accreditation to Spices Board and the accreditation certificate issued is valid for a period of three years.

Spices Board has also developed the guidelines for organic production of export oriented spices. Any grower who has inclination for organic farming can apply for organic certification. The inspectors appointed by the certification agencies will carry out inspection of the farm/ processing operations through records maintained and by periodic site inspections. The producer has to keep all the details with respect to cultivation like input-output details, farm map and organic system plan and reports etc documented. Documentation of farm activities is must for acquiring certification especially when both conventional and organic crops are raised. The documents/records to be maintained are Field map, Field history sheet, Activity register, Input record, Output record,

Harvest record, Storage record, Sales record, Pest control records, Movement record, Equipments cleaning record and Labelling records. (http://www.indianspices.com/html/ind_sp_farm_15.html; accessed on 02.07.2007)

Similarly, processing/ packing and exporting units are also needs to be certified before it deals with organically certified exports. Only those exporters or processors whose produce is duly certified by the official accreditation and certification agencies will be issued license to use the India Organic logo. The tariff on inspection and certification varies between agencies and between the types of farmers. Indian government has taken steps to control the costs of certification by setting the maximum tariff structure for various activities of certification such as travel and inspection, certification and report preparation charges.

Group certification programmes are also in vogue for organized group of producers, processors and exporters with similar production systems located in geographical proximity. The criteria used for group certification is (Kithu, 2005),

Marginal farmers	-	below one hectare
Small farmers	-	between 1 - 1.99 ha
Semi medium farmers	-	between 2 – 3.99 ha
Medium farmers	-	between 4 – 9.99 ha
Large farmers	-	above 10 ha

2.1 Standards for Organic Crop Production

2.1.1 Conversion Requirements

The time between the start of organic management and certification of crops is known as the conversion period. The whole farm, including livestock, should be converted according to the standards over a period of three years and the start of the conversion period is calculated from the date of application to the certification programme. Regular inspection will be carried out at periodical intervals during the conversion period to ensure that standards requirements have been met from that date of conversion. A farm may be converted step by step into organic with a clear plan of conversion. If the whole farm is not converted, the certification programme should ensure that the organic and conventional parts of the farm are separate and inspectable. Simultaneous production of conventional, in conversion and/or organic crops that are clearly distinguished from each other is allowed and the certification programme will inspect the whole production system, where ever appropriate. The full conversion period of three years is not strictly required in farms where de facto full standards requirements have been met for several years which can be verified through records and in such cases inspection can be carried out with a reasonable time interval before the first harvest.

The duration of the conversion period is adapted based on the past use of the land and ecological situation. The crops produced can be certified organic when the national standards requirements are met during a conversion period of at least two years before sowing or in the case of perennial crops at least three years (thirty-six months) before the first harvest of products. The accredited inspection and/or certification agency can decide in cases to extend or reduce the conversion period based on previous status of the land to the minimum of twelve months. During the conversion period of the farm the products can be sold as "produce of organic agriculture in process of conversion" or a similar description.

2.1.2 Choice of Crops and Varieties

The seeds and plant material from certified organic sources, cultivars/varieties that are resistant to pests and diseases and adapted to the soil and climatic conditions are to be given preference. When certified organic seed and plant materials are not available, chemically untreated conventional materials can be used. The use of genetically engineered seeds, transgene plants or plant materials are not allowed under organic farming. The certification programme also sets standards for a minimum percentage of the farm area to facilitate biodiversity and nature conservation. Crop rotation with legumes to maintain or increase soil fertility, microbial activity and general soil health is a recommended practice.

2.1.3 Fertilization Policy

Sufficient quantities of organic material of microbial, plant or animal origin has to be recycled into the soil to increase or maintain fertility and the biological activity. Under organic farming, major strategy of fertilizer management is minimising the nutrient losses and accumulation of heavy metals and other pollutants through nutrient recycling and soil and water conservation measures. Relevant measures should be taken to prevent erosion, salination of soil, excessive and improper use of water and the pollution of ground and surface water. Non synthetic mineral fertilizers and brought in manures should be regarded as supplementary and not a replacement for nutrient recycling. The certification programme sets limitations to the total amount of brought in material onto the farm, considering local conditions and the specific nature of the crops.

The use of manures containing human excreta is not allowed. The certification standards also restricts the use of inputs of mineral origin such as mineral potassium, magnesium fertilizers, trace elements and manures with a relatively high heavy metal content and/or other unwanted substances, e.g. basic slag, rock phosphate and sewage sludge by limiting their quantity (Appendix 1).

2.1.4 Pest, Disease and Weed Management including Growth Regulators

Under organic farming systems, losses from pests, diseases and weeds are minimized with the use of a balanced fertilizing programme, use of crops varieties well-adapted to the environment, fertile soils of high biological activity, adapted rotations, companion planting, green manures, etc. The protection of natural enemies of pests and diseases and their encouragement through proper habitat management forms the basic in plant protection.

The certification standards and the inspecting agency set the guidelines on the products that can be used for pest, disease and weed management (Appendix 1). Priority is given to the products prepared at the farm from local plants, animals and micro-organisms, thermic weed control and physical methods for pest, disease and weed management practices. The use of synthetic herbicides, fungicides, insecticides and genetically engineered organisms or products for pest/ disease management is prohibited.

2.1.5 Processing and Handling

Proper handling and processing of organic products is essential to maintain the quality of the product and to minimise the development of pests and diseases during storage. To avoid contamination, processing and handling of organic products are to be done separately in time or place from that of non organic products. The certification programme sets standards to prevent and control pollutants and level of contaminants. Moreover organic and non-organic products should not be stored and transported together except when labelled or physically separated. Besides storage at ambient temperature, the special conditions of storage like controlled atmosphere, cooling/freezing, drying and humidity regulation are permitted. Similarly, pest control treatments like physical barriers, sound, ultra-sound, light, and UV-light, traps (including pheromone traps and static bait traps), temperature control, controlled atmosphere and diatomaceous earth are permitted. Irradiation is prohibited. A plan for pest prevention and pest control should be developed.

2.1.5.1 Processing

All the ingredients and additives used in processing should be of agriculture origin and certified organic. In cases where an ingredient of organic agriculture origin is not available in sufficient quality or quantity, the certification programme authorises use of non organic raw materials subject to periodic re-evaluation.

Under organic farming, processing methods also should be based on mechanised, physical and biological processes to maintain the vital quality of organic ingredient throughout each step of its processing. The following processing methods are allowed.

- Mechanical and physical
- Biological
- Smoking
- Extraction
- Precipitation
- Filtration

Extraction using water, ethanol, plant and animal oils, vinegar, carbon dioxide, nitrogen or carboxylic acids of food grade quality are allowed. Filtration substances shall not be made of asbestos nor may they be permeated with substances which may negatively affect the product.

2.1.5.2 Packaging

The material used for packaging should be eco-friendly. Recycling and reusable biodegradable packaging systems are encouraged wherever possible to make non contaminated food.

Labelling

Labelling should clearly indicate the organic status of the product as "produce of organic agriculture" or a similar description when the standards requirements are fulfilled. The name and address of the person or company legally responsible for the production/ processing of the product should be mentioned on the label along with all components of additives and processing aids. When a minimum of 95% of the ingredients are of certified organic origin, products must be labelled "certified organic" with the logo of the certification programme. When at least 70% of the ingredients are of certified organic, "made with organic ingredients" may be labelled and where less than 70% of the ingredients are of certified organic origin, such product may not be called as "organic".

All raw materials of a multi-ingredient product should be listed on the product label in order of their weight percentage. If spices constitute less than 2% of the total weight of the product, they may be listed as "spices" without stating the percentage.

2.1.5.3 Storage & Transport

Product integrity should be maintained during storage and transportation of organic products. Organic Products must be protected from contact with any pesticide or materials and substances not permitted under organic farming. Storage areas and transport containers for organic product should be cleaned using methods and materials permitted in organic production.

2.1.5.4 Terms of Quality

The quality of food is generally influenced by the choice of variety, weather, season, stage of maturity, soil, fertilizer and cultivation technique. Through organic farming the quality parameters are influenced especially over a healthy, active soil.

External quality

External characteristics express only the commercial value of agricultural goods with the standards,

- shape
- size
- colour
- disease

In the production of fresh products emphasis is put on the external qualities, as the optical characteristics have the biggest impact on the buying behaviour.

Internal quality

High yields are most often achieved in agriculture at the cost of the internal quality of the produce. The plants contain more water but also higher levels of substances which deteriorate the value or are harmful. Aspired characteristics under organic farming regulation are,

- Highest content of oil, oleoresin, protein and vitamins minerals
- Smallest amounts of fertilizer and pesticide residues and other environmental harmful substances
- Flavor metabolites and its balance

American Spice Trade Association (ASTA) is a premier international agency involved in the upkeep and maintenance of whole spices and related products. International Standards Organization (ISO) also developed standards for all the producing and importing countries. The desired specifications for all the major spices are applicable for organically produced products also (Appendix 2).

2.1.6 Social Justice

Social right and justice are integral part of organic agriculture. The laws relating to labour welfare and rights of children should be honoured. All employees and their families should have access to potable water, food, housing, education, transportation and health services. All employees should have equal wages when doing same job. They must have equal opportunities irrespective of colour, creed and gender. Social security needs (include maternity, sickness and retirement benefits) should be met. Labour conditions regarding noise, dust, light and exposure to chemicals should be within acceptable limits; and they should have adequate protection. The rights of indigenous people should be respected.

Appendix 1

Products allowed for Use in Fertilizing and Soil Conditioning (as per NPOP (2005) guidelines)

In organic agriculture the maintenance of soil fertility can be achieved through the recycling of organic material whose nutrients are made available to crops through decomposition by micro-organisms.

The following inputs are restricted for use under organic production. ("restricted" means that the conditions and the procedure for use shall be set by the certification programme).

The organic inputs that are produced within the farm like farmyard & poultry manure, slurry, cow urine, crop residues, green manure, straw and other organic mulches are permitted.

The organic inputs produced outside and brought in to the farm are limited and their quantity is also restricted to supplement the nutrition and not to replace the on farm inputs. The quantity may vary with the farm conditions and soil fertility and will be specified by the certification programme. The following are some of the allowed inputs of restricted quantity.

- Compost made from any carbon base
- Blood meal, meat meal, bone meal and Guano

- Fish and fish products without preservatives
- By-products from the food and textile industries
- Biodegradable material of microbial, plant or animal origin without any synthetic additives
- Sawdust, wood shavings, wood provided it comes from untreated wood
- Seaweed and seaweed products obtained by physical processes
- Sewage sludge and urban composts from separated sources which are monitored for contamination
- Vermicasts, Animal charcol
- Compost and spent mushroom and vermiculate substances
- By products from oil palm, coconut and cocoa (including empty fruit bunch, palm oil mill effluent, cocoa peat and empty cocoa pods)
- By products of industries processing ingredients from organic agriculture
- Nutrients of mineral origin separated by physical processes like basic slag, calcareous and magnesium rock, calcium carbonate of network origin (chalk, limestone, gypsum and phosphate chalk), mineral potassium with low chlorine content, natural phosphates (e.g. Rock phosphates), sodium chloride, trace elements (boron, Fe, Mn, Mo, Zn), wood ash from untreated wood, magnesium sulphate (Epson salt), aluminium calcium phosphate, sulphur, clay (bentonite, perlite, zeolite)
- Bacterial preparations (biofertilizers), Biodynamic preparations and Plant preparations and botanical extracts

Products for Plant Pest and Disease Control

The products allowed for use in organic agriculture for the control of pests and diseases in plant production are,

- Neem preparations (eg. neem oil), Preparation of rotenone from *Derris elliptica, Lonchocarpus, Thephrosia* spp., preparation of pyrethrins extracted from *Chrysanthemum cinerariaefolium*
- Release of parasite predators of insect pests, Tobacco tea (pure nicotine is forbidden) Lecithin, Casein, sea weed extracts, extract from mushroom, natural acids (vinegar) are restricted
- Use of minerals like chloride of lime/soda, burgundy mixture, clay (e.g. bentonite, perlite, vermiculite, zeolite), copper salts / inorganic salts

(Bordeaux mix, copper hydroxide, copper oxychloride) as a fungicide, diatomaceous earth, light mineral oils, permanganate of potash, silicates, sodium bicarbonate, sulphur (as a fungicide, acaricide, repellent) are also restricted.

- Use of viral, fungal and bacterial preparations of biocontrol agents, herbal and biodynamic preparations and release of sterilised insects is also restricted as per specifications of certification programme.
- Use of chromatic/ mechanical/ pheromones traps are allowed

The products or inputs those are useful and appropriate for use in organic farming and which may or may not fall under these headings can be considered under organic farming standards provided,

- The material is essential for achieving or maintaining soil fertility or specific soil-conditioning and rotation purposes or for the control of a harmful organism or a disease for which alternatives and/or effective management techniques are not available
- The ingredients are of plant, animal, microbial or mineral origin which may undergo the physical, enzymatic, microbial process of decomposition
- Their use does not result in unacceptable effects on the environment

Appendix 2

Quality specifications for spices (Table 1 to 27)

Black Pepper

Specification for black pepper (whole and powdered) is described in table.1 & 2. The cleanliness specification prescribed by ASTA for white pepper is listed in table.3.

Table 1. ASTA cleanliness specifications for whole black pepper

Specification	Suggested Limits
Whole dead insects, by count	2
Mammalian excreta, by mg/lb	1
Other excreta, by mg/lb	5.0
Mold, % by weight	1
Insect defiled/infested, % by weight	1
Extraneous, % by weight	1.00

Contd....

Table 1. Contd....

FDA DALs	
Insect infested and/or moldy pieces by weight	Ave of 1%
Mammalian excreta	Ave of 1mg per lb
Foreign matter pickings and siftings by weight	Ave of 1%
Volatile oil	2.0 % min
Moisture	12.0 % max
Ash	5.0 % max
Acid insoluble ash	0.5 % max
Average bulk index (mg/100g)	165

Table 2. ASTA cleanliness specifications for ground black pepper

Specification	Suggested Limits
Insect fragments or	Ave of 475 or more/50 g
Rodent hair fragments	Ave of 2 or more/50g
Volatile oil	1.5% min
Moisture	12.0% max
Total ash	5.0% max
Acid insoluble ash	0.5% max
Military specifications (EE-S-63IJ, 1981)	
Volatile oil (ml/100g)	2.0 min
Moisture	12.0% max
Total ash	7.0% max
Acid insoluble ash	1.0% max
Crude fibre	12.5% max
Nonvolatile methylene chloride extract	7.5% min
Starch	30.0% max
Granulation	70% min through a U.S.S #30 or 95% min through a U.S.S #16 95% min through a U.S.S #35
Bulk index (mg/100g)	Varies with granulation

White pepper

To prepare white pepper, the berry is picked when fully ripe. The outer layer of shrunken skin is removed, leaving the dried, greyish-white kernel. It has a milder, more delicate flavour than black pepper (Table 3).

Table. 3. ASTA cleanliness specification for white pepper

Specification	Suggested Limits
Whole dead insects, by count	2
Mammalian excreta, by mg/lb	1
Other excreta, by mg/lb	1.0
Mold, % by weight	1
Insect defiled/infested, % by weight	1
Extraneous, % by weight	0.50
FDA DALs	
Insect infested and/or moldy pieces by weight	Ave of 1%
Mammalian excreta	Ave of 1mg per lb
Foreign matter pickings and siftings by weight	Ave of 1%
Volatile oil	1.5% min
Moisture	14.0% max
Ash	1.5% max
Acid insoluble ash	0.3% max
Average bulk index (mg/100g)	150

Sterilised black pepper

Pepper berries are collected farm fresh and after removing all stones and other foreign particles, it is thoroughly washed in boiling water and dried under controlled conditions, till the moisture is brought down to less than 11% (Table 4).

Table 4. Specification for sterilised black pepper

Item	Requirement
Characteristics	125 grains (corns) more than 4 grams weight.
Colour	Black gold
Shape	Uniform
Taste	Spicy
Broken pieces	Max 0.5%
Smell	Strong (pungent)
Discoloured corn	10% by count
Extraneous matter- stem ,leaves	1% by wt.
Infested	Nil
Grit & sand	0.5%
Arsenic	1.0 mg/kg max
Lead	5.0 mg/kg max
Calcium/Magnisium Sulphates or Carbonates	0.1 % max
Copper	15 mg/kg max
E.Coli	Negative/25gm
Salmonella	Absent
Yeast & moulds	1000 cfu/gm
AflatoxinB1	Negative (max2ppb)
AflatoxinB2	Negative (max1ppb)
AflatoxinG1	Negative (max1ppb)
AflatoxinG2	Negative (max1ppb)
	Total sum not to exceed 4 ppb

Table 5: Specification for Dehydrated green pepper

Item	Requirement
Essential oil	3.0 (%)
Piperine	7.0 (%)
Moisture	12.0 (%)
Bulk density	250-400 gm/litre
Infected	0.2 (%)
Grit and sand	0.5 (%)
Copper	15.0 mg/kg max
Zinc	15.0 mg/kg
E.coli	Negative/25gm
Salmonella	absent
Aflatoxin	B1 max.2 ppb
Aflatoxin	B2 max 1 ppb
Aflatoxin	G1 and G2 max. 1 ppb each (Total sum not to exceed 4 ppb)
Yeast and moulds	1000 cfu/g
Aerobic total count	30,000 cfu/g

Source: ipsr solutions2005

Cardamom

Table 6. ASTA cleanliness specifications for whole cardamom

Specification	Suggested Limits
Whole dead insects, by count	4
Mammalian excreta, by mg/lb	3
Other excreta, by mg/lb	1.0
Mold, % by weight	1.00
Insect defiled/infested, % by weight	1.00
Extraneous, % by weight	0.50
FDA DALs	None
Volatile oil	3.0% min
Moisture	12.0% max
Ash	10.0% max
Acid insoluble ash	2.0% max
Average bulk index (mg/100g)	
Bleached	320
Green	250

Table 7. ASTA cleanliness specifications for ground cardamom

Specification	Suggested Limits
FDA DALs	None
Volatile oil	3.0% min
Moisture	12.0% max
Total ash	10.0% max
Acid insoluble ash	2.0% max
Military specifications (EE-S-63iJ, 1981) (Decorticated Cardamom)	
Volatile oil (ml/100g)	3.0% min
Moisture	12.0% max
Total ash	7.0% max
Acid insoluble ash	3.0% max
Granulation	95% min through a U.S.S #40
Bulk index (ml/100g)	190

Ginger

Table 8. ASTA cleanliness specifications for whole ginger

Specification	Suggested Limits
Whole dead insects, by count	4
Mammalian excreta, by mg/lb	3
Other excreta, by mg/lb	3.0
Mold, % by weight	1
Insect defiled/infested, % by weight	1
Extraneous, % by weight	1.00
FDA DALs	
Moldy and/or insect infested pieces	Ave of 3% by weight
Mammalian excreta	Ave of 3mg per lb
Volatile oil	2.0%
Moisture	12.0% max
Ash	5.0% max
Acid insoluble ash	1.0% max
Average bulk index (mg/100g)	N/A

Table 9. ASTA cleanliness specifications for ground ginger

Specification	Suggested Limits
FDA DALs	None
Volatile oil	1.5% min
Moisture	12.0% max
Total ash	5.0% max
Acid insoluble ash	1.0% max
Military specifications (EE-S-63IJ, 1981)	

Contd....

Table 9. Contd....

Volatile oil (ml/100g)	1.5 min
Moisture	12.0% max
Total ash	7.0% max
Acid insoluble ash	1.0% max
Crude fibre	8.0% max
Starch	42.0% max
Granulation	95% min through a U.S.S #30
Bulk index (mg/100g)	210

Turmeric

Table 10. ASTA cleanliness specifications for whole turmeric

Specification	Suggested Limits
ASTA cleanliness specifications:	
Whole dead insects, by count	3
Mammalian excreta, by mg/lb	5
Other excreta, by mg/lb	5.0
Mold, % by weight	3.00
Insect defiled/infested, % by weight	2.50
Extraneous, % by weight	0.50
FDA DALs	None
Curcumin content	5.0% min
Moisture	10.0% max
Ash	8.0% max
Acid insoluble ash	1.0% max
Average bulk index (mg/100g)	N/A

Table 11. ASTA cleanliness specifications for ground turmeric

Specification	Suggested Limits
FDA DALs	None
Curcumin content	5.0% min
Moisture	10.0% max
Total ash	8.0% max
Acid insoluble ash	1.0% max
Military specifications (EE-S-631J, 1981)	
Volatile oil (ml/100g)	3.5 min
Moisture	10..0% max
Total ash	8.0% max
Acid insoluble ash	0.6% max
Crude fibre	9.5% max
Color power, expressed as curcumin	5.0%-6.6%
Granulation	95% min through a U.S.S #40
Bulk index (mg/100g)	185

Cassia and Cinnamon

Table 12. ASTA cleanliness specifications for whole cassia or cinnamon

Specification	Suggested Limits
ASTA cleanliness specifications:	
Whole dead insects, by count	2
Mammalian excreta, by mg/lb	1
Other excreta, by mg/lb	1.0
Mold, % by weight	5.00
Insect defiled/infested, % by weight	2.50
Extraneous, % by weight	0.50
FDA DALs	
Insect infested pieces by weight or	Ave of 5%
Moldy pieces by weight or	Ave of 5%
Mammalian excreta identified as to source when possible	1 mg per lb
Volatile oil	1.5%-3.0%
Moisture	14.0% max
Ash	5.0% max
Acid insoluble ash	1.0% max
Average bulk index (mg/100g)	N/A

Table 13. ASTA cleanliness specifications for ground cinnamon

Specification	Suggested Limits
FDA DALs (6 sub samples)	
Insect fragments	Ave of 400 /50 g
or	
Rodent hairs	Ave of 11 /50 g
Volatile oil	1.5%-3.0%
Moisture	14.0% max
Total ash	5.0% max
Acid insoluble ash	1.0% max
Military specifications (EE-S-63IJ, 1981)	
Korintji cinnamon	
Volatile oil (ml/100g)	1.5 min
Moisture	11.0% max
Total ash	5.0% max
Acid insoluble ash	2.0% max
Granulation	95% min through a U.S.S #60
Bulk index (mg/100g)	165

Clove

Table 14. ASTA cleanliness specifications for whole cloves

Specification	Suggested Limits
Whole dead insects, by count	4
Mammalian excreta, by mg/lb	5
Other excreta, by mg/lb	8.0
Mold, % by weight	1.00
Insect defiled/infested, % by weight	1.00
Extraneous, % by weight	1.00
A 5% Allowance for unattached clove stems over and above the tolerance for other extraneous matter is permitted	
FDA DALs	
Adulteration with stems by weight	Ave of 5%
Volatile oil	16.0% min
Moisture	8.0% max
Ash	5.0% max
Acid insoluble ash	0.5% max
Military specifications (EE-S-63IJ)	5% or more
Average bulk index (mg/100g)	240

Table 15. ASTA cleanliness specifications for ground cloves

Specification	Suggested Limits
FDA DALs	None
Volatile oil	14.0% min
Moisture	8.0% max
Total ash	5.0% max
Acid insoluble ash	0.5% max
Military specifications (EE-S-63IJ, 1981)	
Volatile oil (ml/100g)	15.0% min
Moisture	9.0% max
Total ash	6.0% max
Acid insoluble ash	0.5% max
Granulation	95% min through a U.S.S #30
Bulk index (mg/100g)	180

Nutmeg

Table 16. ASTA cleanliness specifications for whole nutmeg

Specification	Suggested Limits	
Whole dead insects, by count	Broken	Whole
Mammalian excreta, by mg/lb	4	4
Other excreta, by mg/lb	5	0
Mold, % by weight	1.0	0
Insect defiled/infested, % by weight	1	2
Extraneous, % by weight	1	2
	0.50%	0
FDA DALs		
Insect infested and/or moldy pieces by count	Ave of 10%	
Volatile oil	7.0% min	
Moisture	8.0% max	
Ash	3.0% max	
Acid insoluble ash	0.5% max	
Average bulk index (mg/100g)	N/A	

Table 17. ASTA cleanliness specifications for ground nutmeg

Specification	Suggested Limits
FDA DALs	
Insect fragments	Ave of 100 /10 g
or	
Rodent hairs	Ave of 1 /10g
Volatile oil	6.0% min
Moisture	8.0% max
Total ash	3.0% max
Acid insoluble ash	0.5% max
Military specifications (EE-S-631J, 1981)	
Volatile oil (ml/100g)	7.5 min
Moisture	8.0% max
Total ash	3.0% max
Acid insoluble ash	0.5% max
Nonvolatile ether extract	25.0% min
Granulation	95% min through a U.S.S #20
Bulk index (mg/100g)	180

Mace

Table 18. ASTA cleanliness specifications for whole mace

Specification	Suggested Limits
Whole dead insects, by count	4
Mammalian excreta, by mg/lb	3
Other excreta, by mg/lb	1.0
Mold, % by weight	2.00
Insect defiled/infested, % by weight	1.00
Extraneous, % by weight	0.50
FDA DALs	
Insect infested and/or moldy pieces by weight	Ave of 3%
Mammalian excreta	Ave of 3 mg per lb
Foreign matter through a 20 mesh sieve	Ave of 1.5%
Volatile oil	15.0% min
Moisture	8.0% max
Ash	5.0% max
Acid insoluble ash	0.5% max
Average bulk index (mg/100g)	N/A

Table 19. ASTA cleanliness specifications for ground mace

Specification	Suggested Limits
FDA DALs	None
Volatile oil	14.0% min
Moisture	8.0% max
Total ash	5.0% max
Acid insoluble ash	0.5% max
Military specifications (EE-S-631J, 1981)	
Volatile oil (ml/100g)	12.0 min
Moisture	6.0% max
Total ash	3.5% max
Acid insoluble ash	0.5% max
Nonvolatile ether extract	20.0-35.0
Granulation	95% min through a U.S.S #20
Bulk index (mg/100g)	205

Coriander

Table 20. ASTA cleanliness specifications for whole coriander

Specification	Suggested Limits
Whole dead insects, by count	4
Mammalian excreta, by mg/lb	3
Other excreta, by mg/lb	10.0
Mold, % by weight	1.00
Insect defiled/infested, % by weight	1.00
Extraneous, % by weight	0.50
FDA DALs (Condimental Seed):	Ave of 3mg per lb
Adulteration with mammalian excreta	
Volatile oil	0.3% min
Moisture	9.0% max
Ash	6.0% max
Acid insoluble ash	1.0% max
Average bulk index (mg/100g)	285

Table 21. ASTA cleanliness specifications for ground coriander

Specification	Suggested Limits
FDA DALs	None
Volatile oil	0.2% min
Moisture	9.0% max
Total ash	6.0% max
Acid insoluble ash	1.0% max
Military specifications (EE-S-63IJ, 1981)	
Volatile oil (ml/100g)	trace
Moisture	10.0% max
Total ash	7.0% max
Acid insoluble ash	1.0% max
Granulation	95% min through a U.S.S #30
Bulk index (mg/100g)	200

Cumin

Table 22. ASTA cleanliness specifications for whole cumin

Specification	Suggested Limits
Whole dead insects, by count	4
Mammalian excreta, by mg/lb	3
Other excreta, by mg/lb	5.0
Mold, % by weight	1.00
Insect defiled/infested, % by weight	1.00
Extraneous, % by weight	0.50
Ash	9.5% max
Acid insoluble ash	1.5% max
FDA DALs	Ave of 9.5% ash and/ or 1.5% AIA
Adulteration with sand & grit	
Volatile oil	2.5% min
Moisture	9.0% max
Ash	8.0% max
Acid insoluble ash	1.0% max
Average bulk index (mg/100g)	240
Military specification (EE-S-63IJ, 1981)	
Harmless extraneous vegetable matter	5% by weight

Table 23. ASTA cleanliness specifications for ground cumin

Specification	Suggested Limits
FDA DALs	None
Volatile oil	2.0% min
Moisture	9.0% max
Total ash	8.0% max
Acid insoluble ash	1.0% max
Military specifications (EE-S-63IJ, 1981)	
Volatile oil (ml/100g)	2.2 min
Moisture	9.0% max
Total ash	9.5% max
Acid insoluble ash	2.0% max
Bulk index (mg/100g)	185

Fennel

Table 24. ASTA cleanliness specifications for whole fennel

Specification	Suggested Limits
Whole dead insects, by count	1
Mammalian excreta, by mg/lb	1
Other excreta, by mg/lb	1
Mold, % by weight	1.00
Insect defiled/infested, % byweight	1.00
Extraneous, % by weight	0.50
FDA DALs	
Mammalian excreta and/or insects	20% of sub samples
Mammalian excreta	Ave of 3 mg per lb
Volatile oil	1.5% min
Moisture	10.0% max
Ash	9.0% max
Acid insoluble ash	1.0% max
Average bulk index (mg/100g)	210

Table 25. ASTA cleanliness specifications for ground fennel

Specification	Suggested Limits
FDA DALs	None
Volatile oil	1.0% min
Moisture	10.0% max
Total ash	9.0% max
Acid insoluble ash	1.0% max
Military specifications (EE-S-63IJ, 1981)	
Volatile oil (ml/100g)	1.0 min
Moisture	8..0% max
Total ash	10.0% max
Acid insoluble ash	1.0% max
Granulation	95% min through a U.S.S #25
Bulk index(mg/100g)	210

Fenugreek

Table 26. ASTA cleanliness specifications for whole fenugreek

Specification	Suggested Limits
FDA DALs	None
Volatile oil	0.25% min
Moisture	12.0% max
Ash	3.0% max
Acid insoluble ash	0.25% max
Average bulk index (mg/100g)	120

Table 27. ASTA cleanliness specifications for ground fenugreek

Specification	Suggested Limits
FDA DALs	None
Volatile oil	0.25% min
Moisture	12.0% max
Total ash	4.0% max
Acid insoluble ash	0.5% max
Military specifications (EE-S-63IJ, 1981)	
Volatile oil (ml/100g)	-
Moisture	9.5% max
Total ash	5.0% max
Acid insoluble ash	1.0% max
Granulation	95% min through a U.S.S. #25
Bulk index (mg/100g)	190

REFERENCES

International Federation of Organic Agriculture Movements (IFOAM). (2004) *IFOAM – Basic Standards for organic production and processing*, IFOAM, Bonn, Germany, pp. 1-72.

Kithu, C. J. (2005) Certification – the crux in organic farming. *Spice India* 18(8): 7-10.

National Programme for Organic Production (NPOP). (2005) *India Organic*, Ministry of Commerce and Industry, GOI, New Delhi, pp. 1-192.

Yadav, B. S., Yadav, B. R. and Debnath, C. (2007) Organic foods: Production and certification, *Indian Food Industry* 26(2): 62-70.

Glossary

Actinomycetes: filamentous bacteria that have sometimes been classified as Fungi Imperfecti. Actinomycetes typically are saprophytes (especially in soil) but a few are pathogenic to man, animals, and plants.

Adulteration: Mixing of cheap duplicates or similar materials to the original produce to earn more profit.

Aflatoxin: Aflatoxins are fungal metabolite toxins produced by the common moulds *Aspergillus flavus* and *A. parasiticus* during storage and are harmful to humans.

Alluvial soil: Soil deposited by water flow in river basins and are rich in organic matter and nutrients.

Antagonism: The phenomenon in which one element or substance inhibits or reduces the uptake, assimilation or performance of the other. It is not mutually beneficial or performance-enhancing.

Antagonist: An agent or substance that counteracts the action of another

Antibiotic: Substance used to inhibit the growth of micro-organisms, including bacteria and fungi.

Arbuscule: These are fungal growth of VA-mycorrhiza in the form of little tree-like hyphal branching in the host cell. They help in the transfer of nutrients especially phosphates from the soil into the root system. Example: *Glomus, Gigaspora.*

Autotrophic: Self-nourishing organisms capable of utilizing carbon dioxide or carbonates as the sole source of carbon and obtaining energy for life processes from radiant energy or from the oxidation of inorganic elements or compounds such as iron, sulphur, hydrogen, ammonium and nitrate.

Axenic culture: A pure culture e.g. culture with organism belonging to a single species. Example: culture of *Azotobacter chroococcum* or *Azospirillum lipoferum*.

Azospirillum: The spiral-shaped nitrogen fixing bacteria. It is widely distributed in soils and grass roots.

Azotobacter: Non-symbiotic aerobic free living nitrogen fixing soil bacteria.

Bacterial culture: Any medium which is enriched with a particular bacteria. Example: *Rhizobium* culture.

Bacterial species A collection of strains that share many features in common and differ considerably from other strains

Bactericide : A substance that kills bacteria.

Basic Standards : are the minimum requirements that an operation must meet to be certified organic. All of the Standards applicable to the particular farm and enterprise must be met before the certification as organic. Basic Standards use "shall".

Bioassay : A biological analysis performed on living cells or on a living organism, sometimes to detect minute amount of a substance which influences growth. Quantitative estimation of biologically active substances e.g. estimation of Indole Acetic Acid (IAA) produced by *Azotobacter*.

Biodynamic agriculture : It was probably the first comprehensive organic farming system (the apparent beginning of which was a lecture Steiner presented in 1924). The Steiner's philosophy different from organic agriculture principally as it has spiritual, mystical and astrological aspects.

Biofumigation : Use of plants containing biologically active compounds as rotation crops or green manures to suppress soil borne pests and diseases in agricultural production systems or a method of partial disinfection of soil by incorporating leaf residue from the *Brassica* members to release isothiocyanates by the biological transformation of glucosinolate present in these groups of plants under the influence of heat enabled by polythene mulching. Exploited for control of soil borne disease in potato

Biological control : The effect of parasites, predators and pathogens in maintaining another organisms density in an ecosystem at a lower level than would occur in their absence. It is a deliberate use by humans of one species of organism to eliminate or control another.

Bisexual : Bearing both male and female parts by the individual plant.

Blanching : Heating the mace in hot water for few minutes to inactivate the enzymes.

Bordeaux mixture : It is a contact fungicide prepared with copper sulphate and slaked lime at different concentrations.

Botanical pesticides : Pesticides obtained from plant origin .

Certification agency : The body that conducts certification, as distinct from standard-setting and inspection.

Certification program : System operated by a certification agency with its own rules, procedures and management for carrying out certification of conformity.

Chelate: A type of metal-organic complex. A ring like complex in which a metal cation is bound by an organic chelating agent.

Codex Alimentarius : The Codex Alimentarius (Latin for "food code" or "food book") is a collection of internationally recognized standards, codes of practice, guidelines and other recommendations relating to foods, food production and food safety under the aegis of consumer protection. These texts are developed and maintained by the Codex Alimentarius Commission, a body that was established in 1963 by the Food and Agriculture Organization of the United Nations (FAO), and the World Health Organization (WHO). The Commission's main aims are stated as being to protect the health of consumers and ensure fair practices in the international food trade. The Codex Alimentarius is recognized by the World Trade Organization as an international reference point for the resolution of disputes concerning food safety and consumer protection. Codex standards are reference points for developing and harmonizing national standards.

Compatible: Refers to substances which can be tolerated by each other and mixed with each other. Biopesticides such as *Trichoderma* are compatible with N-fixing bacteria such as *Rhizobium*. Only compatible fertilizers are used for making fertilizer mixtures. Opposite of incompatible.

Competition: A more or less active demand on the part of two organisms for some commodity (space, food, etc.) that is inadequate to provide for all organisms present.

Conventional agriculture : The conventional agriculture in present day usages means - agriculture involving external inputs particularly fertilizer and pesticides. Besides these, growth regulators, mechanization, animal medicines are part of the system of modern agriculture.

Conversion (into organic farming) : Conversion of chemical farming or conventional farming into organic farming within a stipulated time. In general, it is two to three years. For annuals or seasonal crops two years and perennial crops three years i.e. after adopting organic farming practices to this specific period, produce from these land is considered as organic.

Conversion period : The time between the start of the organic management and the certification of crops and animal husbandry as organic.

Critical stages (of crop growth) : Refers to the period or stage of development in the life-cycle of a plant when it is most sensitive to the deficiency of a production factor or stress and is most responsive to the inputs. Example: germination, tillering etc.

Crop rotation : The practice of growing a sequence of different crops on the same land in successive years or seasons; done to replenish the soil, curb pests, etc.

Cropping System : A systematic pattern in which crops or follow each other as sole crops or intercrops in a field over a period of time.

Culture: A term referring to the growth of micro-organisms under artificial conditions, or the results of microbial growth on artificial culture media. A microbial culture may be pure or mixed.

Curcumin : (diferuloylmethane), the main yellow bioactive component of turmeric has been shown to have a wide spectrum of biological actions.

Drupe : It is a type of fruit characterized by the presence of pericarp enclosing the seed inside.

East Indian Nutmeg : The nutmeg obtained from Indonesia and contain high levels of myristicin in the seed.

Economic threshold level : This is the most profitable rate of application of an input taking into account the value of the crop and cost of input at a given time. It changes with change in input/output prices at a given response rate.

Ectomycorrhiza: Mycorrhizal fungi, which penetrate in the intercellular spaces of only the root cortex (not inside the roots) and develop on the roots forming fungal mantle and intercellular network of hyphae.

Endomycorrhiza: Mycorrhizal fungi that do not form an external mantle but live within (endo) the root.

Entomopathogenic nematode : A nematode which cause a disease in another organism (insect) in association with a bacteria.

Epicotyl : The first shoot that arise from a germinating seed.

Epidemiology : study of factors affecting the health and illness of population.

Ericoid mycorrhizae: Endotrophic mycorrhizae (biseptate) which occur throughout the fine root system of ericaceous plants

Fallowing: Leaving land unplanted after ploughing, to recover its natural fertility and also to reduce soil-borne pathogens.

Fluorescent pseudomonads: Light emitting bacteria.

Food safety : It is described as the guarantee that food has no negative effects on the health of the consumer when the product is processed, cooked and eaten. Food safety is part of the quality of food. Quality means that the product meets consumer expectations. Food safety considered a minimum standard, which is not negotiable.

Fungicide : chemical compounds used to protect the spread of fungi in crops.

Genetic diversity : The variability among living organisms from agricultural, forest and aquatic ecosystems includes diversity within species and between species.

Genetically Modified Organism (GMO) : A plant, animal, or microbe that is transformed by genetic engineering.

Gibberellic acid/Gibberellin (GA3): A plant growth regulator involved in elongation, enhancement of flower, fruit and leaf size, germination, vernalization and other processes.

Good Agricultural Practices (GAP) : GAP as application of available knowledge to addressing environmental, economic, and social sustainability for on-farm production and post-production processes resulting in safe and healthy food and non-food agricultural products.

Sustainable agricultural methods such as integrated pest management, integrated nutrient management, and conservation agriculture are GAP

HACCP : A specific food safety program to identify contamination risks and actions to prevent exposure to such risks.

Heterotroph: Organisms which cannot produce their own food and depend on complex organic substances for their nutrition and growth. Example: ammonifiers like *Bacillus, Serratia* etc.

Holistic living : This is a way of living in tune with nature considering all aspects without any harm to the individual and environment.

Homestead : House and area of land usually used in a farm.

Hot air treatment : Method of heating rhizome (usually to 48-50°C) by exposing the rhizome to hot air at high RH (usually above 60%) for specific duration of about 10-30min. A kind of rhizome treatment for disinfecting the seed rhizomes from seed borne pathogens.

Hot water treatment : Method of heating rhizome (usually to 50°C) by dipping the rhizome in a hot water for specific duration of about 10min. A kind of rhizome treatment for disinfecting the seed rhizomes from seed borne pathogens.

Indigenous rhizobia: Native rhizobia which are already present in the soil and have not been introduced recently.

Inoculant: Culture of microbes like *Rhizobium, Azotobacter* , blue green algae, phosphate solubilising microbes etc. which is used for inoculation at field level.

Inoculation: Artificial introduction of micro-organisms on or into a medium or living system. Example: treatment of legume seeds with *Rhizobium* inoculant.

Integrated pest management : A sustainable approach to managing pests by combining biological, cultural, physical and chemical tools and host resistance in a way that minimizes economic, health, and environmental risks.

Intercropping: Growing two or more crops on the same field, either simultaneously or in relay with an overlapping period. Simultaneous systems refer to the cultivation of two or more crops either intermingled or with distinct row or strip arrangement. Example: Coconut - black pepper intercropping.

Intercrop : Growing crops in the interspaces of a main crop.

Isolation distance : The distance by which organic crops safely grown to avoid contamination of inorganic materials by drift or water flow etc.,.

Labeling : Any written, printed or graphic representation that is present on the label of a product or accompanies the product.

Land race: In Plant Genetic Resources: An early, cultivated form of a crop species, evolved from a wild population.

Mace : The aril covering the seed of the tree *Myristica fragrance* Houtt and used as spice.

Microbial pathogen : A microorganism which cause a disease in its host.

Micronutrients : These nutrients are required by plants in much lesser amounts than macro/major nutrients but are nonetheless equally essential. Group name for essential plant nutrients boron, chlorine, copper, manganese, molybdenum, zinc, nickel and iron.

Monocrop : A system where only one crop is grown.

Mulch : Material which are placed on soil surface to reduce evaporation, erosion or to protect plant roots from extremely low or high temperature.

Mulching : The practice of placing organic materials such as leaf litter, stones, straw, debris, or plastic sheets on the soil surface around the plant or interspaces to control weeds and prevent soil moisture loss.

Mycelium: A mass of interwoven filamentous hyphae, such as that of the vegetative portion of the thallus of a fungus. Plural is mycelia.

Mycorrhiza: This refers to root fungus (from the greek Myces = fungus; rhiza = root). A symbiotic fungi which form a mutually beneficial association with roots of a plant. Mycorrhizae are of three types (i) Ectotrophic (ii) Endotrophic and (iii) Ectendotrophic.

Mycotoxin : Mycotoxins are fungal metabolites which when ingested, inhaled or absorbed through the skin cause lowered performance, sickness or death in man or animals, including birds.

Myristicin : The active principle present in nutmeg and mace that gives the spicy and medicinal value to nutmeg.

Natural enemy : An organism which causes harm to another organism by its predatory, parasitic or disease inducing habits.

Nematicide: A chemical compound or physical agent that kills nematodes.

Nursery : A place where seedlings are raised for transplanting to the field.

Nutmeg : The dried seed of the tree *Myristica fragrance* Houtt used as spice.

Nutrient Uptake : Refers to the amount of nutrient taken up (absorbed) by plant from the soil or atmosphere. It is the quantitative absorption of nutrients by the plant related to a stage of crop growth.

Obligatory cross pollination : Pollination of flowers of female trees is essentially effected through pollen from separate trees (male in the case of nutmeg).

Organic agriculture (OA) : Organic agriculture as holistic food production management systems, which promotes and enhances agro-ecosystem health, including biodiversity, biological cycles, and soil biological activity. It emphasizes the use of management practices in preferences to the use of off-farm inputs, taking into account that regional conditions require locally adapted systems.

Organic farming : It is a way of agriculture that relies on ecosystem management rather than external agricultural inputs. This approach excludes the use of synthetic inputs, such as synthetic fertilizers and pesticides, veterinary drugs, and genetically modified organisms (GMOs), and usually subscribes to the principles of sustainable agriculture.

Organic principles : The principle of health; The principle of ecology; The principle of fairness and The principle of care. The principles are to be used as a whole. They are composed as ethical principles to inspire action in organic agriculture.

Orthotropic : Shoots that grow erect with the leaves spirally arranged on the shoot.

Paprika like chillies: Chillies having paprika characteristics like high color and low or nil pungency

Paprika: Non-pungent (sweet) and brilliant red ground *Capsicum* powder, derived from the thick flesh dried red pods, with most of the seeds and veins removed

Parasitoid : An organism (insect) which lives at the expense of another organism and kills the host in the process of development, and free living as an adult.

Phosphate solubilising micro-organisms (PSM): Bacteria, fungi and actinomycetes which can solubilise insoluble forms of phosphorus.

Phytotoxic: A substance, usually a chemical, which can cause injury to seeds or growing plants. Example: biuret concentration greater than 2% in urea when used for foliar spray. Excessive concentration of any foliar spray solution can also be phytotoxic.

Pilinut : It is a wild tall evergreen tree naturally occurring in Indonesia. Known botanically as *Canarium ovatum* and the nut is edible.

Plagiotropic : Shoots that grow horizontally from the main stem; also known as fan shoots or lateral branches with leaves arranged in the same plane.

Plant growth promoting rhizobacteria (PGPR): Bacteria capable of playing a useful role in plant growth other than of nutrition through production of extra cellular siderophores, antibiosis against pathogenic bacteria and fungi and production of growth promoting substances.

Pollination: Pollination is an important step in the reproduction of seed plants: the transfer of pollen grains (male gametes) to the plant carpel, the structure that contains the ovule (female gamete). The receptive part of the carpel is called a *stigma* in the flowers of angiosperms and a *micropyle* in gymnosperms. The study of pollination brings together many disciplines, such as botany, horticulture, entomology, and ecology. Pollination is important in horticulture because most plant fruits will not develop if the ovules are not fertilized.

Predator : An organism which feeds (preys) on other organisms and kills it.

Prevention: Any activity which reduces the burden of mortality or damage to plants from disease or pest attack.

Pseudomonas: Important soil bacteria which can solubilise insoluble phosphates (e.g. *Pseudomonas striata*) and act as PGPR as well (*Pseudomonas putida, Pseudomonas fluorescens*).

P-solubilising bacteria (PSB): This includes *Bacillus megatherium* var. *phosphaticum, B. polymyxa, B. subtilis* etc.

Resistance: The ability of an organism to exclude or overcome, completely or in some degree, the effect of a pathogen or other damaging factor.

Rhizome solarization : Method of heating rhizome (usually to 46-48°C) by exposing the rhizome to sunlight for a specific duration under a broad sunlight. A kind of rhizome treatment method disinfecting the seed rhizomes from seed borne pathogens.

Rhizoplane: Refers to the external surfaces of plant roots together with any closely adhering soil particles.

Rhizosphere: Volume or zone of soil immediately surrounding plant roots which is most influenced by root activity and metabolism

Root exudates: Organic and mineral compounds excreted by living plant roots.

Root zone: Soil volume in which most of the active roots of a plant are growing or likely to grow during crop's duration. Usually 25-100 cm deep.

Rootstock : The plant or seedling that provide the root system of any grafted plant.

Salination : Accumulation of salts in the soil.

Scion : Plant part of selected tree which is used for grafting

Seed Spice: These are spice crops where the seed is mainly used as a spice.

Seed Treatment: Most common technique for bacterial biofertilizer application to seeds prior to planting. The process includes preparation of paste or slurry by mixing biofertilizer inoculum in water and pouring paste of biofertilizer on 10-15 kg seeds kept on a gunny bag on the floor and mixing the paste with seeds by hand, spreading inoculated seeds in the shades for 10-15 minutes for drying and sowing.

Shelf life: The period up to which a product (biofertilizer) can be stored while retaining its quality in terms of minimum specified number of viable organisms.

Solarization : A method of disinfecting the soil by exposing the soil to sunlight after wrapping under thin polythene sheets of thickness 100microns. Sterilisation of soil or plant material by cooking the material in a plastic bag using the sun's radiation, which kills off most harmful pathogens.

Spice: A spice is a dried seed, fruit, root, rhizome, bark or vegetative substance or flower bud used in nutritionally insignificant quantities as a food additive for the purpose of flavoring.

Split/ Parallel production: Where only part of the farm or processing unit is certified as organic. The remainder of the property can be (a) non-organic, (b) in conversion or (c) organic but not certified.

Sustainable farming: This is a system of farming in which the yields are maintained with out much purchased inputs and rely on on-farm resources recycling.

Topworking: A method of converting a grown up tree into a new promising one through grafting techniques.

Transgenic crop: a plant that contains a gene or genes which have been artificially inserted instead of the plant acquiring them through pollination.

Tree Spice(s): These are plants yielding spices but grow as trees.

Trichoderma: A naturally occurring saprophitic soil fungus which is used as cellulose decomposer, phosphate solubiliser and biopesticide.

VAM: Vesicular Arbuscular Mycorrhiza. Also abbreviated as VA-mycorrhiza. See mycorrhiza

Vermiwash: A watery extract of compost made with earth worms.

West Indian Nutmeg: The nutmeg obtained from Grenada and the seed contain lower levels of myristicin.

Index

A

B

C

F

I

K

L

M

N

P

Q

R

S

T

U

V

W

X

Y

Z